CW01262354

Projection Displays

Wiley-SID Series in Display Technology

Series Editor:
Anthony C. Lowe

Consultant Editor:
Michael A. Kriss

Display Systems:
Design and Applications
Lindsay W. MacDonald and **Anthony C. Lowe (Eds)**

Electronic Display Measurement:
Concepts, Techniques, and Instrumentation
Peter A. Keller

Projection Displays
Edward H. Stupp and **Matthew S. Brennesholtz**

Liquid Crystal Displays:
Addressing Schemes and Electro-Optical Effects
Ernst Lueder

Reflective Liquid Crystal Displays
Shin-Tson Wu and **Deng-Ke Yang**

Colour Engineering:
Achieving Device Independent Colour
Phil Green and **Lindsay MacDonald (Eds)**

Display Interfaces:
Fundamentals and Standards
Robert L. Myers

Digital Image Display:
Algorithms and Implementation
Gheorghe Berbecel

Flexible Flat Panel Displays
Gregory Crawford (Ed.)

Polarization Engineering for LCD Projection
Michael G. Robinson, **Jianmin Chen**, and **Gary D. Sharp**

Fundamentals of Liquid Crystal Devices
Deng-Ke Yang and **Shin-Tson Wu**

Introduction to Microdisplays
David Armitage, **Ian Underwood**, and **Shin-Tson Wu**

Mobile Displays:
Technology and Applications
Achintya K. Bhowmik, **Zili Li**, and **Philip Bos (Eds)**

Photoalignment of Liquid Crystalline Materials:
Physics and Applications
Vladimir G. Chigrinov, **Vladimir M. Kozenkov and Hoi-Sing Kwok**

Projection Displays, Second Edition
Matthew S. Brennesholtz and Edward H. Stupp

Published in Association with the *Society for Information Display*

Projection Displays
Second Edition

by

Matthew S. Brennesholtz
Insight Media, USA

Edward H. Stupp
Stupp Associates, USA

WILEY

A John Wiley and Sons, Ltd, Publication

This edition first published 2008
© 2008 John Wiley & Sons Ltd

Registered office
John Wiley & Sons Ltd, The Atrium, Southern Gate, Chichester, West Sussex,
PO19 8SQ, United Kingdom

For details of our global editorial offices, for customer services and for information about how to apply for permission to reuse the copyright material in this book please see our website at www.wiley.com.

The right of the author to be identified as the author of this work has been asserted in accordance with the Copyright, Designs and Patents Act 1988.

All rights reserved. No part of this publication may be reproduced, stored in a retrieval system, or transmitted, in any form or by any means, electronic, mechanical, photocopying, recording or otherwise, except as permitted by the UK Copyright, Designs and Patents Act 1988, without the prior permission of the publisher.

Wiley also publishes its books in a variety of electronic formats. Some content that appears in print may not be available in electronic books.

Designations used by companies to distinguish their products are often claimed as trademarks. All brand names and product names used in this book are trade names, service marks, trademarks or registered trademarks of their respective owners. The publisher is not associated with any product or vendor mentioned in this book. This publication is designed to provide accurate and authoritative information in regard to the subject matter covered. It is sold on the understanding that the publisher is not engaged in rendering professional services. If professional advice or other expert assistance is required, the services of a competent professional should be sought.

Library of Congress Cataloging-in-Publication Data

Brennesholtz, Matthew S.
 Projection displays / by Matthew S. Brennesholtz, Edward H. Stupp. — 2nd ed.
    p. cm. — (Wiley series in display technology)
 Includes bibliographical references and index.
 ISBN 978-0-470-51803-8 (cloth)
 1. Information display systems. 2. Liquid crystal displays.
 3. Projectors. I. Stupp, Edward H. II. Title. III. Title: Projection displays.
 TK7882.I6P764 2006
 621.39'87—dc22
                                                          2008024096

A catalogue record for this book is available from the British Library.

ISBN 978-0-470-51803-8 (H/B)

Set in 9/11pt Times by Integra Software Services Pvt. Ltd, Pondicherry, India
Printed and bound in Great Britain by CPI Antony Rowe, Chippenham, Wiltshire, UK

Contents

Foreword	xiii
Preface to the Second Edition	xv
About the Authors	xix

1 Introduction — 1
- 1.1 Overview of Projection Displays — 1
- 1.2 Book Organization — 4
- 1.3 What is not Covered — 4

2 Markets and Applications — 7
- 2.1 Overview — 7
  - 2.1.1 Microdisplays, Light Valves and Light Amplifiers — 7
  - 2.1.2 Emissive Systems — 8
  - 2.1.3 Laser-based Projection Technology — 8
- 2.2 Applications and Performance Requirements — 8
  - 2.2.1 Differentiators among Projectors — 9
  - 2.2.2 Requisite Luminance Levels — 11
    - 2.2.2.1 *Flux requirement for presentation and auditorium applications* — 12
  - 2.2.3 Resolution — 14
  - 2.2.4 Electronic Cinema — 15

3 Emissive Image Sources — 17
- 3.1 Projection CRTs — 17
  - 3.1.1 Luminous Output of Projection CRTs — 18
  - 3.1.2 Phosphors — 19
  - 3.1.3 Resolution of Projection CRTs — 22
  - 3.1.4 Spot Size of Beam — 24
  - 3.1.5 Light Collection/Curvature — 25
- 3.2 Field-emission Devices — 26

4 Liquid Crystal Light Valves and Microdisplays — 29
- 4.1 Active Matrices — 30
  - 4.1.1 Operation of Active-matrix Circuits — 31

CONTENTS

		4.1.1.1	Effects of leakage	33
		4.1.1.2	Charging currents	34
	4.1.2	Technologies		35
		4.1.2.1	α-Si TFTs	35
		4.1.2.2	Poly-Si TFTs	38
		4.1.2.3	Crystalline silicon active matrices	40
		4.1.2.4	Active matrices based on two terminal devices	42
4.2	Liquid Crystal Effects			43
	4.2.1	Liquid Crystal Cells		45
	4.2.2	Nematic Cells		46
		4.2.2.1	Parallel aligned layer cells	46
		4.2.2.2	Twisted nematic cells	46
	4.2.3	Polymer-dispersed Liquid Crystal (PDLC)		50
	4.2.4	Other Liquid Crystal Effects		51
	4.2.5	Liquid Crystal Effects for Reflective Microdisplays		52
	4.2.6	Liquid Crystal Inversion		53

5 Micro-electromechanical Devices   57
5.1	DMD		57
	5.1.1	Device Operation	58
	5.1.2	Gray Scale	60
	5.1.3	Contrast and DLP Pixel Design	61
5.2	Linear MEMS Arrays		62
	5.2.1	Grating Light Valve	62
	5.2.2	GEMS System	66
5.3	MEMS Scanning Mirrors		66

6 Filters, Integrators and Polarization Components   71
6.1	Factors affecting Projector Optical Performance			71
6.2	Component Efficiency			72
6.3	Spectral Filters			73
	6.3.1	Fresnel Reflection at Optical Surfaces		73
	6.3.2	Dichroic Filters		75
		6.3.2.1	Dichroic filters at non-normal incidence	78
		6.3.2.2	Dichroic filters in polarized light	80
		6.3.2.3	Dichroic filters in the imaging path	81
		6.3.2.4	Anti-reflection coatings	82
	6.3.3	Absorptive Filters		84
	6.3.4	Electrically Tunable Color Filters		85
	6.3.5	Mirrors		86
	6.3.6	Total Internal Reflection		89
		6.3.6.1	TIR prisms for angular separation	89
	6.3.7	Filters for UV Control		91
	6.3.8	Filters for IR Control		91
	6.3.9	Indium-Tin Oxide and Other Transparent Electrodes		93
6.4	Integrators			94
	6.4.1	Lenslet Integrators		96
	6.4.2	Rod Integrators		100
	6.4.3	Integrators for Projectors with Laser or LED Illumination		102
	6.4.4	Other Integrator Types		103
	6.4.5	Light Guides		104

		CONTENTS	vii

6.5	Polarization Components	105
	6.5.1 Absorptive Polarizers	106
	6.5.2 Reflective Polarizer Technology	109
	6.5.2.1 Brewster angle reflection	109
	(a) Brewster plate	109
	(b) MacNeille polarizing prisms	111
	6.5.2.2 Birefringent multilayer reflective polarizer	114
	6.5.2.3 Bertrand-Feussner prism	116
	6.5.2.4 Wire grid polarizers	117
	6.5.2.5 Other reflective polarizers	120
	6.5.3 Polarization Conversion Systems	121
	6.5.3.1 Polarization recycling	123
	6.5.4 Polarizing Beam Splitters in the Imaging Path	124
	6.5.5 Compensation Films	126

7 Projection Lenses and Screens — 131
7.1	Projection Lenses	131
	7.1.1 Three-lens Projectors	132
	7.1.2 Single-lens Projectors	132
	7.1.3 Zoom Lenses, Focal Length and Throw Ratio	136
	7.1.4 Projection Lens Offset	137
	7.1.5 Matching the Projection Lens to the Illumination Optical Path	141
7.2	Projection Screens	142
	7.2.1 Projection Screen Gain	144
	7.2.2 Multiple Projectors and Screen Gain	146
	7.2.3 Rear Projection Screens	147
	7.2.3.1 Fresnel lens	147
	7.2.3.2 Fresnel lenses for thin RPTV systems	150
	7.2.3.3 Rear projection CRT screens	152
	(a) Double lenticular screens	152
	(b) TIR screens	155
	7.2.3.4 Microdisplay and light-valve rear-projection systems	157
	7.2.4 Front Projection Screens	159
	7.2.4.1 Light rejecting front projection screens	160
7.3	Speckle in Projected Images	162
	7.3.1 Speckle in Rear Projection Systems	162
	7.3.2 Speckle with Laser Illumination	164

8 Light Sources for Light-valve and Microdisplay Projection Systems — 169
8.1	Lamp Parameters	170
8.2	Types of Projection Lamps	170
	8.2.1 Xenon Lamps	171
	8.2.2 Metal-halide Types	172
	8.2.3 The UHP Lamp	174
	8.2.3.1 Temporal properties of UHP lamps	177
	8.2.4 Tungsten-halogen Lamps	178
	8.2.5 Electrodeless Lamps	179
8.3	Lasers as Projection Light Sources	180
	8.3.1 Choice of Laser Wavelengths	181
	8.3.2 Laser Designs Suitable for Projection Applications	183

viii CONTENTS

		8.3.2.1 Laser architectures	184
		8.3.2.2 Laser wavelength generation	187
	8.3.3	Laser Safety	189
8.4	Light Emitting Diodes as Projection Light Sources		190
	8.4.1	Performance Improvements in LEDs for Projection	191
	8.4.2	Color with LEDs	192
	8.4.3	Thermal Issues with LEDs	194
	8.4.4	LED Drive Issues	195
8.5	Efficacy and Lumen Output		198
8.6	Spectral Characteristics of Lamps		200
	8.6.1	Lamp Spectral Emission Lines	200
8.7	Light Distribution from a HID Lamp		201
8.8	Lamp Life		202
	8.8.1	Lamp Servicing	203
	8.8.2	Failure Mechanisms	203
		8.8.2.1 Measurement of lamp life	203
8.9	Reflectors and Other Collection Systems		206
	8.9.1	Reflectors with Conic Sections	206
	8.9.2	Compound Reflectors	206
	8.9.3	Constant Magnification Reflectors	208
	8.9.4	Refractive Collection Systems	208
	8.9.5	Collection Systems for LEDs	209
8.10	Lamp Ballasts and Ignitors		212

9 Scanned Projection Systems — 217

9.1	CRT Projectors		217
	9.1.1	Three-lens CRT Projectors	218
	9.1.2	One-lens CRT Projectors	220
	9.1.3	Convergence of CRT Projection Systems	221
	9.1.4	Lumen Output of CRT Projectors	223
9.2	Scanned Laser Projectors		225
	9.2.1	Raster Scan Patterns	226
	9.2.2	Laser Projectors with Two-axis Scanning	228
	9.2.3	Laser Projectors with a Single Scanning Axis	229

10 Microdisplay System Architectures — 233

10.1	Microdisplay Systems		233
10.2	Three-panel Systems with Transmissive Microdisplays		234
	10.2.1	Three-panel Equal Path	235
	10.2.2	Unequal Path Systems	237
10.3	Three-panel LCoS Projector Architectures		239
	10.3.1	Three Polarizing Beamsplitters with a Dichroic Combiner	239
	10.3.2	Four-cube LCoS Architectures	240
		10.3.2.1 Four-panel, high contrast LCoS architecture	241
	10.3.3	Three-panel, Three-lens Projectors	242
10.4	Single-panel Projectors		243
	10.4.1	Sub-pixelated Projectors	244
		10.4.1.1 Microfilter projector	244
		10.4.1.2 Angular color separation projectors	245
		10.4.1.3 Resolution of sub-pixelated projectors	246

		CONTENTS	ix

	10.4.2 Color-field Sequential Systems	249
	10.4.2.1 Addressing color-field sequential systems	249
	10.4.2.2 Color wheel and related systems	251
	10.4.2.3 Three-light-source field sequential systems	253
	10.4.2.4 Address-and-flash systems	254
	10.4.2.5 Rotating drum systems	255
	10.4.2.6 Scrolling color systems	256
10.5	Two-panel Systems	259
10.6	Schlieren Optics-based Projectors	259
	10.6.1 Dark Field and Bright Field Systems	260
	10.6.2 Schlieren Light-valve Systems	262
10.7	Stereoscopic 3D Projectors	262
	10.7.1 Separation by Polarization	263
	10.7.2 Stereoscopic 3D with Color Separation	263
	10.7.3 Eye-sequential 3D Systems with Active Glasses	264

11 Modeling Lumen Output — 269

11.1	Simplified Model	269
11.2	Light Collection and Étendue	271
	11.2.1 Definition of Étendue	271
	11.2.1.1 Étendue at a flat surface	273
	11.2.2 Étendue Limited Systems	274
	11.2.3 Lumen vs Étendue Function	274
	11.2.3.1 Étendue conserving transformations	278
	11.2.3.2 Shape conversion	278
	11.2.3.3 Usable étendue	279
	11.2.3.4 Limitations of lumen vs étendue model	281
11.3	Integrators and Lumen Throughput	281
	11.3.1 Overfill Losses	282
	11.3.2 Integrator Étendue and Collection Efficiency	283
11.4	Microdisplay and Light-valve Properties	284
	11.4.1 Panel Transmission	284
	11.4.2 Modulation Efficiency	287
	11.4.3 Duty Cycle	287
11.5	Full Colorimetric Model of the Projector	287
	11.5.1 White Light Throughput prior to Color Correction	289
	11.5.2 Color Correction to the desired White Point	291
	11.5.3 Single-panel Color Sequential Projectors	293
	11.5.4 Colorimetric and Throughput Issues with Projectors with more than Three Primary Colors	294
	11.5.5 Color Separation Efficiency	294
11.6	Problems with Lumen Throughput Calculations	296
11.7	Lumen Output Variation in Production	296

12 Projector Lumen Throughput — 301

12.1	Throughput of a Simple Transmissive Projector	301
12.2	Throughput in a Three-panel Projector	304
12.3	Throughput Estimate using the Full Colorimetric Model	306

13 Characteristics and Characterization — 311

13.1	Characteristics of the Human Visual System	312
13.2	Spatial Characteristics of the Image	313

x CONTENTS

13.2.1 Pixel Count	313
13.2.2 Modulation Transfer Function	314
13.2.2.1 MTF of raster-scanned displays	315
13.2.2.2 MTF of sampled systems	316
13.2.2.3 MTFs of other elements	318
13.2.2.4 Convergence	320
13.2.2.5 MTF of projection screens	322
13.2.2.6 MTF of electronic images vs film	322
13.2.3 Image Quality Metrics	323
13.3 Luminance, Contrast and Color	325
13.3.1 Luminance and Brightness	325
13.3.1.1 Measurement of luminance	326
13.3.1.2 ANSI lumens	327
13.3.1.3 Center weighted lumens	329
13.3.1.4 Luminance patch tests	329
13.3.2 Contrast	330
13.3.3 Colorimetry	332
13.3.3.1 The specification of color	332
13.3.3.2 The gamut of real colors	333
13.3.3.3 White point of displays	334
13.3.3.4 Color gamuts of displays	335
13.3.3.5 Expanded color gamuts and displays with more than three primary colors	337
13.3.4 Measurement of the Luminance and Color of Projection Systems	340
13.3.5 Display Gamma and Gray Scale	342
13.4 Image Content-dependent Adaptive Processes	343
14 Image Artifacts	**347**
14.1 Spatial Artifacts	347
14.1.1 Moiré: Alias and Beat Frequencies	348
14.1.1.1 Origins of Moiré	349
14.1.2 Screen Door Effect or Pixelation	353
14.1.2.1 Depixelation	354
14.2 Temporal Artifacts	355
14.2.1 Flicker	355
14.2.2 Image Smear, Judder and Motion Blur	358
14.2.3 Artifacts in Color-field Sequential Systems	360
14.2.3.1 Color breakup due to motion in the image	360
14.2.3.2 Color breakup due to saccadic motion of the eye	361
Appendix 1 Radiometry and Photometry	**365**
Appendix 2 Colorimetry	**371**
A2.1 Introduction	371
A2.2 CIE 1931 2° Color Matching Functions	372
A2.3 Calculation of Color	373
A2.4 Color Temperature	375
A2.5 The Chromaticity Diagram	375
A2.6 Color-luminance Difference Formulas	375
A2.7 Measurement of Color	377
A2.8 Tabulated CIE 1931 2° Photopic Color Matching Functions	380

Appendix 3 Lumen vs Etendue Parametric Model   **383**
   A3.1 Introduction   383
   A3.2 Definition of Étendue   384
   A3.3 Étendue at a Flat Surface   385
   A3.4 Étendue of a Cylindrical Surface   385
   A3.5 Lamp/Reflector Model   386
   A3.6 Comparison of Measured Data to the Model   389

Appendix 4 Glossary   **393**

Index   **425**

Foreword

When the first edition of this book was published, it was the third volume in this series. Now, nine years later, it is the first to appear as a second edition in a series which its publication will increase to fifteen works.

Much has changed in the intervening years. In my foreword to the first edition, I wrote *"Projection displays have existed since before the invention of the cinema, but it is comparatively recently that there has been rapid development of electronically addressed projection display technologies and literally an explosion in their successful commercialisation."* Well, that last statement was true at the time, but what no-one foresaw was that one of those explosive successes, rear projection TV, would implode as rapidly as it had exploded. Other changes over almost a decade are that new UHP, LED, laser and other light sources have become or are becoming available – some of them at commercial prices. The spectral properties of new light sources have enabled colour gamuts to be increased to much wider values than ever before possible. Liquid crystal on silicon displays and projection screen designs have been developed extensively. Following a successful resolution of the content security issue, electronic cinema technology is finally being rolled out at a significant and increasing rate.

This edition contains substantial revisions and additions to reflect technological progress made and the increased maturity of the topic. Those sections covering light sources for projection displays, liquid crystal on silicon, display colour gamuts, electronic cinema and projection screen design having received the greatest amount of revision. It retains the format of the first edition, apart from the fact that it is not now divided into sections. The book is structured in a way which makes it easy to read straight through or as a source of information into which the reader may selectively dip. Chapter 1 describes the structure. Chapter 2 is an overview of markets and applications which will help those not familiar with the subject to place it in context and perspective. Chapters 3 to 9 cover technical aspects of all the components of projection displays. Chapters 10 to 14 deal with the systems aspects of architecture and performance, display characterisation and artefacts. Four appendices cover the basics of radiometry and photometry, colorimetry, étendue and a glossary.

In its first edition, this book proved to be an invaluable source of information for scientists, engineers and technicians working in the field of projection display technology and projection display product development and also for those who need to understand the strengths, limitations and differences in performance between different projection display technologies for the purposes of specification and purchase.

This second edition will enable this book to remain the most comprehensive work available on projection displays and I thank the authors for making the effort to update it.

Anthony C. Lowe
Series Editor
Braishfield, UK.

Preface to the Second Edition

When the first edition of this book was published, the cinema projector was perhaps the most familiar projection display to the average person. Other common film projectors included slide projectors and viewgraph projectors. The advancement of digital imaging technology and projection displays since the first edition was published has swept away virtually all film projectors except cinema. Even in the cinema, digital projectors are replacing film projectors at a rapid rate.

Direct view displays, when the first edition was published, largely meant cathode ray tube (CRT) displays. CRTs have largely been replaced on the desktop by LCD panels. They are also rapidly loosing market share in TV to LCD-based receivers. In 2007 the worldwide unit sales of LCD displays was larger than the unit sales of CRTs for the first time, with market shares of 47 % and 46 % respectively. This leaves a total of only 7 % for all other display technologies, including plasma and projectors. Yet the total display market is huge, so even 7 % is a very big number in either units or dollars.

Direct view displays are capable of high resolution, luminance and contrast, but they have a serious problem: it is difficult and expensive to make a direct view display large enough so an audience of more than a few can view it simultaneously. While LCD displays up to 108″ and plasma displays up to 150″ have been demonstrated, these systems are extremely expensive and are hard to justify for most applications.

For images larger than about 60″ or so, electronic projection displays can provide a more cost-effective solution than LCD or plasma. Electronic projection displays are capable of generating images as large, bright and high resolution as film projectors can make. The projection display industry has worked hard to build projection displays that meet the needs of large-screen displays. Size, purchase cost and operating costs of projectors have dropped rapidly since the first edition was published. Resolution, image size and luminance have increased rapidly, even while cost and projector size have decreased.

There are also some emerging markets for small-screen projection displays. For example, there are forecasts that pico-projectors embedded into cell phones may someday become almost as universal as cell phone cameras. While these projectors are not expected to make images larger than about 20″, the image size and resolution will be larger than the LCD, OLED or other display built into the cell phone.

The impetus for the development of improved projection displays has come from many sources. Two technologies, neither directly related to projection technology, have had a particular influence on projection displays. First, computers, especially laptop computers, have provided a portable source of a high quality electronic video signal. Over the last 15–20 years, projector manufacturers have rushed to fill this need with portable and semi-portable projectors. A second technology providing impetus for the development of projection displays was high definition TV (HDTV). While LCD and plasma displays dominate HDTV in sizes up to about 60″ diagonal, the price differential between direct view and projection in larger sizes ensures a market for HDTV projectors.

PREFACE TO THE SECOND EDITION

Most of the developments in projection display systems and components have been made in industrial research and development facilities rather than academic settings. Due to the commercial and multi-disciplinary nature of the projection industry, there have been limited opportunities for an individual in an academic environment to specifically study electronic projection technology. One effect of this was the lack of texts and published compendia on many projector components and sub-systems. This happened because of both the proprietary and commercial nature of the components, and the commercial pressures that make it difficult for many workers in industry to publish.

The first edition of this book attempted to fill this vacuum. Technology has advanced considerably since it was published and many new technologies are available to the projection engineer. This text attempts to add to the technologies described in the first edition all the projection related technology developed since the first edition was published. In addition, there were details of a few technologies known to the authors when the first edition was written we were not allowed to disclose because of their Philips proprietary nature. The UHP lamp and liquid crystal on silicon microdisplays are two examples. Many additional details on these two technologies are now public and described in this edition. In addition, discussions covering technologies no longer used in the projection industry have been deleted from this edition.

Many disciplines are required for the successful design of a competitive projection system. These include optical engineering, opto-electronics, illumination engineering, lamp physics, human factors, mechanical engineering, materials science, semiconductor device design and manufacturing, manufacturing engineering, software engineering, and digital, analog and power electronics design. Electronic design must be done at the chip, board and system level, with multiple layers of embedded software. A knowledge of video source formats is also essential. Rapid technological change and competitive pressure mean all these disparate elements must be integrated with severe time and cost pressures. The days when a projection system, such as the now defunct Talaria, could be introduced and remain in production for more than 20 years with minimal changes are gone forever.

Much of the published material on projection systems has been presented in unrefereed publications, such as conference proceedings published by the Society for Information Display (SID) and the SPIE – The International Society for Optical Engineering. Another major source of information on projector components is the patent literature, although by the time a patent becomes public, the information can be more than three years old. Another problem with patents is many patents are issued but not all are important, making it difficult to sort out which patents represent real technology and which represent just ideas.

Texts on optical system design and other related subjects often will use a projection system as an example to demonstrate a particular point of optical science. This example will give an understanding of the illustrated point, without necessarily relating it to other problems encountered in projection displays.

This book, then, is directed at those people with a solid technical background in one or more of the disciplines required in projector design, but with limited experience in projection systems and the many technologies, other than their own, employed in the design and execution of a projection display.

The book concentrates on the optics of projection systems, including the image-generating components. It dwells at length on the system considerations of the light-engine: the subsystem which converts the electrical signal and power into an optical image. The light engine is a complete package, one that can be, and often is, manufactured and sold independent of the electronics, the cabinet, and other system components. The electronics associated with the projector will not be discussed in this book since these subjects are covered in other texts. Inclusion of even an overview would make this book impossibly long.

Any projection system manufacturer depends heavily on a wide array of component vendors. A goal of this book is to elucidate the requirements and limitations of each component and how the components fit into the light engine system. This should allow the projection engineer to ask intelligent questions of vendors and to set specifications for the components. These specifications must not only be achievable by the vendor, but they must ensure the component achieves its goal when integrated into the system.

The discussions of many of the subjects in this book are based on a proprietary Philips Research report entitled *Optical Architectures for Light Valve Projectors*, written in 1995 by Matthew Brennesholtz with guidance and technical inputs from the late William Guerinot, Project Leader of the Philips Research-Briarcliff Advanced Projection project and Edward Stupp, Department Head of the Briarcliff Display Systems Department. Short courses and papers presented by Mr. Brennesholtz at SPIE and SID meetings, and invited papers at SID meetings presented by Dr. Stupp, developed additional material and pointed to the industry-wide need for a book on this subject.

This book combines material from these sources with large amounts of additional published material. The book also includes a significant amount of previously unpublished material from Philips and other companies. In the course of writing this book, it was discovered that there seems to be a large body of knowledge that 'everyone' in the projection business knows but for which no one can provide a citation. If any contributor recognizes material in this book without an appropriate citation, accept our apologies and please contact us with the proper reference. In this way, the bibliographies can be updated to ensure correct future citations.

Many people provided information and assistance while the first edition of this book was being written. Most of this material has been carried forward into this second edition. Geert Lievens of Hitachi America provided data on projection CRTs. Colleagues at Philips Research, Briarcliff, including Jerry Kramer, George Melnik, Christian Hentschel and Caaj Greebe provided important discussions and inputs to this book. The aid of John Fox of the Philips Intellectual Property Department, Betsy McIlvaine and Joan Fuller of the Philips Research Library and Diane McCoach and Antonelle Hay-Passatino of the Philips graphics department was invaluable. The presidents of Philips Research in Briarcliff while the first edition was written, Peter Bingham and Barry Singer, gave the authors the time and resources to produce the first edition.

Additional information for the second edition was provided by Serge Bierhuizen of Philips Lumileds, Tony McGettigan of Luxim, Jens Pollmann-Retsch and Holger Mönch of Philips Research Aachen, Christian Hoepfner of Luminus Devices, Jacob Christensen of DNP Denmark, Charles Bruzzone of 3M, Francis Nguyen of Osram Opto-Semiconductors, William Bleha of JVC North America, Patrick Candry of Barco, Greg Niven of Novalux and Chris Chinnock of Insight Media.

While the first edition was done under the auspices of Philips, the second edition was a more independent effort. One place this can be seen is in the graphics. In the first edition all the figures had a uniform look since they were all redrawn from the source by the Philips graphics department. In the second edition, the figures were often copied directly from the original publication with minor editing, leading to a wider variety of graphical styles.

We received considerable positive feedback on the first edition. We hope the second edition will serve the industry as well.

Matthew S. Brennesholtz
Pleasantville, New York

Edward H. Stupp
Jackson, New Jersey

About the Authors

Mr. Matthew Brennesholtz has worked in the display field since receiving his Masters of Engineering degree from Cornell University in 1978. He has worked on direct view CRT systems and projection systems based on CRTs, Talaria oil-film light valves, DMDs, LCDs and LCoS. In addition to system level development work, he was involved in the design and testing of optical components for projection systems. He helped develop both the imagers and the optical systems for the Philips single-panel LCoS system. He also has experience in the manufacturing of CRTs, Talaria and LCoS systems. Currently he is at Insight Media providing consulting and other services to the display industry. Mr. Brennesholtz is a Senior Member of the SID, plus a member of SMPTE and the SPIE. He can be reached at matthew@insightmedia.info.

Dr. Edward H. Stupp is a display consultant with nearly 30 years' experience in many display-related technologies, including projection and direct-view systems and components. Specific experience includes system architectures, single crystal Si-, polycrystalline Si- and amorphous Si-based microdisplays, liquid crystal materials and phenomena, cathode ray tube projectors and electrophoretic devices and systems. He was previously Department Head, Display Systems at Philips Research, Briarcliff Manor, NY. He is a Fellow of the SID. He can be reached at stuppassoc@optonline.net

1
Introduction

1.1 Overview of Projection Displays

An electronic display is a device or system which converts electronic signal information representing video, graphics and/or text to a viewable image of this information. Displays are categorized as being direct-view, projection or virtual. Direct-view displays produce their images on the surface being viewed. The images from projection displays are formed on auxiliary surfaces, which are physically separated from the image-generating component. With a virtual display, there is no real image in space; the optical signal is brought to a focus only on the retina.

Projection displays produce larger images from electronic signals than are normally achieved by direct-view technologies. The images are created by light-emitting devices, such as cathode-ray tubes or lasers, or by modulating the light from an illumination system with a device called a light valve. Projection displays are traditionally thought of as having large images. The technologies involved are also being applied to very small portable systems, including head-mounted and helmet-mounted systems. These produce images perceived to be large because they occupy a significant fraction of the retinal field-of-view.

This book will only consider projection displays suitable for group viewing, i.e. a display capable of being viewed by two or more people simultaneously. The technologies to be described for group viewing projectors are directly applicable to personal use in, for example, a head-mounted display. However, these personal applications frequently have significantly different performance requirements and unique configurations. Few of these constructions are well documented in the technical literature.

Projection displays were among the earliest electronic information display technologies. Most early video receivers used projection to increase the viewed image size from that produced by the small CRTs used to generate the image (Wolf 1937). The other large screen technology, the Scophony system, used a scanned beam technology remarkably similar to a modern scanned beam laser projector (Okolicsanyi 1937). The first large image size, high luminance projector was described and demonstrated nearly seventy years ago (Fischer 1940). For a more complete review of the history of projection technology,

Projection Displays, Second Edition   M. Brennesholtz, E. Stupp
© 2008 John Wiley & Sons, Ltd

see Brenneshoitz (2007). Today, projection displays are used in a variety of configurations, including consumer products, presentation products for the institutional market, large theater displays, very large field-of-view systems for simulator application, and physically small displays for personal use, e.g. using a head- or helmet-mounted projection display. The luminous outputs of projectors range from just a few lumens to 30 000 lumens, with the higher outputs obviously being applied to larger screen sizes. When 30 000 lumen projector output is not sufficient, two high-output projectors can be stacked and converged at the screen. While projectors with more than 30 000 lumen output are seen as technically feasible, the markets for these extreme projectors are so small that, for the time being, projector stacking provides a more cost-effective solution.

Historically, projectors have served to produce the larger image sizes while direct-view displays have served to produce the smaller images. The correlation between size and technology appears to be breaking down. Ultra-large displays today are most commonly direct-view LED systems. At the 2008 Consumer Electronics Show a 150″ direct-view plasma system was demonstrated by Panasonic. 108″ direct-view LCD displays have also been demonstrated. The extreme high cost of these large direct-view flat panel displays means that most large images will continue to be generated by projection displays.

The reverse appears to be happening for small images. LCD, OLED and a host of other technologies can make cost-effective small displays, but they cannot make 20″ displays from a system that fits in a shirt pocket. So 'pico-projectors', typically the size of a cell phone, are beginning to appear in the marketplace.

The performances of projection displays continue to improve as the costs of manufacturing them continue to decrease. CRT-based projectors represent a declining fraction of the projection market due to availability of the high resolution, high brightness, high contrast and low prices of microdisplay-based projectors. These newer projectors, based on microdisplay and light-valve technologies, now dominate the market, as measured by manufacturers' revenue. The high performance, thinner profile and low price of light-valve projectors is expected to largely drive CRT projectors out except in the few niche applications where CRTs retain advantages.

At the heart of all projectors is the light engine. This is the subsystem within the projector which converts the incoming electrical signal into the intensity-modulated two-dimensional optical image. In the following chapters, the systems, subsystems and component technologies used in the light engines of projection displays will be discussed. One of the objectives of this book is to provide the projector designer with the tools for evaluating the potential performance of a light engine design without having to commit to physical hardware. Models are developed which can evaluate the anticipated luminance and chrominance from component data.

The design of a projector requires key components to execute the light engine. These include the image forming devices, lamps, and optical components. The positive and negative attributes of these components and modules strongly influence the system performance characteristics. The more important of these are discussed in detail before the models are developed.

The design and fabrication of a projector is truly a multidisciplinary activity. Table 1.1 illustrates some of these requisite technologies to implement a projection system. The breadth of the technologies required for execution of a design is belied by the simplicity of the listing given in this table. Within each of the technological categories in Table 1.1, there are far larger sublists of technologies which are required for successful implementation. For some technologies employed in projection systems, dedicated industries are required to provide the technical infrastructure for success. Among these are CRTs, light valves, lamps and some aspects of optics.

A projector based on emissive image sources, such as CRTs, is illustrated in Figure 1.1. This type of projector will typically use separate sources for generation of the three primary colors, red, green and blue. One of the characteristics of emissive image sources is light-on-demand. Light is only generated in response to an appropriate signal. This attribute is valuable in producing images with large white-to-black dynamic ranges. The three images created by systems with emissive devices can be combined internally before the projection lens to form the full color image or the recombination can be done by superposition of the three-color images at the screen plane.

Table 1.1 Projection system technologies.

Subsystem	Technologies
Image generation	• Emissive image sources, such as cathode-ray tubes and lasers, or • Light modulating devices, including liquid crystal devices and micro-electromechanical devices with • Silicon or other semiconductor material backplanes for the imagers and • Very large, very flat transparent substrates.
Optics	• Light collection and utilization, and • Light polarization separation and recovery, and • Spectral filtration, separation, and recombination, and • Lenses, and • Screens
Lamps	• Lasers, • LEDs • Tungsten-halogen lamps, or • High intensity discharge (HID) lamps, with • Lamp ballasts for HID lamps and drive circuitry for lasers or LEDs
System	• System design, incorporating • Human factors and interfaces, including luminance, colorimetry, and resolution, and • Artifact minimization
Electronics	• Signal processing and conditioning, and • Interface electronics between microdisplay, light valve or CRT and signal electronics, and • External video interface circuits, and • Power supplies
Mechanics	• Thermal design, and • Materials, and • Stable opto-mechanical light-engine construction, and • Projector packaging

The lamp or other light source in a microdisplay or light-valve projector, on the other hand, continuously generates internally the peak intensity light. This light is attenuated by the microdisplays or light valves to a level appropriate for the input signal at that image point. The range of this attenuation determines the white-to-black dynamic range. These light modulating projectors are based on one, two, three, or more light valves. Figure 1.2 illustrates a generic light-valve projector, showing different light paths for the primary colors.

The illumination system in these projectors provides high intensity light, using lamps or other light sources normally specifically developed for projector application. Optical components separate this light into three color bands, corresponding to the primary spectral colors. The microdisplay(s) spatially modulate this light by varying the amount of light passing through an image point that enters the entrance pupil of the projection lens. For those systems which use more than one microdisplay or light valve, the color images must be superimposed to form a single full color image projected onto the screen.

4 INTRODUCTION

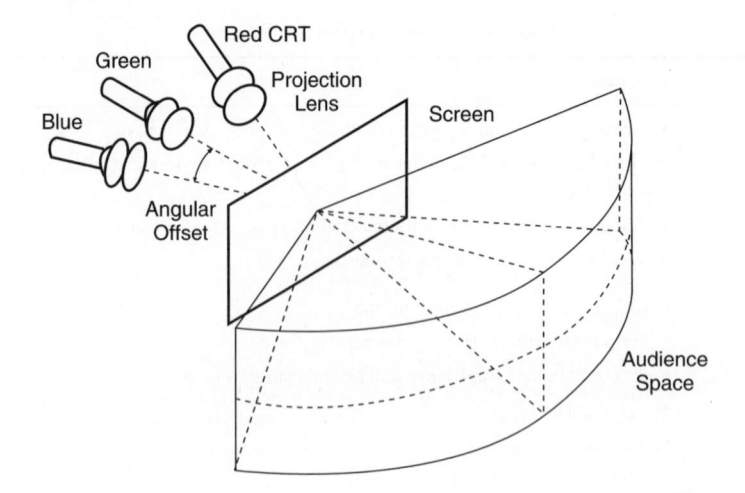

Figure 1.1 Schematic of CRT rear-screen projection system. The screen provides wide-angle horizontal viewing.

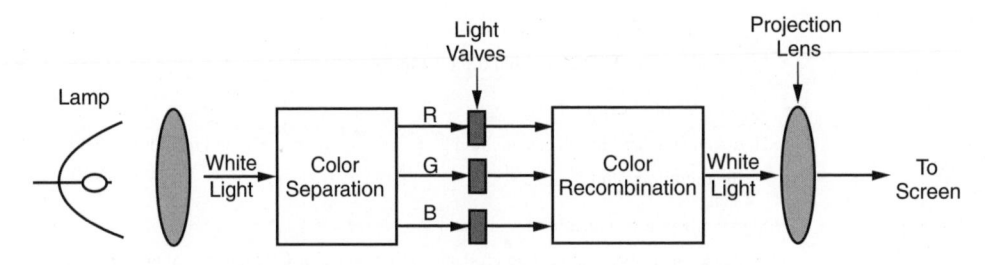

Figure 1.2 Schematic representation of a light-valve projector.

1.2 Book Organization

This book is divided into 14 chapters and four appendices, as described in Table 1.2.

1.3 What is not Covered

Several types of projectors will not be discussed in this book. These include the overhead projector and projector-type systems intended to produce virtual images for personal use. These systems are often called HMDs for Head or Helmet Mounted Displays. Since the technologies used in these systems are similar to the technology used in normal projection systems, the discussions in this book should provide the designers of these systems with requisite information. For more details on the use of microdisplays in near-to-eye systems see Armitage *et al.* (2006).

The overhead projector by itself is nothing more than a projector with very large format static images on film. It can become an electronic projector when coupled with an image generating device such as an active-matrix display like the ones used for laptop computers. Since the displays have been designed and manufactured independently of the illumination source (the overhead projector), it is difficult to discuss the performance of these systems without specific details of both the light modulator and the projector. Another reason for not covering these displays is that they have nearly vanished from conference rooms and have been replaced by full electronic projectors.

Table 1.2 Chapters in *Projection Displays*.

Chapter	Contents
Chapter 1	Introductory material on this book
Chapter 2	Very brief introduction to the markets and market requirements for projection displays
Chapter 3	Cathode ray tubes as an image source for projection displays
Chapter 4	Liquid crystal devices as image sources for projection displays
Chapter 5	Micro-electro-mechanical (MEMs) devices as image sources for projection displays
Chapter 6	Optical components for projection displays, part 1, including filters, polarizers and integrators
Chapter 7	Optical components for projection displays, part 2, including projection lenses and projection screens
Chapter 8	Lamps and solid-state light sources
Chapter 9	The architectures of emissive projection systems
Chapter 10	Architectures for microdisplay, light-valve and light-amplifier projection systems
Chapter 11	Techniques for modeling and predicting the lumen throughput and colorimetry of projection systems
Chapter 12	Numerical examples of the modeling techniques developed in Chapter 11
Chapter 13	The characteristics of projection displays important to their performance and how to measure these characteristics
Chapter 14	The image artifacts present in some projection displays, plus how to evaluate and avoid them
Appendix 1	Background information on photometry and radiometry
Appendix 2	Background information on colorimetry
Appendix 3	Derivation of a semi-empirical model used to forecast the lumen throughput of étendue limited projectors
Appendix 4	A glossary of terms and acronyms related to projection design and the projection industry in general.

Also not covered in this book are the electronics and electronic technologies associated with making a projector. These may include analog and digital signal processing, input signal formats (NTSC, PAL, VGA, ATSC, HDTV, DCI, etc.), signal sources, power supplies, tuners, and others which are required to make a light engine into a projection product. For more information on these topics see Berbecel (2003); Myers (2003) or any of the many other standard texts.

References

Armitage, D., Underwood, I. and Wu, S.-T. (2006) *Introduction to Microdisplays*. Wiley-SID Series in Display Technology. Chichester: John Wiley & Sons, Ltd.

Berbecel, G. (2003) *Digital Image Display: Algorithms and Implementation*. Wiley Series in Display Technology. Chichester: John Wiley & Sons, Ltd.

Brennesholtz, M.S. (2007) The evolution of projection displays: from mechanical scanners to microdisplays. *J. SID* **15(10)** October: 759–74.

Fischer, G. (1940) Auf dem Wege zur Fernseh-Grossprojektion. *Schweiz. Archiv Angew. Wiss. Technik*, **6**: 89–106.

Myers, R.L. (2003) *Display Interfaces: Fundamentals & Standards*. Wiley Series in Display Technology. Chichester: John Wiley & Sons, Ltd.

Okolicsanyi, F. (1937) The Wave-slot, an Optical Television System. *Wireless Engineering*, **14**: 526.

Wolf, M. (1937) The enlarged projection of television pictures. *Philips Technical Rev.* **2**: 249–53.

2
Markets and Applications

Electronic projection displays dominate the large image display market. Film-based projection, including slides and transparencies for overhead projectors have largely been displaced by electronic projection displays. Electronic projection systems are starting to displace film-based cinema systems. The large image area projection systems constitute a large and growing market, especially for institutional applications. On the other hand, large screen consumer projection applications have suffered from competition from flat-panel displays, including both LCD and plasma.

In this chapter, the applications of projection displays, including the performance requirements for the medium to large-size applications, will be discussed. A discussion of various applications will follow. Head-up and helmet-mounted projection applications will not be discussed.

2.1 Overview

Projection systems use a variety of image forming technologies to generate the projected images. These include light modulating image sources, such as microdisplays, light valves and light amplifiers, and emissive image sources, such as CRTs. The current generation of microdisplays is based primarily on either liquid crystal devices (LCDs) or microelectromechanical systems (MEMS). Discussions of the various technologies and systems are the subjects of following chapters.

2.1.1 Microdisplays, Light Valves and Light Amplifiers

Central to the design of most microdisplays or light valves is an active-matrix addressing scheme. This matrix consists of a conductive array of orthogonal columns and rows, insulated from each other at their intersections. The widths of these conductors are small compared with the pitch of the matrix. There is at least one switching device at each intersection. Each switching device is connected to an electrode patterned to maximally fill the space between adjacent columns and rows. This space is called a pixel.

Projection Displays, Second Edition M. Brennesholtz, E. Stupp
© 2008 John Wiley & Sons, Ltd

The switching devices are externally addressable, so that information to be displayed can be directed to specific pixels.

If there are m rows and n columns of pixels in the image forming device, a direct connection to the pixel scheme would require about $m \times n$ external contacts. The active matrix permits the reduction of contacts to the device to about $m + n$. If the addressing circuitry is integrated with the active matrix, the number of external contacts can be reduced to a few tens of contacts since only contacts for the serial input data, clocks, and sources of power are required. Distribution of the signal to the pixels is carried out by the addressing circuitry. The active matrices currently are being executed using amorphous silicon thin-film transistors (α-Si TFTs), polycrystalline thin-film transistors (poly-Si or p-Si TFTs), and single crystal silicon transistors.

Microdisplay and light-valve projectors have been marketed with a very wide range of lumen outputs. On the low end are pico-projectors, typically with 5–25 lumen output and QVGA (320×240) through SVGA (800×600) resolution. On the high end of the performance scale, projectors are being produced with 30,000 lumen output with 2 K (2048×1080) resolution, using a 7000 W lamp. Projectors with still higher resolution are available but at less than 30,000 lumens. A wide range of luminous outputs and resolutions are available for products intermediate between these.

2.1.2 Emissive Systems

This category of projector is dominated by the previously ubiquitous CRT projector. Sales of CRT projectors have declined sharply in recent years as the price of microdisplay projectors has declined and their performance increased. The luminous outputs of CRT projectors are typically less than about 300 lumens, for a full white field. The resolvable information in the image can be as high as XGA resolution, although much higher addressability is possible.

2.1.3 Laser-based Projection Technology

Lasers can be used as either an illumination source or a direct writer of the information in a projection system. Microdisplay-based projection systems which use lasers as an illumination source require a projection lens to form the image at the screen. Alternatively, the laser light can be scanned in a raster pattern and modulated to generate the image directly at the screen. This can be done with either two-dimensional mechanical or opto-electronic scanning or one-dimensional mechanical scanning together with a linear array. There are a wide variety of technologies to scan the laser beam. Mechanical scanning can be done with a polygonal mirrored drum rotating at high speed, a resonant galvanometer mirror or non-resonant mirrors. Some designs use two mirrors or drums while others use a single MEMS mirror capable of bi-axial tilts. In some designs, an acousto-optical modulator is used to scan the beam. While mechanically scanned systems predate the laser (Okolicsanyi 1937), most modern scanning systems use lasers as the light source.

Most lasers suitable for projection applications can be modulated at the video rate, typically 30 MHz or above and no additional modulator is needed. If the laser cannot be modulated at the video rate, an acousto-optic modulator is normally used. Acousto-optical modulators are devices in which an electrically generated diffraction grating is formed. This structure diffracts the laser beam to modulate the light in a Schlieren-type optical system (see Chapter 10).

2.2 Applications and Performance Requirements

The major mass-market applications of projection displays include both front and rear consumer projection systems and front projection products for conference rooms and professional and educational presentations. There are also a number of applications that are small in the number of projectors but these projectors are high-value. These applications include visualization, simulation, large venue and digital cinema.

There are also a number of emerging markets for projectors. These projectors are currently either not sold at all or sold in very small numbers. Examples of this category include pico-projectors, low cost personal projectors and pocket projectors.

Projectors can be broken down into a number of categories based roughly on projector size and lumen output. Table 2.1 shows these various categories of projectors.

Pico-projectors represent an entirely new projector category and none is on the market as of the writing of this book. They are a topic of intensive research and development and are expected to become products in 2008 or 2009. For example, at least 16 companies demonstrated functioning pico-projectors at CES '08. Since there are no products on the market, it is not certain how these projectors will be used. It is expected that they will be used to show images generated by portable devices such as cell phones, PDAs, laptop computers, digital still cameras (DSCs) and digital video cameras (DVCs). These pico-projectors come in two versions. The larger version, a companion projector, is expected to have a total size comparable to that of a cell phone and contain its own battery. Integrated pico-projectors are expected to be small enough to integrate into a portable device such as a cell phone.

2.2.1 Differentiators among Projectors

There are several major differentiators among projectors. These differentiators, and the applications that have special requirements based on these differentiators are:

- *Lumen output.* Lumen output of projectors range from about 5 lumens from the smallest pico-projectors to 30,000 for the largest large-venue projectors. No fundamental technical limit prevents higher output projectors from being built.

- *Resolution.* Resolution in projectors and other displays is currently designated by pixel count. Current projectors range from QVGA (320 × 240) to 4 K (4096 × 2160) and beyond.

    ○ HDTV displays typically are either 1280 × 720 or 1920 × 1080. The trend is strongly toward 1920 × 1080.
    ○ Digital cinema is typically 2 K (2048 × 1080), although some projectors are 4 K.

- *Aspect Ratio.* The ratio of the width to the height of the display, given either as a ratio of whole numbers such as 4:3 or a proportion such as 1.85:1.

    ○ Business projectors tend to be 4:3 or 5:4. Newer ones are also 16:10 or 16:9.
    ○ HDTV projectors are all 16:9.
    ○ Digital cinema projectors are all 1.90:1.

- *Contrast* is the ratio of the brightest white to the darkest black a projector can produce.

    ○ Business projectors can be commercially viable with full-on/full-off contrast ratios in a dark room as low as 400:1.
    ○ Contrast of HDTV projectors are typically specified at >1000:1.
    ○ Some specialty projectors have contrast ratios >30 000:1.

- *Efficiency* is becoming an increasingly important figure of merit for projectors. Values range from a low of about 1 lumen/watt to a high of about 10 lumens/watt. Due to the importance of energy conservation and the development of more efficient light sources, this upper limit is likely to go up in the coming years. Energy conservation in battery-powered projectors is considered especially important because it translates into battery lifetime.

Table 2.1 Categories of projection displays.

Market Segment	Applications	Resolution (Typical)	Output Lumens	Primary Energy Source	Volume	Weight (without battery)	Illumination Source
Pico-projector	Integrated/embedded	QVGA–SVGA	5–15 lm	Battery	<40 cm^3	50–125 g	Laser/LED
	Companion projector	QVGA–SVGA	10–25 lm	Battery	50–300 cm^3	125–250 g	Laser/LED
Head-up display	Professional, consumer	TBD	5–25 lm	Vehicle	NA	NA	Laser/LED/lamp emissive
Pocket projector	Professional	SVGA or more	25–100 lm	Battery/mains	300–500 cm^3	400–650 g	Laser/LED
	Consumer	WVGA or more	25–100 lm	Battery/Mains	300–500 cm^3	400–750 g	Laser/LED
Ultra-portable projector	Professional, crossover	XGA or more	100–1200 lm	Mains	>600 cm^3	900–1500 g	Laser/LED/lamp
	Consumer (family, game, combo)	WVGA or more	100–1200 lm	Mains	>600 cm^3	900–2250 g	Laser/LED/lamp
Rear-projection TV	Consumer	720 p or 1080 p	200–1000 lm	Mains	NA	NA	Laser/LED/lamp
Consumer home theater	Consumer	1080 p	500–2500 lm	Mains	NA	NA	Laser/LED/lamp
Business projector	Professional	XGA or more	1200–5000 lm	Mains	NA	NA	Laser/lamp
Vis/Sim projectors	Professional	SXGA or more	500–7000 lm	Mains	NA	NA	Laser/lamp emissive
Large venue/cinema	Digital cinema	2 K or 4 K	5000–15 000 lm	Mains	NA	NA	Laser/lamp
	Auditorium, rental and staging	2 K	5000–30 000 lm Sometimes stacked for 60 k lm	Mains	NA	NA	Laser/lamp

Source: Reproduced from Brennesholtz (2008) with the permission of Insight Media.

A few of these projector differentiators will be discussed in the following sections, Additional differentiators will be discussed throughout the book but especially in Chapter 13.

2.2.2 Requisite Luminance Levels

Projection systems are used with a wide range of audience and room sizes. The requisite luminance levels and image sizes are very much dependent on the application. As the performance of projectors improves, the goals for the luminance output are increasing. The luminous flux of portable or transportable projectors, typically based on microdisplays, must be sufficiently high to permit viewing with a moderate ambient illumination while using a low enough gain projection screen so the entire audience can see the image clearly. It should be noted that the higher the screen gain, the smaller the viewing angle but the brighter the image within the viewing angle.

The relationship between screen gain, luminous flux onto the screen and the screen luminance is given by Equation (2.1).

$$\Phi = \frac{\pi L W^2}{aG} = \frac{\pi L W H}{G} \tag{2.1}$$

where
- Φ is the luminous flux from the projector (lumens),
- L is the screen luminance (cd/m^2),
- W is the width of the screen (meters),
- H is the height of the screen (meters)
- a is the W:H aspect ratio of the screen (e.g. $16:9 = 1.778$) and
- G is the gain of the screen.

Screen gain is discussed in Chapter 7. It characterizes the increase in luminance produced by the screen by concentrating the output light into reduced horizontal and vertical viewing angles.

Equation (2.1) (most often incorrectly) assumes a uniformly illuminated screen. Industry standards have been adopted for characterizing non-uniformly illuminated screens using a flux unit called the ANSI lumen, which is obtained by averaging nine zones on the screen (ANSI 1992). ANSI lumens are discussed in Chapter 13.

A 300 lumen output projector produces a luminance of about 400 cd/m^2 on a 5-foot diagonal screen with gain 1.5. Portable projection systems with up to about 2500 lumen output are available. The higher lumen output permits larger screens and higher ambient illumination conditions while maintaining a high system contrast.

Projection displays for home TV applications must be bright enough for viewing in a moderately illuminated ambient. Rear projection screens are especially good at rejecting ambient light and are available with screen sizes from 50″ (125 cm diagonal) to as much as 80″ (203 cm) and larger. The typical luminance of a modern consumer HDTV display, whether direct-view or rear projector, is in the range of 400–600 cd/m^2.

Front projection home theater systems are available in virtually any size for which the customer is willing to pay. When these screens are used in a darkened, dedicated home theater, ambient light rejection is less important. In the more common case where there is some uncontrolled ambient light, ambient light rejecting front projection screens are also available.

It is difficult to compare accurately the lumen outputs from CRT projectors and microdisplay projectors. The peak output from the CRT system can vary with signal. Among other parameters, it is dependent on the average signal level and the fraction of the total image area occupied by the peak highlight. A microdisplay projector with about 350 lumens output would produce similar or higher perceived brightness levels for most television scenes when compared with a typical consumer CRT projector with 100 lumen output.

12 MARKETS AND APPLICATIONS

2.2.2.1 Flux requirement for presentation and auditorium applications

There are other requirements for sizable audiences, in settings that include both cinema theaters and auditoriums. For the cinema environment, the relevant standard (ANSI 1995) recommends that film cinema center screen luminance be in the range of 12–22 foot-Lamberts, with 16 foot-Lamberts nominal. This light level is chosen in order to avoid flicker with the 48 Hz presentation rate of cinema film projectors. (The cinema projector normally uses two exposures of each frame of film running at 24 frames/second.) For 60 Hz systems, higher screen luminance is acceptable, due to the reduced human sensitivity to flicker at the higher frequency. The same standard recommends that the luminance at the screen edge be in the range of 75–90 % of the center.

The luminance flux requirement for auditoriums and theaters is a function of the screen size and luminance, which in turn is dependent on the audience size. One often knows the expected audience size for a video projection presentation, and wishes to estimate the minimum required screen size for that audience. Screen width can be estimated from Equation (2.2). This equation is purely geometric, based on how much area each person takes, and how much area is available in front of the screen for seating them. Different assumptions can be made for various viewing conditions and seat pitches. Allowance can also be made for areas with no seating such as aisles or conference room tables.

$$W = \sqrt{\frac{NAP_xP_y}{W_A(D_{Max} - D_{Min})}} \quad (2.2)$$

where
- W is the required screen width,
- N is the audience size,
- A is the aspect ratio of the screen,
- P_x, P_y are the x and y seating pitch of the audience,
- W_A is the width of the audience, relative to the screen width,
- D_{max} is the maximum distance of the audience from the screen, in terms of screen height and
- D_{min} is the minimum distance of the audience from the screen.

For example, assume we have the following values:
- A = 1.778 or 16:9,
- P_x = 24″ or 2.00′,
- P_y = 32″ or 2.67′,
- W_A = 1.5, or the audience occupies a space 1½ times as wide as the screen,
- D_{max} = 6, or the furthest away a member of the audience is 6 picture heights from the screen and
- D_{min} = 1.5, or the closest member of the audience is 1½ picture heights from the screen.

Figure 2.1 shows the calculated screen width, in feet, as a function of the audience size. No resolution or projector lumen output factors enter this calculation.

A typical screen gain for a front projection screen would be about 1.5. For rear projection screens, gains of up to 6 have been used in consumer CRT rear projection systems, although RPTV screen gain is more typically about 4 today. Equation (2.2) can be substituted into Equation (2.1) to get the requisite lumens as a function of audience size, giving:

$$\Phi = \frac{\pi L N P_x P_y}{G W_a (D_{Max} - D_{Min})} \quad (2.3)$$

where

- Φ is the lumens required from the projector,
- L is the target screen luminance (cd/m^2),
- N is the number of people in the audience,
- G is the gain of the screen and
- P_x, P_y, W_a, D_{max}, D_{min} are the same as they were for Equation (2.2).

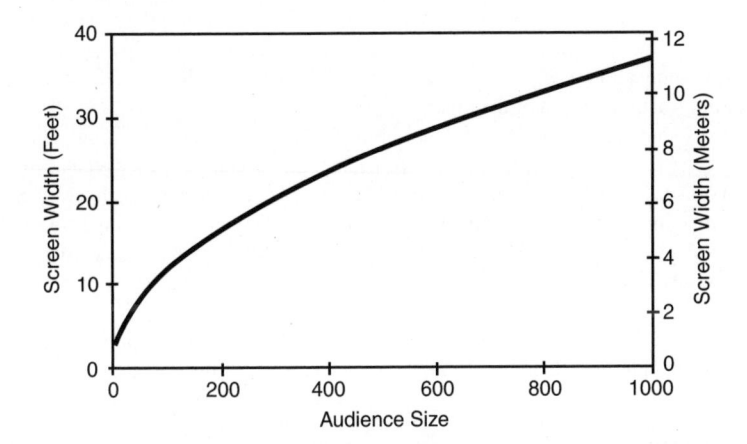

Figure 2.1 Calculated screen size as a function of audience size. Parameters for this example are given in the text.

The projector lumens to produce the requisite luminance for a theatrical presentation L = 16 fL (or about 50 cd/m^2) and for a conference room/auditorium presentation L = 50 fL (or about 150 cd/m^2) are calculated as a function of audience size with Equation (2.3) and are plotted in Figure 2.2.

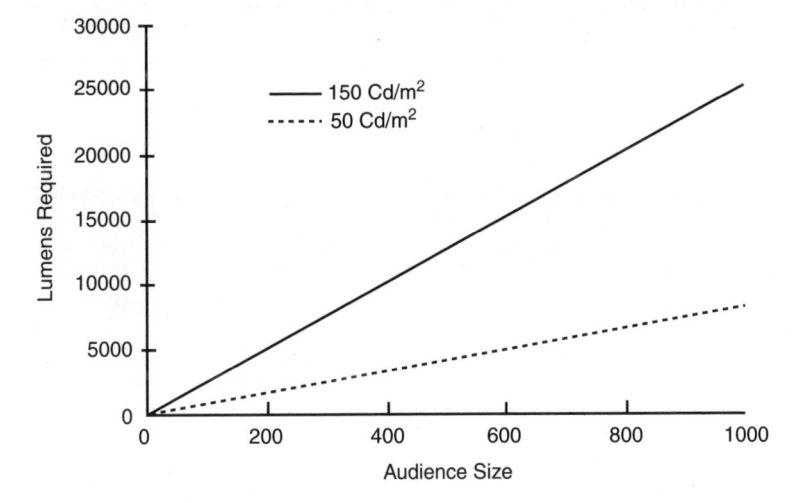

Figure 2.2 Requisite lumen output from projector to achieve luminance levels of 50 cd/m^2 and 150 cd/m^2 on a front projection screen with a gain of 1.4, as a function of audience size.

The highest output projectors available as of this writing have outputs of about 30 000 lumens. This is enough for an audience of about 1100 in a conference room or more than 2500 in a theater, which would be one of the largest movie theaters in the world. A more common large screen in a multiplex seating

14   MARKETS AND APPLICATIONS

1000 people would need about a 12 000 lumen projector. The highest-output projectors are often used for special events such as corporate presentations, product introductions and rock concerts. Most commonly, these projectors are owned by a rental and staging company, not by the end user.

2.2.3   Resolution

Projection displays are made or are being developed with resolutions from below VGA to above UXGA. The appropriate resolution to achieve maximum usable information content in an image is very much dependent on the application. The requisite systems resolutions are functions of the screen sizes, viewing distances and information content of the images.

Table 2.2 gives some resolutions that have been used to build projection systems. This list is by no means comprehensive; non-standard resolution projectors have been built for specialized applications such as military displays. In addition, the microdisplay in a projector may not match the pixel count listed in the table. For example, microdisplays intended for three panel applications often have additional rows and columns of pixels to allow for electronic convergence of the red, green and blue images. In addition, some of the formats such as HVGA are not well established and the pixel count could change as new products are introduced with the same format name.

The aspect ratio for each format is given as either a ratio (1.33) or a proportionality between two integers (4:3). This represents the ratio of the width of the image to the height of the image. Note that

Table 2.2   Pixel counts for projection systems.

Format name	Pixels H	Pixels V	Aspect ratio		MPixels	Application
QCIF	176	144	1.22		0.024	Video conferencing
CIF	352	288	1.22		0.097	Video conferencing
QVGA	320	240	1.33	4:3	0.073	Pico-projectors
HVGA	480	320	1.50	3:2	0.146	PDAs, pico-projectors
VGA	640	480	1.33	4:3	0.293	Pico-projectors
NTSC	720	480		4:3	0.330	SD-TV in US, pixels not square
WVGA	854	480	1.78	16:9	0.391	Low resolution HDTV
PAL/SECAM	768	576	1.33	4:3	0.422	SD in Europe
SVGA	800	600	1.33	4:3	0.458	Computer
WSVGA	1024	576	1.78	16:9	0.563	Low resolution HDTV
XGA	1024	768	1.33	4:3	0.750	Computer
720 p	1280	720	1.78	16:9	0.879	HDTV
WXGA	1280	768	1.67		0.938	Computer
SXGA	1280	1024	1.25	5:4	1.250	Computer
WSXGA	1366	1024	1.33	4:3	1.334	Computer
SXGA+	1400	1050	1.33	4:3	1.402	Computer
1080 p	1920	1080	1.78	16:9	1.978	HDTV
UXGA	1600	1200	1.33	4:3	1.831	Computer
USXGA	1600	1280	1.25	5:4	1.953	Computer
2 K	2048	1080	1.90		2.109	Digital cinema
HDTV+	1920	1200	1.60	16:10	2.197	Computer (also called WUXGA)
QXGA	2048	1536	1.33	4:3	3.000	Computer
4 K	4096	2160	1.90		8.438	Digital cinema
QQXGA	4096	3072	1.33	4:3	12.000	Computer

all these formats are predicated on square pixels except for NTSC. An NTSC video projector would more commonly have a VGA pixel count with reduced horizontal resolution and square pixels. In actual practice, NTSC or other standard definition (SD) video formats are more likely to be shown on a HDTV projector than a dedicated SD projector.

The limiting resolution of the human visual system (HVS) is about 0.35 mrad, with exponentially decreasing response for smaller pitch features. As an example, if the projection display is viewed from a distance of 3 m, the smallest feature that can be resolved is about 1 mm, independent of the screen size. It is necessary for the system designer to understand the environment for the intended applications. Providing the projector with capabilities for higher information content than can be resolved at the expected viewing distances adds economic burden to the product, i.e. makes it more expensive, for this application. Providing it with less resolution can result in annoying fixed pattern artifacts, especially in pixelated displays.

2.2.4 Electronic Cinema

When the first edition of this book was written, electronic cinema was the dream of a few visionaries. The film industry saw it as unlikely. Many film aficionados saw it as a definite threat because they felt it would make the movies look like a projected video while they preferred the 'film look'.

The first tests of electronic cinema took place in 1999 (Bleha and Sterling 1999). By 2002, even some film experts admitted that the image quality of electronic cinema could be better than the image quality of film (Schwartz 2002). The acceptance of electronic cinema became nearly universal in 2005 with the introduction of the Digital Cinema Initiative specification for electronic cinema (DCI 2005). Since then, an increasing number of cinema screens have been converted to digital cinema and it is expected that the eventual penetration of digital projectors into the cinema market will be nearly 100 %.

In addition to providing excellent image qualities, these systems should allow for reduced production costs, distribution costs, and reduced piracy losses to the production and distribution corporations. A full electronic cinema system includes the following subsystems:

- conversion of the film master to a digital data stream; now that digital intermediate (DI) processes are nearly universal in film post-production, the reverse process of converting the digital data stream to film is more commonly needed;
- encryption of the data for transmission;
- transmission of the data via satellite or fiber-optic cable from the production/distribution company to the local cinema;
- local reception and data storage;
- electronic projector which incorporates the final decoding of the encrypted data.

This system, which was considered hypothetical when the first edition was published, is rapidly becoming the mainstream way of delivering films and is illustrated in Figure 2.3.

The encryption of the data stream is intended to minimize piracy when the transmitted signal is in the clear. In the system illustrated in Figure 2.3, the transmission is by satellite or fiber-optic cable. These need not be real time transmissions since local data storage is provided. The decryption occurs during projection to avoid having a non-encrypted signal in the clear.

From a technical perspective, achieving the requisite performance in the projectors to be used in electronic cinemas has been, until recently, the gating item for implementation of these systems. They require very high luminous throughput, high resolution, excellent color rendition, and minimal artifacts.

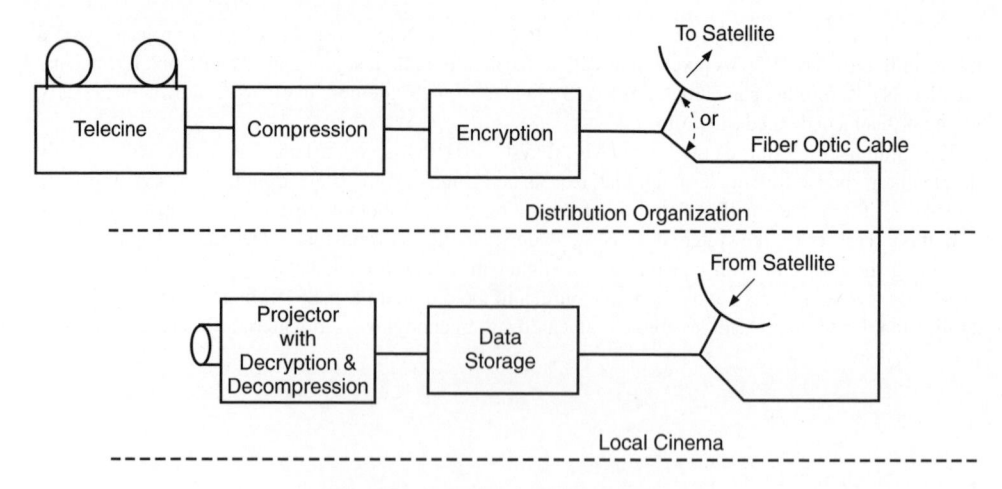

Figure 2.3 Scenario for electronic cinema.

It is seen in Figure 2.2 that, for a 1000-person theater, the projector has to deliver almost 10 000 lumens to the screen. When the first edition was published very few projectors could achieve this output. Today, this represents a medium output professional projector and is available from several vendors using a variety of projection technologies.

Digital cinema projectors enable two additional uses of movie screens that were difficult or impossible with film projectors:

- screening 3D stereoscopic movies;
- display of non-cinema content such as sporting events and concerts.

Most 2D projectors regardless of technology can be converted to 3D by an external or internal add-on that does not affect the projector's ability to function in the normal 2D mode. The lumen efficiency of this conversion is low, however, and the perceived brightness of a digital cinema projector showing 3D material is typically only 14 % of the brightness of the same projector showing 2D material. The adaptation of standard 2D projectors for 3D projection applications will be covered in Section 10.7.

References

ANSI IT7.215-1992. *Data Projection Equipment and Large Screen Displays – Test Methods and Performance Characteristics*.

ANSI/SMPTE 196 M (1995) SMPTE Standard for Motion-Picture Film – Indoor Theater and Review Room Projection – Screen Luminance and Viewing Conditions.

Bleha, W.P. and Sterling, R.D. (1999) Digital cinema using the ILA projector. *Proceedings of the 6th International Display Workshops (IDW-99)*, p. 1009.

Brennesholtz, M.S. (2008) *2007 LED and Laser Projection Systems Market Segment Analysis: Pico-projectors*. Norwalk, CT: Insight Media.

Digital Cinema Initiatives, LLC, Member Representatives Committee (2005) Digital Cinema System Specification V1.0, Issued 20 July 2005.

Okolicsanyi, F. (1937) The wave-slot, an optical television system. *Wireless Engineering* **14**: 526.

Schwartz, C.S. (2002) A review of survey responses to high quality digital and film presentations. *SMPTE Journal* **111**:2, February/March: 108–12.

3

Emissive Image Sources

Among the emissive image sources used, or proposed for use, in projection systems are cathode ray tubes (CRTs), vacuum fluorescent displays (VFDs), field emission displays (FEDs) and electron-beam-pumped lasers (laser CRTs). In this chapter, we will restrict our discussions to devices suitable for higher information content applications. This results in the exclusion of VFDs. While VFDs are employed in head-up displays in automotive applications, they are relatively low information content displays with rudimentary graphics and a limited color gamut and gray scale. Electron-beam pumped lasers will be discussed along with other lasers in Chapter 8.

Television receivers based on CRT projection are among the earliest consumer TV products (Radio-Craft 1937; Wolf 1937). Today, cathode ray tubes are declining rapidly from their position as the dominant light generating devices used in projection systems that they held through the 1990s. Certain properties of CRTs cannot be duplicated by the current generation of microdisplay projection systems, so CRT projectors are still used in some high-end applications, especially visualization and simulation. The discussion in this chapter will concentrate on these CRT devices. Very few, if any, commercial projection systems based on the other emissive technologies have come to market. Nevertheless, some of these have good potential for commercialization, at least in specialized markets such as head-up displays for aviation and automotive applications where VFDs have been used.

3.1 Projection CRTs

Projection CRTs are monochrome tubes optimized for high luminance and small beam width with a relatively narrow spectral output bandwidth. These characteristics are required in projection systems to produce good resolution, chromaticity and high screen brightness. For reasons of geometry and economy, most tubes used for consumer and industrial application employ magnetic deflection and electrostatic focus, although electromagnetically focused tubes are used for some professional applications. The more common tube sizes are designated as 7-inch and 9-inch, which refer to the nominal outside diagonal

---

Projection Displays, Second Edition M. Brennesholtz, E. Stupp
© 2008 John Wiley & Sons, Ltd

18 EMISSIVE IMAGE SOURCES

dimensions of the faceplates. Typically, a 5.25″ diagonal image would be written by the raster on a 7″ tube. Figure 3.1 is a photograph of a production CRT projection tube. The tube shown in the photograph has a concave-curved faceplate. Other tubes are made with flat faceplates. The characteristics of various faceplate configurations will be discussed later in this section.

Figure 3.1 Photograph of a 7″ projection CRT (from Asano *et al.* 1996; reproduced with permission of Hitachi).

3.1.1 Luminous Output of Projection CRTs

The raster diagonal on a 7-inch tube is typically 140–150 mm and it is about 180–190 mm on a 9-inch tube. The 7-inch size is employed in the majority of consumer rear projection TVs (RPTV). The image on the screen in a projector with a 54-inch diagonal image using 7-inch tubes is linearly magnified about 10 times compared with the image on the tube face. The attainment of satisfactory screen luminance requires the tube luminance to be very high. The screen luminance (Kingslake 1983) is the sum of the luminance produced by the three tubes multiplied by the optical and spectral characteristics of the projection system.

$$L_V(screen) = \sum_i \frac{GL_iT_i}{4(f/\#)^2(1+m)^2} \tag{3.1}$$

where
- i refers to red, green, or blue,
- L_i is the luminance of the red, green and blue tubes,
- T_i is the transmission of the optical paths in these spectral bands, excluding the screen,
- m is the linear magnification of the optical system, and
- $f/\#$ is the f-number of the projection lens,
- G is the gain of the screen.

The optics of the screen can increase the luminance of the image by concentrating the light into smaller solid angles. The ratio of the peak system luminance measured from a given screen to that obtained with a Lambertian screen is called screen gain. The transmission of the screen is included in its gain G. Note that screen gain does not imply amplification of the luminous energy. CRT and microdisplay front and rear projection screens will be discussed further in Chapter 7.

With a white field, the green tube produces about 0.6 of the screen luminance. The green tube luminance is obtained by inverting Equation (3.1).

$$L_V(Green) = \frac{[0.6 L_V(screen)]\left[4(f/\#)^2(1+m)^2\right]}{GT(Green)} \qquad (3.2)$$

Consumer projectors can produce a uniform white field screen luminance of $350\,\text{cd/m}^2$ with a $G = 6$ screen and f/1 projection lenses. The optical system transmission is about 85 %. The green tube luminance to produce this screen luminance is then in excess of $1.6 \times 10^4\,\text{cd/m}^2$.

These tubes are usually operated at higher voltages and currents than is common in direct-view CRTs. The shadow mask in direct-view tubes intercepts 75–85 % of the beam. Projection CRTs do not have a shadow mask and, hence, all of the beam current contributes to the high luminous output. The anode voltage can be in the range of 30–35 kV or more and the beam currents can have peaks up to about 7 mA, with a current for a full white field of 600 µA. These high drive levels have implications for system resolution, white-balance as a function of image brightness, and thermal effects in the faceplate.

As beam current is increased, the diameter of the beam increases due to coulombic repulsion of electrons in the beam. This 'blooming' will be discussed in greater detail in Section 3.1.4. White balance is a description of the quality of the white light generated, e.g. cool white or warm white. It is determined by the relative amounts of each of the three primary colors in the white signal. Unfortunately, the relationships of the outputs of the three tubes to the input signals vary between color tube types as well as with beam current. In particular, the phosphor in the blue tube tends to saturate before those in the other tubes, leading to a 'warmer' white at higher beam currents. This is discussed further in Section 3.1.2. The general issue of the white balance of projection systems will be discussed in Chapter 11.

The maximum power in a projection CRT is approximately 20 W, corresponding to an average beam current of 600 µA and an anode voltage of 33 kV. For the case where only a small area is excited, the instantaneous power can be much higher, typically up to 10 x higher. These power levels would be destructive in the absence of efficient cooling. It is now standard practice to use liquid coupling of the faceplate to the rear element of the projection lens (Malang 1989). This provides faceplate cooling and assists in achieving luminance uniformity in the image. The index of refraction and thickness of this coolant, and the index, shape, thickness and curvature of the faceplate are then an integral part of the projection lens design.

3.1.2 Phosphors

The phosphors used in projection CRTs have been undergoing evolutionary improvements over the years. The science and technology of CRT phosphors have long histories (Goldberg 1966). New cathodoluminescent developments are continuing to evolve as the requirements become ever more stringent. Because of the need for high faceplate luminance, the phosphors used in projection CRTs are driven with higher power levels than those in direct-view CRTs. The phosphors used for direct-view application usually cannot be driven to such high luminance, due to saturation and thermal quenching of the luminance. For a complete review of CRT and other phosphors see Yen *et al.* (2006).

While the spectral characteristics, the chromaticity coordinates (see Appendix 2) and the luminous efficacy of the light emitted are paramount in the selection of the phosphor, tradeoffs must be made. This optimization of phosphor characteristics involves tradeoffs among spectral energy distribution, luminous efficacy, saturation, persistence, thermal quenching and degradation under prolonged electron bombardment. Figure 3.2 gives the typical spectral distributions of the light emissions from phosphors used for the three tubes (Raue *et al.* 1989). The actual distributions are manufacturer specific, depending

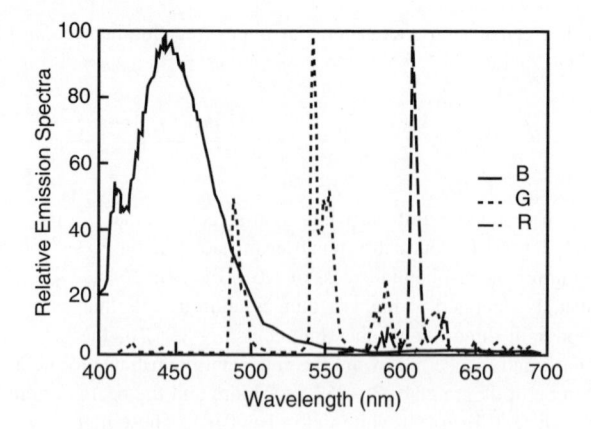

Figure 3.2 Typical projection CRT phosphor spectra.

on the tradeoffs made, and may vary over time as the manufacturer changes the phosphor formulations and/or technologies.

The optimization for each of the phosphors for the three CRTs used in projection systems is based on the characteristics of the inorganic materials from which these phosphors are made. The red phosphor, typically europium-doped yttrium oxide (Y_2O_3:Eu), shows the least amount of problems with respect to luminance, chromaticity and degradation. Y_2O_3 phosphors also have a relatively narrow spectrum giving them good color purity.

The green phosphors have included terbium-doped yttrium aluminum garnet ($Y_3Al_5O_{12}$:Tb or YAG:Tb), Tb-doped yttrium aluminum gallium garnet (YAGaG:Tb), and Tb-doped yttrium silicate (Y_2SiO_5:Tb). YAG:Tb has narrow spectral emissions but other phosphors, e.g. (Zn,Cd)S:Cu, have broad emission bands.[1] Many other green phosphors have characteristics suitable for projection tubes. YAGaG:Tb is one of the more commonly used phosphors in green projection CRTs.

The most common blue phosphor for projection is ZnS:Ag,Al, often called P-55 after its old EIA designation, which is very similar to the blue phosphor used in direct-view tubes. This blue phosphor does not exhibit narrow spectral emissions.

Tube manufacturers will frequently use mixtures of several phosphors to achieve a compromise on characteristics. The emission color of green-emitting, Tb-activated phosphors is yellow green. Filters applied to the tube face or dyes added to the faceplate coolant have been employed to shift the chromaticity into the desired range for satisfactory color reproduction. Alternatively, pigments and/or other phosphors can be admixed with the primary phosphor to improve the chromaticity.

The green phosphors also exhibit a detectable persistence, i.e. the emission can persist for tens of milliseconds after the electron beam excitation has been removed. Under certain conditions, this can lead to visible green 'tails' behind moving objects in the image. Persistence is much less of a problem with the red phosphors and it is not significant in blue tubes.

Saturation of the emission from the blue tube with increasing beam current can be a significant problem. Figure 3.3 shows the luminance of typical R, G, and B tubes as a function of beam current. While some saturation is present in the red and green emissions at the highest beam currents, the stronger saturation of the blue in the range of 300 μA to > 2 mA can result in the limitation of the overall system luminance due to chromaticity errors introduced at higher beam currents.

---

[1] Cadmium-based phosphors have been banned in CRTs in the US because of environmental safety problems. They continue to be used in China, the only remaining significant market for consumer CRT projection systems.

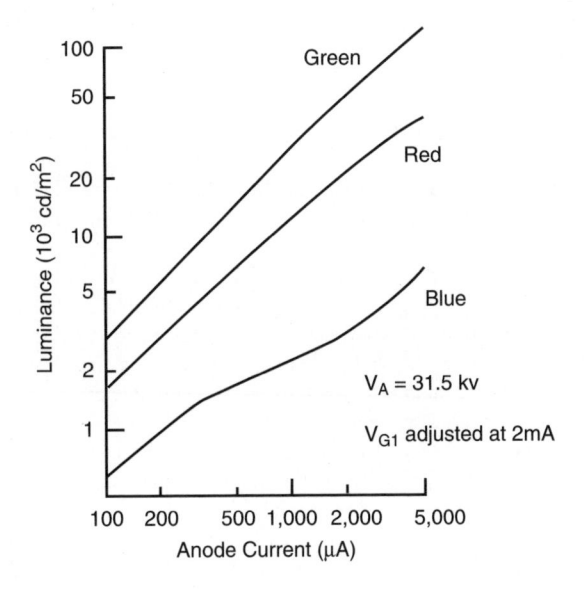

Figure 3.3 Output of the R, G, and B phosphors as a function of beam current.

The output of the blue tube benefits from beam blooming. At higher currents, the beam diameter grows, limiting the current density and minimizing the saturation. Projection systems are sometimes set up with the electron beam in the blue tube slightly defocused, also to decrease the current density. This produces minimal loss of perceived resolution since the human visual system has reduced resolution capabilities at blue wavelengths. While significant progress has been made in reducing the saturation of the blue output, it remains an active research subject (Shiike et al. 1996; Kajiwara et al. 2002).

The luminous output of many phosphors is decreased with increasing temperature. This thermal quenching is most severe in ZnS:Ag,Al and somewhat less in Y_2O_3:Eu. YAGaG:Tb is more resistant. The outputs of the red and blue tubes can decrease by more than a factor of two for phosphor temperatures above 200 °C. These temperatures can be easily reached in the absence of cooling since the phosphor is in a vacuum and it is on a relatively insulating glass substrate. The introduction of liquid cooling of the tube faceplate was a major advance in achieving high luminance since it permitted a significant increase in the electron beam power density without thermal quenching effects.

One proposed solution to the thermal quenching problem is to use CRT faceplates with a higher thermal conductivity than glass, allowing better cooling of the phosphor and higher power densities (Cheng et al. 2002). $MgAl_2O_4$, a transparent, polycrystalline material with a thermal conductivity about 10 x better than glass, was used as the faceplate of a 5″ projection tube. When liquid-cooled these tubes can be driven at a power density sufficiently high to produce 1800 lumens from a set of red, green and blue tubes. As a comparison, the light output of a conventional 9″ tube set with glass faceplates is about 1200 lumens.

Generally it is the blue CRT which limits the lumen output of a CRT projection system, so improvements in the blue phosphor will improve the system performance. Kajiwara et al. (2002) describes an improved blue phosphor that maintains nearly 100 % of its lumen output at 423 K, a temperature at which a conventional P-55 blue phosphor has lost about 40 % of its total output.

Another solution to the blue light-output problem is to defocus the blue gun. By decreasing the current density, the total light output increases significantly without a change in current. In order to avoid the loss of resolution, this can be done dynamically, dependent on image content (Brennesholtz 1998) Content-dependent image enhancement will be discussed in more detail in Section 13.4.

3.1.3 Resolution of Projection CRTs

With perfect convergence of the three tubes, the resolution of the projector is determined by the signal bandwidth, the resolution characteristics of the projection CRTs, the projection lenses and the screen. A measure of the resolution is the MTF, or modulation transfer function, which is discussed in more detail in Chapter 13. To the extent that the system is linear, the system resolution is the product of the MTFs of the various subunits. The resolution performance of the projection tube is one of the principal factors in determining the system resolution.

The electron-optics in projection CRTs uses magnetic deflection and can use magnetic, electromagnetic, or electrostatic focus. Tubes used for consumer application are primarily based on electrostatic focus (Watanabe *et al.* 2000). Projection CRTs are designed with deflection angles from 70° to 100°. The electron gun typically has five elements in addition to the cathode. It is one of the goals of the projection tube designer to produce a high beam current with the smallest diameter spot over the full area of the tube face.

The tube MTF is dependent on several parameters, including beam spot size, discussed further in Section 3.1.4, the dimensions of the raster, including the raster diagonal and raster aspect ratio, beam current, and where on the raster the image is presented (off-axis images typically have lower resolution than on-axis images). Frequently, the beam current distribution is a Gaussian. The MTF of the beam with such a profile in the horizontal direction is given by Equation (3.3). If it is assumed that there is no MTF degradation by the phosphor layer (not necessarily a good assumption), this function will also describe the tube MTF.

$$MTF(f_S, \Delta x) = \exp\left(\frac{\pi^2 \Delta x^2 f_S^2}{4w^2 \ln\left(\frac{i}{i_0}\right)}\right) \quad (3.3)$$

where

- f_S is the spatial frequency in line pairs (l p)/(scan width),
- Δx is the diameter of the beam spot in mm at a current fraction i/i_0 of the peak beam current. It is common to use a 5 % or 10 % fraction in this calculation.
- w is the scan width of the raster in mm

The dimensionless modulation transfer function (MTF) is defined as a function of the spatial frequency. Line pairs per mm (lp/mm) is the common spatial frequency measures outside of television. The MTF in Equation (3.3) includes the width of the display, which is included to adjust the units from 1 p/mm to television lines per picture width (TVL/PW), a common measure of resolution in TV. Two TV lines equal one cycle or one line pair.

The scan width of the raster on the tube face depends on the usable diagonal of the tube and the aspect ratio. The aspect ratio is the ratio of the scan width to the scan height. This ratio is 4:3 for standard definition television and 16:9 for high definition television. Figure 3.4 shows the calculated MTFs for 7″ CRTs with a 4:3 ratio. A spot size of 0.23 mm at a 10 % current is assumed. These calculations describe the contributions to the MTF due to the spot size. The calculations would be valid at the center of the tube; off-axis performance is often somewhat degraded.

The visual resolution of a CRT can be improved beyond what would normally be allowed by Equation (3.3) by a technique called scan velocity modulation (SVM) (Konda *et al.* 2002). SVM, like defocusing for blue luminance increase, is content-dependent. The scan velocity is first slowed down slightly and then accelerated as it passes over an edge in the image, creating a sharper edge. This technique requires an additional coil (Sakurai *et al.* 2003) to be added to the yoke of the tube, plus additional drive circuitry to drive the SVM coil.

The structure of the phosphors, including grain size, packing density, and layer thickness, can also effect the resolution. Light emitted by one phosphor grain can be scattered from other grains, effectively

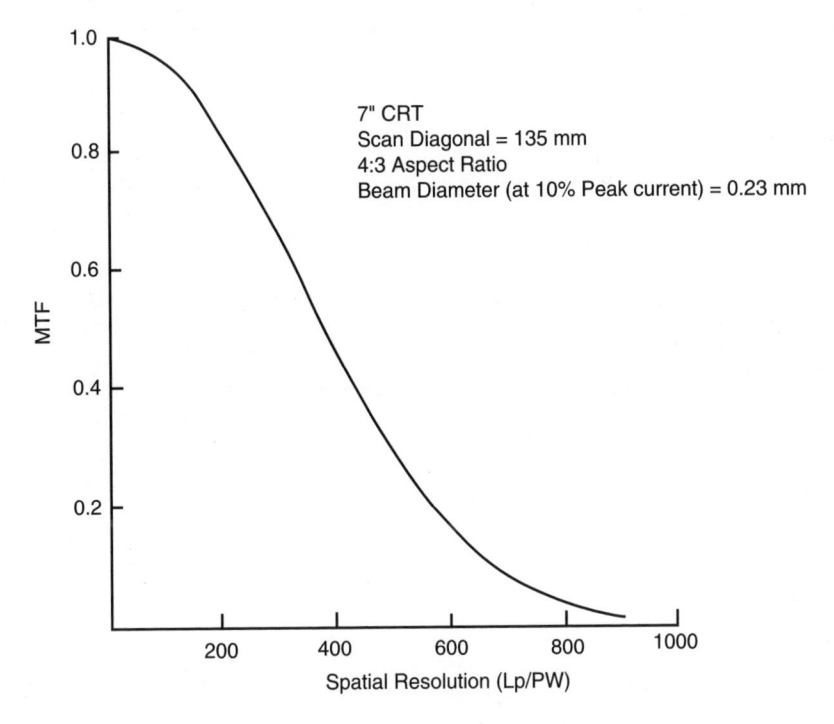

Figure 3.4 Calculated MTF of projection CRTs with 0.23 mm spot size for 4:3 aspect ratio.

increasing the area from which the light is emitted. Scattering can also broaden the electron beam spot size as the beam traverses the phosphor layer before being stopped. This is illustrated in Figure 3.5. Thinner layers with greater packing of smaller grains are important contributors to minimizing this scattering and achieving higher resolution tubes (Asano et al. 1996). The electron beam spot characteristics are discussed in the next section.

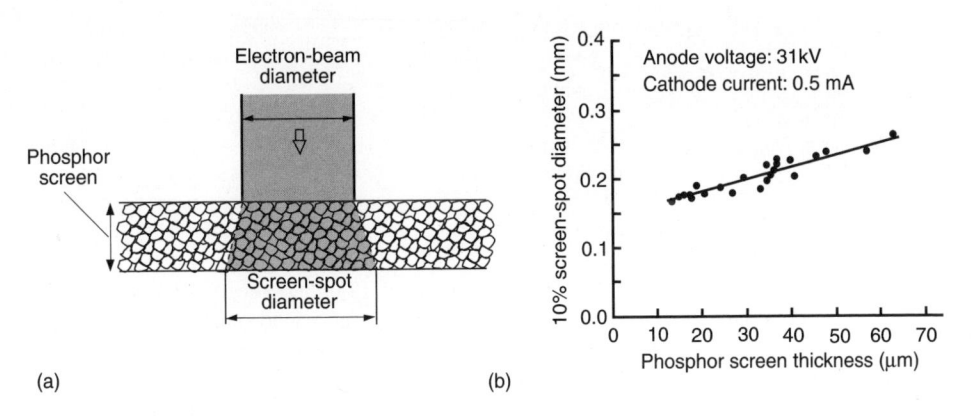

Figure 3.5 Effect of structure of phosphor layer on tube resolution: (a) spot size growth due to scattering in phosphor layer; (b) degradation of spot size with increasing phosphor layer thickness (from Asano et al. 1994; reproduced with permission of Hitachi Review).

3.1.4 Spot Size of Beam

The manufacturer's specified spot size is the composite of the electron-beam diameter and the size-broadening effects of the phosphor layer. The beam produced by the electron-gun can have a Gaussian current distribution across the spot, although other distributions are also observed. The Gaussian distribution with the total current in the spot normalized to unity is given by Equation (3.4).

$$I(r) = \frac{1}{\sqrt{2\pi}\sigma} \exp\left(-\frac{r^2}{2\sigma^2}\right) \tag{3.4}$$

where
- I   is the current density,
- r   is the radial distance from the peak at the center of the spot, and
- σ is the standard deviation.

It is most common to specify the beam size as the full width of the beam measured at a fraction of the peak current, typically 5 % or 10 %. This specification is usually given for the undeflected, on-axis beam at the beam current for which the spot size is smallest (see Figure 3.6). For the Gaussian beam distribution, the full-width (2 r) at 5 % beam current is 4.896 σ.

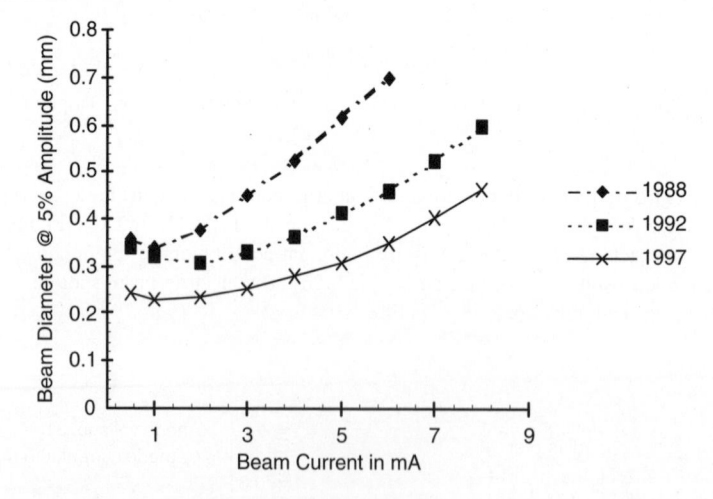

Figure 3.6 Spot size performance evolution of 7-inch faceplate projection CRTs from various manufacturers. (Data courtesy of M. Carpenter, Philips Consumer Electronics Corp.)

Unfortunately, the qualities of the beam frequently degrade when the beam is deflected or when much higher than average beam currents are used. The increase in beam size with current is called blooming. Significant progress has been made in recent years to reduce blooming and off-axis effects. Figure 3.6 illustrates the measured spot sizes as a function of beam current in commercially available tubes. Dramatic improvements were achieved, both in the reduction of spot size at average currents as well as in the reduction of blooming. These reductions have been achieved through improved electron optics as well as through reductions of light scattering in the phosphor layers (Asano et al. 1994).

The average beam current in a 7″ projection CRT is about 0.6 mA. While the best tube shown in Figure 3.6 achieves a 0.23 mm spot size at 1–1.5 mA, the spot size blooms to 0.41 mm at 7 mA. The details of blooming vary with specific tube designs but the general characteristics are seen in all projection CRTs. The system designer has a tradeoff: peak luminance vs resolution.

A higher peak, usually over a limited fraction of the display area, produces a desirable effect called 'punch', which gives the projector a higher perceived contrast and brightness. The current for a full white field is limited by the power supply to control the power dissipation in the faceplate and the phosphor loading. When only a small area of the faceplate is emitting, the peak current can be increased by factors as large as eight or more.

The spot sizes in 9" tubes are typically greater than those in 7" tubes. This is due, in part, to the greater electron-optics magnification of the beam on to the phosphor screen in the 9" tube. Nevertheless, some advantage in resolution is achieved with 9" tubes compared with 7". Most of the advantage of the larger tube format relates to the larger phosphor area. With this increased area, about 60 % more light can be generated compared with the smaller tube, allowing for large screen sizes at comparable luminance.

With constant focus-voltage, the beam would be best focused with the faceplate at a constant distance from the center of deflection of the beam. This would imply that convex-curved faceplates are best, but such shapes are not used because of the optical problems introduced. Some of these are discussed in Section 3.1.5. The electron-optics designer must accommodate flat or concave-curved faceplates in achieving good off-axis performance.

Among the ways the off-axis performance can be improved is to employ dynamic focusing, i.e. varying the focus voltage as a function of deflection angle. However, this technique can require expensive electronics because of the high focus voltages, the high capacitive loads, and the high frequencies involved (15 kHz in consumer projectors, up to 135 kHz in datagraphic products). These high frequencies may lead to the necessity of using more expensive Litz wire in the deflection coils.

3.1.5 Light Collection/Curvature

With a flat faceplate CRT, the angular distribution of the light from any point on the CRT surface is Lambertian-like. Figure 3.7 illustrates the light collection from flat-faced CRTs and curved face CRTs. With the flat faceplate, shown in Figure 3.7(a), vignetting by the aperture of the projection lens limits the amount of light collected with increasing radial position along the tube face. This vignetting contributes to an off-axis drop-off in image brightness on the screen, sometimes by a factor of four or more compared with the center brightness.

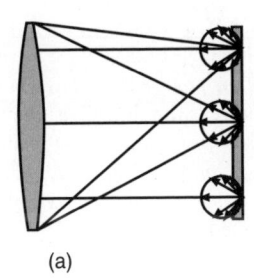

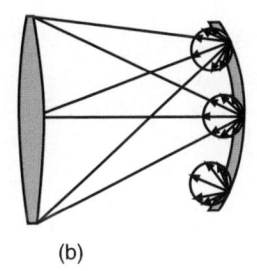

Figure 3.7 Light collection of Lambertian emissions from CRT faceplate (a) with flat faceplate, and (b) with concave faceplates (adapted from Malang 1989; printed with permission of SPIE).

There are other factors which can contribute to this drop-off, including light-trapping in the glass faceplate of the tube. Screen characteristics can further reduce the off-axis luminance. The angular distribution of the light from any point on the CRT surface is only Lambertian-like for angles below the critical angle, i.e. for light entering the faceplate from the phosphor at angles less than the critical angle. Light entering above the critical angle can be trapped within the glass. However, this critical angle effect is eliminated by index matching the faceplate cooling liquid to the faceplate glass.

The use of concave faceplates is an effective method for minimizing the vignetting, as illustrated in Figure 3.7(b). This approach has the additional benefit of providing a curved image field for the projection lens, which eases the lens design issues. This curvature is important in the design of cost-effective, short focal length lenses. These lenses are needed to minimize cabinet depth in rear projection systems. Faceplates having curvatures with constant radius as well as aspheric faceplates have been used in the CRT designs.

3.2 Field-emission Devices

There are several emissive devices and systems in development which are based on other technologies. The field emission device (FED) (Spindt 1968; Fursey, 2005) is a flat, thin CRT device which employs an area array of emitters, shown in cross-section in Figure 3.8. The fields produced by the control grids around the tips is sufficient to produce electron emission. The apertures in the control grids through which the tips emit are less than about 1 μm in diameter and the pitch can be as little as 5 μm. Both high voltage (2–10 kV) and low voltage (0.2 kV–2 kV) versions of FED displays have been in development.

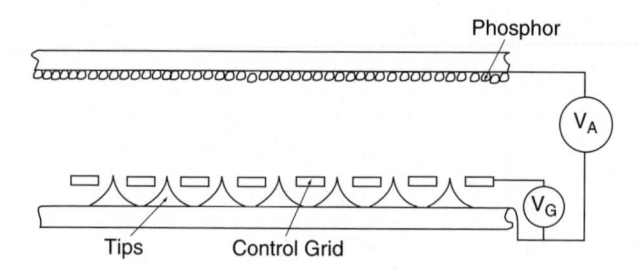

Figure 3.8 Schematic cross-section of FED showing cathode array.

The pixel locations are defined by the intersections of stripe-patterned cathode metallizations (x-direction) and the stripe metallizations in the y-direction on the control grid. Several hundred emitters per pixel are used to average the emissions spatially from many tips to provide uniform current to all pixels. Emitted electrons can be proximity focused onto the phosphor on the anode or the device can employ an auxiliary focus electrode. Alternatively, when lower voltage phosphors are used, anode voltage switching can be employed to change the emitting pixels. Gray scale is achieved with time division multiplexing.

Experimental high-anode-voltage devices have been shown to produce luminance that approaches those required for projection systems application (Palevsky et al. 1994). These had a focus electrode, which complicates the device structure. Stable, vacuum-sealed FED devices with luminance and lives suitable for projection applications have not yet been reported.

References

Asano, T., Hirota, K., Oku, K., Tanaka, Y., Nakayama, K. and Ohsawa, M. (1994) A 6.9-in. high-resolution projection CRT for an HDTV display. *J. SID* **2**: 113–18.

Asano, T., Uehara, Y., Watanabe, T., and Matsumoto, K. (1996) HDTV display with a 16-cm projection CRT. *Hitachi Review* **45**: 203–8.

Brennesholtz, M.S. (1998) Method and apparatus for area dependent dynamic blue de-focusing. US Patent 5,712,691, issued 27 January 1998, Assigned to Philips.

REFERENCES

Cheng, J., Wang, Q.-H. and Lin, Z. (2002) A 5-in. CRT for High Luminance and High Resolution Projection Display. *SID Intl. Symp. Digest of Technical Papers*, Paper 36.4.

Fursey, G.N. (2005) *Field Emission in Vacuum Microelectronics*. New York: Kluwer Academic/Plenum Publishers.

Goldberg, P. (ed.) (1966) *Luminescence of Inorganic Solids*. New York: Academic Press.

Kajiwara, K., Abe, T. and Okada, H. (2002) Blue-emitting ZnS:Ag,Al phosphor with longer lifetime and saturationless performance for projection CRTs, *SID Intl. Symp. Digest of Technical Papers*, Paper 4.1, pp. 6–7.

Kingslake, R. (1983) *Optical System Design*. New York: Academic Press.

Konda, M., Taguchi, Y. Ishikawa, T. and Shimada, K. (2002) Improvement in velocity modulation efficiency for projection CRTs. *Proceedings of the Ninth International Display Workshops (IDW '02), SID and ITE*, Paper CRT2-3, pp. 587–90.

Malang, A.W. (1989) High brightness projection video display with concave phosphor surface. *Proc. SPIE*, **1081**: 101–6.

Nasibov, A.S., Kozlovsky, V.I., Reznikov, P.V., Skasyrsky, Y.K and Popov, Y.M. (1992) Full color TV projector based on A_2B_6 electron-beam pumped semiconductor lasers. *J. Crystal Growth* **117**: 1040–5.

Palevsky, A., Gammie, G. and Koutopoulos, P. (1994) Field-emission displays: a 10,000 fL high-efficiency field-emission display. *SID Intl. Symp. Digest of Technical Papers*, 55–7.

Radio-Craft (1937) The projection kinescope makes its debut, *Radio-Craft*, August, p. 83.

Raue, R., Vink, A.T. and Welker, T. (1989) Phosphor screens in cathode-ray tubes for projection television. *Philips Tech. Rev.* **44**: 335–47.

Sakurai, S., Hisada, T., Watanabe, S. and Asano, T. (2003) A high sensitivity velocity modulation coil for projection CRTs. Paper CRT2-3, IDW '03, Fukuoka, Japan, *Proceedings of the 10th International Display Workshops*, pp. 763–5.

Shiike, M., Komatsu, M., Toyama, H., and Kanehisa, O. (1996) Encapsulated powder screens of blue emitting ZnS:Ag,Al phosphors for projection tubes. *Proc. 16th Intl. Display Res. Conf.*, pp. 504-507, 1996.

Spindt, C.A. (1968) A thin-film field-emission cathode. *J. Appl. Phys*, **39**: 3504–5.

Watanabe, T., Kabuto, N., Hirata, K. and Aoki, K. (2000) Recent progress in CRT projection display. *SID Intl. Symp. Digest of Technical Papers*, Paper 21.2, pp. 306–9.

Wolf, M. (1937) The enlarged projection of television pictures. *Philips Technical Review* **2**: 249–53.

Yen, W.M., Shionoya, S. and Yamamoto, H. (2006) *Phosphor Handbook*, 2nd edn. Boca Raton, FL: CRC Press.

4

Liquid Crystal Light Valves and Microdisplays

There are many technologies used in light valves for projection. The devices can operate in transmission or reflection of incident light. Most of these consist of a matrix of pixels addressed by active devices, such as thin-film transistors (TFTs), diode-like devices or MOS transistors in single crystal silicon (c-Si). The active-matrix array has a switch at the intersection of each row and column drive line. When the switch is opened, normally by a signal on the row electrode, it allows the voltage to be held on the pixel regardless of voltage changes on the column electrode. Passive matrix devices lack this switch at the intersection of row and column lines. While a few early projectors were made with passive matrix light valves, none is currently used in projection systems and passive matrix is falling out of favor even for direct-view displays. Therefore only active-matrix systems will be considered here. For a complete overview of microdisplay technology see, for example, Armitage *et al.* (2006). For a complete review of liquid crystal technology, including both more details on the LC effects discussed in this chapter and information on additional LC effects that may be suitable for projection microdisplays, see Wu (2006).

The voltage put on the pixels by these active matrices (AMs) produces optical effects by a variety of physical phenomena, including liquid-crystal effects, microelectromechanical (MEMs), and piezo-electricity. Liquid crystal effects and MEMs are the principal ones used in commercial projection light valves, with both producing excellent imagery.

Active-matrix technology will be discussed in the first part of this chapter, Section 4.1. Liquid crystal devices are discussed in the second part of this chapter, Section 4.2. MEM-based technologies are covered in the next chapter.

A cut-away view of a transmissive liquid crystal light valve is illustrated in Figure 4.1. It consists of a capacitor defined by transparent electrodes on the two glass or quartz substrates. On the inside of the top glass is a continuous coating. On the substrate containing the active-matrix devices are transparent electrodes patterned to define individual pixels. The patterned electrode is called the pixel electrode and

---

Projection Displays, Second Edition M. Brennesholtz, E. Stupp
© 2008 John Wiley & Sons, Ltd

the unpatterned one is called the counter-electrode. The dielectric liquid crystal in each pixel fills the space between these two sets of electrodes and forms a capacitor with them. The capacitor is electrically connected to the output of the switching device in the active matrix.

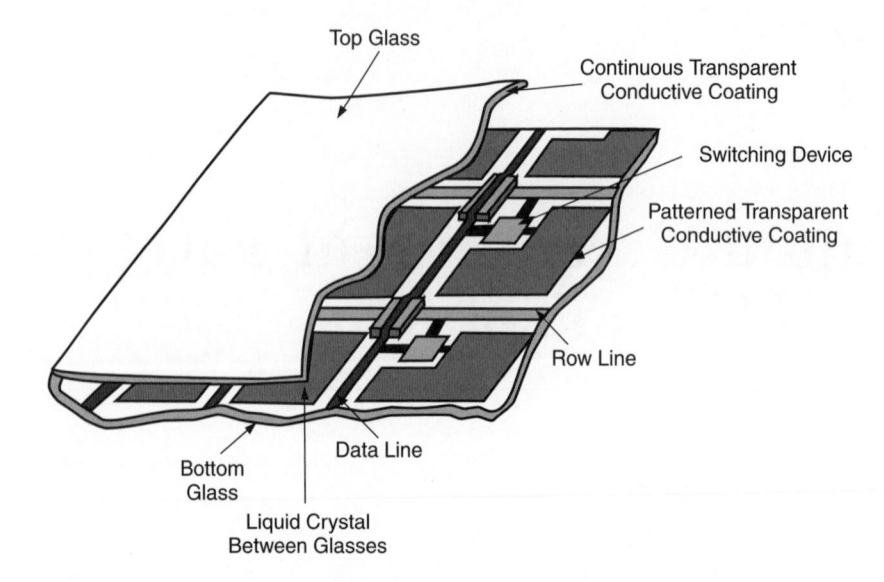

Figure 4.1 Cut-away view of active-matrix liquid crystal light valve.

The vast majority of light-valve projection systems presently use devices made with metal-oxide-semiconductor (MOS) TFT active matrices. The MOS TFTs in projection light valves are made primarily using poly-Si and c-Si, although other materials such as α-Si and CdSe have been demonstrated. The latter material was the first used in a TFT AM application (Brody 1973). While AMs fabricated with this material have many positive attributes, it is rarely used today for projection light valves and it will not be discussed further here. There are also several diode and diode-like technologies that have been used in thin-film AM light valves, including α-Si diodes and metal-insulator-metal (MIM) devices (Hartman 1995). Unless otherwise indicated, the discussion here will pertain to TFT devices.

4.1 Active Matrices

There are many circuit architectures employed in active matrices. The decision as to which one is appropriate is frequently made to optimize one or more characteristics, often at the expense of others. The functionality of AM addressing is illustrated by the electrical equivalent circuit of a generic TFT active matrix in Figure 4.2. In this figure, the liquid crystal light modulator is represented by a capacitor (C_{LC}).

A pixel consists of one or more TFTs located at each crossing in the matrix of row and column conductors. The rows and columns are electrically isolated from each other. The liquid crystal light modulating element is the dielectric of the capacitor. More than one TFT is sometimes used at a crossover for redundancy or reduced leakage currents. An additional storage capacitance (C_{ST}), in parallel with the liquid crystal capacitance and termed the storage capacitor, is normally built into each pixel. Typically $C_{ST} > C_{LC}$ to minimize the effect of the non-linearity of the value of C_{LC}. C_{ST} cannot be too large, however, since it would then require excessive drive from the column driver to charge it for each video field.

ACTIVE MATRICES 31

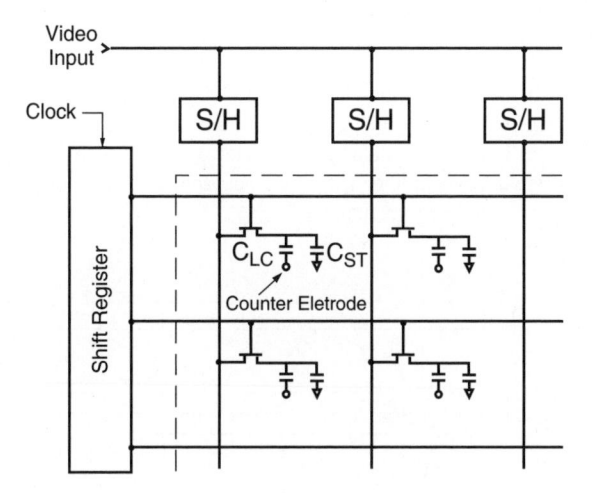

Figure 4.2 Electrical equivalent circuit of TFT AM with LC light modulator.

4.1.1 Operation of Active-matrix Circuits

The gates of the MOS TFTs are connected to the rows, and the TFT sources and drains are connected to the column and pixel capacitances, respectively. The pixel capacitance is comprised of the capacitance associated with C_{LC} in parallel with the storage capacitance C_{ST}. C_{ST} is built into the AM structure to improve the device characteristics, as will be discussed further in Section 4.1.1.1. This capacitor is formed between the poly-Si gate line of the next row and a degenerate region in the poly-Si underneath. This degenerate poly-Si is connected to the pixel electrode. SiO_2 is the most common dielectric for these capacitors. Figure 4.3 shows details of a single pixel.

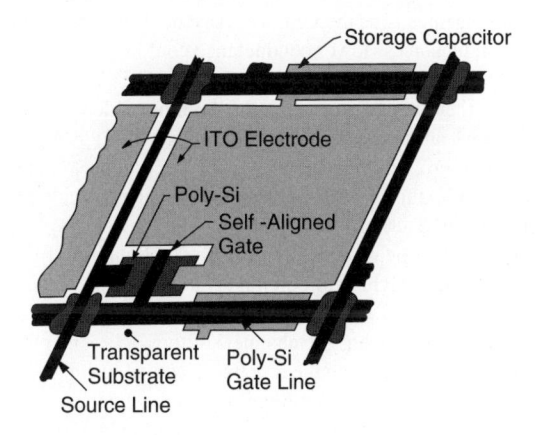

Figure 4.3 Isometric view of a pixel showing TFT and additional storage capacitance. The storage capacitor for the pixel is formed under the next-to-be-active gate line.

The array of TFTs is driven by column and row drivers, which may or may not be integrated with the AM, depending of the technology used. The row line serves normally to open or close the switch

32   *LIQUID CRYSTAL LIGHT VALVES AND MICRODISPLAYS*

(e.g. TFT) in the active matrix and serves a secondary purpose as the ground for the C_{ST} of the previous row. The column line is normally used to load the analog voltage into the pixel capacitance associated with the row selected by the row line. Many designs have been employed for the column driver circuit. One that schematically illustrates the requisite functionality is shown in Figure 4.4. It consists of two sample-and-hold (S/H) circuits for each column plus associated buffers for driving a capacitive load. Each S/H is figuratively illustrated as a sampling switch and a capacitor. A buffer amplifier may be placed on each column to drive the column line capacitance.

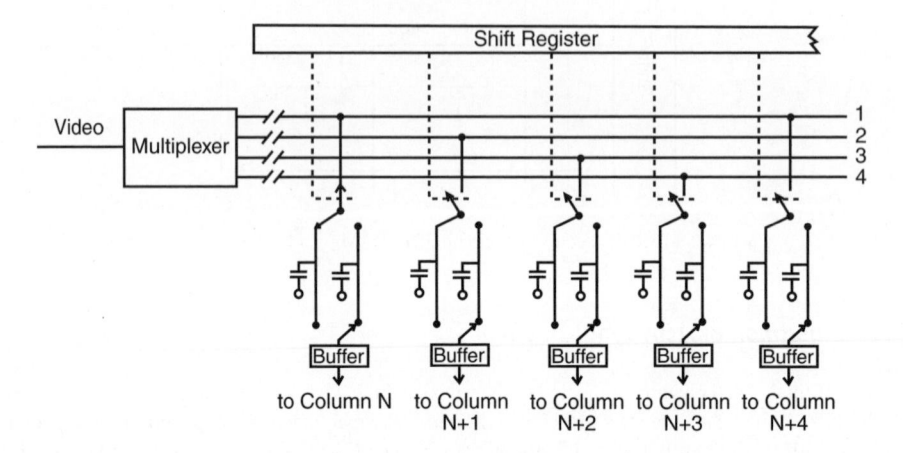

Figure 4.4   Active-matrix column driver circuit configuration using multiplexing onto four video inputs. It is indicated that information is being written to a S/H capacitor on column N.

The input video information is multiplexed onto one or more distribution line(s), depending on the signal bandwidth, the switching speeds of the various active devices used in this column driver circuit, and the gate and row line conductances (lower conductance could require more input lines). Figure 4.4 shows the video being demultiplexed onto four distribution lines. The data rates on each of the distribution lines for this example would then be one-fourth that of the input video signal. In operation, the video information is sampled in temporal succession, controlled by a shift register, from the distribution lines onto the S/Hs on each column, providing successive time-slice samples of the incoming video on the columns. In practice, the multiplexing of the input signal onto separate video lines permits the timings for writing to adjacent columns to overlap.

Figure 4.4 illustrates the sampling being done onto column N. Input line 1 provides signals to column N, N+4,..., input line 2 provides signals to columns N+1, N+5,..., etc. When each column is loaded with the appropriate time-slice of video information for the line to be addressed, a pulse is applied to all the TFT gates in that row, which opens the channels of the TFTs. The pulse is frequently provided by a shift register plus buffer amplifier driving the row lines. With all the gates open in a row, the voltage information is transferred from the respective column lines to the pixel capacitors. After sufficient time has elapsed for the charge transfer to be completed, the gate pulse is removed, opening the switches at the row/column intersections and isolating the charge on the capacitors.

The information written to the pixel capacitor is stored on the pixel until new information is sampled in the next field. Each pixel therefore also behaves as a sample-and-hold element. The isolated charge on the pixel creates a voltage across the liquid crystal, which results in the light-modulating effect. During this process, video information for the next line has continued to arrive and is sampled onto the second S/H for each column. This alternation between S/Hs continues for successive lines.

4.1.1.1 Effects of leakage

The maximum time that each row channel is open depends on the frame rate and the number of rows in the array. Thus, for a VGA (640 columns × 480 rows) display operating at a 60 Hz refresh rate, the maximum line address time is $1/(60 \times 480) = 34.7\,\mu s$. The current devices being produced typically have many more columns and rows, and may be driven at higher field/frame rates. Thus, much less time would be available for addressing. Other circuit configurations can be used which allow for parallel addressing of different lines or regions of the AM to provide more time for depositing the charge in the pixel capacitance.

Continuing with the VGA illustration, the voltage information must be held on the pixel without significant degradation until the pixel is rewritten by the next field information 16.7 ms (1/60 s) later. During a field-time, the voltage change ΔV on the pixel due to charge leakage from all sources is given by Equation (4.1):

$$\Delta V = \frac{i_l \Delta t}{(C_P + C_S)} \qquad (4.1)$$

where
- i_l is the electrical leakage current from all sources
- C_P is the pixel capacitance
- C_S is the storage capacitance associated with each pixel, and
- Δt is the pixel refresh time, 16.7 ms for a 60 Hz display.

The requirements for maximum allowed leakage current in projection light valves can be quite severe. The following example is given to estimate the leakage currents allowed from all sources to keep the voltage on the pixel constant to within 10 % during the frame time.

Consider a representative pixel configuration, with construction outlined in Figure 4.1. The parameters assumed for this example are given in Table 4.1. Figure 4.1 shows the storage capacitor being formed by a stacked structure which includes a degenerate polysilicon electrode connected to the transparent pixel electrode, a dielectric layer and the gate line for the adjacent row (Stupp and Khan 1994), which is the common electrode for all capacitors in a row. The adjacent row line is electrically at AC common, which is the same as that for the counter-electrode. The equivalent electrical circuit is shown in Figure 4.2. It is assumed that the storage capacitor dielectric is SiO_2.

Table 4.1 Pixel parameters.

Pitch	35 μm
Aperture ratio	50 %
LC spacing	6 μm
ε_{LC}	8
SiO_2 thickness	100 nm
Storage capacitor fill factor	20 %
LC dynamic range	2 V
LC p-p voltage	±3 V

In Table 4.1, the fill factor is the percentage of the total pixel area occupied by the storage capacitor. The dynamic range of the LC is the voltage change required to produce a transmission change from maximum to minimum. Several simplifying assumptions are made in this calculation:

1. The area of the liquid crystal contributing to the pixel capacitance is equal to the transmissive area of the pixel.

34   *LIQUID CRYSTAL LIGHT VALVES AND MICRODISPLAYS*

2. The dielectric constant of the liquid crystal material can be represented by a single number. An average value of the dielectric constant of liquid crystal materials is used. The actual dielectric constant can change markedly as the LC state changes under the influence of applied electrical fields.

3. Fringe effects are not a major contributor to the value of the storage capacitor.

Both capacitors, the one with the LC dielectric and the storage capacitor, are plane-parallel capacitors and their values can be calculated using Equation (4.2)

$$C = \frac{\varepsilon \varepsilon_0 A}{d}, \qquad (4.2)$$

where
- ε is the dielectric constant of the LC,
- ε_0 is the permittivity of free space,
- d   is the thickness of the cell, and
- A   is the clear area of the pixel.

As mentioned before, the C_{LC} value is not a constant and depends on the voltage. This occurs because the value of ε depends on the rotation angle of the liquid crystal molecules, which in turn depends on the voltage across the LC.

For the storage capacitor, ε is the dielectric constant of SiO_2, d is the thickness of the SiO_2, and A is the overlap area between capacitor electrodes. The pixel capacitance is then 7.4 fF and the storage capacitance is $\approx$84 fF. The total capacitance is, with rounding, about 90 fF. From Equation (4.1), the leakage current from all sources must be <1 pA for a system with a 60 Hz refresh rate to keep the change in the signal voltage on the pixel to less than 10 % of the dynamic range, or 0.2 V. These sources could include, among others, source-to-drain leakage in the TFT, pixel electrode-to-gate in the TFT, or leakage due to conduction within the liquid crystal material. Increasing the storage capacitance by using a thinner dielectric or a dielectric with higher ε would relax this limit.

On the other hand, smaller pixels or reduced storage capacitance area would make the allowed maximum leakage even less. Were the storage capacitor not included in the pixel, the allowed leakage current from all sources would be less than about 80 fA, a difficult goal to achieve.

Typically the switches (TFTs) in the active matrix are made of photo-sensitive materials such as silicon. In the presence of the very high flux of light in a projection microdisplay, the TFTs would act as phototransistors and rapidly discharge the pixel capacitance. Therefore, the active matrix must be designed to block the incident light and prevent it from falling on the TFT.

4.1.1.2 Charging currents

There are also requirements for the charging currents that the switching device in the active matrix must pass. The liquid crystal cell must be operated with an AC voltage, i.e. successive writes to the pixel must produce the same amplitude signal but with opposite polarity. In addition, there is a threshold voltage before an electro-optic effect is observed. Hence, the maximum voltage swing on successive frames is twice the sum of the threshold voltage plus the dynamic range voltage. This is the maximum peak-to-peak voltage given in Table 4.1. For this, a threshold voltage of 1 V is assumed. There are materials for which the threshold and dynamic ranges are different than those given in Table 4.1.

The average charging current is given by

$$i = \frac{\Delta V (C_P + C_S)}{t_{on}} \qquad (4.3)$$

where ΔV is the change in voltage on the capacitance produced by the charging current i and t_{on} is the time the channel is open for charging ($\leq$ line address time). For an NTSC image converted to 525 lines progressively scanned with a 60 Hz field rate, the maximum charging time is 31.7 µs. The average current to effect this charging is less than 20 nA.

The peak currents, at the point in the charging cycle when the voltage differences between the signal sources and pixels are maximum, are substantially higher. Larger pixels, larger storage capacitors, higher resolution panels or other architectures resulting in shorter pixel address times and/or higher capacitive loads increase the charge current requirements. Nevertheless, except in special circumstances, the charging currents are modest and can be delivered by devices fabricated in a wide range of materials.

4.1.2 Technologies

Only some of the available active-matrix technologies have so far been shown to provide adequately low leakage current suitable for projection application. Among these are amorphous silicon (α-Si) TFTs, polycrystalline silicon (poly-Si) TFTs formed at high temperatures ($\approx$1050 °C), crystalline silicon (c-Si) MOSFETs and two-terminal devices, including α-Si diodes (Hartman 1995). The attributes of each of these technologies will be discussed in the following sections.

4.1.2.1 α-Si TFTs

One of the technologies used to form the active matrix for liquid crystal projection microdisplays is essentially the same as that used for direct-view AM LC displays. These AMs are based on α-Si TFTs. There are several process architectures that have been used to fabricate these AMs but the principal ones employed are called the inverse staggered structures (Oana 1989). Figure 4.5 illustrates one of the several embodiments of the inverse staggered TFT structure.

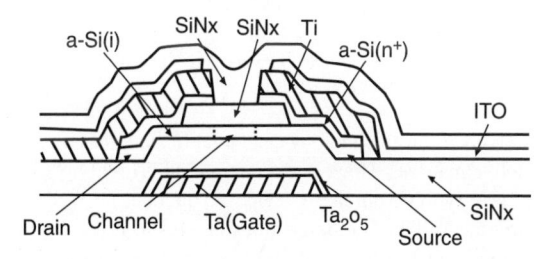

Figure 4.5 Cross-section of inverse staggered (tri-layered) α-Si TFT structure. Conduction path is from α-Si source and drain through the channel region. (From Oana 1989; reprinted with permission of Elsevier Science.)

All of the component layers of the α-Si TFT are deposited at relatively low temperatures, typically less than 450 °C. The substrate on which the device is fabricated is glass (not shown). One of the characteristic features of this inverse staggered structure is the gate electrode being on the bottom of the device. This construction, in combination with the silicon nitride gate dielectric, results a reproducible device with satisfactory stability.

The field effect (F-E) mobilities (Sze 1981) in α-Si TFTs are below 1 cm^2/(V-S). By contrast, the F-E mobilities in c-Si MOSFETs can range from 300 to 1000 cm^2/(V-s) or more. Among the implications of this are limited switching speeds of α-Si TFTs and limited conduction capabilities of these devices. Because of the limited switching speed, active matrices fabricated with α-Si TFTs use external (not integrated) driver circuits. There have been reports of α-Si AM with integrated drivers (Stewart *et al.* 1995; Lebrun *et al.* 1996). This was accomplished by externally multiplexing the video signal onto 40 video input lines to reduce the switching speed requirements of the active devices, since this reduces the data bandwidth for each line by a factor of 40.

The typical electrical characteristics of an α-Si TFT are shown in Figure 4.6. The on-currents are very limited for small channel widths. It is not uncommon to use α-Si TFTs with wide channel widths to increase the maximum current these devices can carry. The off-currents exhibited by this class of AM devices are admirably low, typically below $1\,\mathrm{fA}/\mu\mathrm{m}$ of channel width. Using a wide channel to increase the available on-current still results in off-currents which are well below the leakage limitations for projection applications.

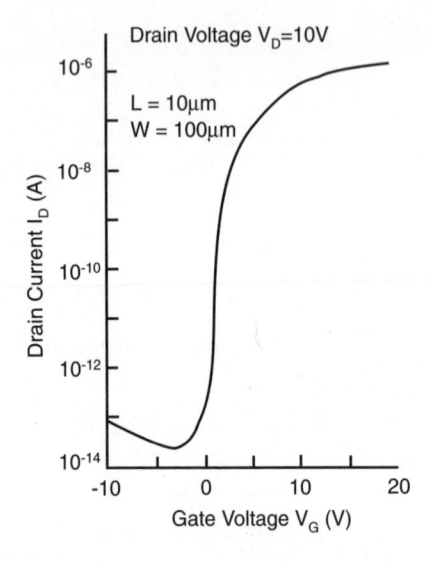

Figure 4.6 Electrical characteristics of α-Si TFT.

Projection light valves based on this technology can be fabricated in essentially the same facilities as are used for direct-view active matrices. However, because α-Si is highly photosensitive, some process and design changes are necessary to permit these devices to operate satisfactorily in the extremely high luminance projector environments. These modifications can result in the requirement for a dedicated facility, possibly a production line no longer used for direct-view devices.

The substrates used for α-Si TFT AM fabrication are large-area glass substrates. The lithography equipment developed for these large substrates is somewhat limiting on the minimum pixel dimensions that can be routinely fabricated. In addition, the process for inverse-staggered TFTs usually does not have a self-aligned gate structure, required for having minimum overlaps between the source and drain metallizations and the gate electrode. The overlap results in undesirable parasitic capacitances, which can negatively impact device operation.

Figure 4.7 illustrates the equivalent circuit of a picture element, including a storage capacitor. In a self-aligned TFT, the source and drain metallizations are defined by the gate metallization, minimizing the overlap. In non-self-aligned structures, overlap must be built into the design to ensure all of the channel is controlled by the gate. The parasitic capacitances C_{gd} and C_{gs} shown in Figure 4.7 increase with increasing overlap.

The consequences of these parasitics are overshoot and undershoot (depending on signal polarity) and DC offset (dV_p) of the pixel voltage relative to the signal voltage. This effect is illustrated in Figure 4.8. It

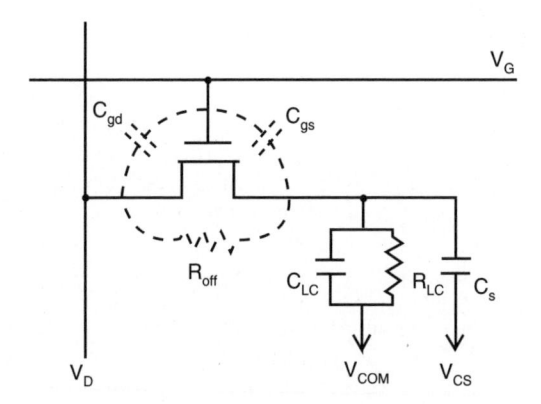

Figure 4.7 Equivalent circuit of a picture element, including a storage capacitor. The pathway for electrical leakage through the transistor is illustrated as the resistance R_{off} and through the liquid crystal as R_{LC}. (From Oana 1989; reprinted with permission of Elsevier Science.)

results from a charge redistribution between the pixel capacitance and the parasitic capacitance. Electronic compensation can be employed to minimize this effect. However, the effects of the parasitics become more apparent, and more difficult to compensate, as the pixel size is decreased.

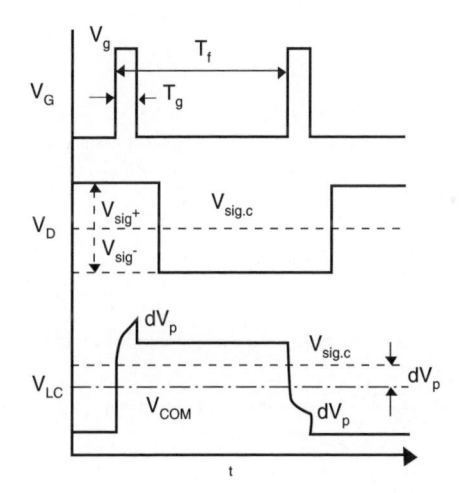

Figure 4.8 Waveform diagram for an ideal pixel with no leakage currents. The LC signal is applied as an AC square wave, and a positive going gate pulse opens the TFT channel. (From Oana 1989; reprinted with permission of Elsevier Science.)

While projection light valves having pixel pitches as small as about 50×50 μm based on this technology have been reported at technical symposia, it is more common that production light valves have pitches 1.5–2× these dimensions. With the exception of single light-valve systems with color filter arrays or angular color separation, discussed in Section 10.4.1, this translates to larger systems with more expensive optical components as compared with systems executed with other AM technologies. As a consequence, this technology is rarely used any more for projection microdisplays.

4.1.2.2 Poly-Si TFTs

There are two generic technologies for forming polycrystalline silicon, or poly-Si, TFTs. The more common one used for projection light valves is called high-temperature poly-silicon (HTPS) (Yamamoto *et al.* 1995). The other is called low-temperature poly-silicon (LTPS). The dividing line between HTPS and LTPS is the softening point of glass. LTPS is produced with process temperatures glass can tolerate while HTPS is formed at higher temperatures and requires a quartz or other temperature-resistant substrate.

LTPS devices are formed on glass substrates using process temperatures typically below 600 °C. While earlier work on fabricating poly-Si at low temperatures was based on furnace annealing of amorphous Si layers (Morozumi 1986), it is more common today for laser annealing to be employed (Im 2007; Itoh 1996). While there are some reports of sub-pA off-currents obtained with laser annealed LTPS TFTs, it is more common that the leakage currents are higher. Thus, despite the excellent mobilities of 150 cm^2/(V-S) or more obtained with this technology, it has not yet been used to any great extent in projection light valves.

HTPS TFTs are fabricated with processes in which the maximum temperature can be in excess of 1000 °C (Morozumi 1984; Mitra *et al.* 1990). The substrates must be quartz because the processing temperatures are above glass softening temperatures. The results of the basic process steps in the fabrication are illustrated in cross-section in Figure 4.9. Many of the production steps and much of the production equipment are common with those used in the fabrication of silicon ICs.

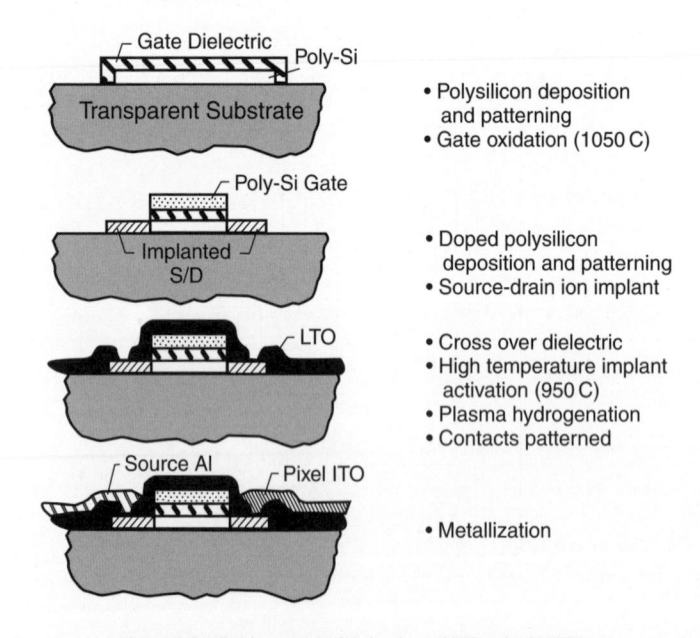

Figure 4.9 Basic steps in fabrication of HT poly-Si TFTs.

The devices have self-aligned gates to minimize parasitics. With standard HTPS processing, electron and hole mobilities of about 50 cm^2/(V-S) and 35 cm^2/(V-S), respectively, are obtained. Much higher mobilities were reported using high-pressure thermal oxidation (Mitra *et al.* 1991a). The mobilities obtained with standard processing have been adequate for full integration of the driver electronics.

The typical electrical characteristics of HTPS TFTs are illustrated in Figure 4.10. The on-currents are about 10 μA/μm of channel width and the off-currents are about 3 fA/μm of channel width. Small devices with channel widths less than 5 μm are frequently used within the active matrix. The on-currents and leakage currents in HTPS TFTs scale with the channel width. For the small devices used in an active matrix, the leakage current would be about 10 fA. The use of a lightly doped drain (LDD) structure in the device has been effective in controlling the increase in current when the TFT is back-biased (Mitra et al. 1991b).

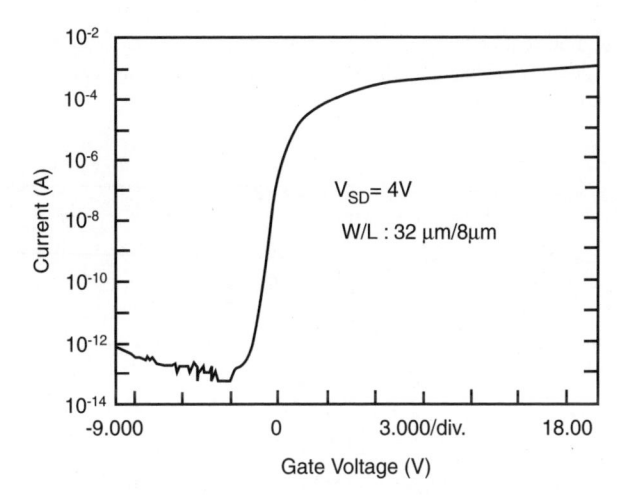

Figure 4.10 Electrical characteristics of HT poly-Si TFT.

HTPS TFTs can have high aperture ratios even with small pixel pitches. Among the factors which contribute to this are self-aligned TFTs, pixel designs which hide most of the storage capacitor under the adjacent row line, and manufacturing technologies suitable for sub-micron design rules. Pixel pitches of about 32 μm and aperture ratios in excess of 60 % were common for the first edition of this book. Pixel sizes in HTPS systems dropped rapidly after that. For example, Shirochi et al. (2003) described an HTPS panel with a 22 μm pitch and a 52 % aperture ratio. Seiko-Epson achieved an aperture ratio of 60 % with 20 μm pixels and an aperture ratio of 45 % with 14 μm pixels in 2001. The effective aperture ratio for these 14 μm pixels could be increased to 70 % if a microlens array was used to focus light on the pixel aperture (Okamoto 2001).

Almost all HT poly-Si and α-Si AM light valves are transmissive devices, i.e. the modulated light leaves from the side opposite to the one where the illumination light is incident. The AM structures such as the row and column metallizations, the TFTs, and some of the storage capacitor structure, require areas not directly involved in producing the optical modulation. These structures can induce electric fields that affect the liquid crystal materials which, in turn, can produce undesirable optical effects. It is common to make these areas where the electric field is not perpendicular to the substrates opaque with a structure called the black mask, or the interpixel mask. This mask has been placed either on the counter electrode or on the active substrate. The latter has been shown to be beneficial to minimize extra area losses resulting from safety margins needed for aligning the black mask on one substrate to the active matrix on the other. This black mask can also serve to prevent the TFT from acting like a phototransistor and discharging the pixel prematurely.

To a reasonable approximation, there are minimum dimensions associated with the structures that must be made opaque. Since the areas of the structures to be hidden by the black mask have only a small dependence on the pixel pitch, decreasing the pixel pitch results in a decrease in the fraction of the total pixel area available for light transmission. As the pitch is decreased, the aperture ratio goes down. This will limit further decreases in pixel pitch of transmissive light valves. While smaller pitch translates to less costly devices and systems, it also reduces the throughput of the optical system. This problem is a key issue in device and system design and will be covered in more detail in Chapter 11.

4.1.2.3 Crystalline silicon active matrices

Systems using active matrices with reflective light valves based on crystalline silicon, or c-Si, are being developed (Sato *et al.* 1997; Alt 1997). Figure 4.11 illustrates a pixel cross-section incorporating this technology and a liquid crystal electro-optic effect. LC effects will be discussed later in this chapter. The TFTs have excellent on-current and off-current characteristics. Obviously, driver electronics are readily integrated. Liquid crystal devices made with active matrices using single crystal silicon are commonly called Liquid Crystal on Silicon or LCoS.

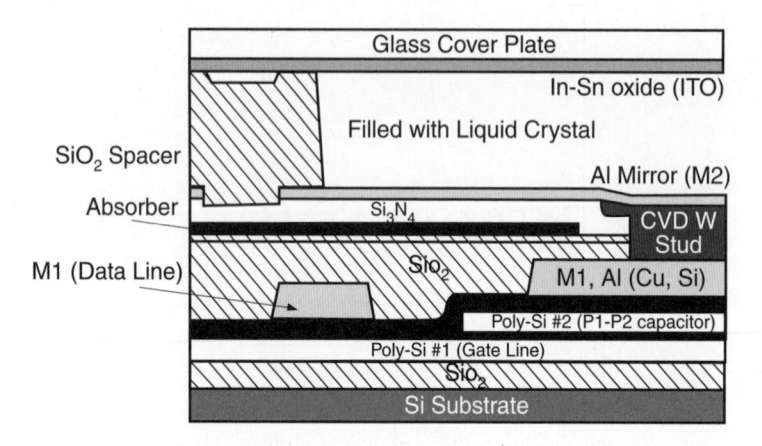

Figure 4.11 Cross-section of a pixel in an LC/c-Si reflective light valve (from Alt 1997; reproduced with permission of SID).

The active matrix is formed in the silicon substrate using conventional CMOS processing. This process is modified so that a highly reflecting metallic mirror is formed atop each TFT and is electrically connected to this active device through conduction paths called vias. One such via, labeled CVD W (for chemical vapor deposited tungsten) stud, is illustrated in Figure 4.11. The via connects the Al mirror (M2) with an underlying Al(Cu,Si) structure. It is necessary to planarize the surface above the TFT but under the Al mirror to within a fraction of a wavelength of visible light. Rougher surfaces result in the mirror not having the requisite reflectivity. While planarization technologies have been developed for multi-metal IC technologies, the planarity requirements of the light valve are more severe.

The light from the illumination system passes through the glass cover plate and the liquid crystal material to reach the mirror. If some of this light enters the silicon substrate, transistors in the ICs and AM can malfunction because of carrier generation by the light. The pixel design must incorporate appropriate structures to shield the active devices from the intense illumination incident on the structure. In Figure 4.11, this is the structure labeled *absorber* under the aluminum mirror. The SiO_2 spacer is a lithographically defined structure to keep the cover plate with its ITO conductor at a well-defined distance from the Si structure. The space between the Al mirror and the ITO is filled with the liquid crystal.

An electron micrograph of a typical LCoS backplane (Sloof and Brennesholtz 2003) is shown in Figure 4.12. Note in this image, the spacers to maintain the correct cell gap are not at every pixel intersection, but only one intersection in four. The pixel pitch for this microdisplay was 13.3 μm square. The small dip in the center of each pixel represents the end of the via where the aluminum mirror, which also served as the pixel electrode for controlling the LC, was connected to the underlying active matrix transistor.

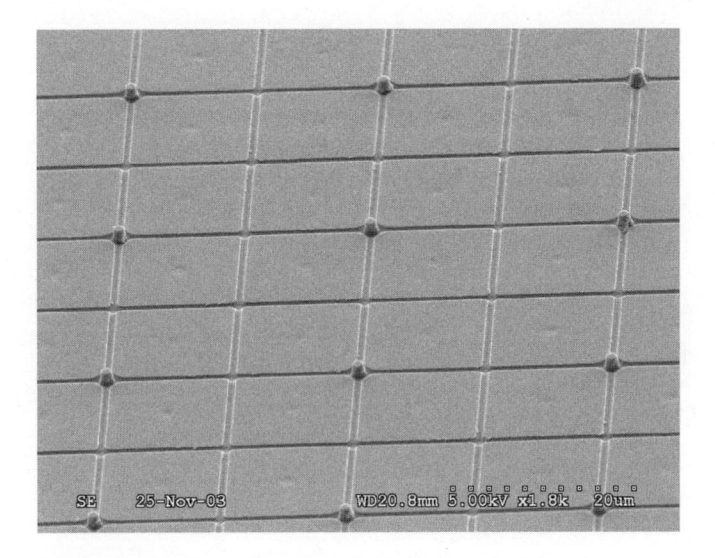

Figure 4.12 Active matrix backplane for a single panel LCoS microdisplay (Reproduced by permission of Royal Philips Electronics).

Reflective projection light valves have the promise of much higher aperture ratios as compared with transmissive ones, as shown in Figure 4.13. All of the structures that would have been in the interpixel areas in a transmissive light valve can now be hidden behind the mirror atop each pixel. It is only necessary to provide a limited space, typically less than 1 μm, between pixels for electrical isolation. This results in a higher aperture ratio and a reduced dependence of the aperture ratio on the pixel pitch. With this increased aperture ratio arrays can be fabricated with smaller pitches. This can be important in the upgrading of a projector to higher resolution without changing the optical system. On the negative side, the optical systems for use with reflective light valves are more complex and expensive.

(a) Transmissive   (b) Reflective

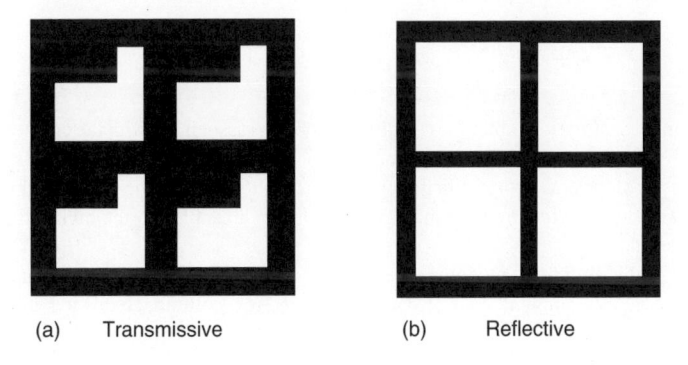

Figure 4.13 Schematic of pixel apertures in (a) transmissive and (b) reflective light valves.

Silicon active matrix backplanes can drive the LC with either analog voltages (Sloof 2003, 2004) or with a digital drive scheme (Shimizu *et al.* 2004; Guttag 2004). These active matrix arrays can also be made with a built-in frame buffer. This frame buffer can greatly simplify the optics of a single-panel system, as will be discussed in Section 10.4.2.

42 **LIQUID CRYSTAL LIGHT VALVES AND MICRODISPLAYS**

c-Si active matrices are also used in light valves based on micromechanics. Some of these will be discussed in Chapter 5.

4.1.2.4 Active matrices based on two terminal devices

Two terminal devices were among the earliest devices proposed for switching an active matrix (Lechner *et al*. 1971). These devices include p-i-n diodes, metal-insulator-metal devices (MIM) (Morozumi 1983), and devices called TFDs which use a semi-insulating material between electrodes (Hartman 1995). All of these exhibit diode-like characteristics with a well-defined threshold for conduction. In theory, all two terminal devices can be produced with fewer mask steps than three terminal devices.

The electrical equivalent circuit of a pixel is shown in Figure 4.14 for one configuration of two terminal device active matrix. In this 'diode-ring' arrangement, the diodes are connected in parallel and are arranged with their polarities reversed. The rows and columns are defined by orthogonal conductive traces, the row electrode and all the diodes being on one inner surface of the liquid crystal cell and the column electrode being on the other.

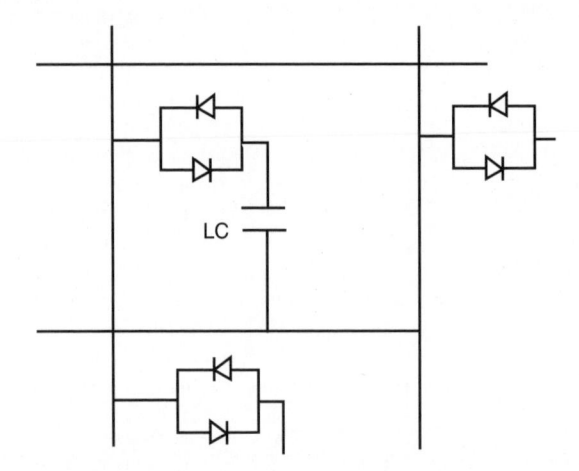

Figure 4.14 Electrical circuit for pixel address by two-terminal devices.

This very simple construction minimizes electrical problems associated with the crossovers when both row and column metallizations are on the same substrate, as is the case with TFTs. In practice, each 'diode' may consist of several two-terminal devices in series to increase the effective threshold voltage to conduction. This may be required for matching the diode thresholds to the higher thresholds of the liquid crystal materials.

The operation of this simplified model is given by the following. The operating voltages are chosen so that the pulse voltages V_R, applied to a row, and V_C, applied to a column, are always in the range given by Equation (4.4).

$$\text{Addressed}: \quad \frac{|V_{Th}|}{2} < |V_R|, |V_C| < |V_{Th}|$$
$$\text{Non-addressed:} \quad V_R \text{ or } V_C = 0 \tag{4.4}$$

To address a specific pixel, a voltage pulse, called the half-select voltage, is applied to the row electrode and simultaneously a pulse of opposite sign is applied to the appropriate column electrode. Since both V_R and V_C are greater than $V_{Th}/2$, the difference between these voltages is greater than V_{Th} and one or the other diode conducts, charging the liquid crystal capacitance to a new voltage level. For the non-addressed pixels attached to either this line or row, either V_R or V_C would be equal to zero. The total voltage due to the half-select voltage being applied to only one terminal of the diode-capacitor series circuit is now less than V_{Th}. No conduction occurs through the diodes and the voltage on the liquid crystal is unchanged. The symmetrical device arrangement permits conduction in both directions for AC operation.

Two terminal devices have had limited application to liquid crystal projection light valves. They have been used in modest resolution direct-view systems. Practical and performance considerations have resulted in constructions and electrical drive sequences (Kuijk 1990) more complex than outlined above. These more complex constructions often require additional mask steps. As a consequence, two-terminal device AMs lose some or all of their economic advantages compared with three-terminal devices.

4.2 Liquid Crystal Effects

Liquid crystal phenomena are among the ones used in projection light valves to produce the spatial light modulation. Projection systems have some unique, sometimes severe, requirements for the electro-optical effects to produce the requisite system characteristics. As a consequence, there are a limited few liquid crystal effects that are actually used in projection light valves. These will be discussed in the following sections, mostly from phenomenological and device structure perspectives. Detailed discussions of the physics and chemistries of liquid crystal materials are beyond the scope of this book. These discussions are available in the literature (Hilsum and Raynes 1983; Bahadur 1984; Bahadur 1990; Blinov 1983; Wu and Yang 2006). Light modulation with microelectromechanical (MEMS) devices will be discussed in Chapter 5.

Liquid crystals can be considered to be an intermediate phase, called the mesophase, between crystalline solids and isotropic liquids. They exhibit some of the longer range order characteristics of crystalline solids and some of the fluidity of liquids. When the liquid crystal material is cooled, a phase transition will be reached where the material becomes solid (Figure 4.15). When it is heated, a phase transition will be reached above which the material is isotropic and liquid. Liquid crystal molecules are generally organic and have an elongated shape. This elongated shape leads to different optical and electrical properties parallel and perpendicular to the molecular long axis.

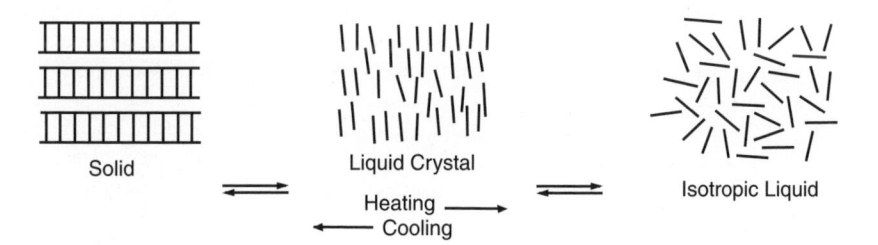

Figure 4.15 Effects of heating or cooling a liquid crystal material (adapted from Bahadur 1984; reprinted with permission of Gordon & Breach Science Publishers).

Within the mesophase phase, liquid crystal materials can have a variety of orderings which result in dielectric and optical characteristics which have been exploited in displays. The materials are classified into three types, depending on the ordering. These are nematic, smectic and cholesteric, illustrated in Figure 4.16. The nematic phase is the simplest of these and, so far, the most important for light-valve application. It consists of rod-like molecules, more or less parallel to each other. The vector that defines the orientation of the long axes of these molecules is called the 'director'.

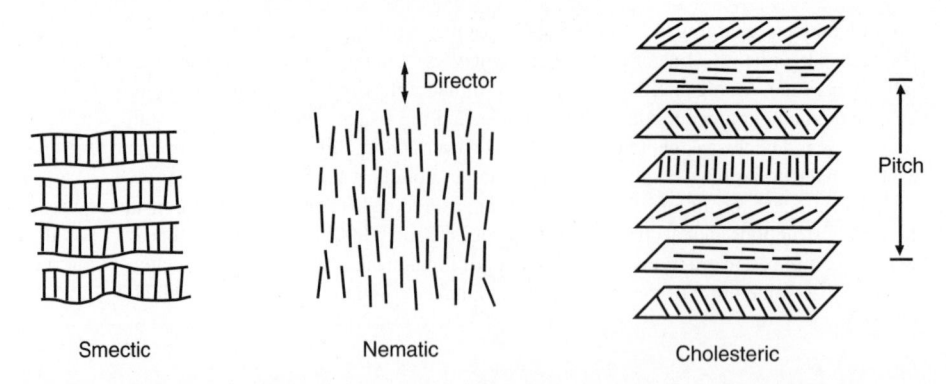

Figure 4.16 Molecular arrangements in various types of liquid crystals (adapted from Bahadur 1984; with permission of Gordon & Breach Science Publishers).

The nematic materials used in light valves are frequently mixtures of many chemical components. These mixtures are varied to adjust the physical characteristics to produce the desired display characteristics. Several of these properties are anisotropic. Among these are birefringence and dielectric anisotropy, i.e. the indices of refraction and dielectric constants parallel to the director are different from those perpendicular, with the parallel values more often being larger (positive anisotropy). The optical characteristics of the material can be varied by manipulation of the orientation of the director in a device by an electric field. This will be discussed further in the next section.

The variation of the dielectric constant of the liquid crystal with applied voltage provides another argument for including a storage capacitor in the active matrix. The switching speed of a nematic liquid, discussed in Section 4.2.2, is typically measured in milliseconds. The time the pixel is having information written, the line address time, is measured in tens of microseconds. During the line address, an amount of charge Q is put on the pixel to achieve the desired voltage $V = Q/C$.

In the absence of a storage capacitor, C is the capacitance of the pixel with liquid crystal dielectric constant appropriate to the molecular alignment at that instant. The liquid crystal material undergoes a change in dielectric constant when subsequently switching to the new applied voltage state. Since Q is fixed and C is changing, the voltage across the liquid crystal must also be changing. Thus, the final voltage state is different from the address voltage. Addition of a voltage-independent storage capacitance, which is much larger than the pixel capacitance, minimizes the effects of the dielectric switching.

The physical characteristics of the liquid crystal in a cell, including the viscosity and the elastic constants of liquid crystal materials, are mixture dependent. The mixtures are adjusted to achieve, within limits, specific microdisplay characteristics. These two parameters are important in determining the rates at which the director direction changes under the influence of external fields and the relaxation back to the undriven state. Lower viscosities and higher elastic forces result in faster display switching speeds.

The transition from the mesophase to an isotropic liquid occurs at the 'melting point' of the liquid crystal, called the 'clearing temperature'. In a projector, the very high light fluxes can result in elevated temperatures. Satisfactory operation of the microdisplay requires a high clearing temperature and good thermal design of the projector to minimize temperature rises. Liquid crystal mixtures having clearing temperatures of around 100 °C are used in projection light valves. The system designer must ensure the temperature of the LC cell does not exceed the clearing temperature for proper operation of the system. If the temperature is exceeded, the LC loses its electro-optical properties and the system most commonly goes to maximum transmission for a full white field. Normally this process is reversible and when the system cools down, the LC molecules realign themselves and proper operation is restored.

4.2.1 Liquid Crystal Cells

In the absence of any ordering of the molecular orientations of the nematic liquid crystal molecules the directions of the ordinary and extraordinary indices of refraction are random. Ordering of these directions can be produced by an external field or, in the absence of such a field, by surface structures with which the liquid crystal molecules have a strong affinity. These surface structures, called alignment layers, are produced by a variety of techniques, including depositing the layer under specific conditions and post-processing the deposited layer.

Among the earliest alignment layer technologies is oblique evaporation of SiO_x (Janning 1972). Evaporation of a few tens of nanometers of SiO_x at an angle greater than 80° with respect to the normal to the cell surface produces an alignment layer structure perpendicular to the evaporation direction. In a liquid crystal cell, the nematic material is in contact with this layer and the directors are all aligned by this structure. This alignment methodology is difficult to implement with the large area substrates used for direct-view α-Si active matrices and, as the substrate size is increased for projection light valves, it is also becoming difficult for these devices.

The current preferred method for producing the alignment layer is the deposition of a thin polymer layer, typically polyimide, followed by a post-processing. This processing consists of a unidirectional rubbing of the polyimide layer (Kahn *et al.* 1973), producing a structure in the rubbing direction to which the nematic molecules align.

Figure 4.17 illustrates the construction of a liquid crystal cell. Each substrate of a transmissive cell has a transparent conducting layer and an alignment layer. The conducting layer is typically indium tin oxide, or ITO. The substrate thicknesses are often less than 1 mm. The interior of the cells is slightly below atmospheric pressure and the external forces would distort the substrates in the absence of internal supports, or spacers. In active-matrix cells, one of the substrates in Figure 4.17 is replaced by the AM circuit substrate, with the alignment layer on the AM.

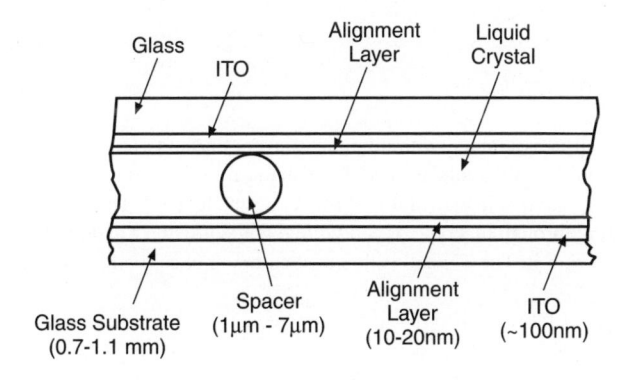

Figure 4.17 Cross-section of a liquid crystal cell.

The cell thickness is controlled by precision spacers between the substrates. Cell thicknesses are typically 3–6 μm, although some LC modes require thinner cells. LCoS cells designed for single-panel operation can be as thin as 1 μm. These spacers can be spheres or rods with well-controlled dimensions which are randomly distributed with a low-concentration density throughout the cell. Alternatively, lithographically defined spacers (Castleberry and Possin 1988), shown in Figure 4.12, are being used more frequently in projection light valves fabricated with poly-Si or crystalline silicon active matrices. This type of spacer has defined locations relative to the clear aperture. These are typically located in non-transmissive regions of the pixel, making the spacers invisible.

4.2.2 Nematic Cells

Nematic cells can be constructed with the orientation of the alignment layers on the two surfaces parallel or at an angle to each other. Cells made with parallel alignment layers are called homogeneously aligned when the LC molecules are parallel to the surfaces and homeotropically aligned when the molecules are perpendicular to the surfaces. Nematic cells made with an angle between the two alignment layers exhibit a twist in the orientation of the molecular axis in the volume between the two alignment directions. Hence, the name given to this type of liquid crystal cell is twisted nematic liquid crystal cell.

4.2.2.1 Parallel aligned layer cells

The nematic molecules are rod-like and have different dielectric constants and indices of refraction along the molecular axis and perpendicular to it. Consider the case of a homeotropically aligned liquid crystal with parallel alignment layers. This mode of operation can be used in reflective devices. Without applied voltage, the indices of refraction for the two orthogonal polarizations are equal. The application of a field to a very thin cell changes the orientation of the molecules, and with it the indices of refraction, and the liquid crystal now behaves as a birefringent material. Since the magnitude of the birefringence is controlled by the applied field, this mode of operation is called electrically controlled birefringence (ECB) (Armitage 1992).

Without applied voltage, the polarization of the light propagates unchanged through the cell and it is blocked by the perpendicularly aligned output analyzer. With the deformation of the LC by the field, light propagates at an oblique angle to the optic axis. This produces elliptically polarized output light, some of which can now pass the output analyzer. These thin ECB cells can provide good contrast, good luminance throughput and faster switching speeds. However, dimensional tolerance requirements can be very severe in this type of cell.

4.2.2.2 Twisted nematic cells

If the liquid crystal cell is assembled with alignment layers on the two substrates perpendicular, or at another angle, to each other, the molecules of the nematic liquid crystal will align with each of these surfaces and will undergo a twist in the volume of the cell (Schadt 1971). The direction of the polarization of the light propagating through this type of cell will be guided by the rotation of the twist in the liquid crystal and will rotate with it. The TN cell is the most common construction used in liquid crystal light valves.

The alignment layer is fabricated so that the LC molecules at the surface are slightly tilted up from the surface. This, together with a chiral dopant added to the nematic liquid crystal, provides a preferred rotation direction. Without this preference, the molecules in some regions would rotate clockwise and in others counterclockwise. Where these different regions meet, a visible defect, called a disclination, may be generated.

For very thin cells, the liquid crystal behaves as a simple birefringent material, with different indices along and perpendicular to the molecular axis. The propagation of light through this material can be described as two counter-rotating elliptically polarized modes (Armitage 1992). For greater thicknesses, large compared with the wavelength of the light, the direction of polarization of the light will be guided by the rotation of the twist in the liquid crystal and rotate with it. At intermediate thicknesses as are typical in LC microdisplays, the situation is more complex and must be evaluated with more sophisticated tools such as Stokes vectors, the Jones Matrix or the Mueller Matrix. These tools are beyond the scope of this book and it is recommended that interested readers see a more detailed reference on LC and polarization such as (Wu and Yang 2006).

LIQUID CRYSTAL EFFECTS 47

The transmission T of the twisted nematic cell with the polarizer and analyzer aligned parallel is given by Equation (4.5), the Gooch-Tarry equation (Gooch and Tarry 1975), which is plotted in Figure 4.18.

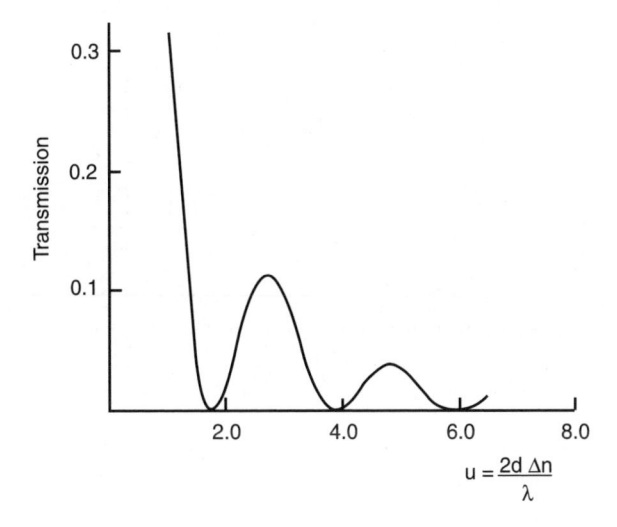

Figure 4.18 Gooch-Tarry curve for parallel polarizer and analyzer.

$$T = \frac{\sin^2[\theta(1+u^2)^{\frac{1}{2}}]}{(1+u^2)}$$

$$u = \frac{\pi d \Delta n}{\theta \lambda}$$

(4.5)

where
- θ is the rotation angle of the cell ($= \pi/2$ for cell with $\perp$ alignment layers),
- λ is the wavelength of the light,
- Δn is the index anisotropy in the liquid crystal between the director direction and the direction perpendicular to the director, and
- d is the cell thickness.

The Gooch-Tarry equation applies to the liquid crystal cell without applied voltage. With the parallel polarizer/analyzer, a minimum in transmission is obtained only for specific values of d, Δn, and λ. For values of u, defined in Equation (4.5), away from the minima, the transmitted light is not linearly polarized, but rather is elliptically polarized. Some of this light can pass the analyzer. The contrast for this parallel polarizer case is determined, in part, by the linearity of the polarization of the output light, i.e. the closeness to zero of the minimum transmission.

With crossed polarizer/analyzer, the Gooch-Tarry equation predicts transmission maxima at the same u values which give the minima for the parallel polarizer/analyzer case. If the values in u in a specific cell do not correspond to those producing the maximum transmission, the effect is to reduce the peak transmission, a relatively small effect.

With crossed polarizer/analyzer, the rotation of the polarization by the cell will result in the maximum transmission of the light (Figure 4.19a). When voltage is applied to the cell, the liquid crystal molecules away from the surfaces align with the field, removing the twist, and the polarization direction is no longer rotated (Figure 4.19b). The incident light is then blocked by the analyzer. For intermediate voltage

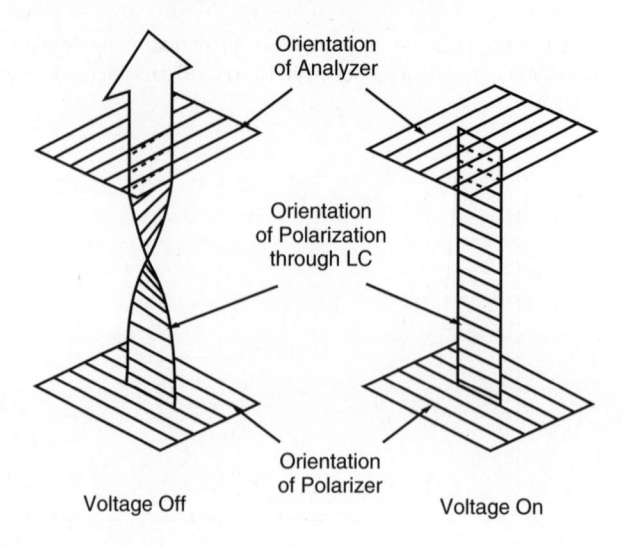

48 **LIQUID CRYSTAL LIGHT VALVES AND MICRODISPLAYS**

Figure 4.19 Passage of polarized light through a twisted nematic liquid crystal cell with crossed polarizer/analyser: (a) no applied voltage; (b) voltage applied.

conditions, the transmitted light becomes elliptically polarized and only a fraction of this passes through the analyzer.

Figure 4.20 illustrates the transmission-voltage characteristics of a twisted nematic liquid crystal cell with crossed polarizer/analyzer. Twisted nematic cells are operated with AC waveforms with minimal DC component. For both polarities, there is a minimum voltage, the threshold voltage, to produce an optical effect. This is the voltage at which the molecules just start to rotate into alignment with the field.

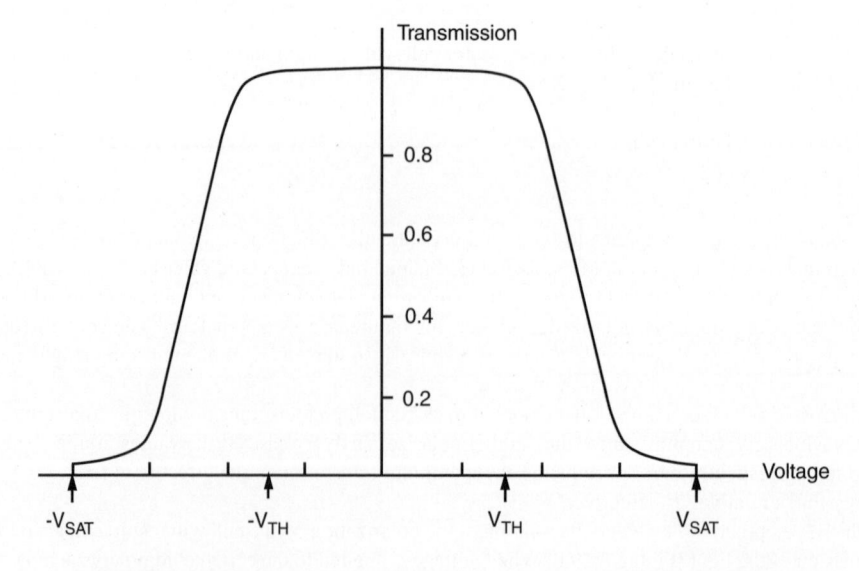

Figure 4.20 Transmission-voltage characteristics of twisted nematic liquid crystal cell.

Above the threshold voltage, the transmission decreases with increasing voltage in a continuous, non-linear manner, as the molecules rotate further. At sufficiently high voltages, the transmission can be reduced to a fraction of a percent of the peak transmission. In a device incorporating an active matrix, the voltages to produce these changes in transmission are applied to the column drive electrodes and are routed to the appropriate pixel under the control of the row voltages, which open or close the channels in the active devices.

This mode of operation is called 'normally white' or sometimes 'driving to black'. By disposing the polarizer/analyzer directions parallel, polarized light traversing the cell will be blocked by the analyzer in the absence of applied voltage. When voltage is applied, the rotation of the polarization is suppressed and light passes the analyzer. This latter mode is called 'normally black' or sometimes 'driving to white'. With this mode, the minimum transmission of the stack is not voltage determined, but rather by the parameters in Equation (4.5).

Except for residual twist near the cell surfaces, the driven state has no optical activity. As the signal voltage is increased, the cell becomes more homeotropic and the direction of polarization less rotated. The transmission through the crossed polarizers is thus minimized by the application of high signal voltages. Because of this advantage, projectors most often are operated in the driven-to-black mode. The range of voltages for achieving minimum transmission is typically in the range of 4–6 volts and the threshold voltages are typically in the range 1–2 volts.

The contrast ratio obtained with this driven-to-black cell is limited by the amplitude of the signal voltage available to achieve the non-transmissive state and the extinction provided by the crossed polarizer and analyzer. It is not uncommon to get contrast ratios of 1000:1, or more, in TN cells with crossed-polarizer/analyzer. TN liquid crystal light valves for projection almost always work in this mode in order to get high contrast.

The switching speed of a twisted nematic is dependent on whether the final state is the driven state, i.e. the lower transmission state for crossed polarizer systems, or the undriven high transmission state (Bahadur 1984). The time to reach the driven state is called the rise time. The decay time to the undriven state, also called the relaxation time, is controlled by relaxation phenomena within the liquid crystal. How fast this process is depends on the material's elastic constants. The parametric relationships for the response times for these two cases are

$$T_{rise} \propto \frac{\eta d^2}{\varepsilon_0 \Delta \varepsilon V^2}$$
$$T_{decay} \propto \frac{\eta d^2}{k}$$
(4.6)

where

- η is the viscosity,
- d is the cell thickness,
- $\Delta \varepsilon$ is the dielectric anisotropy,
- k is the elastic constant, and
- ε_0 is the permittivity of free space.

The rise time is typically much less than the decay time. Both switching speeds of the liquid crystal cell are reduced most by decreasing the cell thickness. A lower viscosity mixture will usually switch faster. Cells for fast switching are often fabricated to operate at the first, or at most the second, Gooch-Tarry maximum for crossed polarizer devices to minimize the cell thickness. Switching speeds suitable for displaying video are now routinely achieved. As mentioned before, LCoS cells can achieve switching speeds under 1 mS and can be fast enough for color sequential operation.

50 LIQUID CRYSTAL LIGHT VALVES AND MICRODISPLAYS

4.2.3 Polymer-dispersed Liquid Crystal (PDLC)

Nematic liquid crystal cells have been developed which require no polarizer/analyzer. These cells operate on a different principle than the twisted nematics. The electro-optic material consists of a dispersion of droplets of nematic liquid crystal in a plastic matrix (Fergason 1985) (Figure 4.21). In the absence of applied voltage, the directors of the liquid crystal molecules are, for the most part, random. There is a mismatch between index of refraction of the plastic and the effective index of the non-aligned liquid. When voltage is applied, the directors align with the field.

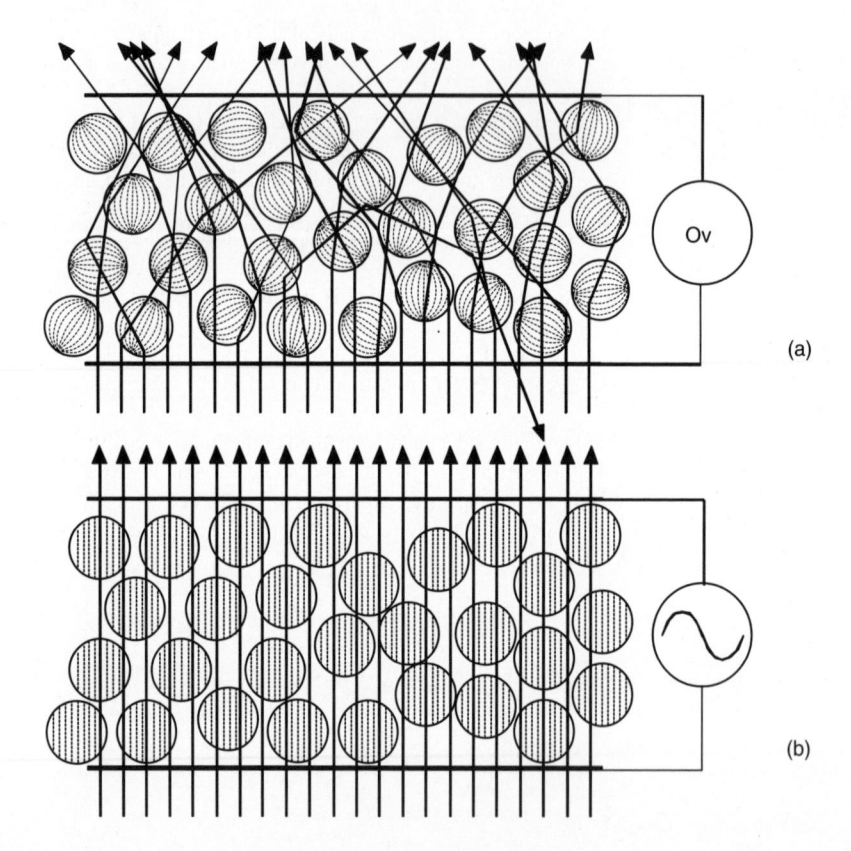

Figure 4.21 Effects on direction of light in passing through PDLC cell: (a) no applied voltage; (b) voltage applied.

The nematic/polymer mixture is prepared by mechanical dispersion (Drzaic et al. 1992) or by polymerizing a mixture of monomers and oligomers in which the liquid crystal is dispersed (Niiyama et al. 1993). The polymer host is chosen so that its index is the same as the liquid crystal along the direction of the director. Without applied voltage, the liquid crystal droplets have a different index in the direction of the light and the light is scattered (Figure 4.21a). When voltage is applied (Figure 4.21b), the indices of the liquid crystal and the polymer are equal and the matrix is transparent. A Schlieren optical system is used to separate the specularly reflected beam from the scattered beam.

PDLC light valves have the advantage of requiring no polarizer or analyzer, with the attendant optical losses these devices impose. However, some of the scattered light can enter the aperture of the projection lens, reducing the contrast. The contrast reduction is minimized by using a higher f/# (smaller aperture) lens. The light available for projection from the clear state is maximized by using a smaller f/# lens. Thus, the design of a system based on PDLC light valves requires a tradeoff between screen illuminance and contrast.

The maximum contrast in a PDLC projector is calculated assuming that the distribution of the output light is Lambertian when the light valve is scattering and the fraction of the specularly reflected light that passes the projection lens when the light valve is in the clear state, which is determined by the f/# of the system. The results are shown in Figure 4.22. Practical system considerations usually produce lower contrast ratios. The voltages required to switch PDLC light valves are higher than those required for nematic devices. A PDLC projection system is shown in Figure 10.31.

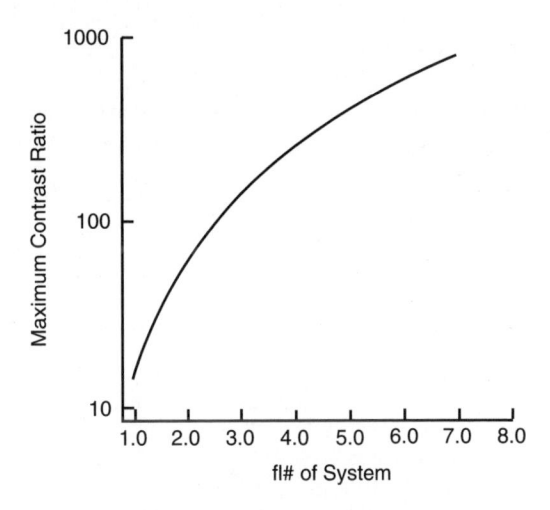

Figure 4.22 Maximum contrast obtainable with PDLC cell as a function of system f/#.

One of the main advantages of PDLC is that it does not depend on the polarization of the light, which makes it especially suitable for use with unpolarized light sources such as lamps and LEDs. Newer LC effects that also work with unpolarized light and show higher contrast are available and may be used in projection systems, especially pico-projectors with LED illumination, in the future (Komanduri et al. 2007).

4.2.4 Other Liquid Crystal Effects

Other liquid crystal effects and configurations are sometimes employed in projection light valves, including hybrid aligned nematics (HAN), ferroelectric liquid crystals (FLC or FELC), and electrically controlled birefringence (ECB). These have characteristics which may be valuable in specific system applications. In HAN displays (Matsumoto et al. 1976), the alignment layer on one surface of the cell is essentially the same as in a twisted nematic cell and the director is parallel to the surface. The alignment layer on the other surface is prepared so that the director direction is perpendicular to the surface. HAN light valves typically are faster switching than nematic cells but require higher drive voltages. They are effective in reflective light-valve applications (Glueck et al. 1992).

FLC is a chiral smectic C phase (S_c^*), which is sometimes considered to be a special case of the nematic phase. They have a lamellar smectic structure with an inclined orientation of the long molecular axes (director direction). Reversal of the applied electric field can cause the director to switch through twice the tilt angle. A consequence of this is that the material exhibits ferroelectricity (Meyer et al. 1975). This structure gives rise to a variety of electro-optic effects (Beresnev and Blinov 1989). Among the attributes of projection light valves using FLC is very fast switching. These materials can switch three or more orders of magnitude faster than twisted nematic. This makes them suitable for use in color-sequential systems.

Special cell constructions and/or materials are required for using the ferroelectric characteristics in a light valve. Included among these are surface-stabilized FLC (SS-FLC) (Clark and Lagerwall 1980) and distorted helix ferroelectrics (DH-FLC) (Fünfschilling and Schadt 1989). SS-FLC devices are bistable, i.e. they either fully transmit or they do not transmit the light, and hence they do not normally provide gray scale. They are fabricated with long pitch FLCs and thin cell gaps. Dithering of the image or operating in a half-tone mode (groups of four or more subpixels employed to form one display pixel) have been employed to generate limited gray scales. It has been shown that if the drive electronics controls the charge put on a pixel rather than the applied voltage that switched and non-switched domains form within the pixel (Hartmann 1989) with the relative areas of these proportional to the charge on the pixel capacitance. As a consequence, continuous gray scales have been demonstrated.

The DH-FLC devices are made with materials exhibiting helical twist pitches shorter than the wavelength of the incident light. Application of an external field distorts the helix. Among the consequences of this is an increase in the birefringence. These effects can be used to produce the electro-optic effect used in light valves.

In electrically controlled bifringence cells (Kahn 1972), the nematic liquid crystal is aligned without a twist, i.e. the alignment layers on the two surfaces are parallel. In the absence of applied voltage, positive dielectric anisotropy liquid crystal molecules would be uniformly aligned parallel to these alignment layers. Incident polarized light would be unchanged in passing through the cell. Crossed polarizer/analyzer will then block the light transmission. When voltage is applied, there is reordering of the orientations of the liquid crystal molecules in the central volume of the cell. As a consequence, the effective index of refraction changes. Polarized light incident on the cell will emerge elliptically polarized and a fraction will pass the analyzer.

The ideal transmission of the ECB cell (Karim 1992) can be expressed in terms of the birefringent phase retardation.

$$T = \sin^2 \frac{\pi d \Delta n_{eff}}{\lambda} \qquad (4.7)$$

where
    d   is the cell thickness
    Δn_{eff} is the effective birefringence (related to the local nematic tilt angle), and
    λ is the wavelength.

$T = 1$ when the argument of the sine function in Equation (4.7) is $(2n+1)\pi$ and $T = 0$ when the argument is $2n\pi$. ECB cells can produce high contrast images when the f/# of the system is not too low. They typically have narrow cell gaps, which results in faster switching times (see Equation (4.6). The transmission is wavelength dependent. However, this wavelength sensitivity is minimized at the lowest usable cell gap. Since the dielectric properties are temperature sensitive, so are the spectral characteristics.

4.2.5 Liquid Crystal Effects for Reflective Microdisplays

Most liquid crystal effects suitable for use in transmissive microdisplays have their counterparts in reflective microdisplays (Pfeiffer 2000). For example, Akimoto *et al.* (1999) describes a FLC LC effect on a silicon backplane and intended for single-panel, high resolution applications. One advantage of reflective microdisplays over transmissive is the higher speed of operation. Response speed of a liquid crystal cell normally proportional to the square of the cell thickness, as shown in Equation (4.6). A reflective LC cell is half the thickness of an equivalent transmissive cell, since the light passes through the cell twice and doubles the effect of the LC on the polarization.

The most common LC effect used for transmissive LC microdisplays is the TN-90 effect. The equivalent effect for reflective microdisplays is the TN-45 effect (Grinberg and Jacobson 1976) With this effect, on the first pass though the LC with no voltage applied, the input linearly polarized light is converted to circularly polarized light. When the light reflects off the mirror at the pixel electrode, there is a 180° phase change for the circularly polarized light. As it passes through the LC again, the circularly polarized light is converted to linearly polarized light that has been rotated by 90° compared to the input state. When a voltage is applied across the LC cell, the LC molecules line up mostly with the electric field except at the boundary layers. Since the electric field in the light is perpendicular to the LC molecules, to first order the polarization of the light is not changed. This creates the dark state of the cell.

The contrast of this basic TN-45 cell is relatively poor, on the order of 50:1, for two reasons. First, in the boundary layers, the LC molecules do not line up exactly with the electric fields. Secondly, since the system has a finite f/#, the electric field of all the light is not exactly perpendicular to the LC molecules. These two problems can be corrected by A and C plate retarders respectively, as will be described in Section 6.5.5. With proper compensation, the contrast of a TN-45 LCoS cell can be increased to 5000:1 or more.

TN-45 and related effects have been used primarily for field sequential LCoS projectors because of the relatively high speed they can achieve (Sloof and Brennesholtz 2003, 2004). A typical transmissive TN-90 LC cell will have a response speed of around 4 mS, depending on its exact design. This is not fast enough for color sequential operation without the introduction of major artifacts. Since the TN-45 is about half the thickness, it would typically have about a quarter the response time, or about 1 mS. This speed is fast enough for color sequential operation.

The other major LC effect used in LCoS is the vertically aligned nematic or VAN effect (Kurogane *et al.* 1998; Pfeiffer 2000). This basic effect is used by both Sony and JVC in their LCoS panels. Since it is not fast enough for color sequential operation, it is generally used in high-performance three panel LCoS projectors. Research into faster materials for the VAN effect is ongoing and it is possible there will be color-sequential VAN microdisplays in the future.

In a VAN cell, the LC molecules are aligned parallel to the light when no voltage is applied, as shown in Figure 4.23a. When the light passes through the cell, the polarization of the light is not changed and the dark state occurs. VAN cells are made with a LC material that has a negative dielectric constant. That means that when a voltage is applied, the LC molecules rotate so they are approximately perpendicular to the electric field, as shown in Figure 4.23b. These molecules now rotate the polarization of the incoming light and create the bright state. Like the TN-45 effect, A-plate and C-plate compensation is needed to get high contrast.

4.2.6 Liquid Crystal Inversion

The voltage applied across most liquid crystal modes must be an AC voltage in order to avoid such artifacts as image sticking. Typically this is done by applying voltages of opposite polarity each time the pixel is readdressed. For example, if a pixel is addressed with a positive voltage, the next time it is addressed, it is done so with a negative voltage and the third time with a positive voltage again. With most LC effects applying a positive or a negative voltage has the same electro-optic effect. For example, if a TN effect requires 6 volts applied to drive the cell to black, the cell will produce black whether $+6$ V is applied or -6 V is applied.

In a normal LC cell, the counter-electrode (the electrode on the side of the LC cell without the active matrix) is held at a constant voltage. If the TN cell requires 6 volts to drive it to black, typically the counter-electrode would be held at a constant 6 V. When the video signal calls for the pixel to be driven to black, the pixel electrode would apply alternately 0 V and 12 V. This would effectively apply -6 and $+6$ V across the LC, providing the correct inversion. Note this requires the active matrix to handle double the maximum voltage across the cell. If a cell requires 6 volts maximum, the transistors or other switching elements in the active matrix must be able to handle 12 V or more without damage. If the active matrix has

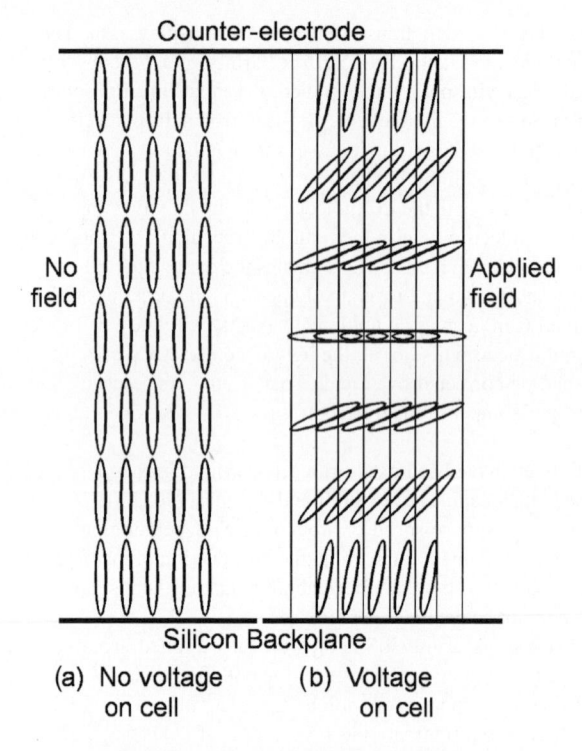

Figure 4.23 VAN Mode Liquid Crystal Effect.

an internal frame buffer, the counter electrode voltage can be switched synchronously with the inversion scheme, reducing the drive voltage required to only the 6 volts again. This type of design in normally only feasible in c-Si backplanes. This will be discussed in more detail in Section 10.4.2.

The pixel inversions can be arranged in various systems. If alternate pixels are inverted out of phase with each other so that one pixel sees a positive voltage and the adjacent pixel sees a negative voltage, it is called pixel inversion. If alternate rows or columns have opposite voltages, it is called row or column inversion, respectively. The most common case, however, is for all pixels in a video field to be addressed with the same polarity, say a positive voltage. The exact voltage applied to each pixel would be dependent on the incoming video signal, but all the voltages would be positive. In the next video field, a negative voltage would be applied to all pixels. This system is called field inversion.

FLC is the main exception to the rule that positive and negative voltages produce the same electro-optic effect. When the negative voltage is applied to a FLC cell, the cell displays the negative image. Therefore, the maximum duty cycle for a FLC cell is typically 50 %, since no light can be applied during the negative portion of the inversion cycle. This 50 % duty cycle compared to the 100 % duty cycle usable in TN and VAN cells is a serious disadvantage for FLC, preventing it from being used in applications where high throughput is required.

In a transmissive LC cell, both electrodes are the same material, typically ITO. This is not true in a LCoS cell where the active matrix side normally has an aluminum mirror/electrode and the counter-electrode is normally ITO. Two different metals with a fluid between them is almost the definition of a battery. Therefore a LCoS cell has a small but noticeable DC bias voltage across it (Melnik *et al.* 2000). This upsets the inversion system and can cause flicker in the system. Adjusting the counter-electrode voltage can cancel this effect, except the DC bias voltage drifts with time, light exposure and illumination. In some cases, this requires active feedback to prevent the flicker from becoming visible.

References

Akimoto, O., Hashimoto, S., Yumoto, A., Mano, M., Yamagishi, M. et al. (1999) UXGA/HDTV Microdisplay using FLC on Si backplane. *The 19th International Display Research Conference (IDRC), (EuroDisplay '99) Proceedings*, 45–8.

Alt, P.M. (1997) Single crystal silicon for high resolution displays. *Intl. Display Res. Conf. Record*, M19–M22.

Armitage, D. (1992) Liquid crystal display device fundamentals. In M. Karim (ed.) *Electrooptic Displays*. New York: Marcel Dekker, 19–67.

Armitage, D., Underwood, I. and Wu, S.-T. (2006) *Introduction to Microdisplays*, Wiley-SID series in display technology. New York: John Wiley & Sons, Inc.

Bahadur, B. (1984) Liquid crystal displays. *Molecular Crystals and Liquid Crystals* **109**: 1–98.

Bahadur, B. (ed.) (1990) *Liquid Crystals: Applications and Uses, Vols 1–3*. New York: World Scientific.

Beresnev, L.A and Blinov, L.M. (1989) Electro-optical effects in ferroelectric liquid crystals. *Ferroelectrics* **92**: 335–43,

Blinov, L.M. (1983) *Electro-Optical and Magneto-Optical Properties of Liquid Crystals*. New York: John Wiley & Sons, Inc.

Brody, T.P., Asars, J.A. and Dixon G.D. (1973) A 6 × 6 inch 20 lines-per-inch liquid crystal display panel. *IEEE Trans. Electron Devices* **ED-20**: 95–101.

Castleberry, D.E. and Possin, G.E. (1988) A 1 mega-pixel color a-Si TFT liquid crystal display. *SID Intl. Symp. Digest of Technical Papers*, 232–4.

Clark, N.A. and Lagerwall, S.T. (1980) Submicrosecond bistable electro-optic switching in liquid crystals. *Appl. Phys. Lett.* **36**: 899–901.

Drzaic, P.S., Gonzales, A.M., Jones, P. and Montoya, W. (1992) Dichroic-based displays from nematic dispersions. *SID Intl. Symp. Digest of Technical Papers*, 571–4.

Fergason, J.L. (1985) Polymer encapsulated nematic liquid crystals for display and light control applications. *SID Intl. Symp. Digest of Technical Papers*, 68–70.

Fünfschilling, J. and Schadt, M. (1989) Fast responding and highly multiplexible distorted helix ferroelectric liquid-crystal displays. *J. Appl. Phys.* **66**: 3877–82.

Glueck, J., Lueder, E., Kallfass, T., and Lauer, H.-U. (1992) Color-TV projection with fast-switching reflective HAN-mode light valves. *SID Intl. Symp. Digest of Technical Papers*, 277–80.

Gooch, C.H. and Tarry, H.A. (1975) The optical properties of twisted nematic liquid crystal structures with twist angles $\leq 90°$. *J. Phys D* **8**: 1575–84.

Grinberg, J. and Jacobson, A.D. (1976) Transmission characteristics of a twisted nematic liquid-crystal layer. *J. Opt. Soc. Am.* **66**: 1003.

Guttag, K.M. (2004) Digital microdisplay backplane with bit serial SIMD processing. *ITE/SID IDW '04*, Niigata, Japan.

Hartman, R.A. (1995) Two-terminal devices technologies for AMLCDs. *SID Intl. Symp. Digest of Technical Papers*, 7–10.

Hartmann, W.J.A.M. (1989) Charge-controlled phenomena in the surface-stabilized ferroelectric liquid-crystal structure. *J. Appl. Phys.* **66**: 1132–6.

Hilsum, C. and Raynes, E.P. (ed.) (1983) Liquid crystals: their physics, chemistry, and applications. *Proceedings of a Royal Society Discussion Meeting*, 27 and 28 October, 1982. The Royal Society.

Im, J.S., Limanov, A.B., van der Wilt, P.C., Chung, U.J. and Chitu (2007) Evolution in LTPS AMLCD manufacturing via advances in laser crystallization techniques and systems. *Information Display* **23** (September): 14–18.

Itoh, M., Yamamoto, Y., Yoneda, H., Yamane, Y., Tsuchimoto, S. et al. (1996) High-resolution low-temperature poly-Si TFT-LCDs using a novel structure with TFT capacitors. *SID Intl. Symp. Digest of Technical Papers*, 17–20.

Janning, J.L. (1972) Thin film surface orientation for liquid crystals. *Appl. Phys. Lett.* **21**: 173.

Kahn, F.J. (1972) Electric-field-induced orientational deformation of nematic liquid crystals: tunable bifringence. *Appl. Phys. Lett.* **20**: 199–201.

Kahn, F.J., Taylor, G.N. and Schonhorn, H. (1973) Surface-produced alignment of liquid crystals. *Proc IEEE* **61**: 823.

Karim, M.A. (ed.) (1992) *Electro-Optical Displays*. New York: Marcel Dekker, 36–9.

Komanduri, R.K., Jones, W.M., Oh, C. and Escuti, M.J. (2007) Polarization-independent modulation for projection displays using small-period LC polarization gratings. *Journal of the SID* **15**(8): 589–94.

Kuijk, K.E. (1990) D^2R, a versatile diode matrix liquid crystal approach. *1990 Intl. Display Res. Conf.* (Euro display '90), 174–9.

Kurogane, H., Doi, K., Nishihata, T., Honma, A. Furuya, M. et al. (1998) Reflective AMLCD for projection displays: D-ILA™. *SID Intl. Symp. Digest of Technical Papers*, paper 5.3, 33–6.

Lebrun, H., Szydlo, N., Maurice, F., Borel, T., Stewart, R.G. et al. (1996) A-Si self-scanned 2.2-in EDTV projection LC light valve. *Society for Information Display International Symposium Digest*, 677–80.

Lechner, B.J., Marlowe, F.J., Nester, E.O. and Tults, J. (1971) Liquid crystal matrix displays. *Proc. IEEE* **59**: 1566–79.

Matsumoto, S., Kawamoto, M. and Mizunoya, K. (1976) Field-induced deformation of hybrid-aligned nematic liquid crystals: New multicolor liquid crystal display. *J. Appl. Phys.* **47**: 3842–5.

Melnik, G., Janssen, P. and Cnossen, G. (2000) Galvanic activity in reflective LCDs. *SID Intl. Symp. Digest of Technical Papers*, 252–5.

Meyer, R.B., Liebert, L., Strzelecki, L. and Keller, P. (1975) Ferroelectric liquid crystals. *J. Phys. Lett.* **36**: 669–71.

Mitra, U., Khan, B.A., Venkatesan, M., Carlson, A., Vaez-Iravani, M. et al. (1990) High performance polysilicon thin-film transistors. *Mat. Res. Soc. Proc.* **182**: 381–6.

Mitra, U., Chen, J., Khan, B. and Stupp, E. (1991a) Low-temperature polysilicon TFT with gate oxide grown by high-pressure oxidation. *IEEE Electron Device Letters* **12**: 390–2.

Mitra, U., Khan, B., Venkatesan, M., and Stupp, E. (1991b) Effect of processing temperature on polysilicon thin film transistors for active matrix LCDs. *Intl. Display Res. Conf Record*, 207–10.

Morozumi, S. (1984) 4.25-in and 1.51 in b/w and full-color LC video displays addressed by poly-Si TFTs. *SID Intl. Symp. Digest of Technical Papers*, 316–17.

Morozumi, S., Ohta, T., Araki, R., Sonehara, T., Kubota, K. et al. (1983) 250 × 240 element LCD addressed by lateral MIM. *Proc. Japan Display*, 404–7.

Morozumi, S., Araki, R., Ohshima, H., Matsuo, M., Nakazawa, T. et al. (1986) Low-temperature processed poly-Si TFT and its application to large area LCD. *Proc. Japan Display*, 196–9.

Niiyama, S., Hirai, Y., Ooi, Y., Kunigita, M., Kumai, H. et al. (1993) Hysteresis and dynamic response effects on the image quality in a LCPC projection display. *SID Intl. Symp. Digest of Technical Papers*, 869–72.

Oana, Y. (1989) Technical developments and trends in α-Si TFT-LCDs. *J. Non-Cryst. Solids*, **115**: 27–32.

Okamoto, N. (2001) Developments in p-Si TFT LCD projectors. *SID Intl. Symp. Digest of Technical Papers*, Paper 46.1, 1176–79.

Pfeiffer, M. (2000) Liquid crystals for LCoS. *SID Intl. Symp. Seminar Lecture Notes*, Seminar M-13.

Sato, F., Yagi, Y. and Hanihara, K. (1997) High resolution and bright LCD projector with reflective LCD panels. *SID Intl. Symp. Digest of Technical Papers*, 997–1000.

Schadt, M. and Helfrich, W. (1971) Voltage-dependent optical activity of a twisted nematic liquid crystal. *Appl. Phys. Lett.* **18**, 127.

Shimizu, S, Ochi, Y., Nakano, A. and Bone, M. (2004) Fully digital D-ILA™ device for consumer applications. *SID Intl. Symp. Digest of Technical Papers*, 72–5.

Shimizu, S., Nakagaki, S., Doi, K. and Bleha, W.P. (2005) Full HD 1920×1080 pixel digital D-ILA™ microdisplay projector technology. *147th SMPTE Technical Conference & Exhibition*, 9–12 Nov.

Shirochi, Y., Murakami, K., Endo, H., Arakawa, S. Kitagawa, H. et al. (2003) 50/60 V Hivision LCD rear-projection TV (GRAND WEGA) with excellent picture quality. *SID Intl. Symp. Digest of Technical Papers*, 114–17.

Sloof, W.A. and Brennesholtz, M.S. (2003) A high resolution LCOS imager for single panel HDTV systems, *Proceedings of the 10th International Display Workshops*, Fukuoka, Japan: 1577–80.

Sloof, W.A. and Brennesholtz, M.S. (2004) An improved WXGA LCOS imager for single panel systems. *Proceedings of ASID '04*, Southeast University Press, 150–3.

Stewart, R.G, Dresner, J., Weisbrod, S., Huq, R.I., Plus, D. et al. (1995) Circuit design for a-Si AMLCDs with integrated drivers. *SID Intl. Symp. Digest of Technical Papers*, 89–92.

Stupp, E.H. and Khan, B. (1994) Active-matrix electro-optic display device with storage capacitors and projection color apparatus employing same. US Patent 5,305,128.

Sze, S.M. (1981) *Physics of Semiconductor Devices*, 2nd edn. New York: John Wiley & Sons, Inc., Chap. 8.

Wu, S.-T. and Yang, D.-K. (2006) *Fundamentals of Liquid Crystal Devices*. New York: John Wiley & Sons, Inc.

Yamamoto, Y., Morita, T., Yamana, Y., Funada, F. and Awane, K. (1995) High-performance low temperature poly-Si TFT with self-aligned offset gate structure by anodic oxidation of Al for a driver monolithic LCDs. *Proc. 15th Intl. Research Conf.* (Asia Display 95), 941–2.

5

Micro-electromechanical Devices

Projectors using microdisplays based on liquid crystal effects led the attack on oil film light-valve projectors and came to dominate the market. The importance of projectors based on micro-electromechanical (MEM) constructions has increased sharply since the first edition of this book. In particular, one of these, the Digital Micromirror Device, or DMD*™ (Hornbeck 1983; Sampsell 1994), has come to dominate certain projector markets including business front projection and electronic cinema. Systems based on this device have also captured a significant portion of the consumer rear projection TV and home theater market. The characteristics of the DMD and other micro-electromechanical systems used in projection displays will be discussed in this chapter.

5.1 DMD

The DMD is part of a complete system that Texas Instruments (TI) calls 'Digital Light Processing' or DLP™. This system included not only the DMD microdisplay, but the drive ASIC, support hardware and software and reference optical designs.

The DMD is fabricated in c-Si using similar processing technologies as are applied to integrated circuit fabrication. It consists of an array of mirrors, one per pixel, suspended over CMOS circuitry. Figure 5.1 shows photographs of the device. In Figure 5.1(a), the micromirrors are shown. In Figure 5.1(b), one of the micromirrors has been removed to show the underlying structure. Figure 5.1(c) shows a magnified view of this underlying structure. When introduced to production, the mirrors were fabricated with a pitch

---

*™ Texas Instruments Corp.

Projection Displays, Second Edition   M. Brennesholtz, E. Stupp
© 2008 John Wiley & Sons, Ltd

58 MICRO-ELECTROMECHANICAL DEVICES

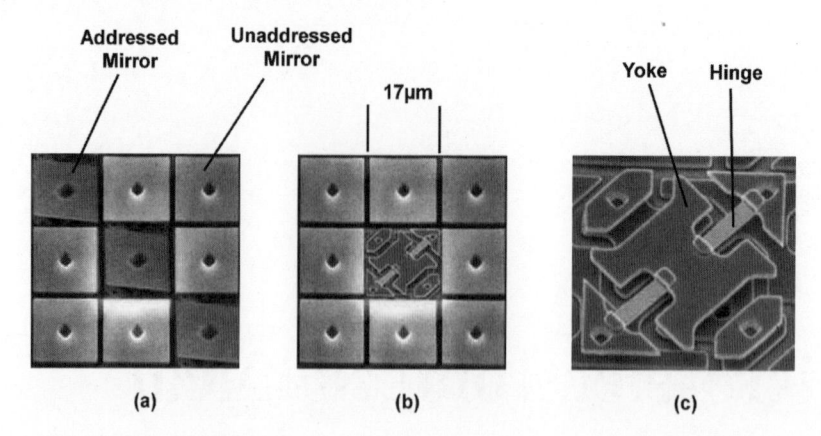

Figure 5.1 Photomicrographs of the DMD: (a) 16 μm mirrors on 17 μm centers, with mirrors shown in both the on and off positions; (b) one mirror removed to show underlying structure; and (c) magnified view of this structure. (Reprinted with permission of Texas Instruments, Inc.).

of 17 μm with 1 μm between mirrors and a maximum mirror tilt of ±10°. The gap between adjacent pixels has since been reduced to 0.8 μm (TI 2005). More recent DMD designs use pixel pitches of 13.8 μm or 10.3 μm and have a tilt angle of ±12°.

The design overhead to change the pixel size in a LCD or LCoS panel is relatively modest and represents straightforward semiconductor design. This is not true for the DLP. While the design of the CMOS circuitry is relatively easy, designing the yoke and mirror assembly, proving its reliability and optimizing the operating parameters is difficult and expensive. Therefore, TI stays with a few proven pixel designs. These pixels can then be grouped in various combinations to get the correct pixel count and aspect ratio.

The idea of a metal mirror suspended by deformable metal hinges was first described in 1970 (Van Raalte 1970). In its original configuration, this reflective device operated in an analog mode. The amount of deflection was controlled by the voltage levels on the address electrodes, and as a result, modulated the amount of light entering the projection lens. It is very difficult to use analog control of the light from all the pixels with the accuracy required for a projector. The present bistable device and its associated circuitry are completely digital (Hornbeck 1989; Gove 1994). Tests indicate that the reliabilities of current DMD devices are satisfactory through thousands of hours of life and environmental tests (Douglass 1996).

5.1.1 Device Operation

Figure 5.2 illustrates a simplified DMD pixel in cross-section. The mirror assembly consists of the square mirror on top of a post attached to a torsion-hinge-suspended yoke. The yoke and hinges are visible in Figure 5.1(b,c). The hinges are not shown in Figure 5.2. The yoke can rotate about ±12° (±10° in older DMDs) before it contacts the mechanical stop. For this deflection, the torsional effects in the hinges are fully elastic.

On the silicon substrate are push-pull address electrodes connected to the CMOS circuitry. The CMOS electronics in the underlying silicon substrate consists of a six transistor SRAM cell per pixel and plus other circuits. With the appropriate bias, the mirrors move in response to the voltages on the address electrodes. In the +12° position, light incident onto the mirror will be reflected into the projection lens, as shown in Figure 5.3. In the −12° position, the light is directed away from the lens input and is internally absorbed. Thus, the light from any position on the mirror array is either present or not present, i.e. the projected intensity is binary. The complete cycle for each mirror of the DLP is complex and tuned by TI

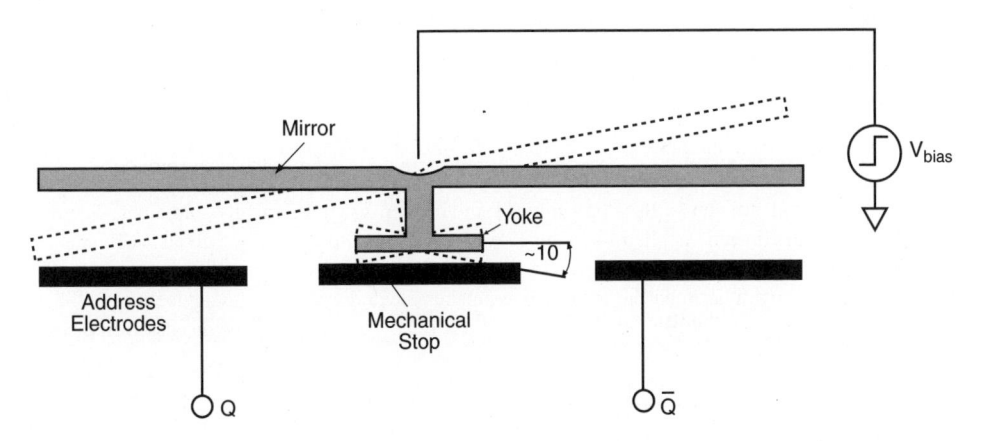

Figure 5.2 Illustration of the cross-section of a DMD pixel.

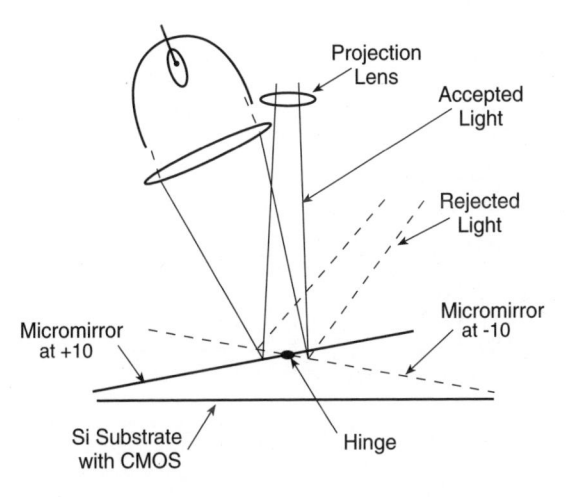

Figure 5.3 Light throughput modulation by a DMD pixel.

to optimize performance. The following description of the cycle is reproduced from Hornbeck (1997) with the permission of the SPIE:

1. *Memory Ready* – All memory cells under the DMD have been loaded with the new address states for the mirrors.
2. *Reset* – All mirrors are reset in parallel (voltage pulse applied to the bias bus).
3. *Unlatch* – The bias is turned off to unlatch the mirrors and allow them to release and begin to rotate to flat state.
4. *Differentiate* – Retarding fields are applied to the yoke and mirrors in order to rotationally separate the mirrors that remain in the same state from those that are to cross over to a new state.
5. *Land and Latch* – The bias is turned on to capture the rotationally separated mirrors and enable them to rotate to the addressed states, then settle and latch.

6. *Update memory array* (one line at a time) – The bias remains turned on to keep the mirrors latched so as to prevent them from responding to changes in the memory, while the memory is written with new video data.

7. *Repeat sequence* beginning at step 1.

This cycle is controlled by an ASIC supplied by TI: the imagers are not normally sold alone. The reset pulse in step 2 can either be applied to all pixels in the DMD or to a group of pixels. (Urbanus and Sampsell 1994, 1998) The group is a collection of rows and can be (in theory) as little as one row. This ability to reset groups rather than the entire device makes the addressing of the device more efficient and prevents dead time, especially when low-order bits are addressed, as will be discussed in the next section.

5.1.2 Gray Scale

The mirrors can switch positions in about 15 µs, which allows achieving gray scales by time-multiplexing the mirror position, as illustrated in simplified form in Figure 5.4. Luminance between the full-on and full-off values is obtained by selecting the appropriate on- and off-times. Thus, to achieve a luminance of 70 % of the peak white, the mirrors would be in the $+12°$ during the intervals 1/2, 1/8, 1/16, 1/128, and 1/256 of the field time and they would be in the $-12°$ positions otherwise. This sequence actually produces a 69.9 % gray, which is the closest a DLP with 8 bit drive can come to 70 %. The eye-brain system cannot temporally resolve the individual light impulses in these five periods and integrates and averages the total output to give the desired relative luminance.

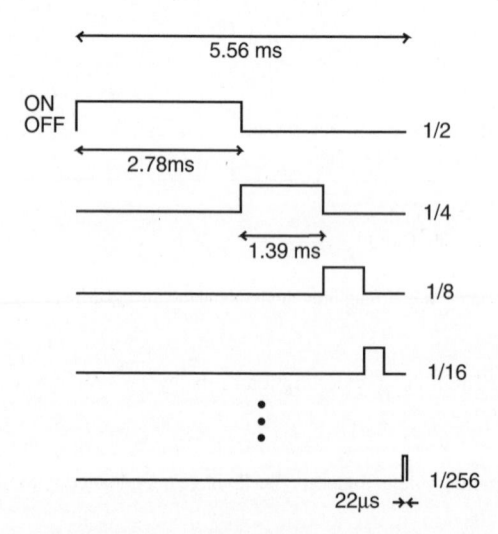

Figure 5.4 Time-multiplexing intervals for achieving 256 gray scales with a DMD for color-field sequential operation. For three-panel operation, the time intervals are three times longer.

This scheme illustrated in Figure 5.4 provides 256 gray levels per color (8 bit). For a single-panel architecture operating at a 60 Hz field rate, i.e. one presenting the R, G and B color fields in succession in a total period of 16.7 ms, the shortest time address interval required is about 22 µs. This is close to the switching speed of the mirrors and may require modified signals to correct errors introduced by this limitation. Eight bits of gray scale (256 levels including black) on a linear device such as the DMD are not sufficient to provide image quality adequate for video applications (Brennesholtz 1998). In single-panel applications, it is necessary for the DLP image processing to dither the signal both spatially and temporally in order to get satisfactory gray scale response. In three panel applications such as electronic cinema where 10 or more bits of gray scale can be used, this is not an issue.

The peripheral electronics accepts serial data on each of the parallel R, G and B inputs. The electronics must reformat these to provide the time-multiplexed outputs. Bit 7 is the MSB and provides output for half of the frame time; bit 0 is the LSB and does so for 1/256 of this time. For three-panel operation, i.e. one light valve per primary color, the panels are written in parallel and the full 16.7 ms is available for each. The low order or LSB bit, bit 0, is on for about 65 μs. For single-panel applications, the R, G and B time-multiplexed outputs must be provided sequentially. Thus, the time for displaying bit 0 in these systems is about 22 μs.

The information for the next bit is written to the panel while the current bit is being displayed. In early DLP systems, it required about 120 μs to load each of the bits of the data onto the address electrodes for all the pixels. For a one-panel system displaying one of the three least significant bits (LSBs), bits 2 through 0, it takes longer to load the information into the CMOS SRAM than the on-times for the bits being displayed. To correct for this, it is necessary to get a minimum time of 120 μs for each bit. For the three LSBs, the mirrors for all pixels were switched to the on-state for the requisite times but they are switched to the off-state so the off time plus the on time totaled 120 μs. The light throughput is zero during these off-periods, resulting in dead time and lost efficiency. Exclusive of system throughput issues to be discussed in Chapters 11 and 12, the efficiency of the DMD was about 65 %

Newer DLP systems can use the group reset to avoid this efficiency loss. In these systems, the low-order bits can be loaded to only a section of the DLP imager and then that section reset. The result is no dead time and increased efficiency. Obviously during some portion of the video field, a complete set of low order bits must be loaded into every pixel in the DLP.

The high order bits, the ones that take more than 2× the array addressing time, more than 240 μS in the early DLP imagers, can be split for display. That is, a bit that needs to be displayed for 240 μS can be split into two sub-bits and each sub-bit displayed for 120 μs with no dead time. These sub-bits can be displayed at different portions of the sequence in order to improve image quality. This is especially important in preventing color breakup, as will be described in Section 14.2.3.

5.1.3 Contrast and DLP Pixel Design

The DLP pixel design has evolved since it was first introduced. In Figure 5.5 (Dewald *et al.* 2002) image B2 represents the original DLP design with a contrast of 49:1. This device was experimental and not introduced into production. The second device, labeled HH for hidden hinge, had a contrast of 220:1. Hidden hinge refers to the fact that the hinge is entirely behind the mirror and does not scatter light into the projection lens pupil. This device was the first DLP introduced into production in 1996. The third device, labeled SRV had a contrast of 310:1. SRV stands for small, rotated via. The contrast-improvement value of making a small via is obvious: there is less area to scatter or diffract light.

The value of rotating the via is less obvious. It relates to the diffraction pattern produced by the array of vias. With the original orientation, this diffraction pattern was oriented so a significant fraction of the diffracted light entered the pupil of the projection lens. By rotating the via, less of the light entered the projection lens, increasing contrast. The final pixel design, designated SRV + SMG had 410:1 contrast. SMG stands for small mirror gap. This reduced inter-pixel spacing not only provides higher contrast but increases the efficiency slightly by increasing the aperture ratio. A final pixel design, not shown because the image would be identical to the SRV + SMG, was designated DM3 for 'Dark Metal 3'. This involves adding a dark organic layer to metal layer three in the structure underneath the mirror. Ray tracing at TI had identified this metal layer as a significant source of scattering. Light would enter this under-mirror region in the inter-pixel gaps and get scattered in such a way it could re-exit through the gaps and enter the projection lens. The DM3 devices are capable of 900:1 to 1000:1 contrast.

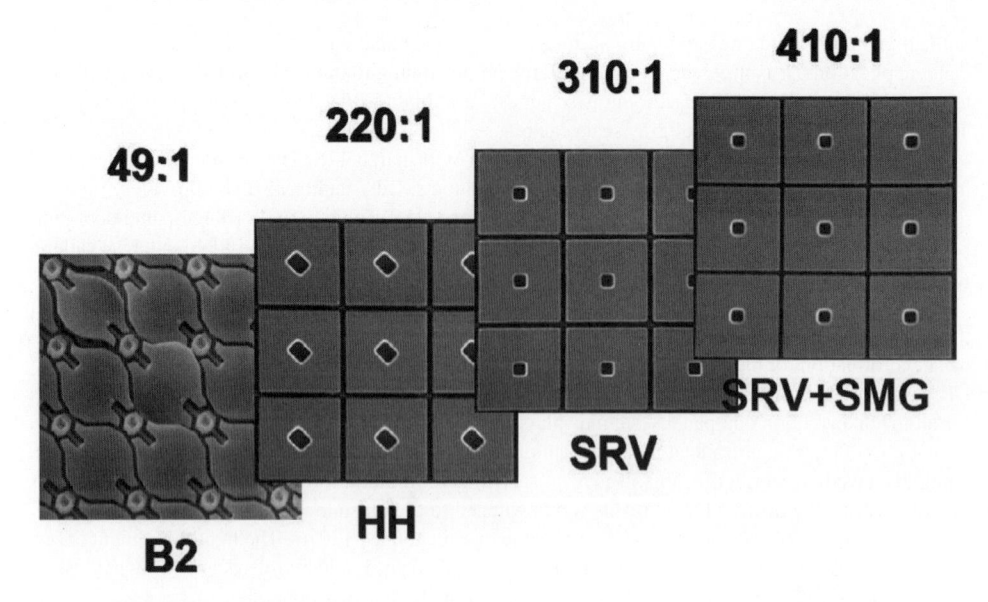

Figure 5.5 Evolution of the DLP Pixel Design (reproduced by permission of © 2002 the SID).

Contrast in a DLP imaging system is very dependent on the design of the optical system. (Dewald *et al.* 2002, 2003; Texas Instruments 2005) The contrast values for the HH, SRV, SRV + SMG and DM3 versions of the DLP cited above were all measured in the same reference optical system. Contrast improvement efforts at TI have focused on the optical system since the introduction of DM3.

The SMG also reduces an image artifact known as the screen door effect, as will be discussed in Section 14.1.2. All DLP imagers since the HH version have a relatively small interpixel region compared to the comparable HTPS active matrices for transmissive LC microdisplays. In these LCD microdisplays the interpixel region can occupy about 35 % or more of the active area, resulting in a much higher visibility of the screen door effect in the image.

5.2 Linear MEMS Arrays

Linear arrays of pixels based on MEMS devices have been developed for display applications. These include the GLV from Silicon Light Machines and the GEMS device from Kodak. These linear arrays are based on diffraction and a dark-field Schlieren optical path, as will be described in Section 10.6. Due to the very low étendue of the linear arrays, they must be illuminated with laser light to get enough throughput to be of interest in projection systems.

These linear arrays produce, typically, a single column of pixels. This column image is then scanned by a moving mirror along the other axis to produce the complete two-dimensional image. This optical system will be described in Section 9.2.3.

Early in the DMD development cycle, TI also developed a linear array of tilting mirrors. Since this device was not used for projection applications, it will not be discussed here.

5.2.1 Grating Light Valve

The deformable grating light valve (Apte *et al.* 1993), or DGLV, was a linear array of pixels fabricated on a silicon substrate. It was based on diffraction of the light by a grating formed on the surface of the device from a series of parallel ribbons that could be moved by application of voltages, as shown in Figure 5.6. Each pixel of this reflective device consists of 2 parallel reflective ribbons, spaced apart from each other. Between the ribbons were reflective regions on the substrate. In the up position, the ribbons are $1/2$ wavelength above the substrate. Reflection from the ribbons and the substrate are in phase and all light is reflected in the 0th diffraction order. Each ribbon can be pulled down approximately one-quarter wavelength to create a diffraction grating. The height difference between the ribbons and the substrate is now $1/4$ wave and 84 % of the light is diffracted into the ± 1st diffraction order.

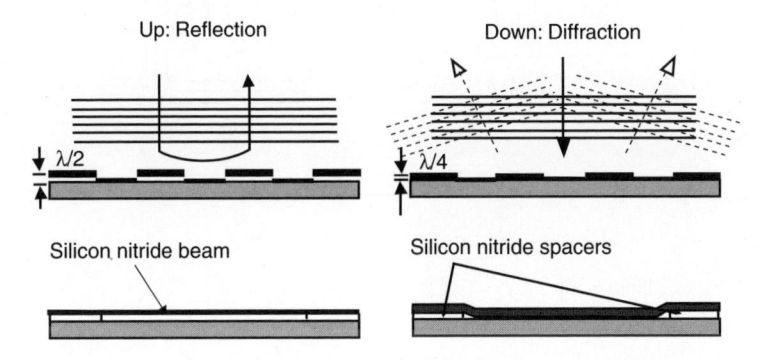

Figure 5.6 Schematic of the GLV operation (from Apte 1993; with permission of SID).

The initial DGLV devices operated as digital systems: the movable ribbons were either up diffracting into the 0th order or down diffracting into the ± 1st order. In these systems, the natural state of the movable ribbons is up. Application of a voltage V_2 between the ribbon and the underlying electrode caused the mirror ribbon to switch to its down position forming the grating structure. The mirror was held down with zero power by a bias voltage V_B, which is less than V_2. Application of a switch-up voltage V_1, which is less than V_B, will cause the mirror to go to its relaxed state. The switching action occurred in 20 ns, allowing this bistable device to be used in a time-multiplexing mode, much like the DMD.

The very poor contrast of this system (20:1) puzzled the authors of (Apte *et al.* 1993). In retrospect we know the condition to get all the light in the 0th order (dark state) is too critical for a structure with two different levels in the dark state to provide satisfactory results. Later versions (Bloom 1997) inserted fixed ribbons between the moving ribbons so when the moving ribbons were in the up state, the system presented a flat surface to the incident light broken only by the gaps between the fixed and moving ribbons. With this arrangement, a contrast of 200:1 was achieved. This newer system was designated the grating light valve or GLV[1]™. A set of 4 ribbons, 2 fixed and 2 moving, constituted a pixel 20 μm square. This system, like the earlier version, was driven digitally and used temporal modulation to achieve a 1000:1 dynamic range.

With newer versions of this device it is more common to drive each pixel with an analog drive. With no voltage applied, the moving ribbons are in the up position and in the same plane as the fixed ribbons and all light goes into the 0th order. As a voltage is applied, the ribbons are drawn down proportional to the applied voltage as long as the applied voltage is less than the V_2 voltage that causes the ribbons to go suddenly to the full down position. Note that as the ribbons are drawn down, an increasing proportion of the light goes into the ± 1st order and a decreasing portion of the light goes into the 0th order. No light is ever diffracted into angles between the 0th and ± 1st orders.

The GLV device is used in a Schlieren optical system, as shown in Figure 5.7. When the mirrors are in the relaxed, or dark, state incident light is specularly reflected from the surface. This light is intercepted by the Schlieren stop and does not reach the screen. When the mirrors form a grating (bright state) incident light is diffracted at angles sufficiently large to avoid interception by the Schlieren stop. This diffracted light enters the projection lens for imaging on the screen.

---

[1]™ Silicon Light Machines Corp.

64 *MICRO-ELECTROMECHANICAL DEVICES*

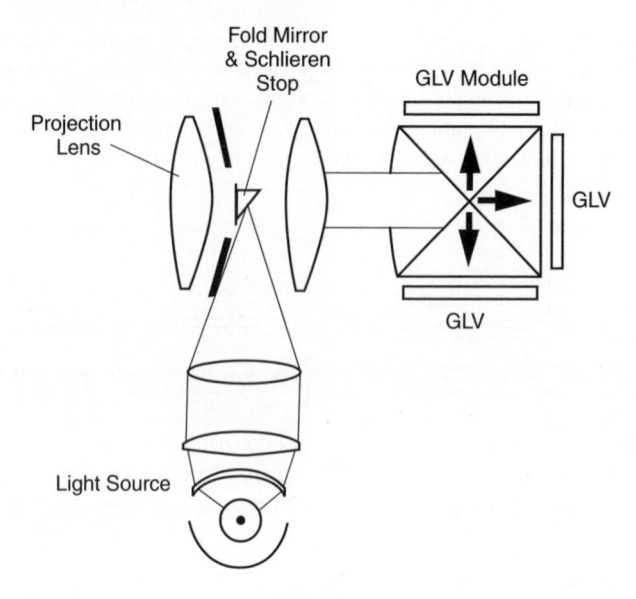

Figure 5.7 Schematic of schlieren optical system for GLV projector (reproduced from Apte 1993; with permission of SID).

Currently there are two versions of this device. The first, now called the grating light modulator (GLM), is made in a linear array of 4096 pixels, with each pixel consisting of a fixed/moving ribbon pair. The ribbons in the pixel array are each 3.7 μm wide, with a pixel size of 7.4 μm for one pixel. The ribbons are 200 μm long and the vertical motion of the aluminum-coated silicon carbide moving ribbon is 150 nm, about a quarter-wave of green light. A SEM photograph of a GLM is shown in Figure 5.8.

Figure 5.8 SEM photograph of a GLM from a Sony laser projector (reproduced by permission of © 2007 the SID).

LINEAR MEMS ARRAYS   65

Unlike the early versions of the GLV, the GLM is driven with analog voltages. The 16 bit DAC drive chips and the GLM are mounted together in a multi-chip module (MCM). This arrangement is needed to support the very high data rates demanded by the Evans & Sutherland 4 K × 5 K resolution projector.

The other development of the original GLV is from Sony, which calls the device the GxL (Tamada 2007). One of the key differences between the GLM and the GxL is the GxL has blazed ribbons, as shown in Figure 5.9. In the off state, shown in Figure 5.9(a) with the movable ribbons in the up position, all the light is diffracted into either the 0th or the −2nd diffraction order. The Schlieren optical system is designed so these two orders are blocked and do not reach the projection screen. With the ribbons in the on-position (down), shown in Figure 5.9(b), all the light is diffracted into the −1st diffraction order. The light in this diffraction order is passed through the Schlieren stop and eventually to the screen. With the movable ribbons in an intermediate state, as shown in Figure 5.9(c), some of the light is diffracted into the −1st order and the rest of the light is diffracted into either the 0th order or the −2nd order. This produces an intermediate gray level since only the light in the −1st order reaches the projection screen.

For this system to get good contrast, the height of the fixed and moving ribbons in the dark state must be exactly the same. According to Tamada, if there is a 10 nm difference between adjacent ribbons, the

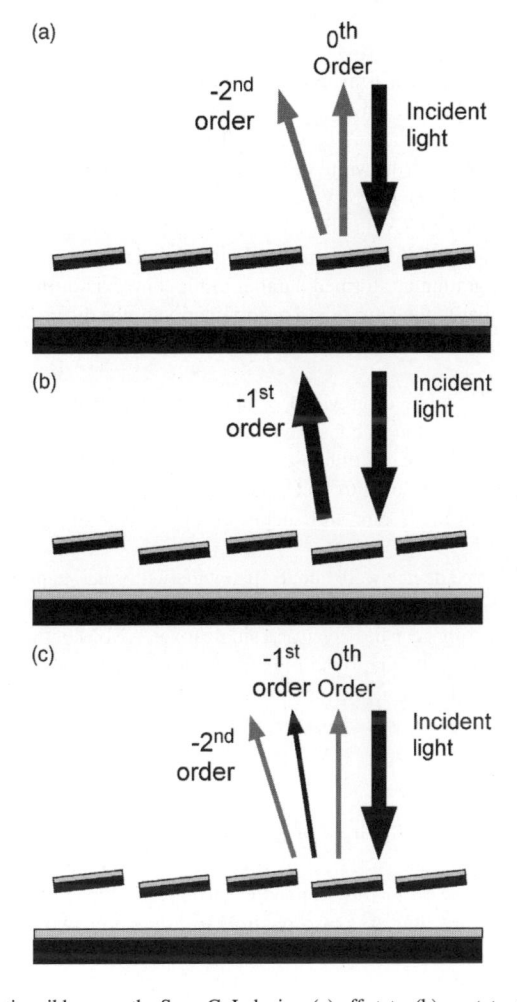

Figure 5.9 Blazed diffractive ribbons on the Sony GxL device: (a) off state; (b) on state; (c) intermediate gray level.

contrast is less than 100:1 and if there is a 1 nm difference, the contrast is less than 10 000:1. Tamada reports Sony has achieved 34 700:1 contrast in the system. An earlier version of the system was publicly shown at the 2005 Aichi World Expo using a 2005" diagonal screen with 10 000:1 contrast (Eguchi 2006). Contrast is also affected by other design details of the device, including bow of the ribbons (which would show up as a curve of the ribbon in Figure 5.9) and the exact tilt angle. The 1080 line device used at the Aichi World Expo had 6 ribbons/pixel for a total length of 1.09" (27.7 mm). The current generation of the GxL device uses 4 ribbons per pixel and is 0.72" long (18.3 mm). Due to the very high power densities (10s of watts of laser light on the 1.09" long device) Sony uses a Al−Cu film as a mirror surface rather than a pure aluminum film.

The GxL is driven by an analog system instead of the time division multiplexing used in the early GLV systems. The 1.09" device at the Aichi World Expo used 10 bit drivers and the current device uses 12 bit drivers. The resonant frequency of the ribbons is about 1 MHz. While the 15 GxL projectors used at Aichi operated at 60 Hz with 1920 pixels per line, this resonant frequency can support 90 Hz operation with 4000 pixels per line.

5.2.2 GEMS System

Eastman Kodak has developed a linear array independent of the GLV called the grating electromechanical system (GEMS) device (Brazas and Kowarz 2004; Kowarz *et al.* 2006). The GEMS device, like the original GLV, is a digital device switching between a flat mirror corresponding to the dark state and a diffraction grating, corresponding to the white state. Gray scale is achieved using pulse width modulation.

One important difference between the GLV and its derivatives and the GEMS system is the orientation of the diffraction gratings. In the GLV, the lines in the diffraction grating are perpendicular to the long axis of the device. The leads to the diffracted orders tilting parallel to the long axis. In the GEMS system, the lines of the diffraction grating are formed parallel to the long axis of the device and the diffracted orders are tilted perpendicular to the long axis. This orientation of the diffracted orders allows a simpler optical design. In fact, the original GEMS demonstrations were given using standard, off-the-shelf cinema projection lenses.

The device, like the GLV, is based on MEMS ribbons stretched perpendicular to the long axis of the device. In the GEMS device, the ribbons have intermediate supports, typically at 25 μm to 50 μm intervals, and all ribbons are movable ribbons. When a voltage is applied between the ribbon and the substrate, the ribbon is drawn down to touch the substrate. The intermediate supports prevent the ribbon from moving down in that region, producing a ridge oriented parallel to the long axis of the device.

The ribbons in the prototype device were about 5 μm wide with small gaps between adjacent ribbons to relieve stress in the ribbons and allow removal of sacrificial material during fabrication. There is no limit to the number of ribbons that can be driven as a single pixel. A SEM of the ribbon structure is shown in Figure 5.10. For scale, the supporting ribs in Figure 5.10(b) are 10.6 μm apart. One thing to note in Figure 5.10(a) is the very small distance the moving ribbon actually travels, about 170 nm in this device, compared to the 1 μm thick ribbon.

5.3 MEMS Scanning Mirrors

The linear arrays discussed in Section 5.2 scanned electronically in one dimension and use a moving mirror to scan in the other direction. It is also possible to use one or two moving mirrors to scan an optical beam to create a picture. For example, a rotating mirror array and a galvanometer mirror can work together to generate a full raster scanned image. It is also possible to use MEMS scanning mirrors to generate the raster-scanned light beam. This section will discuss these MEMS mirrors. All of these

MEMS SCANNING MIRRORS 67

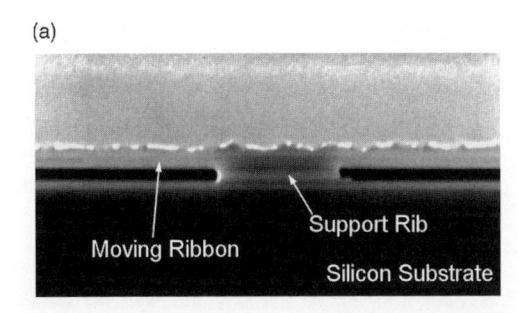

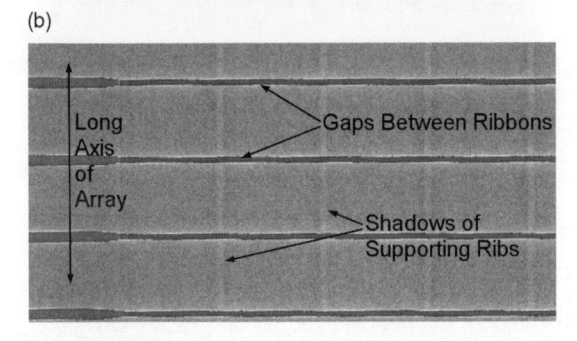

Figure 5.10 Ribbon structure of the Kodak GEMS linear array: (a) cross-section of a ribbon; (b) several ribbons (reproduced by permission of © 2004 SPIE).

single axis and two axis optical scanning technologies will be discussed at a system level in more detail in Section 9.2

Figure 5.11(a) shows the basic layout of a biaxial MEMS scanning mirror. In this system, a single mirror, typically about 1 mm to 1.5 mm in diameter, has a double-yoke arrangement to allow it to tilt in both directions. The low speed flexures typically allow vertical scanning at 60 Hz or above while the high-speed flexures typically allow horizontal scanning at 20 kHz or above. Figure 5.11(b) shows a photo of a typical bi-axial MEMS mirror from Microvision.

The mirror drive can be either electrostatic (James *et al.* 2007) or electromagnetic (Sprague *et al.* 2005). The electrostatic drive described by James uses interleaved combs attached to both the moving mirror and the fixed frame. Multiple fingers are used on the combs to amplify the electrostatic interaction and minimize the voltage required. The electromagnetic drive involves fixed magnets near the moving mirror and an electromagnetic coil on the moving frame, as shown in Figure 5.11(b). Applying a force to the moving frame can obviously tilt the mirror vertically by tilting both the frame and the attached mirror. In fact, the magnetic drive can apply enough force to the mirror to drive it vertically in a non-resonant mode, i.e. with a linear scan. Force applied to the frame by the magnetic coil at the horizontal resonant frequency of the mirror couples into the mirror and excites the resonance of the mirror, with the frame tilting slightly horizontal 180° out of phase with the mirror. Due to the added mass of the frame, the vertical resonance frequency of the mirror is much lower (typically 20 Hz vs 20 kHz) so this high-frequency component has little effect on the vertical motion of the mirror.

It is also possible to add a 3rd motion to the mirror: focus (Shao and Dickensheets 2004). In this arrangement, the mirror is a membrane instead of the normal silicon carbide. This allows the mirror not only to aim the laser beam but focus it as well to produce a 3D scan for applications such as confocal microscopy.

MEMS mirrors can also be designed to have a single scan axis. In addition, conventional, non-MEMS single axis galvanometer mirrors can be used to scan laser beams in projectors.

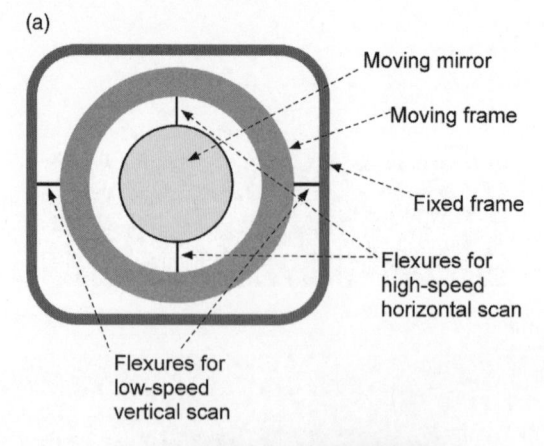

Figure 5.11 Biaxial MEMS mirror design: (a) layout; (b) photo of Microvision magnetic drive device (reproduced by permission of © 2005 SPIE).

References

Apte, R.B., Sandejas, F.S.A., Banyai, W.C. and Bloom, D.M. (1993) Grating light valves for high resolution displays. *SID Intl. Symp. Digest of Technical Papers*, 807–8.
Bloom, D.M. (1997) The Grating Light Valve: revolutionizing display technology. In Ming H. Wu (ed.), *Projection Displays III, SPIE Proceedings* **3013**: 165–84.
Bloom, D.M. and Tanner, A.H. (2007) Twenty megapixel MEMS-based laser projector. *SID Intl. Symp. Digest of Technical Papers*, Paper 3.3, 8–11.
Brazas, J.C. and Kowarz, M.W. (2004) High-resolution laser-projection display system using a grating electromechanical system (GEMS). *MOEMS Display and Imaging Systems II, SPIE Proc.* **5348**: 65–75.
Brennesholtz, M.S. (1998) Luminance contouring at the display in digital television. *Projection Displays IV (Ming Wu, ed.), SPIE Proceedings* **3296**: 62–70.
Dewald, D.S., Segler, D.J. and Penn, S.M. (2002) Advances in contrast enhancement for DLP projection displays. *SID Intl. Symp. Digest of Technical Papers*, Paper 46.1, 1246–9.
Dewald, D.S., Bartlett, T. and Iyengar, A. (2003) Optical model of DMD device in DLP™ projection system. *Proceedings of the 10th International Display Workshops*, Paper LAD2-1, 1557–60.
Douglass, M.R. and Malemes, C.G. (1996) Reliability of displays using Digital Light Processing™. *SID Intl. Symp. Digest of Technical Papers*, 774–7.

REFERENCES

Eguchi, N. (2006) GxL laser dream theater at the Aichi Expo (equivalent to 2005-in. TV). *Proc. of the 13th International Display Workshops*, 9–12.

Gove, R.J. (1994) DMD display systems: The impact of an all digital display. *SID Intl. Symp. Digest of Technical Papers*, 673–6.

Hornbeck, L.J. (1983) 128 × 128 deformable mirror device. *IEEE Trans. Elect. Dev* **ED-30**: 539–45.

Hornbeck, L.J. (1990) Deformable-mirror spatial light modulators. Presented at *SPIE Conference on Spatial Light Modulators and Applications III*, San Diego, CA, 1989. Text in *Proc. SPIE*, **150**: 86–102.

Hornbeck, L.J. (1997) Digital Light Processing™ for high-brightness, high-resolution applications. *Projection Displays III, Ming H. Wu, Ed., SPIE Proceedings* **3013**: 27–40.

James, R., Gibson, G., Metting, F., Davis, W. and Drabe, C. (2007) Update on MEMS-based Scanned Beam Imager, Paper 64660J. In D.L. Dickensheets, B.P. Gogoi and H. Schenk (eds), *MOEMS and Miniaturized Systems VI, Proc. of SPIE* **6466**: 11 pp.

Kowarz, M.W., Phalen, J.G. and Johnson, C.J. (2006) Line-scanned laser display architectures based on GEMS technology: from three-lens three-chip systems to low-cost optically efficient trilinear systems, *SID Intl. Symp. Digest of Technical Papers*, **XXXVII**, Paper 67.2, 1908–11.

Sampsell, J.B. (1994) An overview of the performance envelope of Digital-Micromirror-Device-based projection display systems. *SID Intl. Symp. Digest of Technical Papers*, 669–72.

Shao, Y. and Dickensheets, D.L. (2004) MEMS 3D scan mirror. In H. Urey and D.L. Dickensheets (eds), *MOEMS Display and Imaging Systems II, Proceedings of SPIE* **5348**: 175–83.

Sprague, R., Montague, T. and Brown, D. (2005) *Bi-axial Magnetic Drive for Scanned Beam Display Mirrors*. In H. Urey and D.L. Dickensheets (eds), *MOEMS Display and Imaging Systems II, Proceedings of SPIE* **5721**: 13 pp.

Tamada, H. (2007) Blazed GxLP™ light modulators for laser projectors. *J. SID* **15**: 817–23.

Texas Instruments (2005) Single-Panel DLP™ Projection System Optics, Application Report Discovery DLPA002, March 2005, 32 pp. Can be downloaded from http://focus.ti.com/download/dlpdmd/Discoverydlpa002.pdf

Urbanus, P.M. and Sampsell, J.B. (1994) DMD architecture and timing for use in a pulse-width modulated display system. US Patent 5 339 116, issued 16 Aug. 1994, assigned to TI.

Urbanus, P.M. and Sampsell, J.B. (1998) DMD architecture and timing for use in a pulse-width modulated display system. US Patent 5 745 193, issued 28 April 1998, assigned to TI.

Van Raalte, J.R. (1970) A new Schlieren light valve for television projection. *Applied Optics* **9**: 2225–30.

6

Filters, Integrators and Polarization Components

This chapter will cover the key optical components of a video projector, excluding the microdisplay or other image generating component, the projection lens and the lamp. The physics and function of each component in the system will be covered. The effect of each component on color and lumen throughput will be discussed. Included in this chapter are spectral filters, integrators and polarization components. Projection lenses and projection screens will be discussed in the next chapter.

For those readers unfamiliar with the concept of étendue as it applies to microdisplay projectors, it is recommended Section 11.2 be studied before this chapter.

6.1 Factors affecting Projector Optical Performance

Every component of a projector light engine affects the optical performance in some way. For virtually all components, there is a trade-off between cost of the component and performance of that individual component. For example, anti-reflection coatings on optical surfaces come in various grades. A high efficiency AR coating on a component will increase the throughput of the component, but will cost more than an AR coating that is less efficient. The trade-off between throughput and cost must, of course, be made by the system designer.

For some components, there is a trade-off between throughput and some other system property. For example, in TN-LCD projectors there is often a trade-off between throughput and contrast. With a higher f/# optical path, étendue limited projectors will have reduced throughput but can have better contrast compared to a lower f/# system.

72 *FILTERS, INTEGRATORS AND POLARIZATION COMPONENTS*

6.2 Component Efficiency

Component efficiency is the percentage of the light[1] falling on or passing through an element that can be used by the projector. For most components, efficiency is a function of the wavelength of the light. For example, a simple, uncoated lens of a high transmission glass, has very little loss except for Fresnel reflection losses at the surfaces. The typical reflection loss at a surface is about 4 %. Therefore, for both surfaces of a lens, the loss would be 8 % and the efficiency 92 %. An anti-reflection coated lens, on the other hand, will typically reflect about 0.75 % per surface, for a total loss of 1.5 % or an efficiency of 98.5 %. Generally speaking, all optical surfaces in commercial projectors are anti-reflection coated.

Component efficiency for most passive components is just the measured transmittance or reflectance, as appropriate, as a function of wavelength. The symbol $\eta(\lambda)$ is often used for this spectral efficiency. In polarization sensitive projectors, i.e. most LCD projectors, the spectrum must be measured in light of the correct polarization.

It is often helpful to the projector designer to characterize the efficiency of a component with a single number, η. For components that are nearly spectrally neutral, i.e. $\eta(\lambda) \cong$ constant, this is a reasonable thing to do. There are several methods in use to calculate this single efficiency value, and all give the same number for optical elements that have neutral density.

In the following equations, colorimetric units are used. The reader is encouraged to read Appendix 2 if these are not familiar concepts. There are three distinct ways to define efficiency of a component:[2]

1. Photometric efficiency using CIE Illuminant E:[3]

$$\eta_{Photometric} = \frac{\sum_{380}^{780} \eta(\lambda) \bar{Y}(\lambda)}{\sum_{380}^{780} \bar{Y}(\lambda)} \qquad (6.1)$$

2. Photometric efficiency using the spectrum of the lamp of interest:

$$\eta_{Lamp} = \frac{\sum_{380}^{780} \eta(\lambda) E(\lambda) \bar{Y}(\lambda)}{\sum_{380}^{780} E(\lambda) \bar{Y}(\lambda)} \qquad (6.2)$$

3. Average efficiency over the wavelength range of 432 to 680 nm, or some other range:

$$\eta_{Average} = \frac{\sum_{432}^{680} \eta(\lambda)}{N}. \qquad (6.3)$$

In these equations:

$\eta(\lambda)$ is the component efficiency as a function of wavelength,
$\bar{Y}(\lambda)$ is the CIE 1931 2° luminance matching function,

---

[1] Here the meaning is the fraction of the energy of the light, not the amplitude of the electric field. Unless otherwise specified, this will be the meaning throughout the book.
[2] These equations are given in numerical form suitable for computation with tabular data in spreadsheets or other computer programs.
[3] Illuminant E is a fictitious light source where there is equal energy at all wavelengths, i.e. $E(\lambda) =$ constant.

$E(\lambda)$ is the spectrum of the lamp in the projector and
N is the number of wavelength steps. For example, N = 63 for the range of 432 nm to 680 nm, taken in 4 nm steps.

If the component of interest is exactly neutral density, i.e. $\eta(\lambda)$ = constant, then Equations (6.1), (6.2) and (6.3) will all give the same efficiency value. With components that are not spectrally neutral, such as color filters, one must be very cautious about calculating a single number efficiency. The issue of non-neutral components will be covered in detail in Section 11.5. Obviously, the lamp and color filters are not neutral and must be handled accordingly. Normally the goal in component design is to make the rest of the components neutral.

6.3 Spectral Filters

Optical filters come in two main types: dichroic and absorptive. A dichroic filter reflects some wavelengths of light and transmits other wavelengths with very little absorption. An absorptive filter will transmit some wavelengths while absorbing others. The purpose of any filter is to modify the spectrum of the light passing through or reflected by the filter.

6.3.1 Fresnel Reflection at Optical Surfaces

Electromagnetic theory predicts that, at any interface at which there is a change in the index of refraction, there will be both reflection and refraction of the energy at the interface. The amount of reflection, predicted by the Fresnel equations (Smith 1990), is dependent on the polarization of the light:

$$R_s = \frac{\sin^2(\theta_i - \theta_r)}{\sin^2(\theta_i + \theta_r)} \quad \text{(a)}$$

$$R_p = \frac{\tan^2(\theta_i - \theta_r)}{\tan^2(\theta_i + \theta_r)} \quad \text{(b)}$$

(6.4)

where:
  R_S is the reflectivity of the component of the light polarized normal to the plane of incidence and
  R_P is the reflectivity for the light polarized parallel to the plane of incidence.
  θ_i is the incident angle, and
  θ_r is the refracted angle.

These two angles are related by Snell's law:

$$n_i \sin(\theta_i) = n_r \sin(\theta_r) \qquad (6.5)$$

where:
  n_i is the index of refraction of the material on the incident side of the interface and
  n_r is the index of refraction of the material on the refracted side of the interface.

The geometry of these angles is shown in Figure 6.1. In this figure, the paper is the plane of incidence because it contains both the incident light and the vector normal to the optical surface.

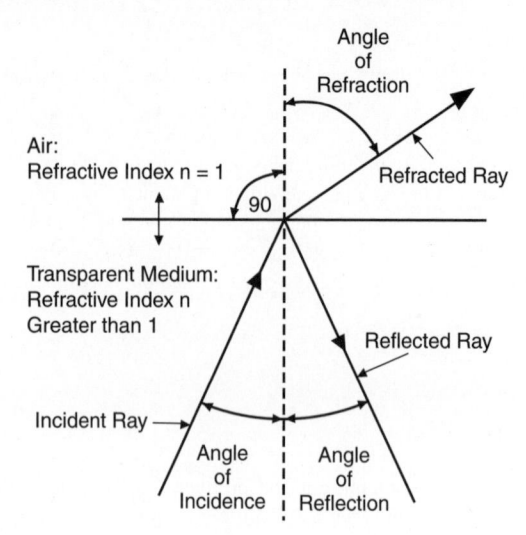

74 FILTERS, INTEGRATORS AND POLARIZATION COMPONENTS

Figure 6.1 Geometry of Snell's Law.

For unpolarized light, the average reflectance is:

$$R = \frac{1}{2}(R_S + R_P) \tag{6.6}$$

Note that after the first surface that is not at normal incidence, the light will be partially polarized and Equation (6.6) cannot be used: Equations (6.4(a,b)) must be used for each component separately.

At normal incidence, Equations (6.4) and (6.5) can be combined to get the equation:

$$R_S = R_P = R_A = \left(\frac{n_i - n_r}{n_i + n_r}\right)^2. \tag{6.7}$$

For $n_r = 1.52$ (BK–7), $n_i = 1.00$ (air) Equation (6.7) gives $R_A = 4.3\%$. Notice that the value is the same if $n_r = 1.00$ and $n_i = 1.52$, 4.3%. This means the same amount of light is reflected at both sides of an optical element, as the light enters the element and again as it leaves.

As mentioned before, these coefficients R_s and R_p deal with the reflected power. There are similar equations that are related to the electric field of the reflected and transmitted beam. The projector designer is primarily concerned with the power, not the field, in designing a projector. For design of dichroic filters, it is necessary to use the Fresnel equations in electric field form.

Equation (6.4(a,b)) and Equation (6.6) are plotted in Figure 6.2 for a typical glass/air interface. The index of refraction of air was taken to be 1.00, the index of the glass was taken to be 1.57 and the angle of incidence was allowed to vary from 0° to 90°. There are several interesting things to note about this plot. First, at 0° (normal incidence) there is no difference between s- and p-polarized light. For typical glasses, the normal incidence reflection is about 4% per surface. For increasing angle, the reflectivity of the s-polarized light increases monotonically with angle, to 100% at 90° incidence.

The behavior of p-light is entirely different, however. Here, the reflectivity *decreases* with increasing angle until one reaches what is called *Brewster's Angle* where the reflectivity of the p light is 0.0%. Brewster's angle occurs when the condition:

$$\theta_i + \theta_r = 90° \tag{6.8}$$

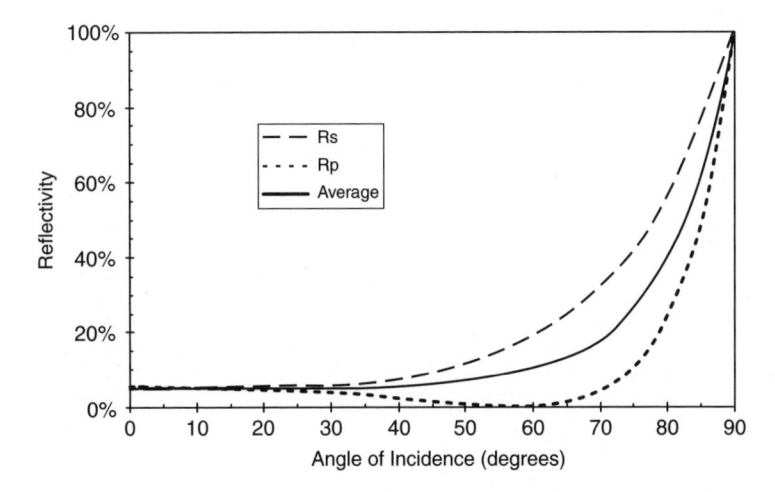

Figure 6.2 Fresnel reflection from a glass/air Interface. Glass index of refraction = 1.57.

is satisfied by the angle of incidence and the angle of refraction in Snell's law, Equation (6.5). For the glass in Figure 6.2, Brewster's angle is 57°. For non-absorptive media, all p-polarized light at Brewster's angle is transmitted; none is reflected. At still higher angles, the reflection of p-light begins to increase and also reaches 100 % at 90° angle of incidence.

While the Fresnel equations are nominally independent of wavelength, the wavelength of light enters into the surface reflection off objects in two different ways. First, the indices of refraction of all optical materials including glass, plastic, crystalline materials and liquid crystals, depend on wavelength. As the index of refraction varies with wavelength, the Fresnel reflections vary. Secondly, and also normally more importantly, if there are two or more interfaces separated by a distance comparable to the wavelength of light, interference patterns are established between the two interfaces. These interference patterns are very wavelength- and angle-dependent and therefore the reflection and transmission are strongly wavelength- and angle of incidence-dependent.

6.3.2 Dichroic Filters

Dichroic filters are critical components in most modern projectors. The term 'dichroic' is most commonly used to signify a filter that splits a beam into two or more separate color bands, but it is also used with polarizing filters (Section 6.5) and hot or cold mirrors (Section 6.3.7).

Dichroic filters can be designed to split an incident optical beam into two exiting optical beams, as shown in Figure 6.3(a), or recombine two beams into a single beam, as in Figure 6.3(b). Note that in order to recombine two beams, the beams must be distinguished from each other by wavelength or polarization. For example, the red or green dichroic filter in Figure 6.5 could be used to recombine a red beam with a green beam with relatively high efficiency, but neither filter could not combine two red beams. If the polarizations of the two red beams differ, they could be combined by a polarizing coating, such as the MacNeille prism discussed in Section 6.5.2.1(b).

Dichroic filters are made of many layers of dielectric material. These are arranged in alternating layers of high and low index of refraction material. The most commonly used materials are transparent and non-absorbing. Common materials for visible light are MgF and SiO_2 for the low index layers and TiO_2, Ti_2O_3, and Ta_2O_5 for the high index layers The thickness of each layer is typically on the order of one-eighth to one-half of a wavelength of the light of interest. The total number of layers is typically on the range of 15–40 although sometimes more layers than this are used. While these filters normally

76 FILTERS, INTEGRATORS AND POLARIZATION COMPONENTS

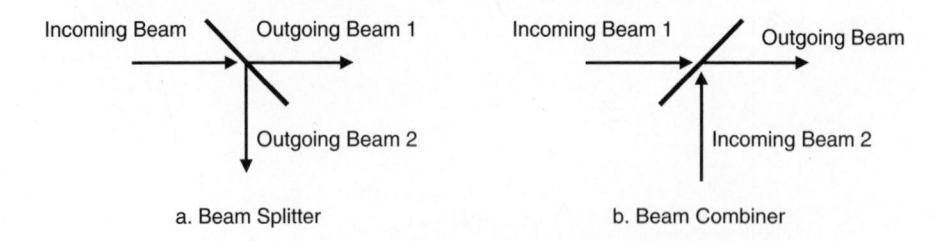

Figure 6.3 Dichroic filter used as a beam splitter or combiner. a) beam splitter, b) beam combiner.

only use two types of material, one low index and one high index, designs with three or more different materials are sometimes necessary to get the desired optical properties.

The structure of a typical dichroic filter is shown in Figure 6.4. The most common manufacturing technique for dichroic filters is to deposit all the layers of the coating in a vacuum chamber on one of the substrates. This sequence of layers is commonly called the stack. If the filter design calls for a top substrate, an optical cement is used to bond the second piece of glass on top of the stack.[4] If the top substrate is relatively thin, the coating is called an embedded coating. If the two substrates are thick, the system becomes a prism. Many common dichroic designs omit the second substrate and are designed to have air as the second boundary material.

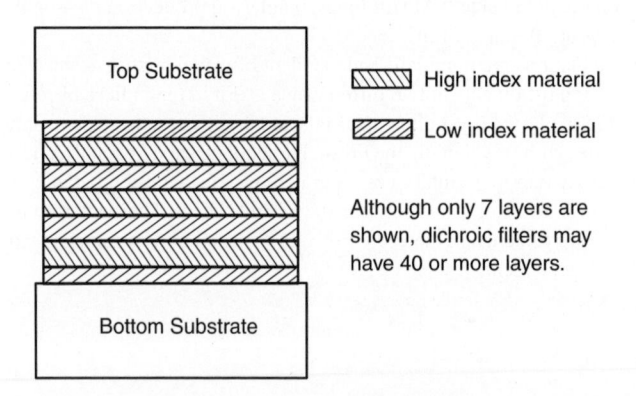

Although only 7 layers are shown, dichroic filters may have 40 or more layers.

Figure 6.4 Structure of a dichroic filter.

There is a small amount of Fresnel reflection at each surface between the high index and low index materials in Figure 6.4. According to Equation (6.7), at an interface between SiO_2 (n = 1.48) and Ti_2O_3 (n = 2.35), the reflectance at normal incidence would be 5.2 %. The multiple reflections from all of the layers sets up interference patterns in the filter. Depending on the wavelength of the light, the polarization of the light, the thickness of the various layers and the angle between the light beam and the filter, this can result in either destructive or constructive interference. Destructive interference results in reflection of the light from the filter while constructive interference results in transmission through the filter.

---

[4] If the index of refraction of the cement is different from the index of the top substrate, it must be allowed for in the dichroic design. This is a special problem if the top substrate is a high index glass and no index-matching cement is available.

SPECTRAL FILTERS

In an ideal dichroic filter with non-absorbing materials including the substrate, there is obviously no absorption in the filter. All light is either reflected or transmitted. Realistic dichroic filters from a variety of manufacturers approach this ideal very closely. This is a particularly important property when dealing with light beams that carry a very high energy density. Since little energy is absorbed, there is little heating and little thermal stress on dichroic filters. The slight amount of absorption in a real dichroic filter is not normally an issue except in high power laser applications.

When there is no absorption or scattering in a dichroic filter, reflectivity can be calculated by the equation:

$$R(\lambda) = 1 - T(\lambda) \qquad (6.9)$$

The design of dichroic filters is much too complex an issue to go into in detail in this book. The reader is referred to one of the many texts on filters (Rancourt 1996; Macleod 2001). Most projector manufacturers do not in fact design or manufacture their own dichroic filters. If stock filters available from one of the many specialized filter manufacturers will not perform satisfactorily, the projector manufacturer would provide a performance specification to a filter maker and let the filter maker design the filter to match their manufacturing process. In high volume, custom dichroic filters are not significantly more expensive than standard filters with the same number of layers.

Small quantities of custom filters can be expensive, however. Dichroic filters are most commonly made in large vacuum chambers. The cost of running the chamber is roughly the same whether there is one glass substrate in the chamber or 1000. Because of this fixed cost of a coating run, running only a small number of glass substrates for custom filters is not very cost effective, although it is sometimes necessary in the research or development stage of a projector design.

Figure 6.5 shows measured transmission curves for typical red, green and blue filters at normal incidence as a function of wavelength.

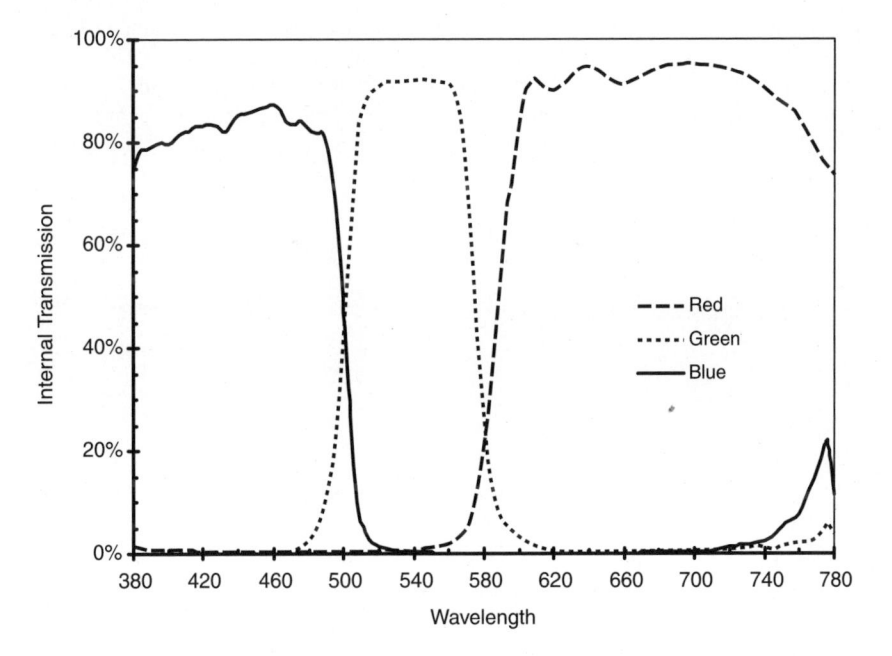

Figure 6.5 Transmissions for typical red, green and blue dichroic filters at normal incidence.

Since it is much easier and more accurate to measure the transmittance of a filter than the reflectance of a filter, it is common to measure transmission and use Equation (6.9) when the reflectance of a dichroic

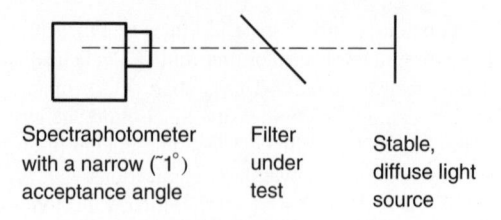

Figure 6.6 Simple apparatus for measuring filter transmission.

filter is desired. If it is believed that there is some absorption in a filter, or the filter acts as a mirror on an opaque substrate, it is necessary to measure reflectivity directly. For details on how to measure reflectivity accurately, see, for example, Palmer (1995). This paper also gives a good overview of the measurement of transmission and absorption.

The apparatus shown in Figure 6.6 can be used to measure the transmission of most filters. The light source can be as complex as an integrating sphere light source or as simple as a tungsten-halogen bulb[5] driven by a regulated power supply reflecting off a diffuse surface such as a piece of white paper. The key to making accurate transmission measurements is stability in the light source. The light source is first measured with the filter removed from the system, giving a spectrum L(λ). The filter is inserted into the system and without moving the detector or the source and another spectrum F(λ) is measured. The filter transmission is then given by:

$$T(\lambda) = \frac{F(\lambda)}{L(\lambda)}. \tag{6.10}$$

Many computer controlled spectrophotometers or spectroradiometers have Equation (6.10) built in as an internal software function. Otherwise, the spectra F(λ) and L(λ) must be downloaded into a desktop computer for the computation. This technique gives the transmission of the filter over a cone angle that matches the acceptance angle of the detector used. The facility for accurate optical measurements should include a darkened room with light baffling to minimize unwanted light paths.

6.3.2.1 Dichroic filters at non-normal incidence

As mentioned earlier, the optical properties of dichroic filters vary with both polarization and angle of incidence. The shift with angle of the center wavelength of a bandpass filter is described by Equation (6.11) (Melles Griot 1997):

$$\lambda(\theta) = \lambda_0 \sqrt{1 - \left(\frac{n_0}{n_{eff}}\right)^2 \sin^2\theta}. \tag{6.11}$$

In this equation:

- λ is the center wavelength at the angle θ,
- λ_0 is the center wavelength at normal incidence,
- n_o is the external index of refraction ($n_o = 1.00$ for air),
- n_{eff} is the effective index of refraction of the coating and
- θ is the angle of incidence at the coating.

---

[5] An ordinary non-halogen incandescent light bulb is not satisfactory because it does not produce enough blue or near UV light (380–450 nm) to produce satisfactory, low noise results.

The effective index of refraction is a complicated function of the index of the high and low index materials in the dichroic stack and of the design of the filter. Equations for n_{eff} for various filter designs are given by Macleod (2001).[6]

Note that for an angle of incidence, θ, greater than 0, the term under the square root sign in Equation (6.11) is always less than 1.0. Therefore, $\lambda(\theta)$ is always less than λ_0. This is a universal property of multi-layer dielectric filters: as the angle of incidence increases, the cut-off and cut-on, as well as the center wavelength, shift to shorter wavelengths.

Unfortunately, the assumptions made in the derivation of Equation (6.11) rarely match the conditions under which the projection engineer would like to use an equation of this type. First of all, this equation applies to the center wavelength of a band pass filter. A projection engineer is more commonly concerned with the cut-on or cut-off wavelength of a filter. Secondly, Equation (6.11) is intended to be used at small angles, while the projection engineer is frequently interested in angles up to 45° or more. Next, the projector designer will rarely have enough information to calculate n_{eff} from the equations given by Macleod and this value is rarely available from the dichroic filter manufacturer. Finally, the value for n_{eff} is different for s- and p-polarized light. If the projector designer is interested in unpolarized light, Equation (6.11) would need to be applied separately to the two polarizations.

In practical terms, however, Equation (6.11) is still very useful. A filter of interest can be measured at several different angles and the measured cut-off or cut-on data fit to the equation. For example, Figure 6.7 shows the measured optical properties of a typical dichroic filter as it tilts from 0° to 45°. This filter was intended to be a yellow[7] transmitting filter at normal incidence and was measured in unpolarized light. The cut-on data was then fit to Equation (6.11) and n_{eff} derived empirically. The measured value of the cut-on wavelength at 0°, 516.1 nm, was used for λ_0. The value of $n_{eff} = 2.970$ was then the best fit to the cut-on wavelength at higher angles.

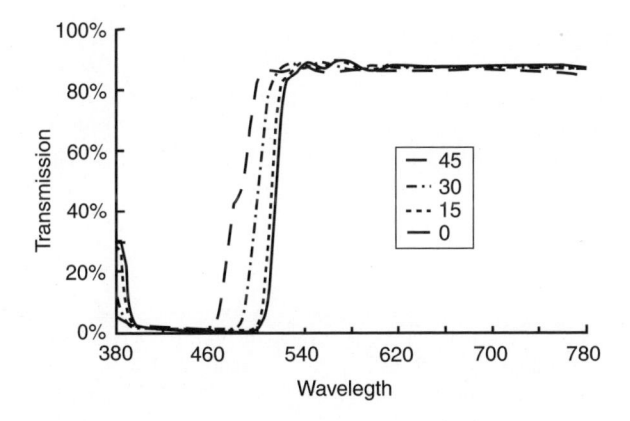

Figure 6.7 Dichroic filter transmission as a function of angle.

These values, when used in Equation (6.11), predict the cut-on wavelength of the filter at non-normal incidence to an accuracy of ±0.6 nm. This accuracy is comparable to the accuracy of the measurement of the cut-on wavelength and is perfectly satisfactory for most projector colorimetry work. For instance, this empirical relationship between the angle of incidence and the cut-on wavelength can be useful in colorimetric modeling of projectors with finite cone angles or non-telecentric dichroics.

---

[6] Macleod also gives a derivation of Equation (6.11). His final form for this equation is much less convenient for the projector designer.

[7] Yellow dichroic filters reflect blue light and transmit red and green light.

80 FILTERS, INTEGRATORS AND POLARIZATION COMPONENTS

The change in optical properties with angle do not become significant to the projector designer until the angle of incidence reaches about 20°. At 45°, however, there is a noticeable change in the optical properties of dichroic filters compared to the properties at normal incidence. This change in properties must be designed into the filter for correct operation.

Note that in addition to a shift in the cut-off wavelength of this filter, there is a change in the shape of the transmission curve at increasing angles. Part of this is due to a real change in the shape of the curve as the angle changes. In addition, as the angle increases the effect of polarization becomes increasingly important. Since n_{eff} is different for s- and p-polarized light, the shift in cutoff wavelength is different for the two polarizations. This factor accounts for the step in the transition region near the 50 % transmission value when the filter is measured at high angles in unpolarized light. The effect of polarization will be discussed further in Section 6.3.2.2.

This wavelength shift affects the colorimetry of a video projector if the light passing through a tilted dichroic is not telecentric. The angle of incidence of light that comes from one side of the panel is different from the angle of incidence of light from the other side of the panel. This means the left and right side of the panel see different cut-off or cut-on wavelengths. This results in a color gradient across the image, which can be noticeable and objectionable. This effect is sketched in Figure 6.8. Here ray 1, from the left side of the panel, sees an angle of incidence of $(45 + \theta)°$, while ray 2 from the right side sees an angle of incidence of $(45 - \theta)°$. Since dichroic filters always shift to shorter wavelengths with increasing angle, the left side of the panel will appear bluer than the right side in this example.

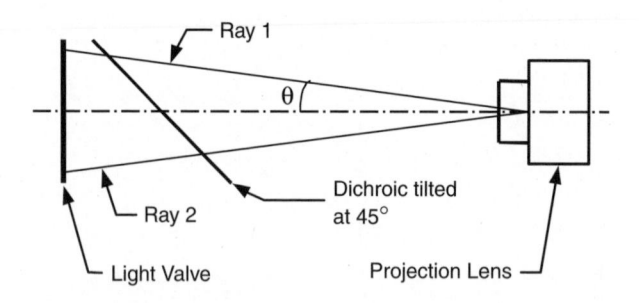

Figure 6.8 Dichroic filters in non-telecentric light.

Generally, non-telecentric systems can be made more compact and use smaller and less expensive components than systems that are telecentric at the tilted dichroics. Therefore, it is a problem that will become increasingly important in a competitive projector market. One possible solution is to build a gradient into the dichroic filter to compensate for the gradient caused by the non-telecentric light, as described by Bowron et al. (1996).

6.3.2.2 Dichroic filters in polarized light

Figure 6.9 shows the effect of polarization on the optical properties of a dichroic filter. This is the same filter as shown in Figure 6.7, measured in the apparatus of Figure 6.6, with the addition of a polarizer between the filter and the detector. This polarizer could rotate about the optical axis and be set to pass either s-polarized or p-polarized light. As can be seen, the transmission of the filter is significantly different for the two types of light. If Equation (6.11) is to be applied to s- and p-polarized light separately, different values of n_{eff} must be used. For this example dichroic, a value of $n_{eff} = 3.400$ is appropriate for s-polarized light and $n_{eff} = 2.473$ is appropriate for p-polarized light. Since there is no difference between s and p at normal incidence, $\lambda_{0s} = \lambda_{0p} = 516.1$ nm for this filter.

SPECTRAL FILTERS 81

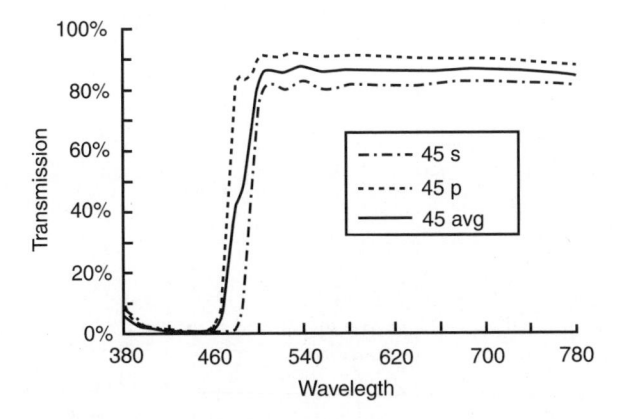

Figure 6.9 Filter properties as a function of polarization.

For this filter at 45°, light in the wavelength range of 460–495 nm is reflected if it is s-polarized light but is transmitted if it is p-polarized light. This means that the projector designer must specify what the polarization of the light will be when it encounters a tilted dichroic filter: unpolarized, s-polarized or p-polarized. The dichroic designer can then optimize the filter appropriately.

One way to make a polarizing beam splitter is to design a dichroic coating that maximizes the difference s and p polarizations. Coatings like this are normally embedded in a prism for maximum efficiency over the range of wavelengths of operation. Section 6.5 discusses polarization components in more detail.

6.3.2.3 Dichroic filters in the imaging path

In three panel, single-lens projectors, the three separate modulated color beams must be recombined with dichroic filters. This dichroic recombination can cause imaging problems. The simplified two color projector, as shown in Figure 6.10, is sufficient to illustrate the problem.

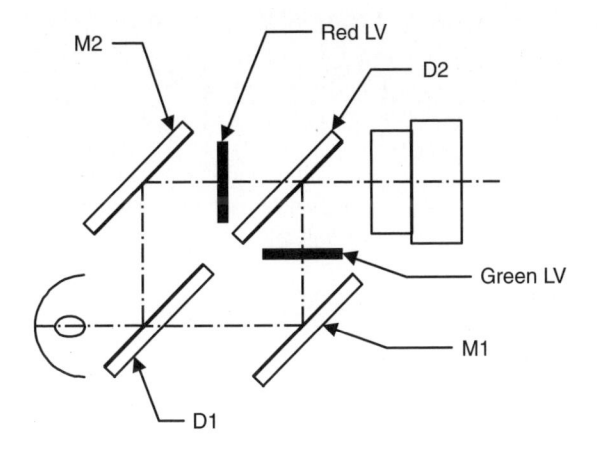

Figure 6.10 Simple 2-panel projector.

82 FILTERS, INTEGRATORS AND POLARIZATION COMPONENTS

Light from the lamp is split into red and green beams by the first dichroic D1. These beams are redirected to the two microdisplays LV1 and LV2 by mirrors M1 and M2. After the light is modulated, the beams are recombined by D2 and projected onto the screen. Note that the red beam passes through dichroic D2. Since the dichroic is tilted and has a finite thickness, this induces coma into the red beam. The green beam, on the other hand, is reflected off the first surface of D2 and never passes through the tilted plate. Therefore, no coma is induced into the green image.

The projection lens designer can largely correct coma in the lens design, but here he is faced with two beams: one with coma and one without. If he corrects for the coma in the red beam, then he will induce coma in the green beam. Additional optical elements may be required in the red channel for compensation (Shimizu and Janssen 1996). If the coating on D2 is made, a second surface coating for the green path the problem still occurs. Here the red light passes through the substrate once while the green light passes through the substrate twice, inducing more coma in the green path than the red path.

There is a second problem with the green beam. If the dichroic D2 is not perfectly flat, it will distort the image in the green beam as it reflects off the dichroic. The solution to this problem is to make the dichroic thick enough so it will remain flat in the projector. Unfortunately, making the dichroic thicker increases the amount of coma in the red beam.

One solution is to make an embedded coating, as shown in Figure 6.11. In Figure 6.11(a), the dichroic layer is embedded between two substrates of the same thickness. In this case the same amount of coma is induced in each beam. The red light will pass through the first substrate, transmit through the coating and pass through the second substrate. The green beam passes through the second substrate, reflects off the coating and passes again through the second substrate. The lens designer then has the same amount of coma to correct in each beam, while the dichroic filter can be made thick enough to maintain it's flatness.

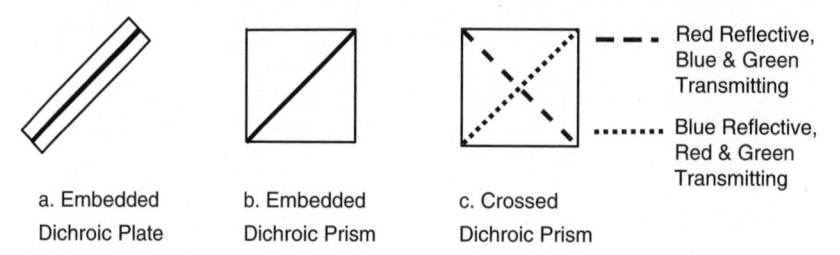

Figure 6.11 Embedded dichroic filters. a) embedded dichroic plate, b) embedded dichroic prism, c) crossed dichroic prism (X-prism).

In the second solution, as shown in Figure 6.11(b), the faces of the prism are normal to both beams and no coma is induced at all. Flatness is not normally an issue with prisms. While the prism will introduce spherical aberration and chromatic aberrations, these problems are relatively easy to correct in the projection lens design. This prism solution (with two crossed dichroics as shown in Figure 6.11(c)) is commonly used by the designers of unequal illumination path projectors, as described in Section 10.2.2.

This problem also occurs in the illumination path of projectors, at dichroic D1 of Figure 6.10. Here the problem is much less serious because the image quality requirements on the illumination side of the microdisplays are much less critical than on the image side.

6.3.2.4 Anti-reflection coatings

One very important class of multilayer thin film coatings is anti-reflection coatings at glass-air interfaces in projectors. If a lens or other optical element in a projector is uncoated, the Fresnel equations (Equation (6.4)) apply and about 4 % of the light incident on the surface is reflected and 96 % is transmitted. Since every lens has two surfaces, the losses per lens are about 8 %. This loss represents not only a loss of projector throughput, but an increase in stray light that if not controlled might degrade system contrast. While 96 % may sound like fairly high throughput, if there are five optical elements in the system with a total of ten surfaces, the system throughput would then be $0.96^{10} = 66\%$. Since modern projectors can have many more than ten optical surfaces between the lamp and the projection screen, reflection off these surfaces must be reduced.

In theory, it is possible to eliminate all reflection, at least for a single wavelength at normal incidence with a single layer coating. Ideally, the index of refraction of this coating should be $n_i = \sqrt{n_1 n_2}$, with a thickness d such that $n_i d$ equals a quarter wavelength. For $n_1 = 1.00$ (air) and $n_2 = 1.52$ (BK-7 glass), $n_i = 1.23$. Unfortunately, there are not any suitable materials with an index of 1.23: the lowest index of a common optical material is $n = 1.38$ for MgF. Nevertheless, a single layer of MgF is an effective and low cost anti-reflective coating, typically reducing the reflectivity per surface from about 4% to about 1%. The performance of a single layer MgF coating at normal incidence and 45° is shown in Figure 6.12(a).

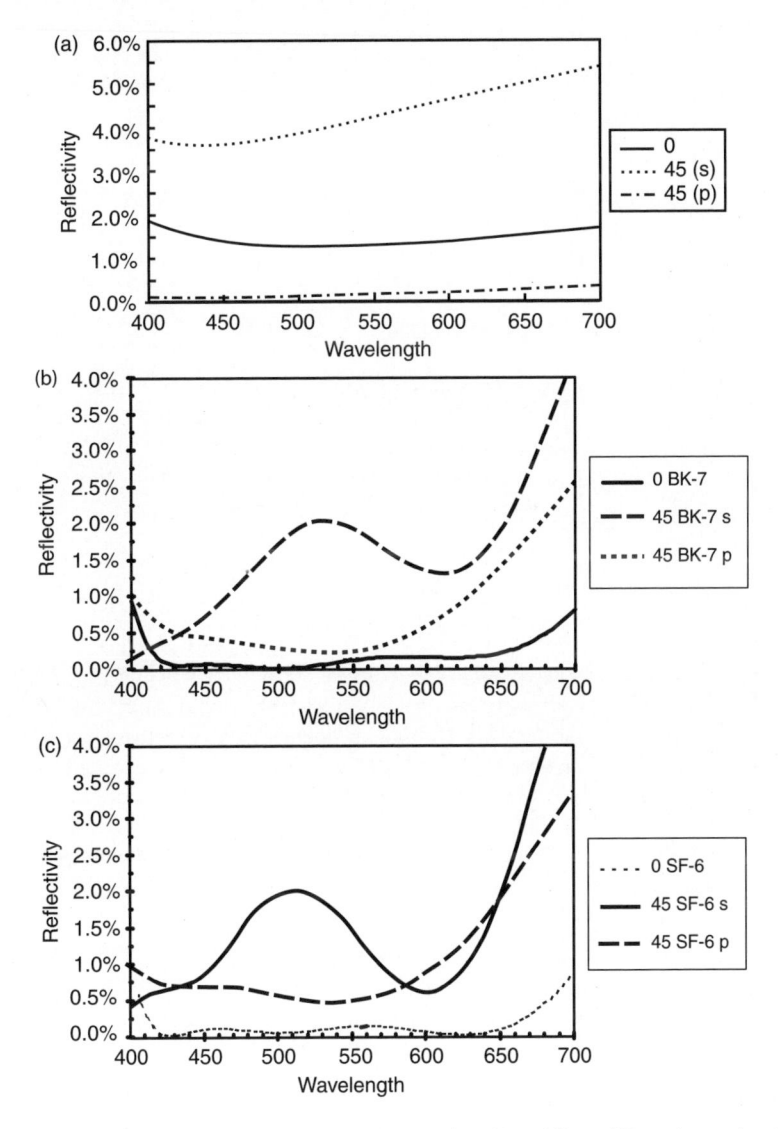

Figure 6.12 Antireflection Coatings. a) single layer MgF_2 coating, b) multilayer AR coating on low index glass (BK-7) c) multilayer AR coating on high index glass (SF-6). (Data Courtesy of JML Optical Industries, Inc.).

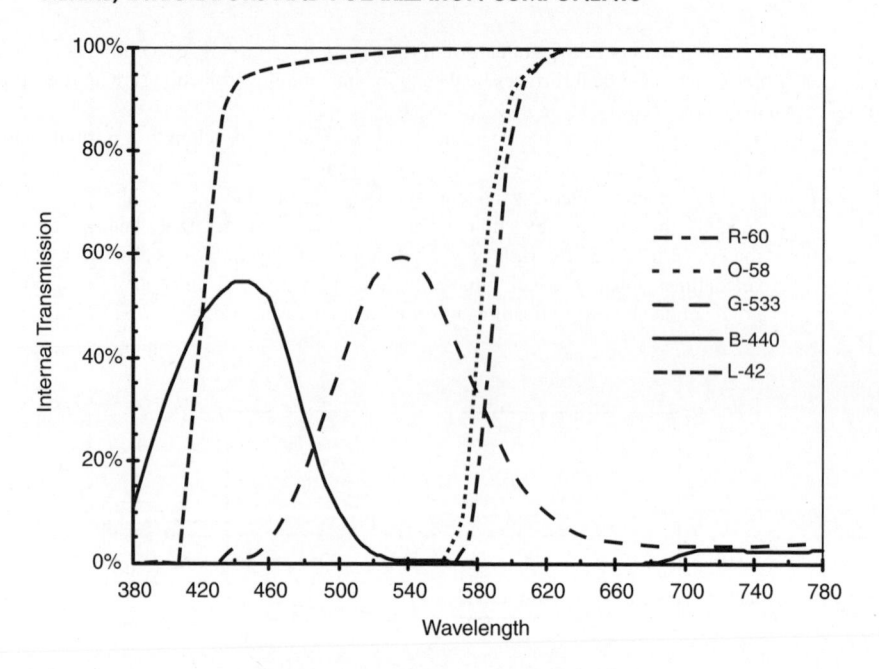

84 FILTERS, INTEGRATORS AND POLARIZATION COMPONENTS

Figure 6.13 Transmission of absorptive filters.

More effective and more expensive AR coatings are formed from multiple layers, often 3 or 4 layers. The price/performance ratio of modern video projectors is such that these more efficient coatings are normally justified for all optical surfaces. The performance of these coatings is also shown in Figure 6.12(b). Reflection at normal incidence and at 45° are shown for coatings on both low index glass in Figure 6.12(b) (BK-7, $n_g = 1.52$) and high index glass in Figure 6.12(c) (SF-1, $n_g = 1.72$).

6.3.3 Absorptive Filters

The other common type of filter used in projectors is the absorptive filter. These operate by absorbing the undesired light rather than reflecting it. Transmission curves for sample red, green, blue and UV filters are shown in Figure 6.13. These absorptive filters are from Hoya Optics. Filters with similar optical properties are available from a variety of other vendors. Similar spectral properties are available in either glass or plastic based filters, or as dye additives to liquids.

Note that relatively efficient long pass filters[8] are available. Two versions of a red filter (R-58 and R-60) and a UV absorbing visible transmitting filter (L-42) of this type are shown. Absorptive filters with cut-on wavelengths anywhere in the visible or IR range are available from a variety of vendors. This type of filter has nearly 100 % transmission at long wavelengths, transmission near 0 % at short wavelengths and a relatively steep and smooth transition between the two regions.

Band pass and short pass absorptive filters are much less efficient than long pass filters. They also have much poorer color saturation than the long pass filters. The G-533 filter has its primary transmission in the green region, but it has a long tail in the red region that significantly desaturates the color of the light passing through the filter. The blue filter, B-440, transmits in both the blue and the blue-green regions. It blocks most of the green and red light, but begins to transmit again in the near IR region. Both blue and

---

[8] A long pass filter is one that transmits long wavelengths but blocks with absorption or reflection shorter wavelengths.

the green filters have peak transmissions in the range of 55–60 %. Fresnel reflection occurs at the surface of absorptive filters. For this reason, these filters are often AR coated to increase the transmission of the desired wavelengths.

Since an absorptive filter operates by absorbing energy from the light beam passing through it, the temperature of the filter can rise appreciably in operation. This can lead to cracking the filter from thermal stress, especially if the light is distributed unevenly over the filter, creating a center to edge thermal gradient. At high power densities, bleaching is also a possibility, especially if the dye used in the filter is an organic material. The projector designer must ensure the power density limitations of the filter are not exceeded and it is mounted in such a way as not to put undue stress on the filter. Opto-mechanical design is a science of its own that will not be covered in this text (see, for example, Ahmad (1997)).

6.3.4 Electrically Tunable Color Filters

Effects have are reported which can be used to produce electrically tunable color filters. One of these is based on voltage control of surface plasmon resonance frequencies (Wang 1997). The filter incorporates a liquid crystal layer in contact with a very thin metal layer. The device absorbs incident photons at the surface plasmon resonance frequency and photons out of resonance are reflected. Device switching times as low as 0.2 ms have been observed, making it possible to build a color-field sequential projector using this approach (see Chapter 10).

Another approach to electrically tunable color filters is based on cholesteric color filters (Schadt and Schmitt 1997). The device makes use of the unique wavelength selective transmission and polarization properties of stacked cholesteric layers. Figure 6.14 schematically illustrates the structure and operation of the device for one of the three color bands. Operation over the entire visible spectrum requires three tuned band-modulation filters in series in a color subtractive mode.

The device operates using wavelength and polarization selective transmission of the cholesteric layers. The transmission of the filter changes according to the voltage-dependent polarization state of the twisted nematic layer. The center wavelength and bandwidth of the filter are independently adjustable. Good contrast is attainable throughout the visible region.

Other cholesteric LC devices have been designed (Li *et al.* 2000) to cover the entire visible range with a variable reflective bandpass filter with in-plane drive voltages. These devices are capable of producing full color reflective displays without color filters. Unfortunately, they cannot be turned 'off' and can

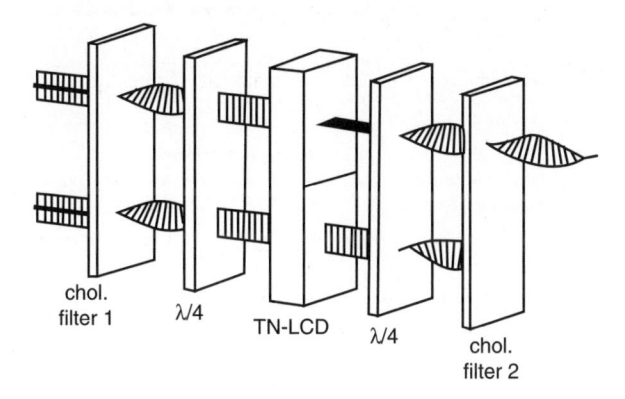

Figure 6.14 Schematic of a band modulation filter for a single color consisting of two identical cholesteric filters, two quarter wave plates, and a TN-LCD modulator in its on-state (top) and its homeotropic off-state (bottom). (reproduced from (Schadt and Schmitt 1997), with permission of SID).

86 FILTERS, INTEGRATORS AND POLARIZATION COMPONENTS

only produce colors around the periphery of the color space: they cannot produce white. In addition, the prototype device required 0–300 V drive voltages

Electrically tunable filters can have broad wavelength adjustability with modest drive voltages, suitable for adjusting color temperature of a display, as discussed in Brennesholtz (2003). This filter design is based on a variable retarder between polarizers. Typically, the first polarizer would be the PCS and the second would be the clean-up polarizers on LCD panels or the PBS in a LCoS system. The optical properties of this system are shown in Figure 6.15. This filter could adjust the color temperature of a UHP-type light source, or a projector based on a UHP-type lamp, over the range of 4228–9246 K.

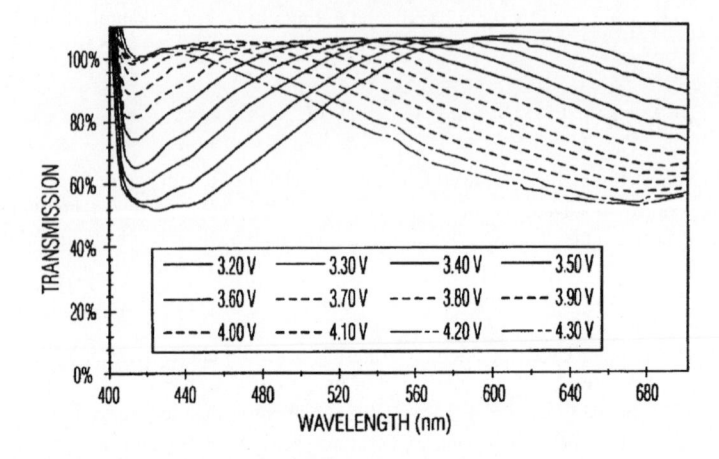

Figure 6.15 Electrically tunable color filter for color temperature control (Brenneshaltz 2003; United States Patent and Trademark Office).

A more sophisticated version of color adjustment using wavelength sensitive retarders and non-pixelated liquid crystal cells is discussed in Sharp et al. (2000). In this technology, multi-layer retarders are designed to have distinct wavelength bands, as discussed in Sharp and Birge (1999). These retarders can have very sharp cut-off values with relatively small numbers of layers, typically in the order of 3–9. When placed between crossed polarizers, a sample red/cyan retarder of this type with between 1 and 9 layers has the properties shown in Figure 6.16. This retarder and polarizer stack can be combined with a liquid crystal (LC) cell. This polarizer/LC/retarder stack/polarizer will transmit either red or cyan light, depending on the voltage applied to the LC cell. More complex arrangements can be designed to transmit red, green or blue for use in a color sequential projector system, as will be described in Section 10.4.2. These retarders can also be used in three-panel LCoS systems to separate the colors, as will be described in Section 10.3.2.

6.3.5 Mirrors

Mirrors are a common feature of video projectors. For example, they are used in Figure 6.10 to steer the illumination beams to the microdisplays. Most commonly, the mirrors in a projector are first surface mirrors. This means that the light beam encounters the mirror coating before it encounters the substrate, as shown in Figure 6.17(a). In this case, flatness is the only important optical property of the substrate.

Second surface mirrors have the coating on the back surface of the substrate, as shown in Figure 6.17(b). With this type of mirror, the beam passes through the substrate, reflects off the mirror and then passes through the substrate again. This does two things, both undesirable in the imaging portion of a projector.

First, there are ghost reflections due to Fresnel reflection on the first surface, shown as a dotted line in Figure 6.17(b). While this can be minimized with a good AR coating, it cannot be eliminated completely.

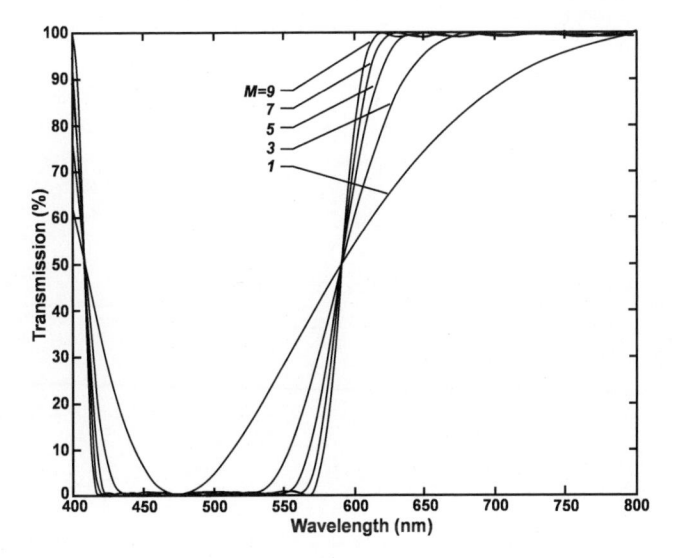

Figure 6.16 Red/cyan retarder from ColorLink between crossed polarizers (Reproduced by permission of © 2000 the Society for Information Display (Sharp and Birge 1999)).

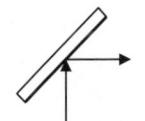

a. First surface mirror

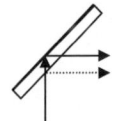

b. Second surface mirror

Figure 6.17 First and second surface mirrors. a) first surface mirror, b) second surface mirror.

Secondly, the beam passes through the tilted substrate, inducing coma in the beam. Both of these are less of a problem in the illumination path before the microdisplay than they are in the imaging path after the microdisplay.

Enhanced metallic mirror coatings are metallic coatings with a dielectric overlay of several layers to enhance the reflection and prevent tarnishing of the metal. The most common metals used in visible light are silver and aluminum. The reflectivity of enhanced aluminum and silver mirrors are shown Figures 6.18(a) and (b) respectively. Enhanced metallic coatings are always used as first surface mirrors. Protected metal mirrors have a transparent dielectric overlay, typically SiO_2, to prevent tarnishing of the metal. The spectral characteristics of a protected aluminum mirror are shown in Figure 6.18(c).

Metallic coatings without enhancement are sometimes used where cost is the primary concern. One example of this is in a rear projection consumer CRT set, where the final fold mirror can be quite large and comparable in size to the area of the projection screen. An aluminum coating is most commonly

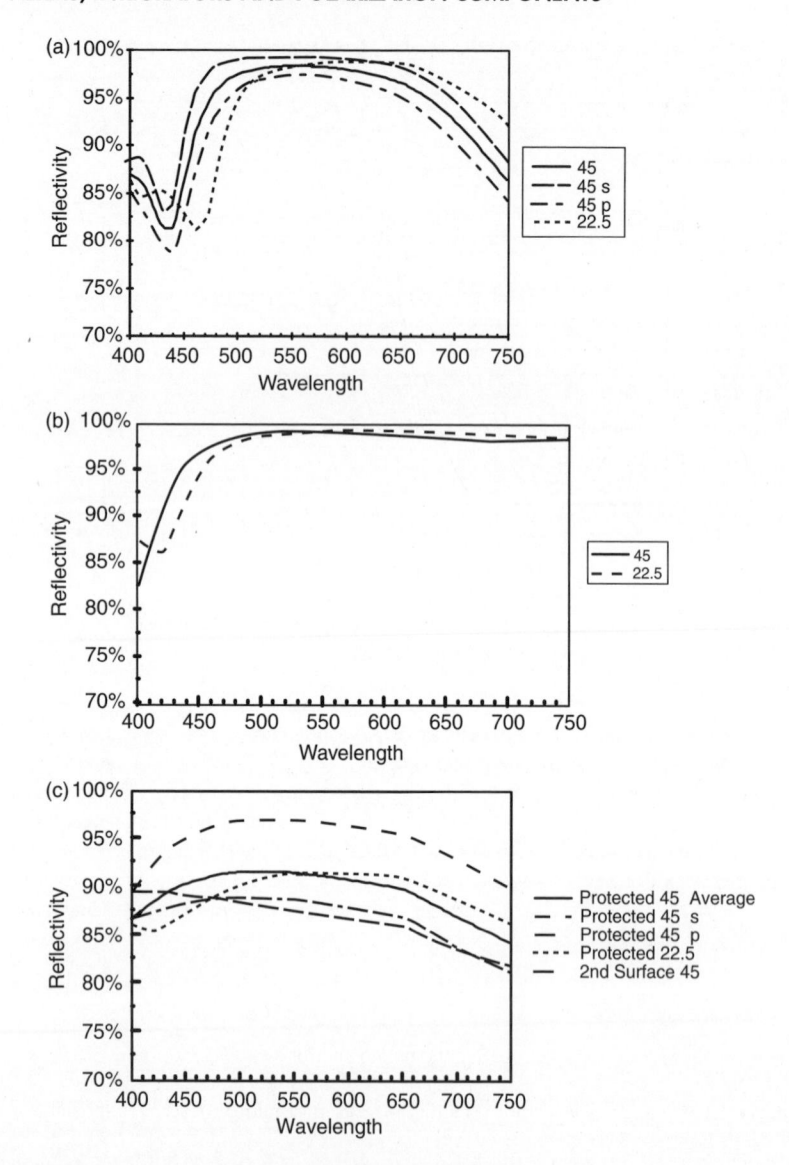

88 FILTERS, INTEGRATORS AND POLARIZATION COMPONENTS

Figure 6.18 Reflectivity of various mirror coatings. a) enhanced aluminum, b) enhanced silver, c) protected aluminum. (Data courtesy of JML Optical Industries, Inc.).

used in this type of mirror. In order to prevent the mirror from tarnishing, the aluminum is deposited on the second surface. The exposed side of the aluminum can be protected by an inexpensive lacquer while the side used as a mirror is protected from the air by the glass substrate. Since the fold mirror is far from the image plane and the resolutions of consumer CRT projectors are limited as much by the video signal as by any projector component, the resolution degradation from the ghost reflection and the coma due to the second surface mirror is not too significant. Second surface aluminum reflectance without enhancement is also shown in Figure 6.18(c).

The most efficient and most expensive mirror coatings are multi-layer dielectric coatings. Efficiencies in the range of 95–99 % across the visible range are readily achievable. These coatings would normally

SPECTRAL FILTERS

be applied as first surface coatings, although they could be designed as second surface if desired. The cold mirror coating shown in Figure 6.23(a) is a typical example of a dielectric mirror coating.

6.3.6 Total Internal Reflection

Total internal reflection (TIR) is another method to achieve a mirror effect. Snell's law, Equation (6.5), is symmetric with respect to the high and low index of refraction materials, and applies whether the incident beam is in the high index material or the low index material. When the light is within the high index material, there is an angle of incidence for which Snell's law predicts an angle of refraction of 90°. This angle is called the critical angle or θ_C and its value is given by:

$$\theta_C = \sin^{-1}\left(\frac{n_{low}}{n_{High}}\right). \tag{6.12}$$

For angles of incidence larger than this all of the light is reflected. For example, for $n_{high} = 1.51$ and $n_{low} = 1.00$, $\theta_C = 41.5°$. In non-absorbing media, the beam totally reflects for incidence angles greater than θ_C, the TIR angle.

A right angle TIR prism is shown in Figure 6.19. Light enters the prism at normal incidence through face #1, so the light is not refracted. When it encounters face #2 at 45°, the angle is larger than the critical angle, so the light is completely reflected. The light continues and exits the prism at face #3, again at normal incidence.

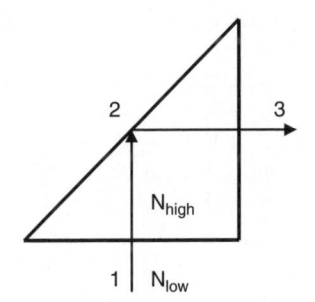

Figure 6.19 Right angle TIR prism.

For all optical materials with an index $n_{high} > 1.414$, the critical angle is less than 45°. This condition is true for most optical materials, so TIR prisms with a 90° bend can be made out of most optical glasses and plastics. Since the beam encounters both air/glass interfaces (surfaces #1 and #3 in Figure 6.19) at normal incidence, there is no coma induced in the beam.

6.3.6.1 TIR prisms for angular separation

The phenomena of total internal reflection can be used as an angular analyzer for microdisplays that separate the image and the waste light by angle, especially if that angle is relatively large. The DMD (Section 5.1) falls into this category, and TIR prism-based DMD projectors have become the standard architecture for these projectors (Florence and Yoder 1996).

This prism, illustrated in Figure 6.20, is used to separate the incoming and outgoing beams. The illumination beam encounters the air gap at angles larger than the critical angle. Therefore, all the light in the illumination beam reflects and illuminates the device. The reflected light exits at normal incidence to the DMD surface and enters the beam-separating prism, where it is incident on the air-gap at an angle

90   FILTERS, INTEGRATORS AND POLARIZATION COMPONENTS

below the critical angle. Therefore, the light transmits through the air gap. After passing through the second half of the prism, it reaches the projection lens and eventually the screen.

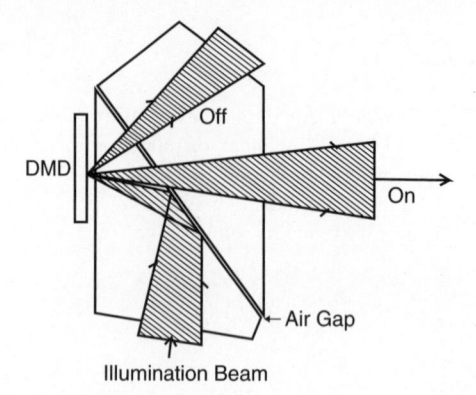

Figure 6.20   TIR prism for angular separation in a DMD system.

The DMD mirrors switch from either $-10°$ to $+10°$ or $-12°$ to $+12°$, depending on the design of the device (TI 2005), in going from the off-state to the on-state, producing a deviation of the off-beam direction of either 40° or 48° with respect to the on-beam. If the DMD mirrors are in the 'off' position, the light also passes through the air gap, exits the prism but its optical path does not intersect the entrance pupil of the lens. It eventually reaches a beam stop where it is absorbed to prevent it from reaching the projection lens as contrast-reducing stray light.

One problem with this TIR prism approach is the tilted air gap in the prism, which results in coma being induced in the image beam. Since the gap is relatively close to the image plane, and the DMD with its 10.3, 13.8 or 17 µm pixels is inherently a high spatial resolution device, this coma can significantly affect the image quality. It is necessary to design the projection lens carefully to correct for this.

Reflective LCD projection systems have also been designed based on TIR prisms, as shown in Figure 6.21 (Brandt 1991). A conventional reflective LCD system can use a polarizing prism to separate the illumination and image containing beams, as described below in Section 6.5.3.1. The polarizing

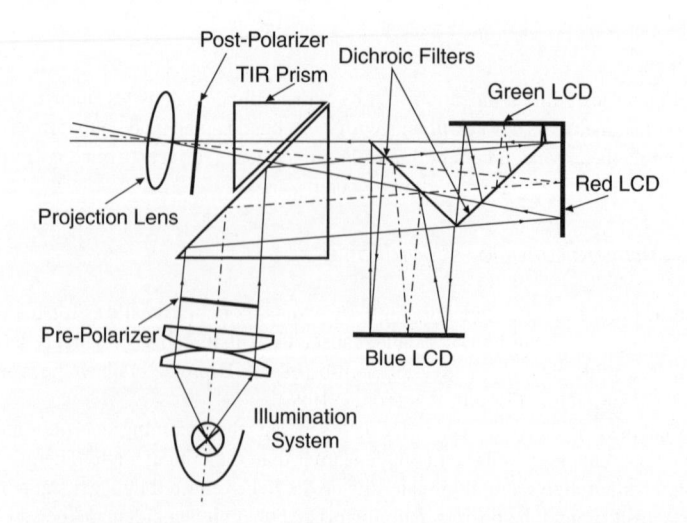

Figure 6.21   TIR prism for angular separation in a reflective LCD system.

prism also does at least part of the analysis job as well. When a TIR prism is used with a reflective LCD, the prism only separates the illumination beam from the reflected beam: it does nothing to separate the image from the waste light in the reflected beam. This must be done by separate polarizers.

6.3.7 Filters for UV Control

Arc lamps put out significant quantities of ultraviolet (UV) and infra-red (IR) light. In general, UV and violet light with a wavelength shorter than about 420 nm and IR light with a wavelength longer than 700 nm contribute nothing to the luminance or colorimetry of a projector. As can be seen in the spectra of common projection lamps, shown in Chapter 8, there can be significant energy in these wavelength bands.

While all arc lamps emit some UV, lamps containing mercury, including most metal-halide lamps, put out significant amounts of UV radiation. There is no significant UV flux from a tungsten-halogen lamp. Most optical glasses are opaque at wavelengths shorter than 365–380 nm, so the wavelength window where UV is a problem is fairly narrow, from about 365–420 nm.

UV and violet light has two effects on a system. First, the energy in the UV light can be absorbed and will heat the component where it is absorbed. A more serious problem is that many organic materials are significantly degraded by UV light. Organic materials in a typical projector include molded plastic lenses and other optical elements, optical cements, polarizers, compensation films, LC materials and LC alignment layers. In order to protect these components, it is necessary to block the UV before it reaches the first organic optical component. Short wavelength blue light in the 400–420 nm band contributes little to color or luminance but like UV it can cause lifetime problems and should also be blocked (Yakovenko et al. 2004).

One way to block the UV is with an absorbing glass filter, such as the L-42 long pass filter shown in Figure 6.13. Filters of this type are very effective at blocking UV without blocking a significant amount of visible light. In high power projectors, or projectors with very small optical beam diameters, the UV power density may be high enough to cause thermal cracking in an absorptive UV filter. In this case, a dichroic UV filter can be used, and the UV is reflected back toward the lamp. Most of this UV is eventually absorbed somewhere in the lamp/reflector system. Unfortunately, dichroic filters do not normally reject UV as completely as absorptive filters. In a high power density projector, it may be necessary to combine both dichroic and absorptive filters to prevent damage to both the UV filter and the other components in the projector. This arrangement is shown in Figure 6.22.

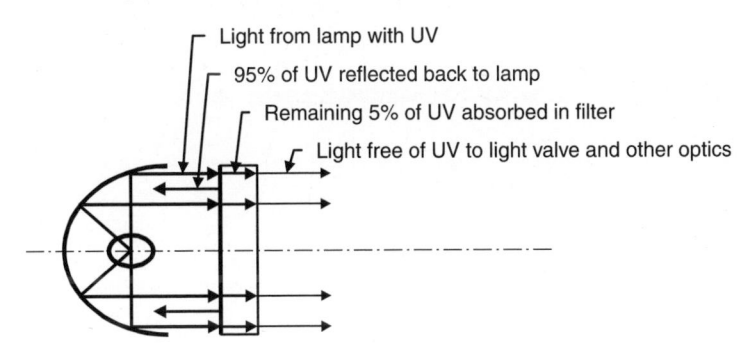

Figure 6.22 Composite UV filter.

6.3.8 Filters for IR Control

Incandescent lamps, including the tungsten-halogen lamps used in some projection systems, put out very large quantities of infra-red (IR) radiation. Arc lamps can also put out significant quantities of IR light, depending on the fill material. IR light is only a thermal problem for the projector designer since IR light

92 *FILTERS, INTEGRATORS AND POLARIZATION COMPONENTS*

does not degrade organic components the same way UV does. Components that are transparent at visible wavelengths may be opaque at IR wavelengths, absorb the IR and lead to thermal problems. Therefore, it is necessary to prevent the IR from entering the projection engine.

This is often done with a cold mirror or a hot mirror. A cold mirror coating is one that transmits IR and reflects visible light. The transmission of a typical cold mirror is shown in Figure 6.23(a). Cold mirrors are normally designed for a 45° angle of incidence as shown in this figure but could be designed for any angle of incidence needed. A hot mirror is a mirror that reflects infrared but allows visible light to be transmitted. Figure 6.23(b) shows a typical hot mirror. Hot mirrors can be designed for either normal incidence as shown in the solid curve in Figure 6.23(b) or for 45° angle of incidence, as shown by the dashed curves.

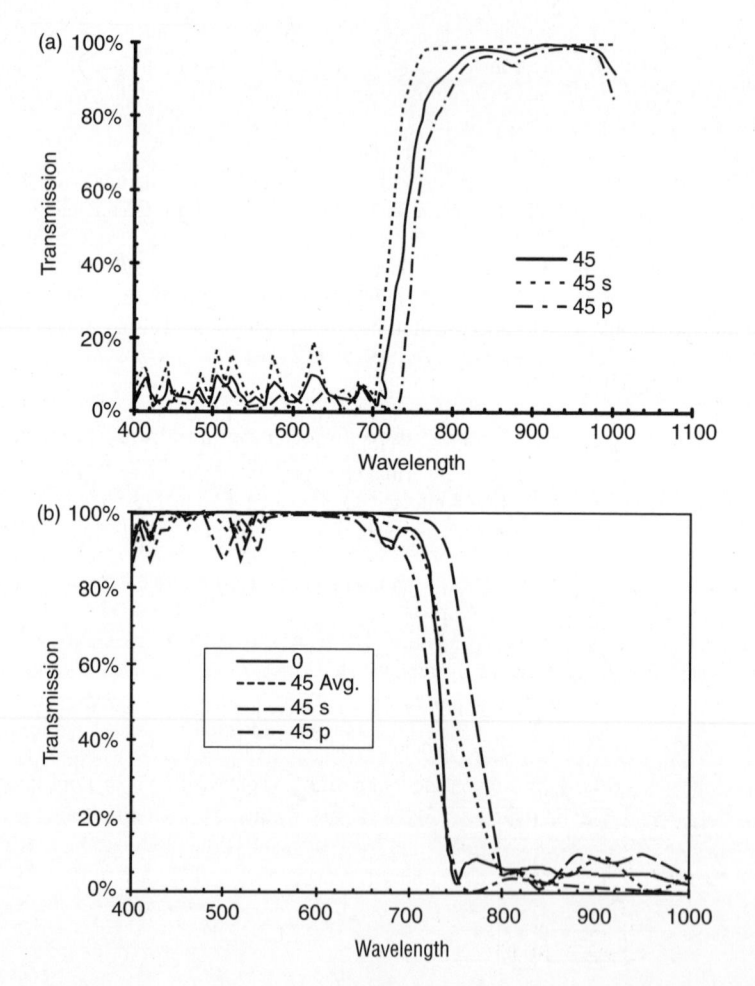

Figure 6.23 Hot and cold mirror coatings. a) typical cold mirror coating, b) typical hot mirror coating. (Data courtesy of JML Optical Industries, Inc.).

The first line of defense against IR light in a projector is often a cold mirror coating on the lamp reflector. Figure 6.24 shows how cold mirrors are used in a projection system. The arc emits both visible and IR light. When the light reaches the reflector, the visible light is reflected and concentrated in a beam while much of the IR light is transmitted through the coating and the glass reflector and absorbed in a heat sink.

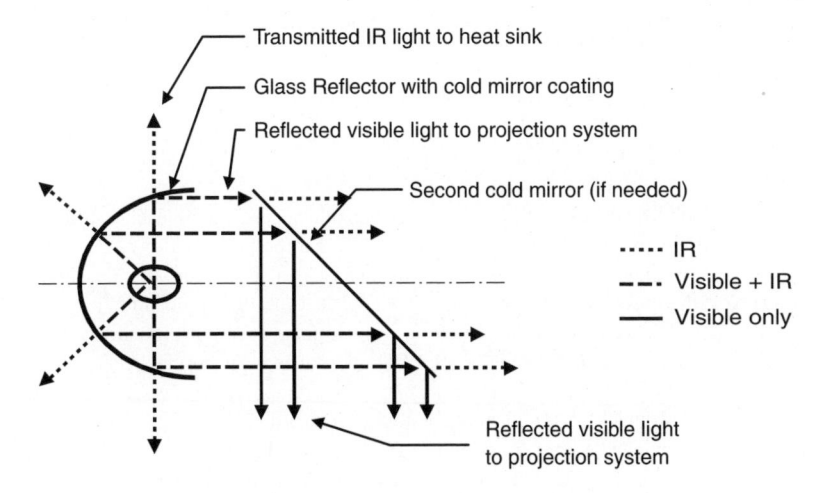

Figure 6.24 Cold mirrors in a projection system.

If there is still too much IR in the beam, a second cold mirror coating can be applied to a fold mirror immediately following the reflector. If there is a fold mirror anywhere in the illumination path, it is normally a cold mirror regardless of whether thermal management requires it or not. This can be done economically because a cold mirror coating is just as efficient at reflecting visible light and costs no more than an ordinary high efficiency mirror coating.

If there are thermal problems that require more IR protection than is given by the reflector coating alone, and the architecture of the projector does not allow folding the optical path after the reflector, a hot mirror at normal incidence can be inserted between the reflector and the next optical element. This hot mirror coating at normal incidence would reflect the IR back into the reflector. There much of the IR would be transmitted through the reflector coating and the reflector itself to be absorbed in the heat sink. Some of the IR, however, would be reflected off the coating and be refocused on the arc, affecting the thermal balance and possibly the life of the lamp.

While hot mirrors are most commonly used at normal incidence, sometimes it is necessary to use one at non-normal incidence The transmission of a 45° hot mirror is shown in Figure 6.23(b). Note that the 0° and 45° data in this figure represent different coatings: one optimized for each angle of incidence. If a hot mirror optimized at 0° were used at 45°, the cutoff wavelength would shift in accordance with Equation (6.11).

Absorptive IR filters are rarely used in projectors. First, there are few efficient IR absorbers suitable for projectors. As mentioned in Section 6.3.3, there are few efficient short pass absorptive materials of any type, as would be required for IR absorption and visible transmission. Secondly, an absorptive IR filter would heat up significantly, and would require cooling to prevent thermal stress and cracking.

Since water absorbs almost all IR energy at wavelengths longer than 1.0 μm, some very high powered projectors use water filters to absorb some of the IR. If heating of the water is excessive, convective or forced circulation of the water can be used to stabilize the temperature.

6.3.9 Indium-Tin Oxide and Other Transparent Electrodes

Indium-tin oxide (In_2O_3 : Sn), better known as ITO, is a transparent conductive coating. It is used in LCD microdisplays to provide the transparent electrodes to control the state of the LCD material. ITO is an alloy of indium oxide with tin oxide. Tin represents about 9–10 % of the final mix. The main losses in an ITO coating are the Fresnel reflections.

94 FILTERS, INTEGRATORS AND POLARIZATION COMPONENTS

The index of refraction of a conductive layer such as ITO is a complex number. One of the side effects of the conductive coating and consequential complex index of refraction is that an ITO coating will have some absorption of the optical energy. If the ITO will be on a glass/air interface, it can be AR coated with MgF_2, aluminum oxyfloride or indium-tin-oxyfloride (Hamberg and Granqvist 1986; Jiang et al. 1988). The best available AR coated ITO films have a photometric transmission of about 95 %. The 5 % loss represents about 1 % absorption and 4 % reflection.

If ITO is deposited on low-cost soda-lime glass and then annealed, Na will leach out of the glass and affect the optical and electronic properties of the ITO. This can be prevented by inserting a blocking layer of TiO_2-SiO_2 between the ITO and the glass. This layer can also serve as an index-matching layer between the ITO and the glass (Tsai 1997). If borosilicate glass is used as a substrate, there is much less sodium to leach into the ITO and there is no need for the blocking layer. While borosilicate glass is higher cost than soda-lime glass, its use in projection α-Si LCD microdisplays is normal because the cost of the glass is a very small part of the cost of the microdisplay. High-temperature poly-Si (HTPS) is typically fabricated on quartz substrates. Blocking layers are not needed for ITO on quartz.

Figure 6.25 shows the measured transmission of ITO on cover glass intended for the construction of LCD cells. This transmission curve was measured in an instrument similar to the one shown in Figure 6.6, with the ITO optically coupled with a liquid with an index of refraction of 1.52, to simulate the coupling effect of liquid crystal. In some cases it is cost effective to AR coat the ITO to minimize reflections at the ITO/LC and ITO/substrate interfaces.

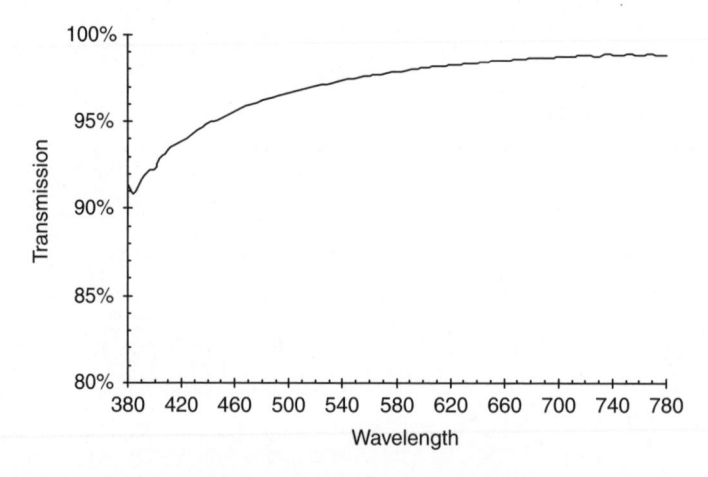

Figure 6.25 Indium-tin oxide transmission.

Indium is not a particularly common element and the demand for it is outstripping supply. Other materials that may some day be suitable for transparent electrodes in microdisplays include zinc-tin oxide, tantalum-tin oxide (Mihara et al. 2007) and transparent conductive organic films. Fluorine and antimony have also been suggested as tin oxide dopants to produce transparent electrodes.

6.4 Integrators

Integrators have two functions. The first is for producing uniform illumination of the microdisplay, and therefore, nearly uniform illumination of the projection screen (Michaloski 1994). The second function is to convert the circular cross-section of the lamp output beam to the rectangular format and size of the microdisplay.

Due to the effect of the projection lens and other optical elements in a projection system, even with perfectly uniform light at the microdisplay, there is normally a decrease in illumination from the center to the edge of a projection screen. Factors causing this fall off will be discussed in Sections 6.5.5 and 7.1.5.

Microdisplay projectors normally use critical illumination, as shown in Figure 6.26(a) (Inoué 1995). For critical illumination, the reflector or condenser lens images the arc or filament of the lamp onto the microdisplay. If no integrator is used, this leads to a very strongly peaked light distribution at the microdisplay and the projection screen, as shown in the solid curve of Figure 6.27.

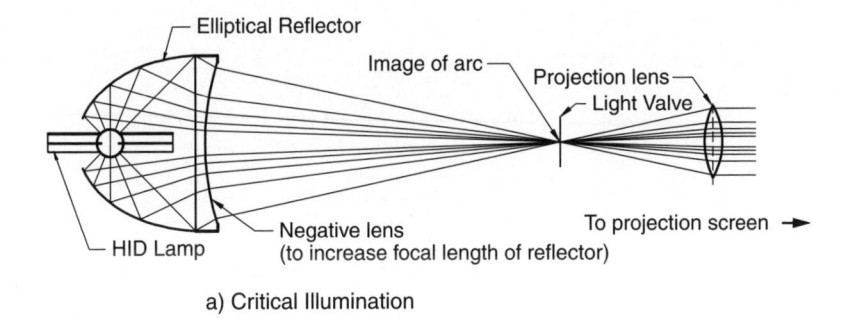

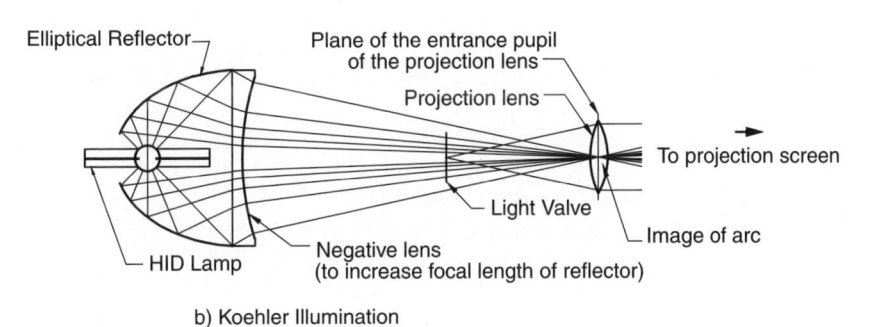

Figure 6.26 Critical and Köhler illumination: (a) critical illumination; (b) Köhler illumination.

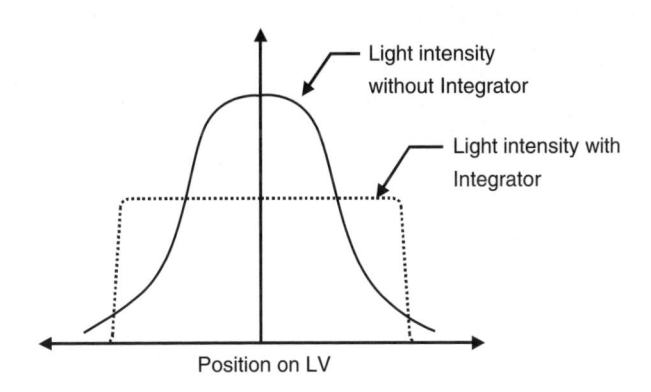

Figure 6.27 Light distribution with and without an integrator.

Koehler illumination, as shown in Figure 6.26(b), is when the light source is imaged onto the entrance pupil of the projection lens. This can achieve improved uniformity without an integrator, at a significant cost in efficiency (Kingslake 1983). Koehler illumination has a second problem: the illumination of the microdisplay is no longer telecentric. For an LCD microdisplay, this would mean the contrast would vary with position on the microdisplay. By using an integrator, uniform illumination across the microdisplay can be achieved with critical illumination, as shown in the second curve of Figure 6.27.

Integrators preserve the cone angle through the system, but spatially average over the area of the light body.[9] This means that light that is near the center of the distribution must be moved to an area near the edge of the distribution, without changing the cone angle of the light. Since both total area and angle are preserved through a properly designed integrator, étendue (see Chapter 11) is preserved. Since light that was originally near the center of the microdisplay is moved to a location nearer the edge of the microdisplay, the center brightness at the screen always decreases when an integrator is used. Due to the extra optical surfaces in the integrator, the total system lumen throughput also normally decreases slightly as well. The effect of an integrator on lumen throughput is discussed in more detail in Chapters 11 and 12.

While an integrator will normally decrease lumen throughput, it can in some cases increase the throughput (Chang et al. 1996). This occurs primarily in systems where Koehler illumination had originally been used to get satisfactory uniformity. If the integrator allows the Koehler illumination to be replaced by critical illumination, it is possible to get a net gain in throughput.

There are two main types of integrators in use in projection systems: lenslet integrators (Section 6.4.1) and rod integrators (Section 6.4.2). Due to their high efficiency, relatively compact design and ability to work well with polarization conversion systems, lenslet integrators are common in production LCD and LCoS projectors. DLP and other projectors not concerned with polarization are more commonly designed with rod integrators.

Since the tooling cost for a lenslet integrator/PCS combination is quite high and a custom design is normally required for each projector design, lenslet integrators are often not used in experimental projector designs. Custom lenslet integrators are required because the integrator must be matched to the lamp, microdisplay and projection lens in the projector. Due to their low single-piece tooling costs and greater versatility, rod type integrators are often used in these types of research projectors.

6.4.1 Lenslet Integrators

A lenslet integrator (Wang and Ronchi 1988; Brandt and Timmers 1992; Itoh et al. 1997), also called a 'fly's eye' lens, is illustrated in Figure 6.28. An integrator for an actual projector would commonly have more lenslets than are shown. A lenslet integrator consists of two integrator plates, which may or may not be identical to each other. Integrators with two different plates can be significantly more efficient than integrators with two identical plates, but due to tooling costs and design and alignment problems, integrators with two identical plates were initially more common in production. With the introduction of smaller microdisplays and projectors, it has become common to make the second integrator plate smaller than the first to reduce the overall size of the system. The example array has a 3 × 4 array of lenslets where each element has a 4:3 aspect ratio. This gives a square overall shape to the array. The aspect ratio of the lenslets in the first integrator plate must always match the aspect ratio of the microdisplay in the system.

---

[9] The light body is the area in the plane of the light valve illuminated by the lamp. It is normally slightly larger than the usable area of the light valve to minimize alignment problems and edge effects.

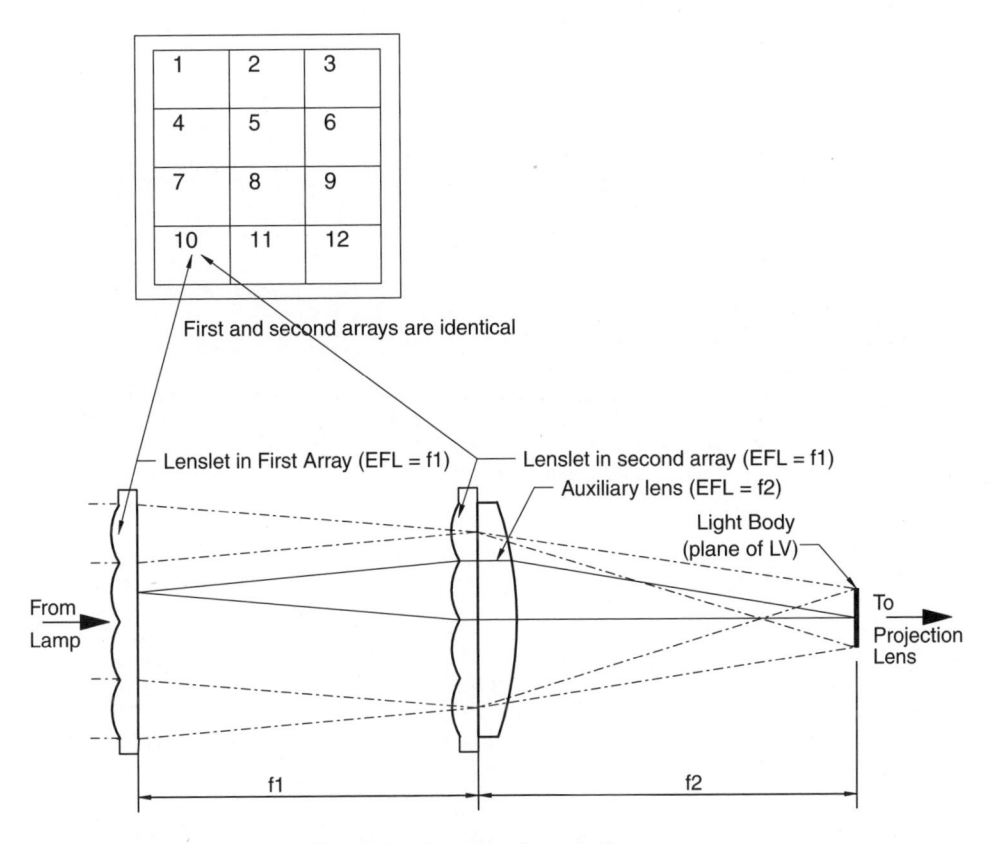

Figure 6.28 Operation of a lenslet integrator.

The most common form of lenslet integrator takes collimated light as input, such as light produced by a parabolic reflector. Each of the 12 lenslets in the first array images the source onto the center of the corresponding lenslet in the second array. This forms an array of 12 virtual sources. Each lenslet in the second array acts as a field lens and, in conjunction with the auxiliary lens, images the corresponding lenslet in the first array on the microdisplay. Therefore, the light distribution at the light body is the sum of the light distributions at all of the lenslets in the first array. For this system to work correctly, the focal length of the lenslets in each array must be equal to each other and the separation of the two lenslet arrays must be equal to that focal length.

Figure 6.29 shows how two lenslets work together to modify the light distribution. With a normal light distribution in critical illumination, the light decreases monotonically from the center of the overall distribution to the edge, as shown by the luminance contour lines in Figure 6.29(a). For lenslet #1, which is in the corner of the integrator as shown in Figure 6.29(b), this means the lower right corner is brighter than the upper left corner. For lenslet #12 from the diagonal corner, as shown in Figure 6.29(c), the converse is true: the lower right corner is dimmer than the upper left corner. When the light distributions from these two lenslets are added together in the light body, the asymmetries largely cancel each other and the sum is much more uniform than either lenslet alone. This is shown in Figure 6.29(d). When the light from a large number of lenslets is added together, the asymmetries in diagonal pairs of lenslets tend to cancel and the distribution becomes very uniform.

Lenslet integrators for production projectors are always molded, although sometimes integrators for experimental projectors are assembled from small lenses sawn into rectangular shapes. While either glass or plastic could be used for the lenslet integrator, glass is normally required for thermal and UV reasons.

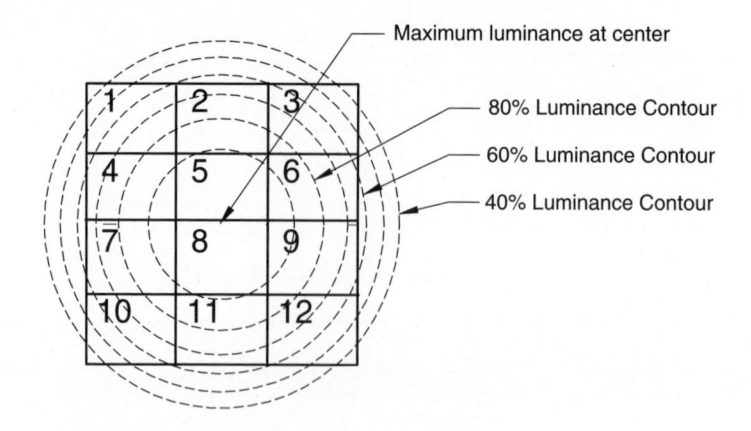

98 FILTERS, INTEGRATORS AND POLARIZATION COMPONENTS

a) Luminance distribution on the first integrator plate

b) Distribution from Lenslet 1 alone

c) Distribution from Lenslet 12 alone

d) Distribution when light from Lenslets 1 and 12 are added together

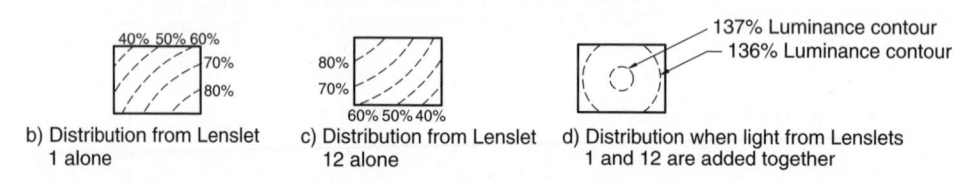

Figure 6.29 Summing of two lenslets in a lenslet integrator: (a) luminance distribution on the first integrator plate, (b) luminance distribution from lenslet 1 alone; (c) Luminance distribution from lenslet 12 alone; (d) luminance distribution when the light from lenslets 1 and 12 are added together.

This is necessary because the integrator is typically the first element in the optical path after the lamp and reflector.

Typical lenslet integrators in projectors have between 25 and 80 lenslets in each array. If there are too few lenslets in each array, the distribution may not be sufficiently uniform. If there are too many lenslets in each array, the efficiency will be reduced and the alignment of the two arrays becomes more critical. Figure 6.30 shows why the efficiency decreases with an increasing number of lenslets. In the ideal cases (Figures 6.30(a,b)), there is no transition region between two adjacent lenslets. The simple integrator design shown in Figure 6.28 would have no step between adjacent lenslets. In this case, all light falls on either one lenslet or the other, and is imaged correctly onto the second array or the microdisplay. In the real case, there is a transition region between two lenslets. In the case where there is no step between lenslets, as shown in Figure 6.30(c), this transition region is typically 0.2–0.3 mm.

a) Centered Lenslets Ideal Case

b) Offset Lenslets Ideal Case

c) Centered Lenslets Realistic Case

d) Offset Lenslets Realistic Case

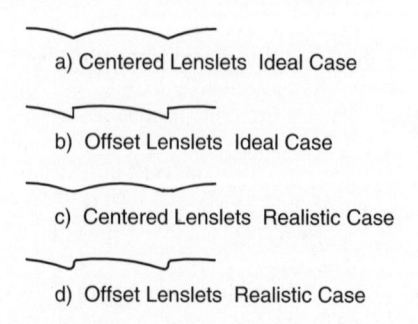

Figure 6.30 Junction between two lenslets: (a) centered lenslets, ideal case; (b) offset lenslets, ideal case; (c) centered lenslets, realistic case; (d) offset lenslets, realistic case.

When there is a step between lenslets, as shown in Figure 6.30(d), the situation is even worse. More complex integrator designs, including the one in Figure 6.31 discussed below, require a step in the glass between adjacent lenslets. Here the transition region is at least 0.3–0.4 mm, but the vertical step must also have a 2–4° draft angle to allow the part to release properly from the mold. These transition regions represent dead space where the light is not properly imaged onto the microdisplay. Therefore, the more lenslets, the more dead space and the less light imaged on the microdisplay.

a) First Integrator Plate

Images of the Arc formed by the Reflector and the First Integrator Plate

b) Second Integrator Plate

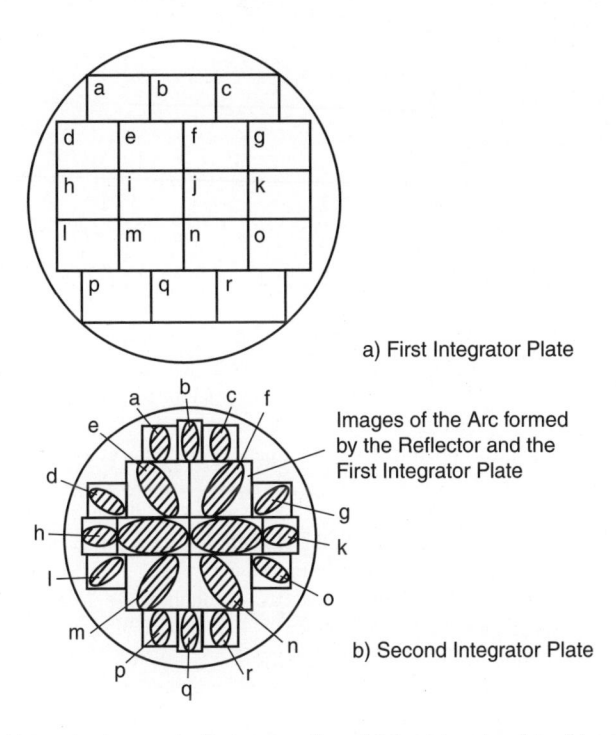

Figure 6.31 Lenslet integrator to correct reflector aberrations: (a) first integrator plate; (b) second integrator plate.

Lenslet integrators also affect projector lumen throughput by changing the collected lumens vs étendue curve. Discussion of this will be deferred to Section 11.3, after the necessary concepts have been discussed.

Figure 6.28 shows the lenslet arrays on a plane surface and operating in collimated light from a parabolic reflector. It is also possible to design an integrator where the lenslets are on spherical surfaces and operating in converging light from an elliptical reflector. This configuration is shown in Figure 6.32. In this design, the focal point of the ellipse with no integrator plates present is at the center of the microdisplay. Both lenslet arrays are on spherical surfaces that are concentric with the center of the microdisplay. Each lenslet in the first array images the arc on the center of the corresponding lenslet in the second array. The lenslet in the second array imaged the lenslet in the first array onto the light body.

One advantage of this configuration is that no auxiliary focusing lens is required, simplifying the optical design and reducing the part count. This type of integrator was used in the Talaria oil film light-valve projector, now out of production. In the Talaria, the electron gun that addressed the light valve had to be on the optical axis. The integrator array was designed so the electron gun was in the shadow of the lamp electrodes and ferrules. Not only did this mean there was no light loss associated with the electron gun on the center line, but the gun was not heated up by the illumination beam.

One disadvantage of the configuration in Figure 6.32 is that the light is not telecentric at the light body. This did not affect the Talaria but it would affect the performance of a LCD microdisplay. If a

100 FILTERS, INTEGRATORS AND POLARIZATION COMPONENTS

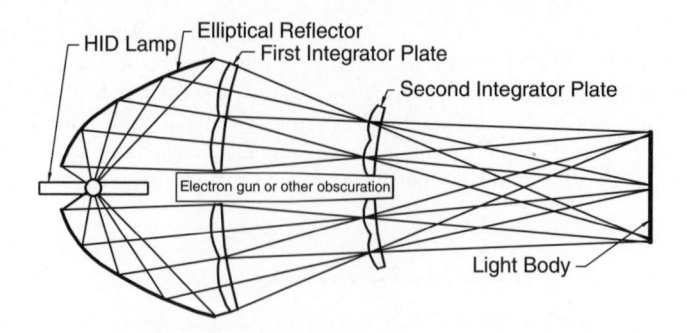

Figure 6.32 Lenslet integrator in converging light.

negative lens is added after the second integrator array, the illumination path can be designed to produce telecentric illumination. This combination of an elliptical reflector, two unequal integrator plates and a negative lens is common in LCD projectors.

Lenslet integrators with more elaborate designs than the rectilinear patterns shown in Figures 6.28 and 6.32 and have been reported in the patent literature. One such design (Masumoto 1995) is shown in Figure 6.31. Note that the lenslets in the first integrator plate are rectangles with the same aspect ratio as the microdisplay. Each lenslet in the second plate is also a rectangle, but is sized and shaped to match the image of the arc more closely as formed by the corresponding lenslet in the first integrator plate.

Note that for these plates to work together properly, it is necessary for each lenslet in the first array to be decentered so it correctly images the arc on the element in the second array. It is this decentering that causes the steps in the glass contour shown in Figure 6.30. The lenslets in the second array must be correspondingly decentered to bring the principal rays through each lenslet back into collimation. Still more elaborate designs are possible, where the lenslets in the second array are no longer rectangles (Brandt and Timmers 1992; Shimizu and Janssen 1997).

The main goal of this type of design is to improve the collection efficiency by correcting some of the aberrations inherent in parabolic or elliptical reflectors. Due to the difficulties associated in designing, tooling and aligning these types of integrators, they are not commonly used in production projectors.

6.4.2 Rod Integrators

Rod integrators (Michaloski and Tompkins 1994), also called tunnel integrators, are another class of optical device intended to produce a uniform light body at the microdisplay. A rod integrator is a rectangular element with the same aspect ratio as the microdisplay to be illuminated. The walls of the rod are highly reflecting. The rod may be hollow (air filled) and depend on high quality first surface mirrors for the reflection, or it may be solid and depend on total internal reflection. Therefore, as light passes down the length of the rod it is reflected multiple times off the walls of rod with very little loss.

The operation of a rod integrator is shown in Figure 6.33. Light from the lamp/reflector combination is focused on the entrance end of the rod integrator. Off-axis light will hit a wall and reflect back into the rod. Since the walls are parallel, the cone angle of the light is preserved at each reflection along the length of the rod. Since the area of the input and output ends are the same and cone angle is preserved, the rod integrator preserves the étendue of the light that passes through it. If the rod is long enough and there are enough reflections, the position of each ray at the output face is uncorrelated with the position of the ray at the input face. This means the light at the output will be spatially uniform across the face of the rod.

INTEGRATORS

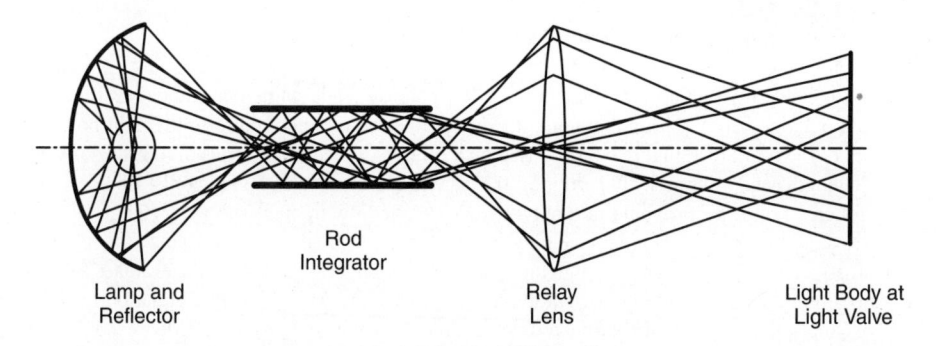

Figure 6.33 Rod integrator in an optical system.

The uniformity of light as it passes through the rod integrator increases with length. This effect is shown in Figure 6.34. In this figure, uniformity is defined by:

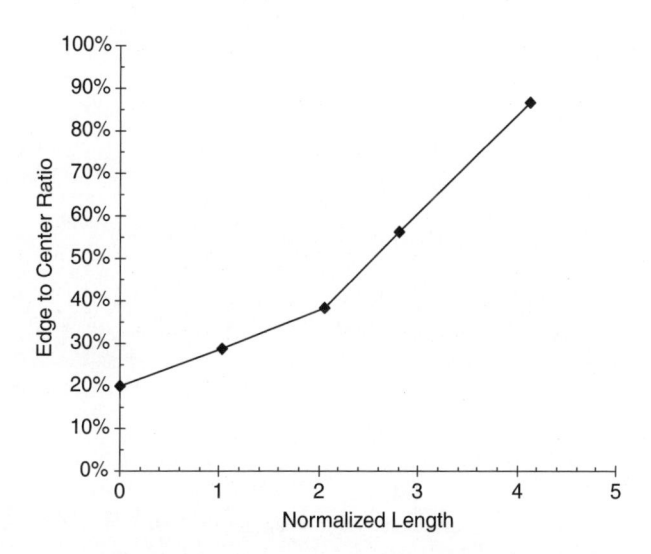

Figure 6.34 Uniformity with a rod integrator.

$$U = \frac{\text{Edge Illumination}}{\text{Center Illumination}} \tag{6.13}$$

and normalized length is defined by:

$$L_n = \frac{L \cdot NA}{n_1 \sqrt{A}}, \tag{6.14}$$

where:
- L_n is the normalized length,
- L is the geometrical length of the rod,
- NA is the numerical aperture of the light entering the rod ($\sin(\theta_{1/2})$),
- n_1 is the index of refraction of the material filling the rod and
- A is the cross sectional area of the rod.

102 FILTERS, INTEGRATORS AND POLARIZATION COMPONENTS

For hollow rods using first surface mirrors, $n_1 = 1.00$. For solid rods that depend on TIR, n_1 is the index of refraction of the glass or plastic used in the rod. The normalized length corresponds to the number of reflections made by a light ray at the periphery of the cone angle while that ray passes through the rod. Figure 6.34 indicates, that up to a point, the longer the rod the more uniform the light body at the output end of the rod. Since Figure 6.34 is based on a limited data set for a series of 4:3 rods of varying geometric lengths and a numerical aperture of 0.17, it should be used as a guide only. The design of a rod integrator for a projector would normally include detailed ray tracing with a very large number of test rays to assure uniformity at the image plane (Park *et al.* 2004).

The cross-sectional area of a rod integrator is generally much smaller than the area of the light body to be illuminated, and the numerical aperture is higher. As can be seen in Equation (6.14), this minimizes the geometric length of the rod for a given normalized length and degree of uniformity. Therefore, a relay lens must be used to image and magnify the output end of the rod onto the light body.

In a rod integrator the shape of the rod defines the area but there is normally nothing to define the angle. Therefore, a rod integrator can be used over a wide range of étendue without affecting the system throughput, as long as the microdisplay has the same aspect ratio as the rod and the maximum input acceptance angle of the rod is not exceeded. The focal length of the relay lens, as shown in Figure 6.33, will need to be changed if a microdisplay with a different area is used. The numerical aperture of the relay lens also must be at least as large as the NA of the illumination system and the microdisplay, in order to prevent the relay lens itself from becoming the étendue limiting component.

Rod integrators are the integrator design of choice for DLP and other polarization-independent microdisplay technologies. They are rarely used with polarization-dependent microdisplays such as LCD and LCoS designs. While attempts have been made to design polarization conversion systems for integrating rods, none has been fully satisfactory. If the polarization conversion system is after the rod integrator, the uniformity of the light may be lost. If the polarization conversion is before the rod, skew rays that reflect off two orthogonal sides of the rod may have their polarization rotated, and some of the benefit of polarization conversion is lost. Polarization conversion systems will be discussed in more detail in Section 6.5.3.

6.4.3 Integrators for Projectors with Laser or LED Illumination

LEDs normally have a relatively high étendue, comparable to the étendue of a lamp. In addition, they do not produce polarized light. Therefore, they can use the same integrator types used by lamps, including the polarization conversion system. LEDs can also use special collection systems that combine the functions of light collection and light integration. These systems will be discussed in Section 8.9.5.

Laser projectors based on microdisplays need an integrator to spread the light in the laser beam uniformly over the microdisplay area. They can use either lenslet integrators or rod integrators, although the use of a rod integrator would be more common. Typically, lasers are polarized and no polarization conversion is needed.

Special integrator types are beginning to emerge that are suitable for laser illumination but would not typically be suitable for lamp or LED illumination. For example, Kappel *et al.* (2005) describes an integrator based on a beam expander, a holographic diffuser and a pair of field lenses, as shown in Figure 6.35. Note that this integrator, unlike the lenslet or rod integrators, does not preserve the étendue of the laser beam. The small diameter, low divergence input beam would typically have an étendue on the order of 10^{-4} mm^2 steradians. The output of this integrator would have a much higher value, depending on the output area and output cone angle. By tailoring the diffuser properties, the output étendue may be tailored to match the microdisplay to be used.

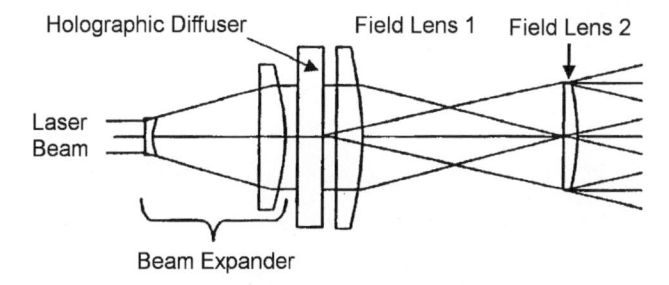

Figure 6.35 Integrator for use with a laser light source (Kappel 2005; United States Patent and Trademark Office).

6.4.4 Other Integrator Types

The fiber-optic integrator is a variation on the rod integrator and is shown in Figure 6.36 (Cheng and Chen 1995). In a normal fiber-optic bundle, the position of a fiber at one end of the bundle is uncorrelated with the position at the other end of the bundle. This spatially averages the light, at least on a scale larger than individual fibers. The individual fibers preserve cone angles through the fiber. Since the area (# of fibers) and f/# are preserved, étendue is preserved. Like a rod integrator, the output end of a fiber-optic integrator is normally imaged with demagnification onto the microdisplay. In order to avoid graininess, the image would be slightly defocused.

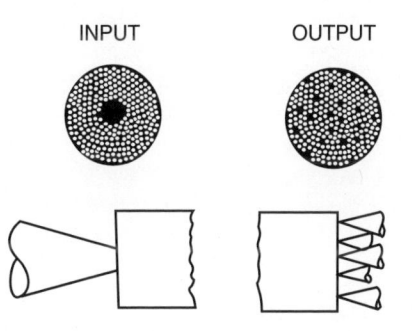

Figure 6.36 Fiber-optic integrator (From Cheng 1995; with permission of SPIE).

One advantage of a fiber-optic integrator is the ease with which the fiber bundle can be reshaped to change the aspect ratio of the light beam. Normally, the input end of the bundle would be round, to match the spot of light from the lamp. The output end can be formed into the appropriate rectangle to match the display. Another advantage of a fiber-optic integrator is that the light can be carried into an inaccessible location from a remote lighthouse. This may simplify the design of a complex optical illumination path.

The fiber-optic integrator is not used in production projectors for two reasons. First, the dead space between fibers gives the system a relatively low efficiency. It can also cause a 'grainy' light distribution at the output if there are not enough fibers in the bundle. Secondly, a fiber-optic integrator is significantly more expensive in volume production than either a rod integrator or a lenslet integrator.

Another optical element that performs some of the functions of an integrator is an Axicon or cone prism, as shown in Figure 6.37 (Ooi *et al.* 1995). The cone prism is not a true integrator in that there is no spatial averaging of the light. The cone prism operates by rearranging angles to give a more uniform lumen vs angle distribution. For example, light ray A in Figure 6.37 just misses the lamp ferrule. Without a cone prism, it would continue along the dotted line, and strike the panel at position A'. Similarly, ray B

104 FILTERS, INTEGRATORS AND POLARIZATION COMPONENTS

would strike at B'. This would leave a dark region in the center of the display in the area shadowed by the lamp. The cone prism, however, refracts these rays so they strike the center of the panel. The contour of the prism can be designed to give uniform illumination at the panel. This is shown in Figure 6.37 by the equal spacing of the rays across the panel.

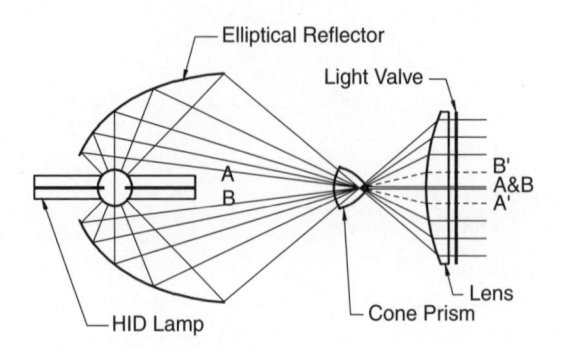

Figure 6.37 Cone-prism integrator.

The cone prism has a number of advantages: it is compact, low-cost and efficient. Unfortunately, the cone prism approach has a major drawback preventing its widespread use. Since true averaging does not take place, the prism must be designed for the light distribution given by the nominal lamp. If an individual lamp varies from the nominal, the distribution at the screen will be different than desired. This problem is especially severe in the center of the screen. If the lamp location in the reflector is off slightly, rays A and B will either overlap at the center of the screen, leading to a bright spot, or not quite converge, leading to a dark spot.

6.4.5 Light Guides

Light guides are closely related to rod integrators except their main function is to transport light from the source to a destination. A light guide system can replace a series of relay lenses, field lenses, mirrors and filters. This can both simplify the system and reduce its size. Often the functions of light transport and integration are combined.

Since light guides depend on total internal reflection for light transport, they can be very efficient. Light guides can be made from injection molded plastic, made from pressed or molded glass or made from ground and polished glass. Even ground and polished glass light guides are relatively low cost since all surfaces are flat and no exotic glasses are needed. In addition, only the relatively small input and output faces of a light guide need AR coatings.

One problem with using light guides for light transport is turning corners. At a turn in a conventional light guide, light can be lost, the étendue can be increased and uniformity can be affected (Park *et al.* 2004; Janssen 2006). One solution to this problem is to insert a low-index layer between the light guide and the turning prism at the bend in the light guide, as shown in Figure 6.38. This low-index layer can either be air or a low index cement to hold the light guide to the turning prism. This low-index layer ensures that the light would normally be lost will make it around the turn and continue to the next optical element. Since this loss occurs preferentially at one side of the fold prism, it will affect light uniformity unless the light guide after the fold is long enough to re-integrate the light.

An optical illumination path that uses light guides to replace the conventional integrator, PCS, relay lenses, dichroic filters and fold mirrors is described by Shimizu (2005).

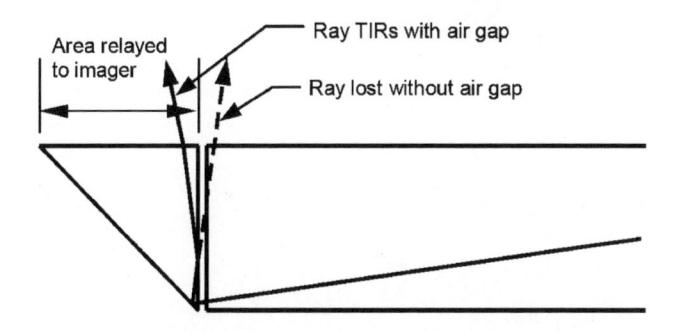

Figure 6.38 Light-loss at a light-guide fold prism (reproduced by permission of © 2004 the Society for Information Dispaly (Park *et al.* 2004)).

6.5 Polarization Components

Most liquid crystal projection microdisplays require polarized light in order to operate. In general, this section does not apply to projectors that use diffractive or micro-mechanical microdisplays, or light valves which use polarization-independent liquid crystal effects. For designers of these types of projectors, the main issue related to polarization is the effect of tilted dichroic filters on polarization and throughput, as covered in Section 6.3.2.1. An excellent but dated general reference on polarizers is Bennett and Bennett (1978). The physics of polarized light is well covered in Born and Wolf (1980). The issues of polarization in LCD projection systems is are covered in considerable detail in Robinson *et al.* (2005).

Absorptive polarizers, as will be discussed in the next section, are used in the vast majority of commercially available LCD projection systems. Even in systems where the majority of the polarization is done with a reflective polarizer, an auxiliary absorptive polarizer is normally required in order to obtain adequate system contrast ratio. Absorption polarizers are also called dichroic polarizers. This use of the word 'dichroic' should not be confused with its use in 'dichroic filters', discussed in Section 6.3.2.

Reflective polarizers have two main applications in LCD microdisplay projection systems:

1. Separation of incoming and reflected light in reflective LCD systems and
2. Polarization conversion.

There are several ways to quantify the properties of a polarizer in the literature, but this book will use what is the simplest and most intuitive. Two spectral curves are necessary to fully quantify a polarizer to be used with visible light: $T_{\parallel}(\lambda)$, the transmission of the polarizer when the polarization vector of the light is parallel to the axis of the polarizer, and $T_{\perp}(\lambda)$, the transmission of the polarizer when the polarization vector is perpendicular to the axis of the polarizer.

When examining published $T_{\parallel}(\lambda)$ and $T_{\perp}(\lambda)$ curves for a polarizer, one must determine if the curves are normalized relative to unpolarized light or relative to light of a single polarization. If the value of $T_{\parallel}(\lambda)$ is in the range of 80–90 %, the normalization was done relative to polarized light. If the value is in the range of 40–45 %, the normalization most likely was done relative to unpolarized light. All the polarization transmission curves presented in this book are normalized with respect to light of a single polarization unless noted otherwise.

The extinction ratio is a measure of the efficacy of a polarizer in generating one polarization and it is defined by:

$$\text{Extinction Ratio} = \frac{T_{\parallel}(\lambda)}{T_{\perp}(\lambda)}. \tag{6.15}$$

FILTERS, INTEGRATORS AND POLARIZATION COMPONENTS

For a 'perfect polarizer $T_{\parallel}(\lambda) = 100\%$, $T_{\perp}(\lambda) = 0\%$ and the extinction ratio is ∞. Another way to express the extinction in a polarizer is by the polarization efficiency $\varepsilon(\lambda)$, defined by:

$$\varepsilon(\lambda) = \frac{T_{\parallel}(\lambda) - T_{\perp}(\lambda)}{T_{\parallel}(\lambda) + T_{\perp}(\lambda)}. \tag{6.16}$$

Realistic polarizers can have extinction ratios $>10^4$, or polarization efficiencies above 99.9 %. As will be discussed in Section 6.5.3.1, the extinction ratio cannot be used as the only factor in judging a polarizer for inclusion in a projector. Extinction ratio will influence system contrast but other factors, such as transmission of the required polarization state, must be taken into consideration when calculating efficiency and throughput. Polarizer lifetime is also a major factor in the choice of a polarizer for a system.

6.5.1 Absorptive Polarizers

Absorptive polarizers operate by absorbing one polarization and allowing the other polarization to transmit largely unattenuated. No absorptive polarizer is perfect, but extinction ratios of 1000:1 are readily attained.

The basic manufacturing of all absorbing polarizers is similar. First, a thin film of a polymeric material is made. Polyvinyl alcohol (PVA) is normally used for this purpose. Next this film is stretched in one direction. This process tends to line up the long chain polymer molecules along the axis of the stretch. Next, the material is modified to absorb light polarized parallel to the molecules and transmit light polarized perpendicular to the molecules.

In the most common type of absorptive polarizer iodine is added to the PVA film. Iodine polarizers are also called type H polarizers, after the original designation given to them by Edwin Land of the Polaroid Corporation. The iodine forms polymer chains that align themselves with the PVA chains and provide the absorption required. Iodine polarizers are the best absorptive polarizers in terms of extinction ratio. The typical transmission curve of this type of polarizer is given in Figure 6.39.

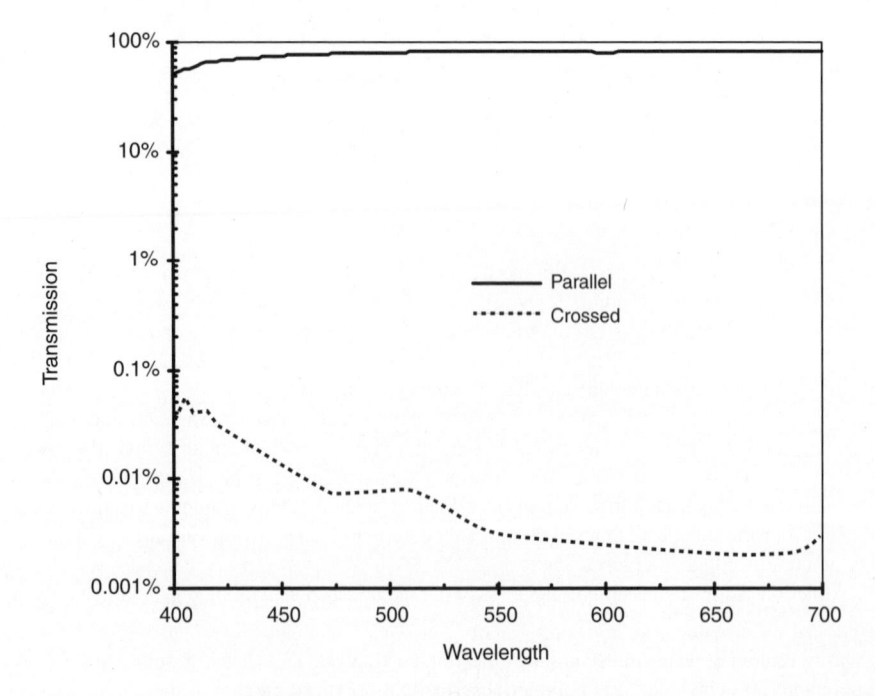

Figure 6.39 Transmission of an iodine polarizer.

Unfortunately, the durability of iodine polarizers with respect to heat and humidity is relatively poor. High ambient light level, high temperature and high humidity tend to bleach the iodine out of the PVA matrix. This occurs because iodine has a relatively high vapor pressure and can sublimate at a significant rate at the temperatures encountered in projectors. As the iodine is removed from the film, the iodine chains become shorter and less effective in absorbing the unwanted light.

In dye-type polarizers an absorptive dye is added in place of iodine. The dye molecules used are long chain dyes that exhibit a large difference in absorption for light polarized parallel and perpendicular to the chain. The dyes used do not normally absorb across the entire visible range, so two or three different dyes are used in combination. For some applications, only a single dye is added so only one wavelength band is absorbed. A polarizer of this type will polarize the color light absorbed by the dye, but leave the other colors of light unpolarized. The transmission of a dye polarizer is illustrated in Figure 6.40.

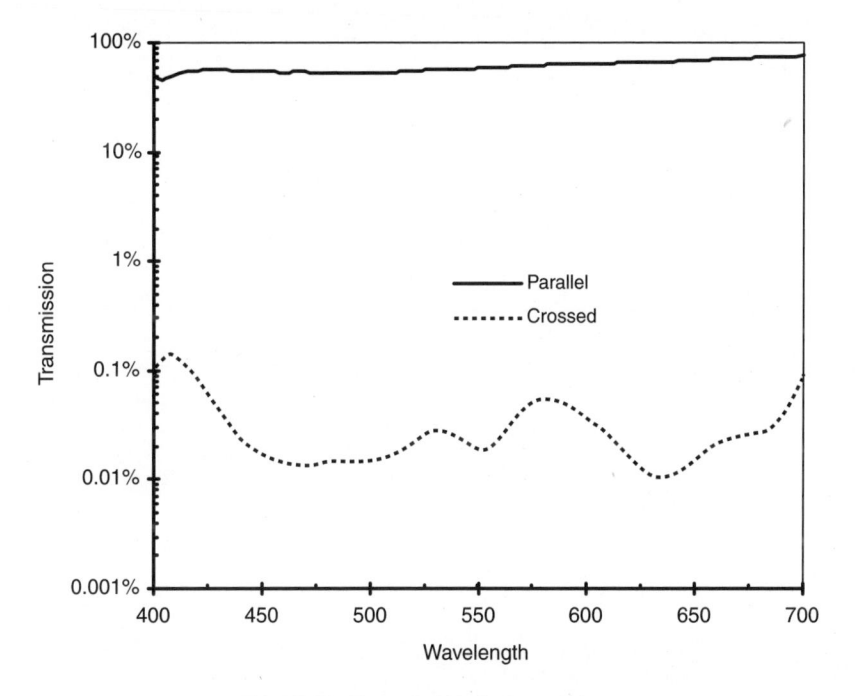

Figure 6.40 Transmission of a dye polarizer.

In general, dye type polarizers are more durable than iodine type polarizers, but have lower extinction ratios (Mohri and Matsuo 1995). While modern dye type polarizers have improved, so have iodine polarizers. In general, it remains true that iodine polarizers are more efficient than dye type polarizers in two ways: higher transmission of the desired polarization and lower transmission of the undesired polarization. The durability of dye polarizers may make them appropriate for microdisplay projectors with very high flux densities at the polarizer.

A third type of polarizer is commonly called a Type K polarizer, after the original 1942 Polaroid Corporation designation. The high-efficiency version of the polarizer was developed in the 1990s and is designated KE by Polaroid's successor, 3M (Ralli *et al.* 2007). In this polarizer the PVA film is formed and stretched in the same way as an iodine polarizer. Then the chemical structure of the PVA film is modified by removing water molecules with heat and a catalyst. This converts some of the PVA into dichromophore polyvinylene. This polymer absorbs light polarized parallel to the polymer chain and transmits light polarized perpendicular to the chain. Thus, the film itself becomes the absorbing medium.

108   FILTERS, INTEGRATORS AND POLARIZATION COMPONENTS

Polarizers of this type are very durable, but they typically have lower polarization efficiency than other absorptive polarizers (Cael et al. 1997). Transmission of this type of polarizer is shown in Figure 6.41.

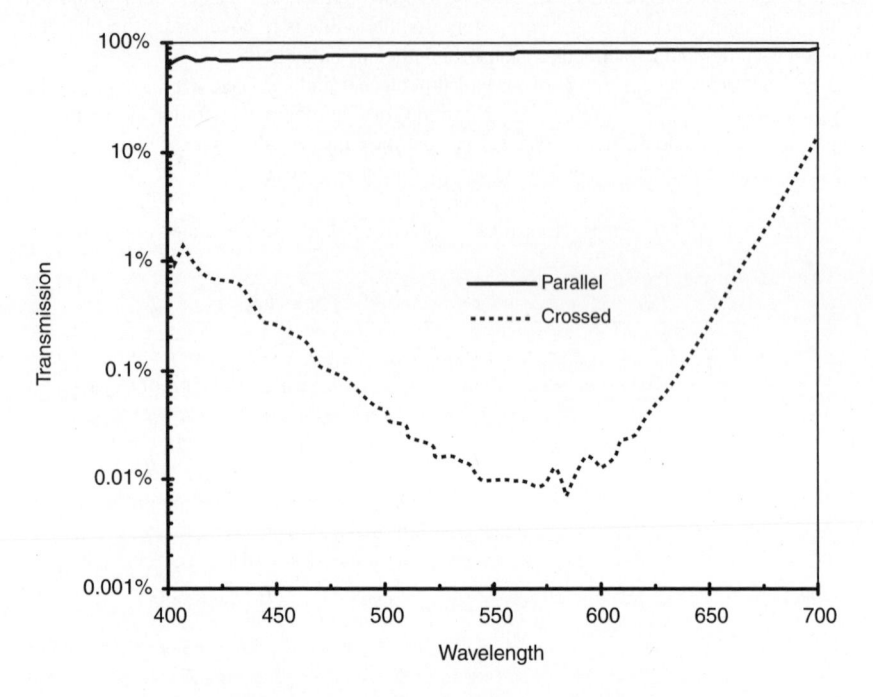

Figure 6.41   Transmission of Polaroid Type K polarizer.

In all absorptive polarizer types there is a trade-off between transmission of the desired polarization and blocking of the other polarization. This effect is shown in Figure 6.42, which is generated from data taken from (Cael et al. 1997). Note that for both polarizer types, as the transmission of the polarizer in unpolarized light (K_V) increases, the polarization efficiency decreases. This indicates that the higher transmission polarizers are not only transmitting more of the desired polarization, they are transmitting more of the undesired polarization as well.

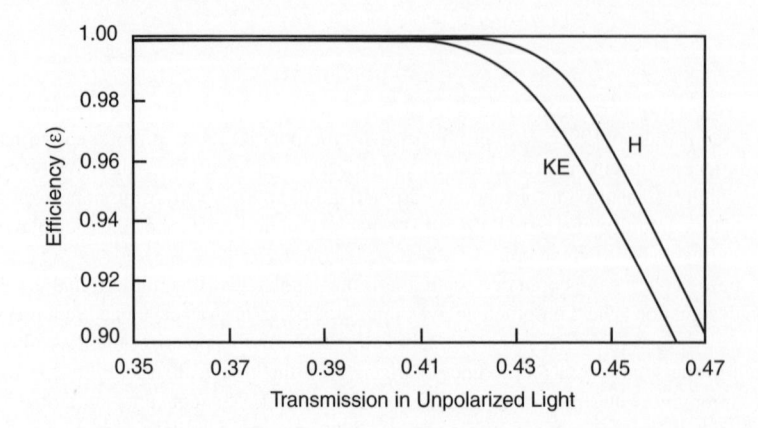

Figure 6.42   Polarization efficiency vs polarizer transmission.

6.5.2 Reflective Polarizer Technology

A reflective polarizer is a polarizer that transmits one polarization of light and reflects the other polarization, with little absorption of the light. There are several physical effects that can be used to produce reflective polarizers suitable for use in projection systems.

1. Fresnel reflection at the Brewster angle. This can be implemented with non-birefringent materials.
2. Fresnel reflection at normal incidence where birefringent and non-birefringent materials are used.
3. Total internal reflection at an interface between birefringent and non-birefringent materials, or between two birefringent materials. The basics of total internal reflection are given in Section 6.3.6.
4. Wire grid polarizers.

How these physical effects are used in reflective polarizing components will be described in the following sections.

6.5.2.1 Brewster angle reflection

At the Brewster angle, the Fresnel equations predict that the reflected light has only a single polarization, as was discussed in Section 6.3.1. This effect can be used to produce efficient reflective polarizers.

(a) Brewster plate
A Brewster plate (Figure 6.43) is an air-spaced plate with a high index of refraction material inserted into the optical beam at the Brewster angle. The transmitted light through the Brewster plate is a mixture of s and p light and the reflected light off the plate is pure s-polarized light, for an ideal plate with a high f/# incoming beam. The polarization efficiency as defined by Equation (6.16) of the transmitted beam is relatively poor.

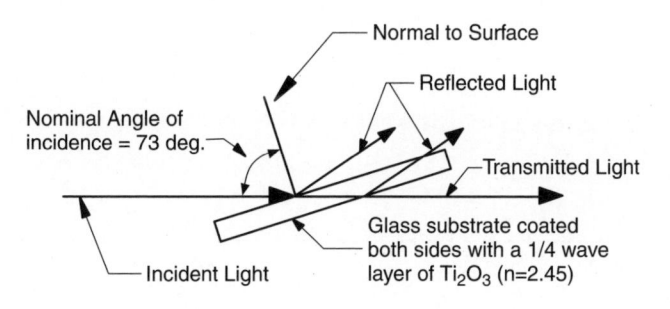

Figure 6.43 Brewster plate with a Ti_2O_3 coating.

If desired, this purity can be increased by using an air-spaced stack of Brewster plates (Bennett 1978). Note that even with a single Brewster plate, there are reflections off both the front and back surface of the plate. The situation is similar to the case of a second surface mirror as shown in Figure 6.17(b), except the reflections off the two surfaces are more nearly equal. Therefore, the reflection path cannot be used in the imaging path of a projection system. The transmission path, however, can be used if allowance for the significant amount of coma induced by the highly tilted plate is made in the projection lens. The transmission and reflection for this structure is shown in Figure 6.44.

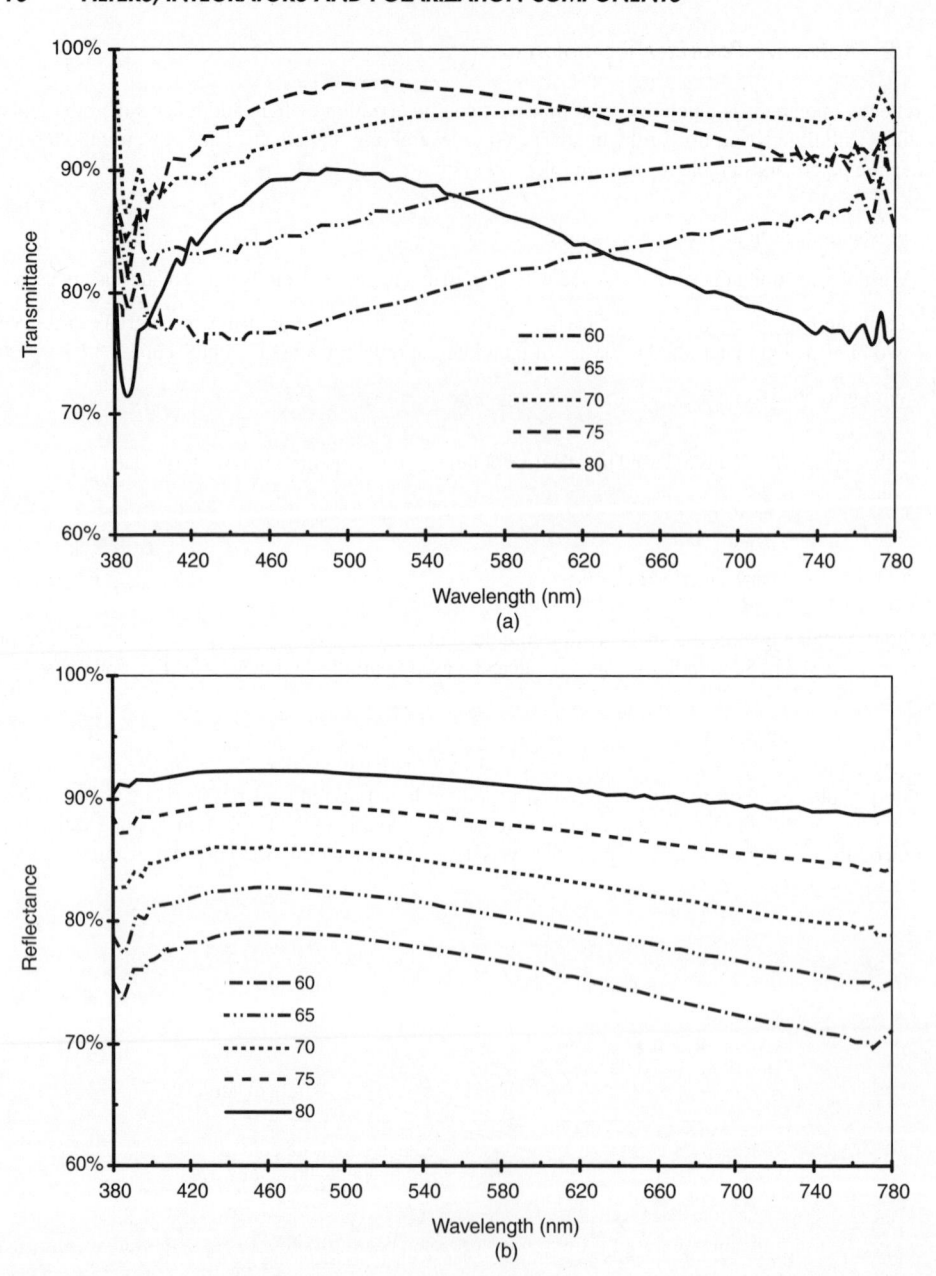

110 **FILTERS, INTEGRATORS AND POLARIZATION COMPONENTS**

Figure 6.44 Brewster plate transmission & reflection: (a) transmission of p-polarized light; (b) reflection of s-polarized light.

A Brewster plate is relatively simple to manufacture, low cost and readily available. Its main disadvantages are: (1) the very high angle of incidence required, about 73° and (2) the low polarization efficiency of a single Brewster plate. For these reasons, Brewster plates are not used as polarizers in modern projectors.

(b) MacNeille polarizing prisms

A MacNeille prism uses a multilayer dichroic coating as its polarizing element. This coating is embedded in a prism at an angle related to the Brewster angle. The layout of a 45° MacNeille cube is shown in Figure 6.45. For maximum performance, the index of refraction of the high index coating material (n_h), the index of low index coating material, the index of the glass prism (n_g) and the angle of the prism interface (θ_p) are related by the equation:

$$\sin^2(\theta_G) = \frac{n_H^2 n_L^2}{n_G^2 (n_H^2 + n_L^2)}. \tag{6.17}$$

where
- θ_G is the angle of the prism interface,
- n_H is the index of refraction of the high index coating material,
- n_L is the index of refraction of the high low coating material and
- n_G is the index of refraction of the glass prism.

When this condition is exactly satisfied, the coating will be at the Brewster angle and only light of a single polarization will be reflected. A derivation of this equation is given by Macleod (2001). For example, the MacNeille angle for BK-7 glass ($n_G = 1.52$), SiO$_2$ ($n_L = 1.46$) and TiO$_2$ ($n_h = 2.35$) the MacNeille condition calls for a 54.7° prism angle. If a MacNeille cube with a 45° prism angle is desired with these same coating materials, it should be made out of glass with a $n_G = 1.75$, such as SF-4. By modifying the design of the dichroic stack, it is possible to make a satisfactory MacNeille prism even if Equation (6.17) is not exactly satisfied.

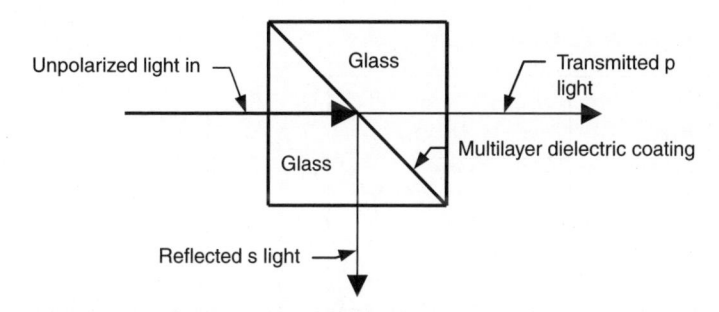

Figure 6.45 MacNeille polarizing cube.

A MacNeille prism always reflects primarily s-polarized light and transmits primarily p-polarized light. For light with small cone angles and a narrow spectral range, the polarizing efficiency of the prism can be very high. In general, the reflectivity of s-polarized light can be made very high, with very little s-polarized light in the transmitted beam. The transmission of p-polarized light is typically not so high: there can be a significant amount of p-light in the reflected beam. In order to get high system contrast, a clean-up polarizer is needed to remove this p-polarized light, especially at low f/#.

The transmissions of p- and s-polarized light for a commercially available MacNeille cube are shown in Figure 6.46 for a f/2.6 optical system (Foreal Spectrum 2006). Properties of a MacNeille prism vary with the numerical aperture of the system, as shown in Figure 6.47. Over the range of cone half-angles examined, there is a modest decline in T_p with increasing angle, and almost no variation in R_s. On the other hand the T_p/T_s shows a strong decline corresponding mostly to an increase in T_s. There is also a decline in the R_s/R_p ratio, corresponding to an increase in the R_p value. The third graph, Figure 6.47(c), is the $T_p{}^*R_s$ efficiency value as a function of cone angle. The prism has very high efficiency (96 %) at small cone angles corresponding to a f/# of f/7.1. At large cone angles (f/0.9), the efficiency

112   FILTERS, INTEGRATORS AND POLARIZATION COMPONENTS

declines to about 67 %. Contrast would also decline, although the data in these graphs is not sufficient to determine by how much, since it is determined by other system considerations in addition to the PBS.

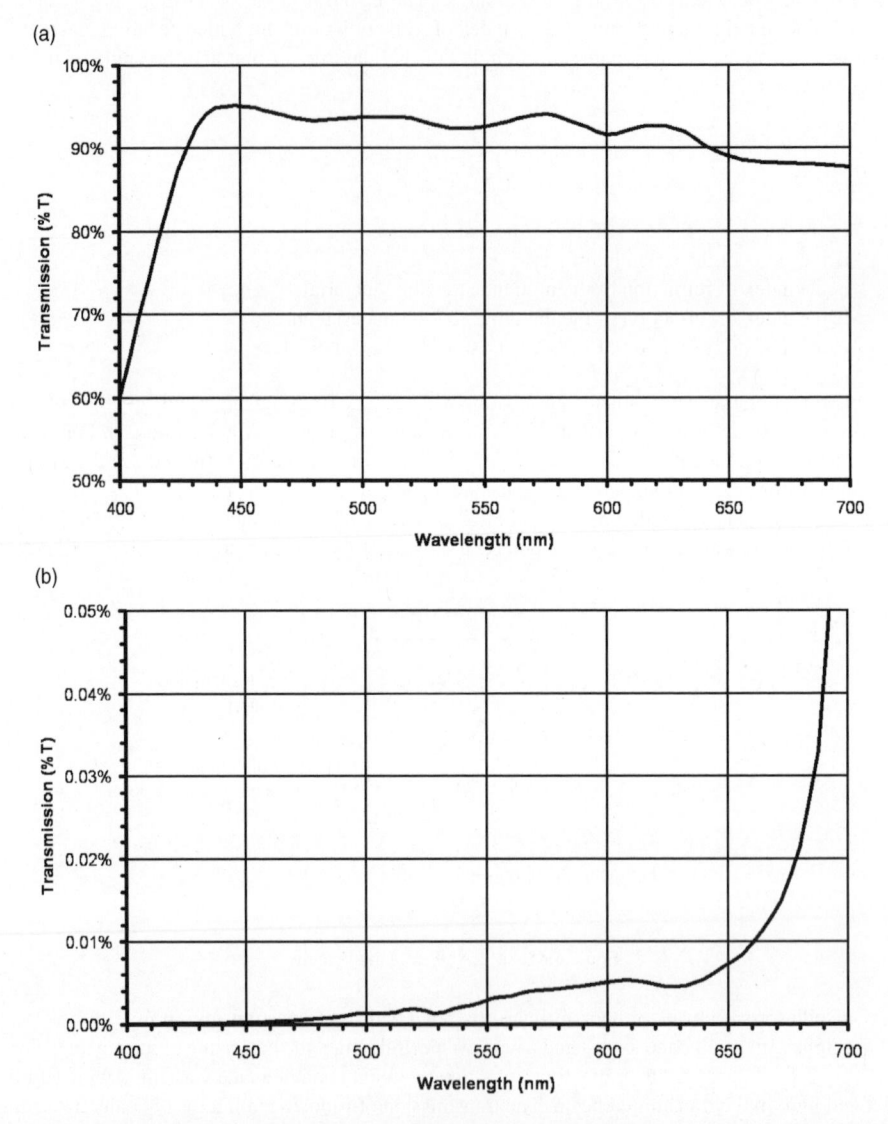

Figure 6.46 Typical spectral performance at f/2.6 of MacNeille prism PBS optimized for LCoS applications (data courtesy of Foreal Spectrum); (a) transmission of p-polarized light; (b) transmission of s-polarized light.

Prisms such as the MacNeille prism that are based on Fresnel reflection have a significant contrast problem at high cone angles (low f/#) when used in the imaging path of a LCoS projector. This is a geometrical problem related to the difference between s- and p-polarization and horizontal and vertical polarization and will be discussed in Section 6.5.4. When this problem is corrected and a clean-up polarizer is used between the PBS and the projection lens, a MacNeille PBS can provide good contrast and high throughput even at a f/1.25 aperture.

POLARIZATION COMPONENTS 113

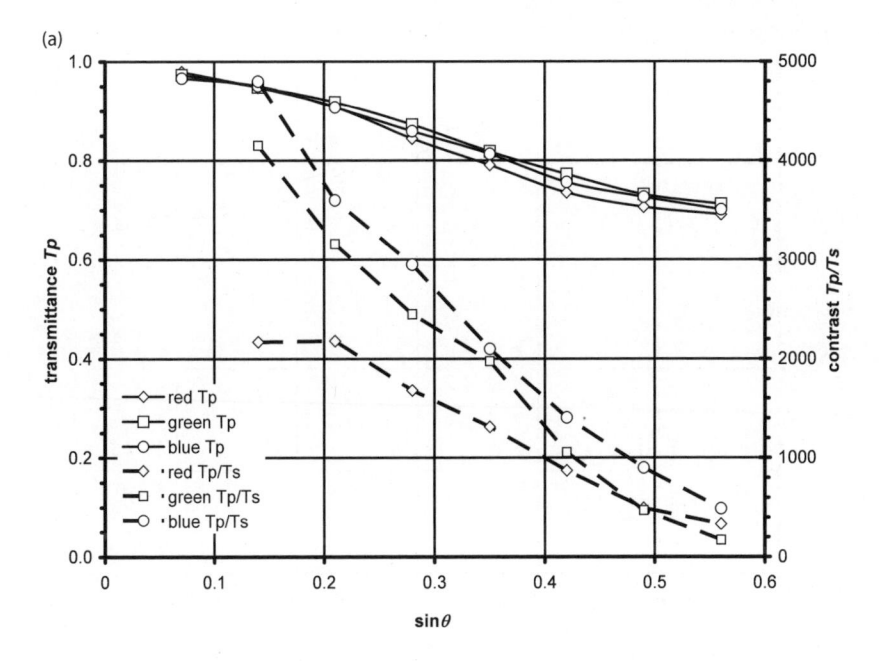

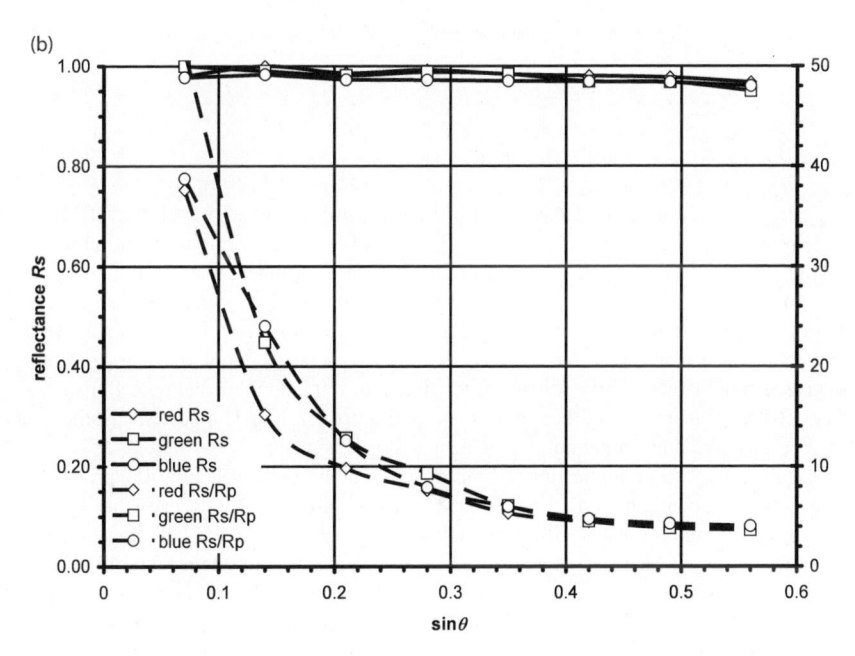

Figure 6.47 Variation in MacNeille prism properties with cone angle (data characterizing Foreal Spectrum PBS courtesy of Mark Handschy of Displaytech).

114 FILTERS, INTEGRATORS AND POLARIZATION COMPONENTS

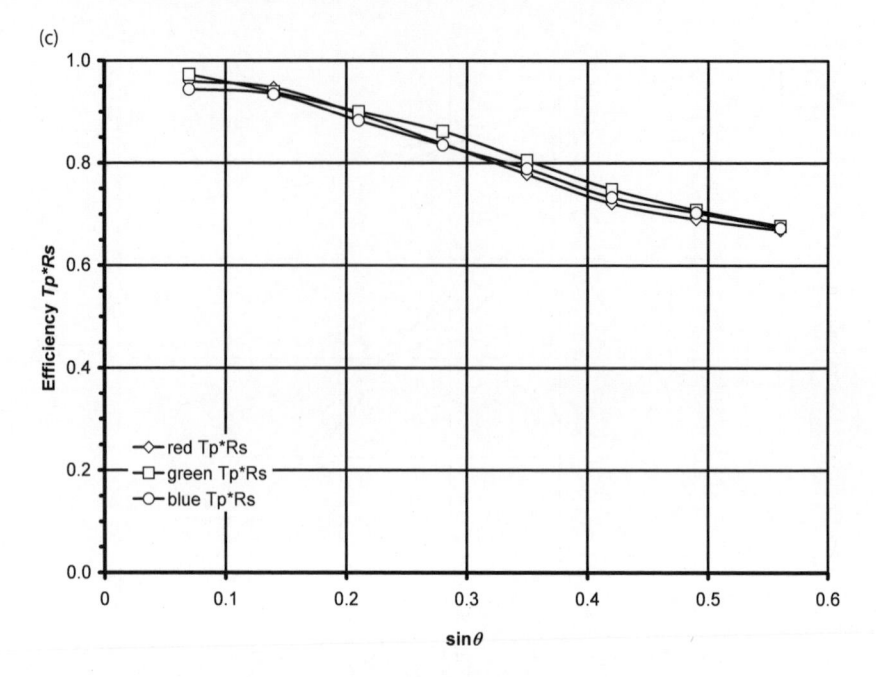

(c)

[Graph showing Efficiency Tp*Rs vs sinθ with curves for red Tp*Rs, green Tp*Rs, and blue Tp*Rs]

Figure 6.47 (Continued).

6.5.2.2 Birefringent multilayer reflective polarizer

A new type of reflective polarizer is called the birefringent multilayer reflective polarizer (BMRP) (Wortman 1997; Weber *et al.* 2000; Eckhardt *et al.* 2003). It consists of alternating layers of birefringent and non-birefringent materials. The index of refraction of the non-birefringent material is matched to one of the indices (n_e or n_o) of the birefringent material. This matched index is designated n_1 in Figure 6.48. For light polarized parallel to the optical axis of the material, the layers have the same index of refraction and no reflection takes place at the surface between the two layers. Light polarized perpendicular to the optical axis sees alternating layers of high and low index of refraction. Fresnel reflection, as described in Equation (6.4), occurs at each of these interfaces for this polarization. Unlike a MacNeille prism, this design can be used as a polarizer at normal incidence as well as off-axis.

When used for projection applications, the film is normally embedded in a 45° prism, as shown in Figure 6.48. This configuration is sometimes called a Cartesian PBS because it separates light based on horizontal and vertical polarization, not s- and p-polarization. This eliminates the need for the corrective quarter-wave plate typically used in MacNeille prism systems. The high index plate with index n_3 shown in Figure 6.48 is needed because the BMRP material typically has a finite thickness, about 1 mm, and has a relatively low index of refraction. The prism halves have a relatively high index of refraction, n_2, that is not as high as n_3. This tilted plate of low index material (n_1) when sandwiched between the prism halves induces coma in the imaging path. The high index plate (which must be higher than the prism halves) induces coma in the opposite direction and corrects this coma problem. The plate n_3 needs to be much thicker than the organic multilayer stack n_1 because normally there is a much higher index difference between n_1 and n_2 than there is between n_2 and n_3. Developments in prism glass types would be needed to reduce n_2 to the point where the

POLARIZATION COMPONENTS   115

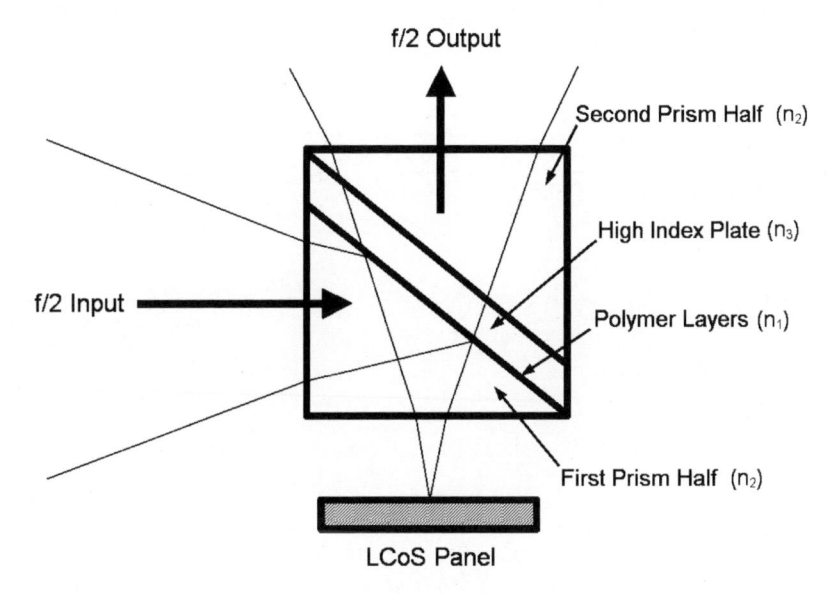

```
                    f/2 Output
                        ↑
                       ╱│╲ Second Prism Half (n₂)
                      ╱ │ ╲
                     ╱  │  ╲ High Index Plate (n₃)
    f/2 Input →     ╱   │   ╲ Polymer Layers (n₁)
                   ╱    │    ╲
                  ╱     │     ╲ First Prism Half (n₂)
                  ▬▬▬▬▬▬▬▬▬▬▬
                    LCoS Panel
```

Figure 6.48 Design of a BMRF Prism for a LCoS application (reproduced by permission of © 2003 3M Company (Eckhardt 2003)).

thickness of the coma-correcting plate can be reduced, or perhaps the plate can even be eliminated completely.

Because of the finite film thickness and the coma correction, a BMRF prism can only be used with the image transmitted though the prism, as shown in the figure. The coma and other distortions induced in the input illumination beam have little effect on the overall system performance.

The optical properties of a sample of a BMRP polarizer film not embedded in a prism are illustrated in Figure 6.49. These properties were measured in an apparatus similar to the one shown in Figure 6.6, with the addition of a polarizer between the light source and the sample of BMRP. The transmission for p-polarized light over the range of 0–45° is shown in Figure 6.49(a) while the reflectivity of s-polarized light is shown in Figure 6.49(b). There are two warnings necessary related to these figures. First, the s and p polarizations are not defined at 0° angle of incidence. S- and p-polarization for the 0° curve refer to light polarized parallel to s- or p-polarized light in the off-axis curves.

Secondly, if the sample of BMRF is rotated 90° about the optical axis of the system shown in Figure 6.6, the material will now reflect p-polarized light and transmit s-polarized light. If the optical properties were to be remeasured, the transmission of s-polarized light would not be quite as high as the transmission of p-polarized light as shown in Figure 6.49(a) and the reflection of p-polarized light would not be quite as high as the reflection of s-polarized light shown in Figure 6.49(b). This occurs in part because the Fresnel reflection at the front and rear surfaces reflects s-polarized light and transmits p-polarized light regardless of the orientation of the BMRP. This tends to improve the properties of the material slightly when the orientation or the material is correct to reflect s-polarized light and degrade the properties slightly when the orientation is off by 90° so p-polarized light is reflected.

Since the polarizing layer in this system is an organic film, the lifetime of the layer is an issue in a projection system. Bruzzone *et al.* (2004) shows the lifetime of preliminary versions of this film could be up to 20,000 hours in a 400 lumen light engine.

One projection application of this material is a 3-PBS optical engine for LCoS projection (Bruzzone *et al.* 2003). This engine design will be discussed in Section 10.3.1.

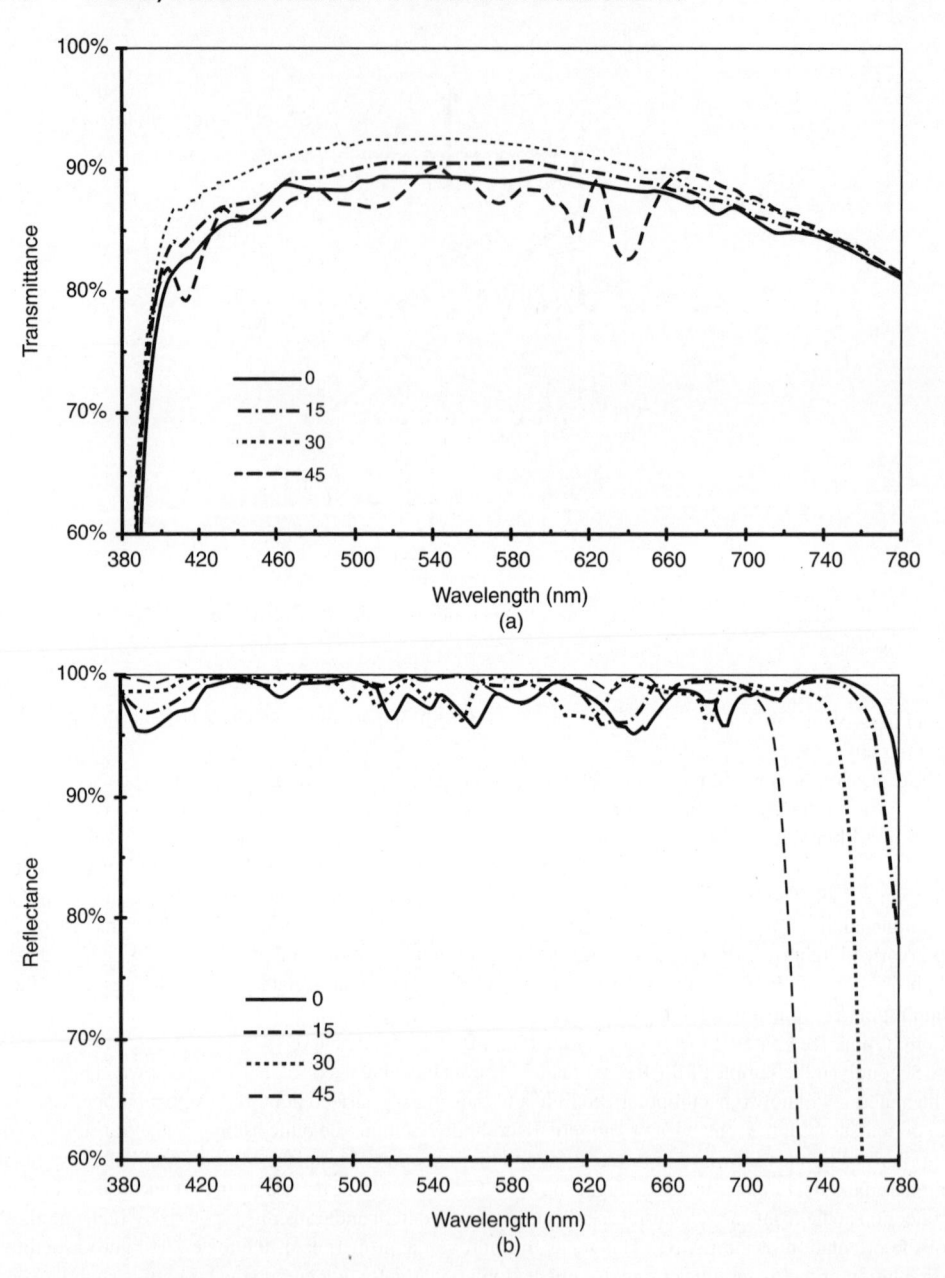

Figure 6.49 Transmission of a birefringent multilayer reflective polarizer: (a) transmission of p-polarized light; (b) reflection of s-polarized light.

6.5.2.3 Bertrand-Feussner prism

The Bertrand-Feussner prism is a polarizing prism based on a thin layer of birefringent material within a two-part prism made of non-birefringent glass, as shown in Figure 6.50. A prism of this type is described by Vinokur et al. (1998). The birefringent material ideally has its high index of refraction (n_e) matched to the index of the glass (n_g) and its low index (n_o) as low as possible. The optic axis of the birefringent material would be perpendicular to the plane of the paper in Figure 6.50.

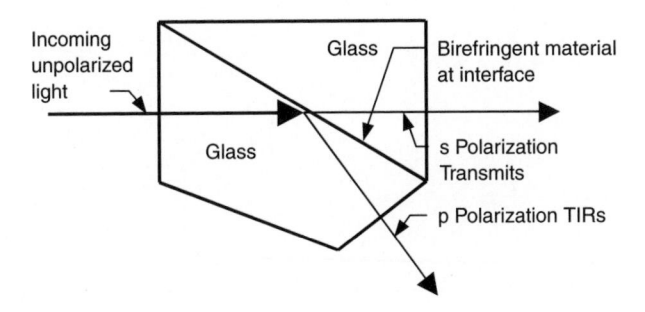

Figure 6.50 Bertrand-Feussner prism.

When light reaches the interface, the p polarized light sees an interface between low and high index materials. The angle of the interface is greater than the TIR angle, and the p light undergoes total internal reflection. The s-polarized light, on the other hand, sees the n_e or high index of the birefringent material, which is matched to the glass. If the match is perfect, the light sees no interface at all and is completely transmitted. Note that in this prism the role of s and p is reversed compared to other polarizers, since s reflects and p transmits in a polarizer based on Brewster angle Fresnel reflection.

This reversal of s and p does not cause a problem if there is a perfect match between n_e and n_g. If there is a mismatch, however, there would be some Fresnel reflection of s-light at the interface, reducing the polarization purity of the reflected light. This effect could be reduced or eliminated by AR coating the interfaces between the glass and the birefringent material. Other designs of Feussner-type prisms can be made to reflect s and transmit p. However, with common birefringent material types on the interface, this reduces the range of angles over which the prism will work satisfactorily.

The transmission and reflection of a Bertrand-Feussner prism at normal incidence at the entrance face and ±8° (in air) is shown in Figure 6.51. The poor performance of this Bertrand-Feussner prism in the blue region is due to absorption of light by the SF-1 glass used in the prism.

Due to the availability of improved MacNeille prisms, BMRP prisms and wire grid polarizers, there is little current interest in Bertrand-Feussner prisms for projection systems.

6.5.2.4 Wire grid polarizers

The wire grid polarizer started as exactly that: an array of stretched wires to polarize microwaves. It is necessary for the spacing of the wires in a wire grid polarizer to be less than the wavelength of light so obviously the stretched-wire approach would not work for projection systems that must deal with 440 nm blue light. Using photolithographic techniques,[10] Moxtek has succeeded in making wire grid polarizers suitable for use with all wavelengths of visible light (Arnold et al. 2001; Hansen et al. 2002).

Wire grid polarizers reflect light that has its electric field vector parallel to the wires and transmits light that has its electric field vector perpendicular to the wires, as shown in Figure 6.52. This occurs regardless of the angle of incidence, so wire grid polarizers can be used at normal incidence. Figure 6.53 shows an electron microscope image of the aluminum wires that make up the polarizer material.

---

[10] The exact technique used is proprietary at Moxtek and is not public knowledge.

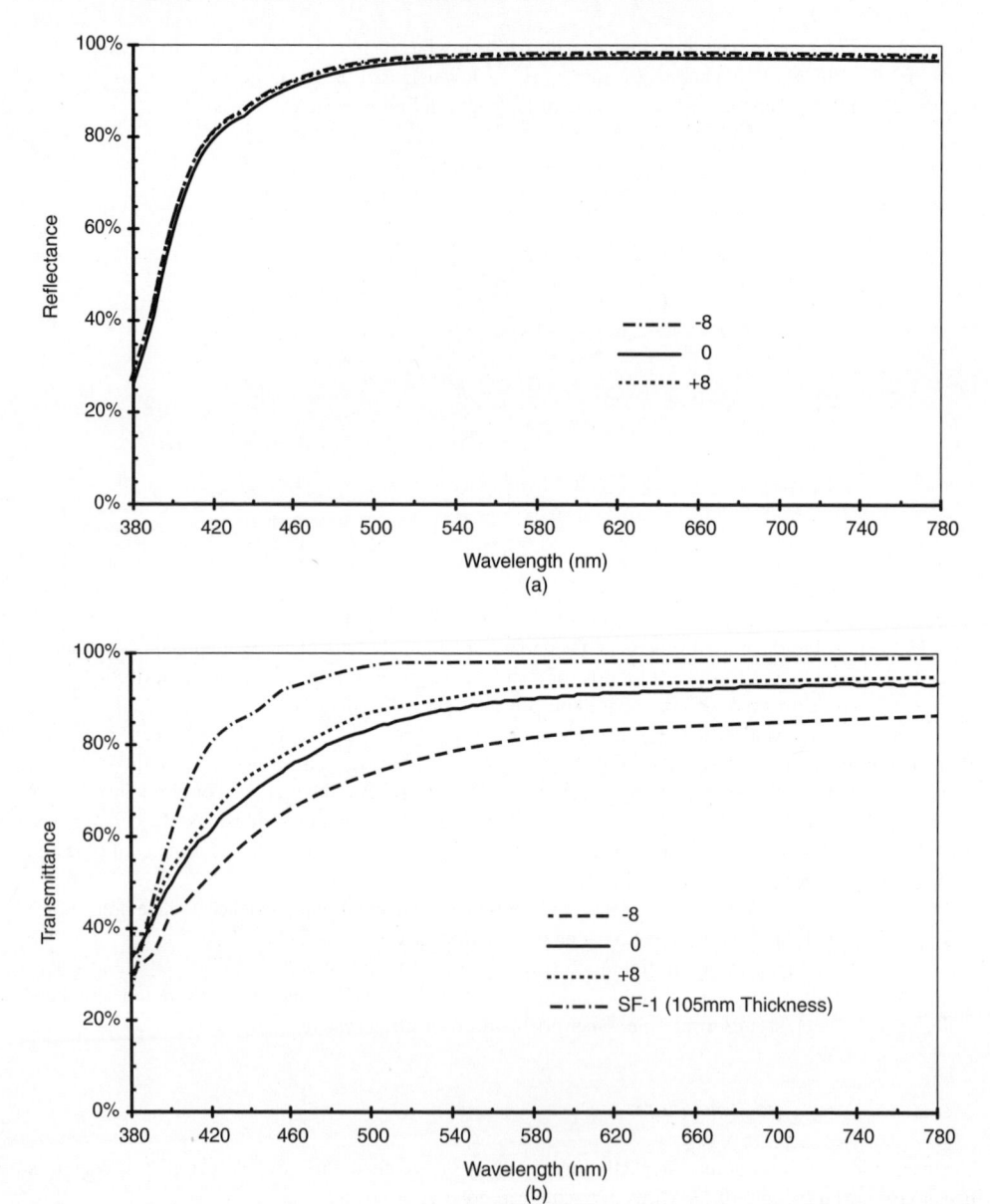

118 FILTERS, INTEGRATORS AND POLARIZATION COMPONENTS

Figure 6.51 Transmission and reflection of a Bertrand-Feussner prism: (a) reflection of p-polarized light; (b) transmission of s-polarized light.

Figure 6.54 shows the optical properties a typical wire grid polarizer from Moxtek. Note that the efficiency, defined by $R_s * T_p$ is high. Due to the relatively high R_p value for this PBS, it would always need to be used with an auxiliary post-polarizer (analyzer) in order to get acceptable contrast in an imaging system.

Like the BMRP polarizer described in Section 6.5.2.2, the wire grid polarizer is a 'Cartesian' polarizer, separating light on the basis of horizontal vs vertical polarization, rather than s- and p-polarization. Therefore, the wire grid polarizer does not suffer from the skew angle problems described in Section 6.5.2.1(b) on the MacNeille prism.

POLARIZATION COMPONENTS   119

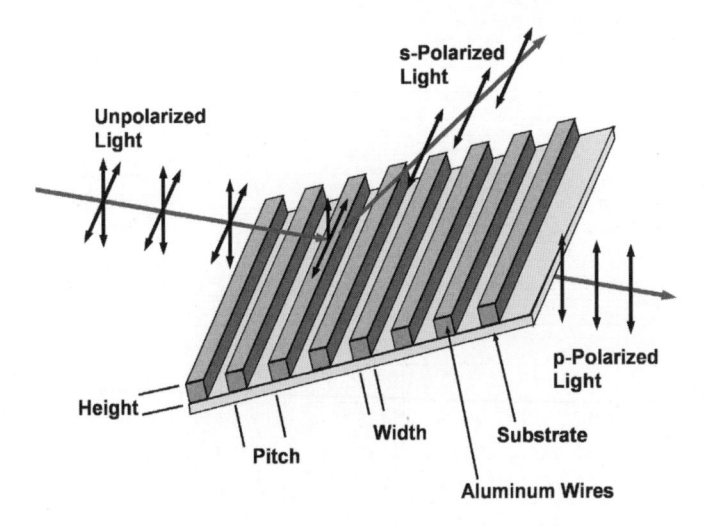

Figure 6.52 Operation of a wire grid polarizer (reproduced by permission of Moxtek Inc.).

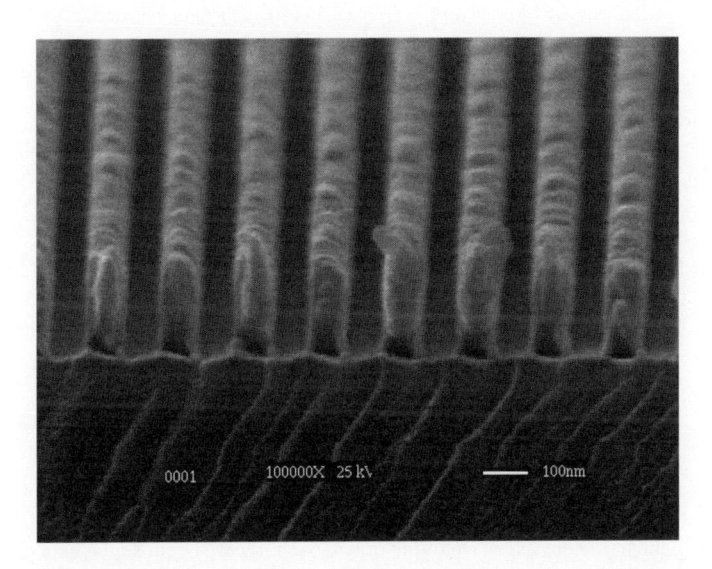

Figure 6.53 Scanning electron microscope image of a wire grid polarizer (reproduced by permission of Moxtek Inc.).

Wire grid polarizers are typically made in plate type of configurations, rather than embedded in a prism. One reason for this is the space between the wires visible in Figure 6.53 must be filled with a low index material such as air for proper operation of the polarizer. If they were used in the imaging path, they would normally be used to reflect the image to avoid inducing coma in the image by the tilted plate. In this configuration, PBS contrasts up to 20,000:1 have been reported using dummy LCoS cells (Gardner and Hansen 2003). With real cells, typically the contrast would be much lower than this.

In addition to their use as polarizing beam splitters, an application that will be discussed further in Section 6.5.4, wire grid polarizers can be used at normal incidence as pre-polarizers. For example, they

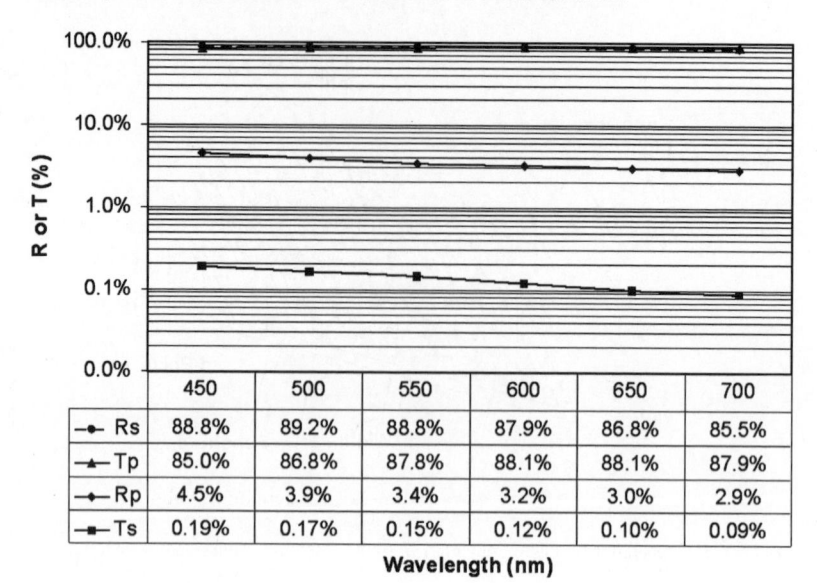

120 FILTERS, INTEGRATORS AND POLARIZATION COMPONENTS

	450	500	550	600	650	700
Rs	88.8%	89.2%	88.8%	87.9%	86.8%	85.5%
Tp	85.0%	86.8%	87.8%	88.1%	88.1%	87.9%
Rp	4.5%	3.9%	3.4%	3.2%	3.0%	2.9%
Ts	0.19%	0.17%	0.15%	0.12%	0.10%	0.09%

Figure 6.54 Properties of a Moxtek PBS-02A wire grid polarizer.

can be used before the PBS in an LCoS light engine to increase the system contrast. They can also be used before the LCD panel in a 3-panel LCD (3LCD) projector. In the pre-polarizer application, the lifetime of the wire grid polarizer is sometimes important. As an all-inorganic system, lifetime can be quite long compared to organic absorptive polarizers.

In general, wire grid polarizers cannot be used after the microdisplay in either transmissive or reflective systems. Since the wire grid reflects rather than absorbs the unwanted light, this reflected light could return to the microdisplay as stray light and reduce contrast.

6.5.2.5 Other reflective polarizers

Dozens, perhaps hundreds, of polarizing prisms have been described in the literature. The most important types are covered in Bennett and Bennett (1978). Two of the most important classes of prisms are the Nicol types and the Glan types. In general, Glan type prisms have the optical axis of the birefringent material in the plane of the entrance face of the prism. Nicol types are cut so the optical axis is neither parallel nor perpendicular to the entrance face. Most of these prisms are not useful in a projector due to their high cost and small aperture.

Unlike a Bertrand-Feussner prism where only a thin layer of birefringent material is used, the birefringent material forms the body of both halves of a Nicol or Glan type prism. The birefringent material used normally is calcite. The cutting and polishing of the calcite, and the rarity of the natural crystals, leads to the high cost of these prisms. The size of the available crystals limits the size of the prisms that can be made.

On the other hand, a Nicol or Glan type prism is capable of very high extinction ratios over the entire visible spectrum, with extinction ratios $> 10^4$ easily obtained. This makes them excellent reference polarizers to be used in polarization measurement instruments. These instruments can then be used to measure the s- and p-polarization performance of the desired reflective or transmissive polarizers.

Reflective polarizers based on diffraction are possible (Tyan et al. 1997), They have not, however, entered the commercial market and will not be covered in detail this book.

6.5.3 Polarization Conversion Systems

In most types of LCD projectors, polarized light is required at the microdisplay. One way to achieve this with light from an unpolarized source is to absorb all the light of the wrong polarization with absorptive polarizers. This reduces the amount of light available by more than 50 %, depending on the transmission of the polarizer. This potentially gives a significant advantage to technologies that do not require polarized light. A polarization conversion system (PCS) can significantly reduce the advantage of microdisplays that do not require polarized light compared to microdisplays that do. A detailed numerical example of the effect of polarization conversion on luminance and collection efficiency is given in Chapter 12.

A projector incorporating a basic polarization conversion system is sketched in Figure 6.55. Light from the lamp encounters a polarizing beam splitter and is divided into two beams with orthogonal polarizations. One beam passes through a half-wave plate oriented at 45° to the polarization vector. The effect of this wave plate is to rotate the polarization vector by 90°. At this point, the two beams have parallel polarization vectors. Both beams are then used to illuminate the LCD panel.

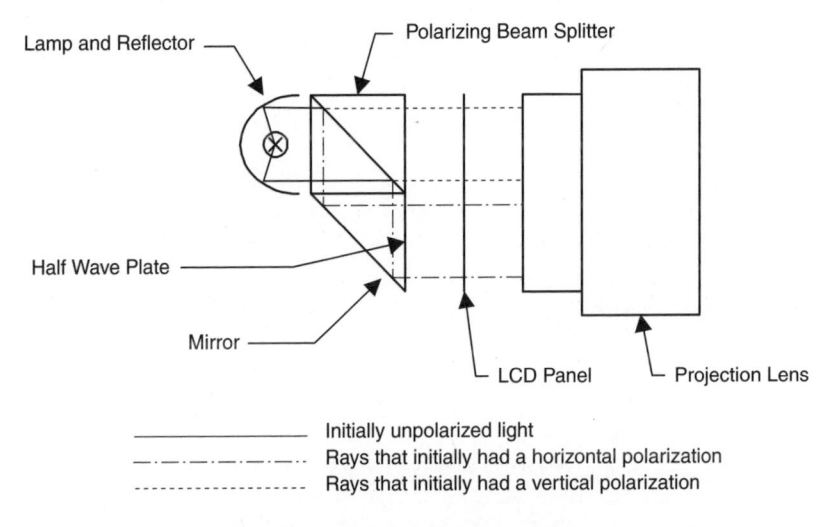

Figure 6.55 Basic polarization conversion system.

This projector in Figure 6.55 is not a practical design as sketched: it would have severe luminance uniformity problems. This would be due to the presence of two bright regions, one from each polarization and centered on the top and bottom halves of the panel. In most cases, a projector with polarization conversion will also include a lenslet integrator as described in Section 6.4.1 to eliminate these hot spots and provide uniform illumination. It is also possible to build satisfactory PCS designs without integrators (Shikama et al. 1990).

A discreet PCS such as the one shown in Figure 6.55 can be used in miniature form with a rod integrator and lamp illumination (Duelli and Taylor 2003) as shown in Figure 6.56. While in theory rod integrators are not polarizing preserving, in practice the relatively short rod integrators used in projection systems preserve polarization well enough for this system to be useful.

Figure 6.57 shows a more common PCS system, which includes a lenslet integrator that combines the two beams into uniform illumination at the LCD panel. Each path of the PCS illuminates half of the

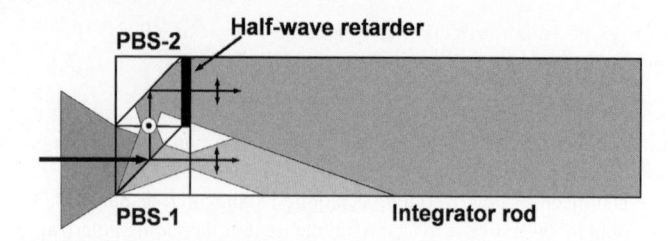

122   *FILTERS, INTEGRATORS AND POLARIZATION COMPONENTS*

Figure 6.56   Rod integrator with a PCS at the input (reproduced by permission of © 2003 the Society for Information Display (Duelli and Taylor 2003)).

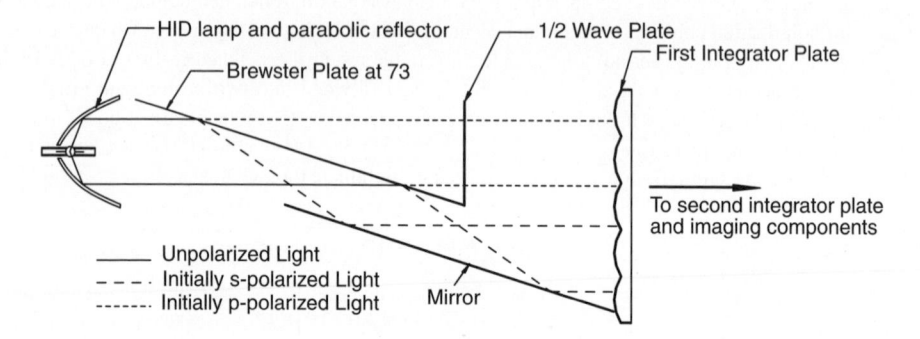

Figure 6.57   Polarization conversion system with a Brewster plate PBS.

first integrator plate. The second integrator plate and the auxiliary lens then image each lenslet from the first plate onto the LCD panel. This type of arrangement can be used to illuminate one and three panel projectors (de Vaan *et al*. 1993).

Another arrangement for a polarization conversion system is given by Itoh *et al*. (1997). The illumination path from a projector with this type of integrator is shown in Figure 6.58. Note that in this case, the PCS is after the integrator, rather than before the integrator, as shown in Figure 6.57.

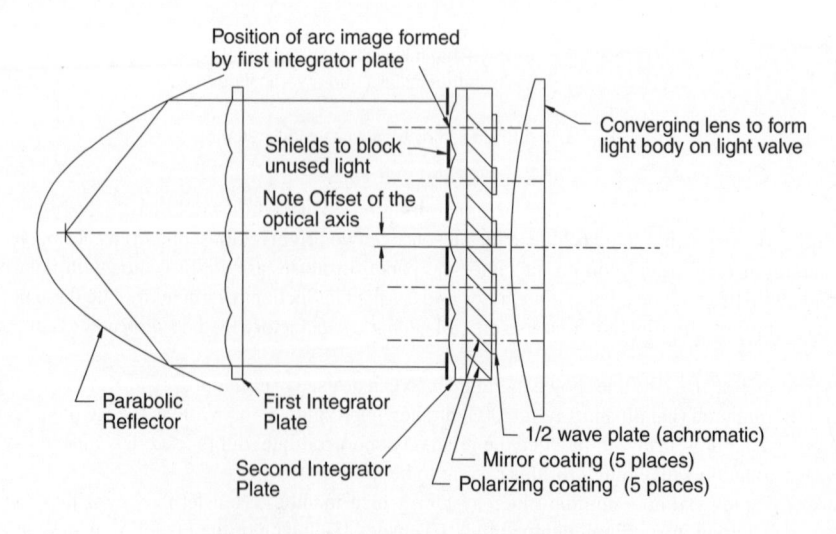

Figure 6.58   Flat polarization conversion system.

In this arrangement, the first integrator plate produces an array of images of the arc at the second integrator plate. The tall, narrow polarizing beam splitters associated with each column of lenslets split these images into two images: one consisting of s-polarized light and the other consisting of p-polarized light. A half-wave retardation foil rotates the light from one of these sets of images by 90° so all the light is polarized in the same direction. There is now an array of virtual light sources, all polarized in the same direction, with each virtual source occupying half the area of a lenslet in the second array. This array of virtual sources defines the pupil of the microdisplay.

Typically the PBS array used in this system is based on MacNeille prisms. Due to the very small range of the angle of incidence at the PBS array, the design of the MacNeille prism can be optimized to overcome the normal limitations of this technology.

This flat PCS arrangement can be made to be very compact. While this arrangement is theoretically no more efficient that a PCS based on a discrete PBS, as shown in Figure 6.57, it is relatively easy to design a flat PCS where the actual efficiency comes close to matching the theoretical maximum. Past discrete PCS designs have sometimes been designed in such a way that the full theoretical efficiency for a PCS system has not been achieved. Another advantage of the flat PCS is their small form factor fits more easily into modern compact projectors.

For a projection system with a high enough étendue such that the system is not étendue limited, polarization conversion can nearly double the light throughput, subject only to the light losses at the extra optical surfaces. For étendue limited systems, some of the efficiency gained by using both polarizations is lost through reduced collection efficiency. The net gain is typically in the range of 30–60 %, or more, over a system without polarization conversion. Typically, the smaller the microdisplay, the smaller the gain from a PCS. The smallest modern microdisplays, which can be 0.25″ diagonal or smaller, might show no gain at all from a PCS, or even a net loss due to the extra optical surfaces. The projector designer needs to balance the projector cost, complexity and size increases against the lumen output increase before deciding to incorporate polarization conversion into the illumination system design.

6.5.3.1 Polarization recycling

The polarization conversion systems described above are deterministic: light passes through the system once and p- and s-polarized light follow distinctly different paths. Polarization recycling involves a more statistical approach. In this system, there is at least one randomizing step in the polarization conversion process, such as passage down an integrating rod or reflection off a diffuse surface. This polarization recycling process is particularly suitable for LCD or LCoS systems with LED illumination because of the high étendue of the LEDs (Sakata et al. 2006).

Figure 6.59 shows a typical integrating rod based polarization recycling system. The lamp is focused on a circular aperture at the entrance to the rod. Typically, the aperture will cover about half of the rod end, and the rest of the end has a mirror coating. Somewhere in the system is a quarter wave plate. Figure 6.59 shows it directly behind the aperture, but it can be anywhere between the aperture and the wire grid polarizer. The unpolarized input light is integrated as it passes down the length of the rod and is spatially uniform by the time it reaches the output. At this point, the desired polarization is transmitted through the wire grid polarizer and the other polarization is reflected back toward the aperture. As the light passes through the quarter wave plate, it is converted to circular polarization. When the light reaches the end of the rod, if it hits the aperture, it is lost to the system with a lamp illumination system. If it hits the mirror portion of the input face, however, it is reflected back through the quarter-wave plate that converts it to linearly polarized light of the desired polarization. When it reaches the wire grid polarizer again, it passes through it and on to the microdisplay.

With LED illumination, the condenser lens will reimage the light exiting the aperture onto the LED. Here it can be scattered by the surface roughness of the LED. If it is scattered into the acceptance cone of the condenser lens and is reimaged onto the aperture, it will pass down the rod, through the polarizer and become illumination light for the microdisplay.

124 FILTERS, INTEGRATORS AND POLARIZATION COMPONENTS

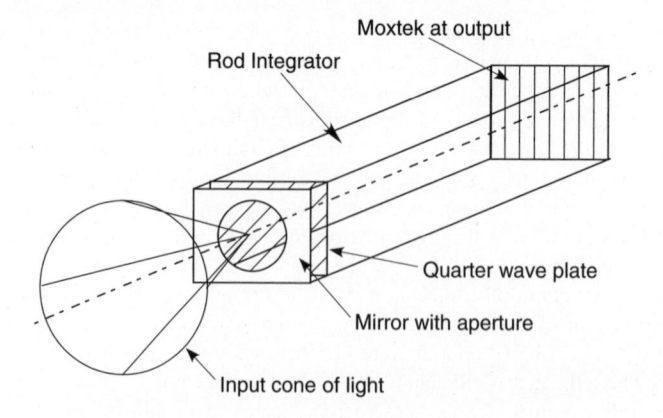

Figure 6.59 Polarization recycling with a rod integrator.

The most common quarter-wave plates are organic materials and are subject to optical degradation from the very high light fluxes typical in a polarization recycling scheme. Duelli et al. (2002) has described a system where an all-inorganic, thin film retarder is used on an embedded surface inside the light pipe to rotate the polarization without the organic wave plate.

Polarization recycling can be done without a rod integrator, as discussed for a lamp by Pentico et al. (2001). Recycling can also be used with LEDs as well as lamps (Sakata et al. 2006; Beeson et al. 2006).

6.5.4 Polarizing Beam Splitters in the Imaging Path

One application for a reflective polarizer, also known as a polarizing beam splitter (PBS), is to separate the input beam from the modulated and waste output beams in a reflective LCD projection system. Reflective LCD systems are most commonly built as liquid crystal on silicon (LCoS) systems, as were discussed in Sections 4.1.2.3 and 4.2.5. A typical arrangement for a reflective LCD and polarizing beam splitter is shown in Figure 6.60. In this system, the light passes through the PBS twice. In the first pass, the s-polarized light is reflected off the splitter coating and emerges from the bottom of the cube to illuminate the reflective LCD. Therefore, high reflectivity for s light leads to an efficient PBS. For the second pass through the prism, the p-polarized light must be transmitted to the projection screen, therefore a high transmission for p light leads to high efficiency.

LCoS systems that use a BMRP prism must match the configuration shown in Figure 6.60, with the illumination light reflected and the image light transmitted. LCoS systems that use a wire grid polarizer as a polarizing beam splitter must use the reverse configuration, with the illumination light transmitted through the PBS and the image light reflected off the PBS. If the configuration of Figure 6.60 is used with a wire grid polarizer, there will be significant degradation of the image. Systems that use a MacNeille prism can use either configuration.

The efficiency of a PBS in a reflective LCD application in the configuration shown in Figure 6.60 is given by:

$$\eta_{PBS}(\lambda, \theta, \phi) = R_s(\lambda, \theta, \phi) T_p(\lambda, \theta, \phi). \tag{6.18}$$

In this equation, the efficiency has been left as an explicit function of the wavelength λ, the in-plane angle θ, and the out-of-plane angle ϕ, both defined in Figure 6.60. This was done because the properties of a typical MacNeille prism PBS are a function of these three parameters. Therefore, these dependencies must be considered when evaluating the overall efficiency of a prism.

The efficiency, η_{PBS}, calculated by integrating Equation (6.18) using Equation (6.1), (6.2) or (6.3), with additional integrations over θ and ϕ, can make a good merit function for the design and evaluation

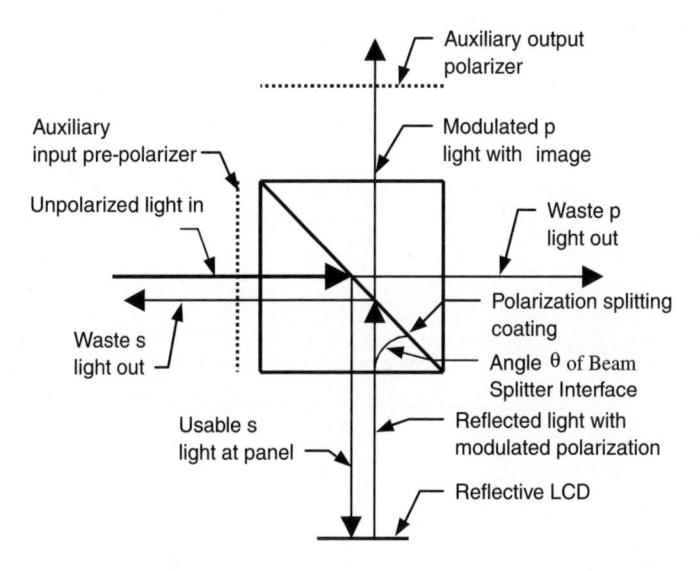

Figure 6.60 Schematic of a typical reflective LCD system using a reflective polarizer.

of PBS prisms. Note that the conventional merit functions for polarizers, the extinction ratios given in Equation (6.19), do not make very good merit functions for prisms to be used in reflective LCD systems:

$$\text{Extinction Ratio} = \frac{T_p(\lambda)}{T_s(\lambda)} \text{ or } \frac{R_s(\lambda)}{R_p(\lambda)} \tag{6.19}$$

This is especially true if only the on-axis properties are used and the dependence on θ and ϕ are ignored. It is possible to have high extinction ratio but low efficiency, and vice versa. In a reflective LCD system, if a PBS has low efficiency, there is nothing that can be done to improve the system efficiency except replace the PBS with one of a different design. If, on the other hand, extinction ratio is low and system contrast is poor, auxiliary polarizers as shown in Figure 6.60 can be used to improve overall extinction ratio and contrast. The auxiliary polarization at the input, for instance, may be done by a polarization conversion system.

In calculating system throughput for a projector, reflective polarizers have a couple of pitfalls for the unwary projector designer. First, the angular acceptance angle of the reflective polarizer is often quite limited, perhaps as small as ±5°. If the polarizer is in a prism, the designer will attempt to minimize prism size and cost. Therefore, the reflective polarizer often has a rather small étendue, and can be the limiting element in the projector.

Another major problem with reflective polarizers is that they are rarely neutral density, and their spectral properties can change rapidly with angle. This makes estimating system throughput and colorimetry significantly more difficult. Calculation of the color and throughput of a projector, given the spectral properties of the individual components, will be covered in Chapter 11.

Another problem with a polarizing beam splitter in a reflective LCD system is that the PBS, even with auxiliary pre- and post-polarizers, may limit the maximum contrast the system can achieve. This can occur if the PBS mixes polarization states, and converts some horizontally polarized light into vertically polarized light or vice versa. This is a particularly severe problem in MacNeille prisms, where the polarization is split on the basis of s- vs p-polarization instead of horizontal vs vertical polarization in Cartesian polarizers. This is a geometrical effect that is caused by the difference between s and p polarization vectors and horizontal and vertical polarization vectors for skew rays at a tilted interface. This effect is covered in more detail in Rosenbluth et al. (1997) and Sannohe and Miyatake (1995). This problem is particularly severe with a MacNeille prism, which is based on Fresnel reflection and s- and p-polarization axes play a strong role. It has also been observed in BRMF prisms.

Correction of this problem involves the insertion of a quarter-wave plate between the PBS and the LCoS imager. Sophisticated, wavelength and angle independent quarter-wave plates can be used (Sharp *et al.* 2002) but even a single layer waveplate can increase the contrast by a factor of 10. If this problem is properly corrected, it is rarely the limiting factor in reflective LCD contrast.

Another geometric effect occurs in wire grid polarizers (Shahzad and Shimizu 2006) if a field lens is used with the wire grid polarizer in an attempt to reduce the size or complexity of the illumination or imaging path. If no field lens is present, the polarization errors induced in the transmitted beam are canceled out by the mirror image errors induced in the reflected beam, leading to no polarization mixing and very high contrast. If a field lens is used, the reflected and transmitted polarization errors do not exactly cancel each other and contrast can be degraded significantly. The solution to this problem is to only use the wire grid polarizer in telecentric light and not use a field lens between the wire grid and the LCoS microdisplay.

6.5.5 Compensation Films

Two polarization problems were discussed in the previous section. These two problems are not the only polarization problems associated with high contrast LCD and LCoS systems. Additional problems are caused by the pre-tilt of the directors of the LC molecules in the liquid crystal cell, the finite cone angle of the illumination light and the inability of the applied voltage to drive the LC molecules to exactly parallel or perpendicular states, as would be required to get perfect black states. Different types of compensation plates are used to correct for these different problems. Compensation plates are also called compensation films, retarder plates or simply retarders.

The detailed mathematics and geometry of compensation films are beyond the scope of this book. For more details on the topic see, for example, Robinson *et al.* (2005). The compensation of liquid crystal devices is generally the responsibility of the microdisplay vendor. The microdisplay vendor preferably builds the compensation into the LCD microdisplay. Alternatively, he can specify exactly how the light engine manufacturer must compensate the cell in order to get maximum contrast. The microdisplay vendor cannot expect the light engine manufacturer to develop the compensation films required by the microdisplay. In either case, however, it is useful for the projector designer to understand the basic principles.

Compensation plates are normally designed to affect the dark state of the LCD to get the best possible contrast (Tan *et al.* 2005). Since the films are present in the white state as well, they often affect the white state efficiency. Typically, this is of little importance. After all, a 0.2 % of maximum brightness in the dark state can have a critically important effect on observed contrast, but a 0.2 % change in the modulation efficiency of the white state is difficult if not impossible to measure, let alone see with the eye.

Compensation films are generally described as A-plate, C-plate or O-plate. An A-plate film uses a uniaxial birefringent material with the optical axis in the plane of the retarder and generally perpendicular to the light passing through the retarder. A C-plate also uses a uniaxial material but the optical axis is perpendicular to the film and generally parallel to the light passing through the film. An O-plate has an oblique optical axis that is neither parallel nor perpendicular to the surface or the light passing through the surface. Each of these types of compensation films comes in two flavors, a positive plate with $n_e > n_o$ and a negative plate with $n_o > n_e$.

While it is possible to replace an A-plate/C-plate pair with a properly designed O-plate with a biaxial birefringent material (Mi *et al.* 2005; Chen *et al.* 1998), managing three different indices of refraction of a biaxial material and orienting them correctly is a difficult problem. Therefore, it is more common to use a separate A-plate and C-plate in microdisplays that require both types of compensation. This simplifies design, construction and alignment of the compensation films.

In most liquid crystal cells, the black state is produced by leaving the polarization state of the input light unchanged. For example, an ideal TN-90 cell when driven to a black state would have all its LC molecules aligned parallel to the electric field of the cell. Light on axis would see no birefringence and the polarization of the light would be unchanged. In practice, at the substrates the LC molecules are tilted slightly rather than perpendicular to the substrate. This slight residual tilt converts the input linearly polarized light into slightly elliptically polarized light. A small amount of this elliptically polarized light will leak through the output linear polarizer, degrading the dark state. An A-plate compensator with the correct retardance value and azimuthal angle can convert this elliptically polarized light back into linearly polarized light, improving the dark state.

Let us look at this same ideal TN-90 system again and examine what happens when off-axis light passes through the TN cell driven to the dark state. As mentioned before, the light on-axis is parallel to the LC molecules and in the ideal case sees no retardance, so it is not affected by the molecules. Off-axis light, however, is not parallel to the LC molecules and sees some residual retardance. Again this converts some of the linearly polarized light into elliptically polarized light. Since all real projection systems have finite cone angles in the illumination, this off-axis contrast degradation degrades overall system contrast, which is the weighted average of on-axis and off-axis contrast. If a C-plate is inserted with a retardance parallel to the LC molecules in the dark state and with a birefringence exactly opposite to the LC birefringence, off-axis light will see a net birefringence of 0 and the polarization of the light will not change. Therefore, the linearly polarized input light will still be linearly polarized at the output polarizer, increasing the contrast.

The most common compensation plates are made from organic polymer materials such as stretched films of polyvinyl alcohol (PVA), polycarbonate (PC) or triacetyl cellulose (TAC). Sometimes they are made from liquid crystal polymer molecules themselves (Koch *et al.* 1997), frozen into place by cross-linking. For very high power compensation plates, it is possible to use inorganic birefringent crystalline materials such as quartz.

In addition to organic and crystalline compensation plates, it is possible to make all-inorganic compensation films with non-birefringent materials (Tan *et al.* 2006). These films are based on form birefringence. Like wire grid polarizers, these require very fine lines to be made, typically a fraction of the wavelength of light. Under an electron microscope, A-plates made with this technology will look very much like the wire grid polarizer shown in Figure 6.53. In fact, a wire grid polarizer can be thought of as a form birefringent retarder with a n_e that is a complex number, due to the conductivity of the aluminum wire. Normal form birefringent materials are made with the same sorts of dielectric materials used to make dichroic filters.

To a first approximation, the indices of refraction of a form birefringent A-plate are given by:

$$n_o = \sqrt{f(n_1)^2 + (1-f)(n_2)^2}$$
$$n_e = 1 \bigg/ \sqrt{f/(n_1)^2 + (1-f)/(n_2)^2} \quad (6.20)$$

In Equation (6.20), n_1 and n_2 are the two dielectric materials making up the grating and f is the duty cycle of the two materials. In the most common version of form birefringence, one of the materials is air, with n = 1.00. More sophisticated analysis with techniques such as Rigorous Coupled Wave Analysis can be used to give a more accurate answer than Equation (6.20), although this equation is satisfactory for most purposes.

A form birefringent C-plate is made up of multiple thin dielectric layers, in precisely the same manner as a dichroic filter. Tan describes a form birefringent AR coating, or FBAR, which serves as both the AR coating and the C-plate for a VAN LCoS cell. This FBAR coating along with a form birefringent A-plate dramatically increased the contrast of the VAN cells, as shown in Table 6.1.

Table 6.1 VAN microdisplay contrast with form-birefringent A- and C-plate compensation.

Color	Red	Green	Blue
Microdisplay contrast without compensation	1200:1	1700:1	1900:1
Microdisplay contrast with compensation	4400:1	5600:1	7500:1

Source: Tan *et al.* (2006).

References

Ahmad, A. (ed.) (1997) *Handbook of Optomechanical Engineering*. Boca Raton, FL: CRC Press.
Arnold, S., Gardner, E., Hansen, D. and Perkins, R. (2001) An improved polarizing beamsplitter for LCoS projection display based on wire-grid polarizers. *SID Intl. Symp. Digest of Technical Papers*, Paper 52.3, 1282–5.
Beeson, K., Zimmerman, S., Livesay, W. Ross, R., Livesay, C. *et al.* (2006) LED-based light-recycling light sources for projection displays. *SID Intl. Symp. Digest of Technical Papers*, Paper 61.5, 1823–6.
Bennett, J. and Bennett, H. (1978) Polarization. In W. G. Driscoll (ed.), *Handbook of Optics*, New York: McGraw-Hill, Inc.
Born, M. and Wolf, E. (1980) *Principles of Optics*, 5th edn. Oxford: Pergamon Press.
Bowron, J. Baker, J. and Schmidt, T. (1996) New high-brightness compact LCD projector. In Ming H. Wu (ed.) *Projection Displays II, SPIE Proceedings*, **2650**: 217–24.
Brandt, H.J. van den (1991) *Projection System*. US Patent 4,983,032.
Brandt, H.J. van den and Timmers, W.A.G. (1992) *Optical Illumination System and Projection Apparatus Comprising Such a System*. US Patent 5,098,184.
Brennesholtz, M.S. (2003) *Projector Color Correction to Target White Points*. US Patent 6,631,992 issued 14 October 2003, assigned to Philips.
Bruzzone, C.L., Ma, J., Aastuen, D.J.W. and Eckhardt, S.K. (2003) High-performance LCoS optical engine using Cartesian polarizer technology. *SID Intl. Symp. Digest of Technical Papers*, Paper 10.4, 126–9.
Bruzzone, C.L., Schneider, J.J. and Eckhardt, S.K. (2004) Photostability of polymeric Cartesian polarizing beam splitters. *SID Intl. Symp. Digest of Technical Papers*, Paper 6.1, 60–3.
Cael, J., Nkwantah, G., Trapani, G. (1997) High-durability KE polarizer for LCD applications: Design considerations. *SID Intl. Symp. Digest of Technical Papers*, 809–12.
Chang, C.-M. Lin, K.-W., Chen, K.-V., Chen, S.-M., Shieh. H.-P.D. (1996) A uniform rectangular illuminating optical system for liquid crystal light valve projectors. *Eurodisplay '96 (16th Intl. Display Res. Conf)*, 257–60.
Chen, J., Kim, K.-H., Jyu, J.-J., Souk, J.H., Kelly, J.R. *et al.* (1998) Optimum film compensation modes for TN and VA LCDs, *SID Intl. Symp. Digest of Technical Papers*, Paper 21.2.
Cheng, D.Y. and Chen, L.-M. (1995) Survey of light sources for image display systems to achieve brightness with efficient energy. In Ming Wu (ed.), *Proc. SPIE: Projection Displays* **2407**: 12–22.
Duelli, M., McGettigan, T. and Pentico, C. (2002) Integrator rod with polarization recycling functionality. *SID Intl. Symp. Digest of Technical Papers*, Paper 36.2, 1078–1100.
Duelli, M. and Taylor, A.T. (2003) Novel polarization conversion and integration system for projection displays. *SID Intl. Symp. Digest of Technical Papers*. Paper 16.3, 766–9.
Eckhardt, S., Bruzzone, C., Aastuen, D. and Ma, J. (2003) *3M PBS for High Performance LCOS Optical Engine*, 3M whitepaper provided by Charles Bruzzone, 5 pp.
Florence, J.M. and Yoder, L.A. (1996) Display system architectures for digital micromirror device (DMD™) based projectors. In M.H. Wu (ed.), *Proc. SPIE: Projection Displays II* **2650**.
Foreal Spectrum, Inc. (2006) Polarization Management Coatings Data Sheet, Rev 1.2, issued 13 February 2006. San Jose, CA.
Gardner, E. and Hansen, D. (2003) An image quality wire-grid polarizing beam splitter. *SID Intl. Symp. Digest of Technical Papers*, Paper 7.1, 62–3.
Hamberg, I. and Granqvist, C.G. (1986) Evaporated Sn-doped In_2O_3 films: Basic optical properties and applications to energy-efficient windows. *J. Appl, Physics* **60**: R123–R159.
Hansen, D., Gardner, E., Perkins, R., Lines, M. and Robbins, A. (2002) The display applications and physics of the ProFlux™ wire grid polarizer. *SID Intl. Symp. Digest of Technical Papers*, Paper 18.3, 730–3.

REFERENCES

Inoué, S. and Oldenbourg, R. (1995) Microscopes. In *Handbook of Optics*, 2nd edn, Vol. II. New York: McGraw-Hill, Inc.

Itoh, Y., Nakamura, J.-I., Yoneno, K., Kamakura, H. and Okamoto, N. (1997) Ultra-high-efficiency LC projector using a polarized light illuminating system. *SID Intl. Symp. Digest of Technical Papers*, Santa Ana, CA: Society for Information Display, 993–6.

Janssen, P.J. (2006) *Loss-less étendue-preserving light guides*, US Patent 7,006,735 Issued 28 February 2006, 10 pp.

Jiang, S.-J., Jin, Z.-C. and Granqvist, C.G. (1988) Low-refractive-index indium-tin-oxyfloride thin films made by high-rate reactive dc magnetron sputtering. *Applied Optics* **27**: 2847–50.

Kappel, D., Fischer, R.E. and Tadic-Galeb, B. (2005) *Illumination device and method for laser projector*, US Patent 6,870,650 issued 22 March 2005.

Kingslake, R. (1983) *Optical System Design*. New York: Academic Press.

Koch, G.C., Winker, B.K. and Gunning, III, W.J. (1997) *LCD splay/twist compensator having varying tilt and/or azimuthal angles for improved gray scale performance*, US Patent 5,619,352 issued 8 April 1997.

Li, Z., Desai, P., Atkins, R. and Ventouris, G. (2000) Electrically Tunable Color (ETC) for full-color reflective displays. *Conference Record of the 20th IDRC, SID*, Paper P.41, 265–8.

Macleod, H.A. (2001) *Thin Film Optical Filters*, 3rd edition. Bristol: Institute of Physics Publishing.

Masumoto, Y. (1995) *Optical Illumination System and Projection Display Apparatus Using the Same*. US Patent 5,418,583,

Melles Griot Inc. (1997) 1997–1998 Catalogue. Irvine CA, p. 199.

Mi, X.-D., Kurtz, A.F. and Kessler, D. (2005) *Display apparatus and method*, US 6,909,473 issued 21 June 2005.

Michaloski, P. and Tompkins, P. (1994) Design and analysis of illumination systems that use integrating rods or lens arrays. In *International Optical Design Conference 1994 Technical Digest*. Optical Society of America, 229–31.

Mihara, Y., Satoh, R., Usui, R., Morinaga, E. and Iwata, Y. (2007) Ta-doped SnO_2 thin films for PDP, *SID Intl. Symp. Digest of Technical Papers*, Paper P-210, 399–402.

Mohri, S. and Matsuo, T. (1995) Highly durable dyed polarizer for use in LCD projection. In M.H. Wu (ed.), *Proc. SPIE: Projection Displays* **2407**: 62–72.

Ooi, Y., Hirai, Y. Kunigita, M., Niiyama, S., Kumai, H. et al. (1995) Reflective-type LCPC projection display. *SID Intl. Symp. Digest of Technical Papers*, 227–30.

Palmer, J.M. (1995) The measurement of transmission, absorption, emission and reflection. In M. Bass (ed.), *Handbook of Optics*, 2nd edn, Vol. 2, 25.1–25.25.

Park, J.-M., Yoo, E.-H., Kim, Y.-W. and Brennesholtz, M.S. (2004) A 1080 line, single panel LCOS rear projection system with a rotating drum. *SID Intl. Symp. Digest of Technical Papers*, Paper 12.4, 178–81.

Pentico, C., Gardner, E., Hansen, D. and Perkins, R. (2001) New, high performance, durable polarizers for projection displays. *SID Intl. Symp. Digest of Technical Papers*, Paper 52.4, 1287–9.

Ralli, P.J., Gunnarsson, G.H., Nagarkar, P.V. and Trapani, G.B. (2007) Intrinsic polarizers – ultra durable dichroic polarizers for LCD Projection. *SID Intl. Symp. Digest of Technical Papers*, Paper 9.1, 101–4.

Rancourt, J. (1996) *Optical Thin Films User Handbook*. Bellingham, WA SPIE Press.

Robinson, M.G., Chen, J. and Sharp, G.D. (2005) *Polarization Engineering for LCD Projection*. SID Series in Display Technology. New York: John Wiley & Sons, Inc.

Rosenbluth, A.E., Dove, D. and Doany, F. (1997) Contrast losses in projection displays from depolarization by tilted beam splitter coatings, *Conference Record of the 1997 International Display Research Conference*, SID, 226–9.

Sakata, H., Aruga, S., Egawa, A., Morikuni, E., Akahane, Y. et al. (2006) A high-efficiency HTPS projection engine with LED light sources. *SID Intl. Symp. Digest of Technical Papers*, Paper 55.4, 1724–7.

Sannohe, S. and Miyatake, Y. (1995) *Polarization Beam Splitter and Projection Display Apparatus*. US Patent 5,453,859 issued 26 Sep. 1995.

Schadt, M. and Schmitt, K. (1997) Flat liquid crystal projectors with integrated cholesteric color filter/polarizers and photo-aligned optical retarders. *Intl. Display Research Conf. Record*, 219–22.

Shahzad, K. and Shimizu, J.A. (2006) Contrast behavior of a field lens with a wire-grid PBS. *Journal of the SID* **14**(3): 293–301.

Sharp, G.D. and Birge, J.R. (1999) Retarder stack technology for color manipulation. *SID Intl. Symp. Digest of Technical Papers*, Paper 52.01, 1072–5.

Sharp, G.D., Birge, J., Chen, J. and Robinson, M.G. (2000) High throughput color switch for sequential color projection, *SID Intl. Symp. Digest of Technical Papers*, Paper 9.4, **31**: 92–5.

Sharp, G.D., Chen, J., Robinson, M.G. and Birge, J.R. (2002) Skew-ray compensated retarder-stack filters for LCoS projection. *SID Intl. Symp. Digest of Technical Papers*, Paper 51.2, 1338–41.

Shikama, S., Toide, E. and Kondo, M. (1990) A polarization transforming optics for high luminance LCD projector. *Eurodisplay* '90 (IDRC), pp. 64–7.
Shimizu, J. (1995) Brightness distributions in projection displays. *SID Intl. Symp. Digest of Technical Papers*, 66–9.
Shimizu, J. and Janssen, P. (1996) *Compensation Plate for Tilted Plate Aberration*. US Patent 5,490,013.
Shimizu, J. and Janssen, P. (1997) *Integrating Lens Array and Image Forming Method for Improved Optical Efficiency* US Patent 5,662,401, issued 2 Sep. 1997.
Shimizu, J. and Janssen, P. (2005) A light guide engine for single-panel LCoS. *SID Intl. Symp. Digest of Technical Papers*, Paper 64.1, 1806–9.
Smith, W.J. (1990) *Modern Optical Engineering: The Design of Optical Systems*, 2nd edn. New York: McGraw-Hill, Inc.
Tan, K.L., Hendrix, K.D., Duelli, M., Shemo, D.M., Ledeur, A. et al. (2005) Design and characterization of a compensator for high contrast LCoS projection systems. *SID Intl. Symp. Digest of Technical Papers*, Paper 64.2, 1810–13.
Tan, K.L., Hendrix, K.D., Hruska, C., Mache, A. Duncan, J. (2006) Inorganic trim retarder compensator for VAN-mode LCoS projection systems. *SID Intl. Symp. Digest of Technical Papers*, Paper 50.3, 1610–13.
Texas Instruments (2005) *Single-Panel DLP™ Projection System Optics*, Application Report Discovery DLPA002. This report can be downloaded from http://focus.ti.com/download/dlpdmd/Discoverydlpa002.pdf
Tsai, R.-Y., Ho, F.-C. and Hua, M.-Y. (1977) Annealing effects on the properties of indium tin oxide films coated on soda lime glasses with a barrier layer of TiO_2-SiO_2 composite films. *Optical Engineering* **36**: 2335–40.
Tyan, R.-C., Salvekar, A.A., Chou, H.-P., Cheng, C.-C., Scherer, A. et al. (1997) Design, fabrication, and characterization of form-birefringent multilayer polarizing beam splitter, *Journal of the Optical Society of America* **14**(7), July: 1627–36.
Vaan, A.J.S.M. de, van de Brandt, A.H.J., Karsmakers, R.A.M., Stroomer, M.V.C. and Timmers, W.A.G. (1993) Polarization conversion system for LCD projection. *1993 Eurodisplay (IDRC), SID*, 253–6.
Vinokur, K., Shariv, I., Gorbatov, N., Pilossof, N. and Cheifetz, E. (1998) High contrast ratio broadband liquid crystal polarizing beam splitter, *SID Intl. Symp. Digest of Technical Papers*, 690–3.
Wang, S. and Ronchi, L. (1988) Principles and design of optical array. In E. Wolf (ed.), *Progress in Optics*, Vol. 25, Amsterdam: North-Holland, 279–347.
Wang, Y. (1997) Surface plasmon tunable color filter and display device. *SID Intl. Symp. Digest of Technical Papers*, 63–6.
Weber, M.F., Stover, C.A., Gilbert, L.R., Nevitt, T.J. and Ouderkirk, A.J. (2000) Giant birefringent optics in multilayer polymer mirrors, *Science* **287**, 31 March, 2451–6.
Wortman, D.L. (1997) A recent advance in reflective polarizer technology. *Conference Record of the 1997 International Display Research Conference*, M98–M106.
Yakovenko, S., Konovalov, V. and Brennesholtz, M. (2004) Lifetime of single panel LCOS imagers, *SID Intl. Symp. Digest of Technical Papers*, Paper 6.2, 64–7

7

Projection Lenses and Screens

This chapter will cover the projection lenses used in microdisplay, light-valve, light-amplifier and CRT projection systems. These lenses are a key factor to the image quality, throughput and cost of projection systems. The chapter will also cover front and rear projection screens. Strictly speaking, screens, unlike all of the other optical components and systems discussed in this book, are not part of the light engine of a projector. Since screen properties can significantly enhance or degrade the projected image, they are covered in order to allow the projector designer to understand the total performance of his system.

This chapter will also discuss speckle in projected images. Speckle first became a problem in microdisplay-based rear projection systems. In these systems, modifications to the screen design proved to be sufficient to reduce speckle to acceptable levels. When laser illumination is used, speckle is an issue in all projector designs for both front and rear projection.

7.1 Projection Lenses

The design of projection lenses is both an art and a science. Fortunately, there are many good books on optical design, including Smith (1990, 1992), Kingslake (1983) and Fischer and Tadic (2000). There are two prerequisites for getting a good lens design: a talented and experienced lens designer and considerable computer support, primarily in the form of modern design, optimization and evaluation software (Betensky 1993). However, there are a few special considerations that apply to projection lenses of which the projector designer must be aware of, even when the lens design work is left to a specialist.

132 PROJECTION LENSES AND SCREENS

7.1.1 Three-lens Projectors

The lenses associated with three-lens projectors are the simplest type of design. Three-lens CRT architectures are discussed in Section 9.1.1 and three-lens light-valve designs are discussed in Section 10.3.3. There are three main reasons why these designs are simple:

1. Since each lens transmits only a single color (red, green or blue), color correction of the lens is normally either unnecessary or relatively simple, at least for lower resolution (e.g., consumer) systems.
2. Three-lens projectors rarely have zoom lenses because of convergence issues.
3. Since there are no color combining optics between the image source and the lens, the required back focal length of the lens is short.

An example of a projection lens for a CRT is given by Clarke (1988a) and is shown in Figure 7.1(a). This is a four-element lens with all plastic elements that is intended for consumer CRT projectors. Higher resolution professional projectors typically use more elements and a glass-plastic hybrid design. One advantage of plastic lens elements is the ease with which aspherical elements can be manufactured. In glass, aspherical elements are significantly more expensive than spherical elements of similar size. In molded plastic elements, once the mold is made, there is little cost difference between a spherical and an aspherical element of the same size.

CRT projection lenses have a very low f/#, typically in the range of f/0.9-f/1.1. The lenses must have large diameters, about 3.5 inches for a typical consumer CRT system with 7 inch tubes. While the details may differ, one of the standard features of a CRT projection lens is a large negative element as a field flattener near the CRT, as shown in Figures 7.1(a) and 7.1(b). This element is typically liquid coupled to the CRT faceplate. This liquid coupling serves two functions. First it suppresses Fresnel reflection off both the outer surface of the faceplate and the inner surface of the field flattener. Secondly, the liquid serves to cool the CRT and allow it to operate at higher power and higher brightness. Obviously, this field flattener must be larger than 3.5 inch in diameter: it typically has a rectangular shape with the same size as the tube face plate.

Another typical feature of modern CRT projection lenses shown in Figures 7.1(a,b) is that most of the positive power of the lens is in a single element. In the case of Figure 7.1(b), taken from (Moskovich 2003), the element is labeled L2. The additional thin elements L1, CR and L3 before and after this strong element correct the aberrations introduced by the main element. These corrective elements can have either simple or compound curvatures, as shown in elements L1 and L3 in Figure 7.1(b). In a modern high-performance lens, the element with power would typically be a glass element and all the additional elements would be plastic (Watanabe 2000).

Rear projection CRT lenses must have very wide field angles, in order to minimize the depth of the projector cabinet. Currently, CRT projection lenses for consumer applications are limited to about ±45° field angles by the Fresnel lens used as a field lens at the projection screen. This effect will be covered in more detail in Section 7.2.3. Another problem limiting CRT projection lens field angles to about ±45° is the $\cos^4(\theta)$ law, which will be discussed in Section 7.1.5.

7.1.2 Single-lens Projectors

Single-lens projectors use significantly more complex lenses. These lenses must be well corrected for color, since all three colors are projected by the same lens. In any of the transmissive single-panel

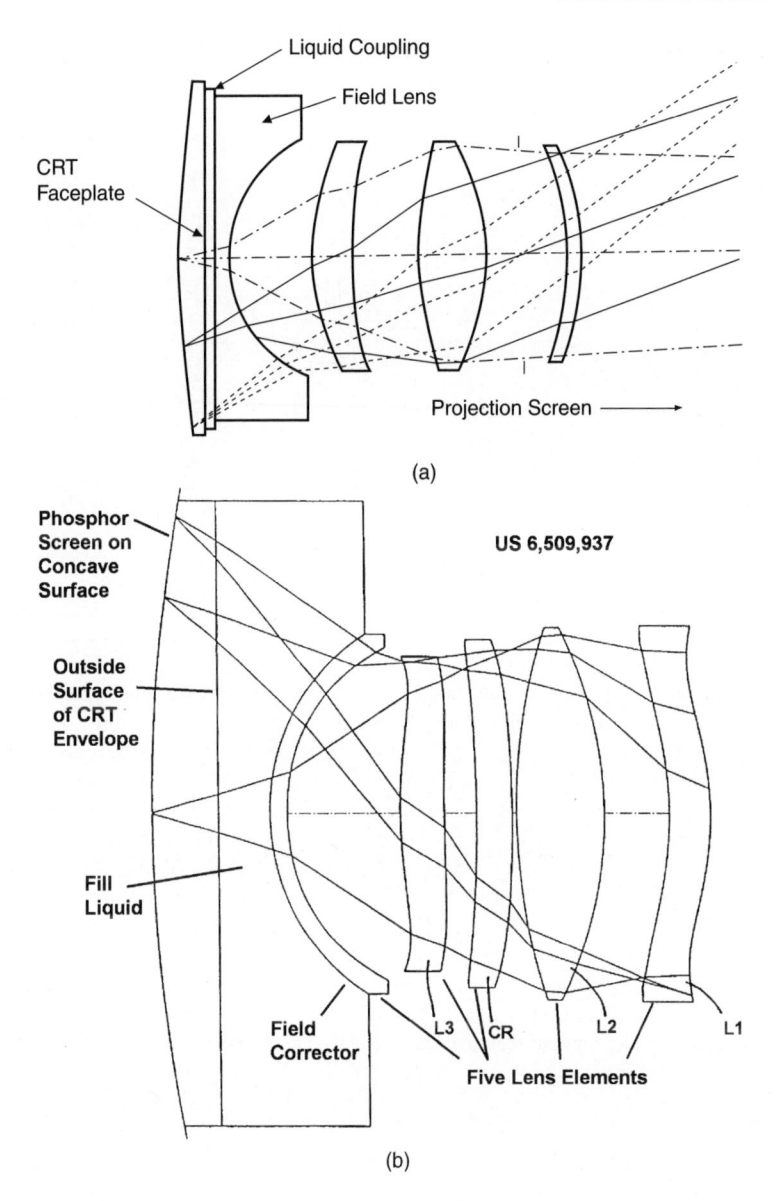

Figure 7.1 Examples of CRT projection lenses: (a) moderate performance lens for a consumer rear projection TV (from Clarke 1988a; with permission of the SPIE); (b) high performance lens for a professional projection system (Moskovich 2003; United States Patent and Trademark Office).

color projectors, discussed in Chapter 10, the back focal distance[1] is not a problem because there is no color combining optics between the panel and the lens, and the distance can be made relatively short.

However, for a three-panel architecture where there are dichroics between the lens and the panel, relatively long back focal length lenses must be used. The most common design is the reversed telephoto

---

[1] The back focal length is the distance from the rearmost lens element in a design to the image plane of the microdisplay or light valve.

lens. An example of a reversed telephoto design (Moskovich 1997) is shown in Figure 7.2. Another example of a similar lens is given by Chen *et al.* (1995), where extensive use has been made of plastic elements and aspherical surfaces.

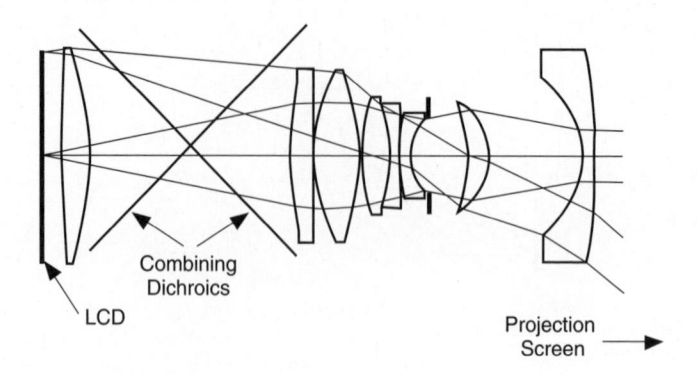

Figure 7.2 Typical three-panel LCD projection lens.

Note that this design requires an optical element relatively close to the panel. The position of this element is before the dichroic that combines the blue and green beams with the red beam. Therefore, this element must be replicated in each channel. Similar designs put a field lens directly adjacent to the LCD itself. In any lens where there are duplicated optical elements, it is necessary to ensure that the magnification, distortion, etc. of all three paths are the same.

Back focal length is a design problem when the back focal length must be greater than the focal length of the lens. This is rarely a problem in front projectors, but rear projection systems typically need very short focal length lenses in order to fit the optical path into a reasonably sized cabinet. In single-panel DLP projectors and three-panel LCD projectors where current design practice has only a single prism between the projection lens and the microdisplay, back focal length is a relatively modest problem, except where the design goal is to achieve a very shallow cabinet. For three-panel LCoS, where it is necessary to have two prisms in the back focal length region between the lens and the microdisplays, the design issue is more serious. These three optical architectures will be discussed in more detail in Chapter 10.

When it comes to converging a three-panel system, mechanical x-y adjustment of two of the panels can align the screen image of a single point on each panel with the corresponding point on the third reference panel. The reference panel is normally the center panel, which is typically green in three -LCD and LCoS designs. In practice, the two adjustable panels must have full six-axis adjustments, three linear (x, y and z) and three angular (θ, ϕ and ψ). If the three optical paths have the same magnification and distortion, then the entire image can be brought into convergence. If there is a magnification or distortion difference between the three optical paths, it will be impossible to achieve convergence everywhere in the image.

An alternative approach to convergence is to produce a lens with good MTF for all three colors but poor convergence properties. The lens can have differential magnification, lateral color, differential distortions and other problems normally not acceptable in imaging systems. Convergence is then achieved in the video signal through image processing and geometry correction of the three colors separately (Ramachandran and Prior 2007). This can significantly reduce the cost of the projection lens in single-panel and three-panel systems and simplify convergence in three-panel systems.

This convergence issue is one of the reasons why three-lens pixelated light-valve projectors are not common. The raster in CRT projectors can be distorted electronically to correct for differences between the three lenses. In fact, this must be done, in order to correct the trapezoidal distortion of the off axis tubes. Raster addressed light valves such as the Hughes/JVC ILA, the GE Talaria or the Eidophor[2] could also correct for the difference between lenses by adjusting the raster geometry. For a pixelated light valve or microdisplay, which dominate both the current commercial market and current R&D efforts, this raster distortion is not possible and the lenses would have to be very closely matched for a three-lens projector to be acceptable. Alternatively, electronic image geometry correction would need to be used.

The rear projection TV industry, in an attempt to compete with flat panel displays such as LCD and Plasma, have designed very thin systems. These thin systems have very wide angle projection lenses, with very short focal lengths. These projection lenses can be of all-refractive design, as shown in Figure 7.3(a) (Gohman 2005) or can be a combination of a moderate focal length refractive lens with an

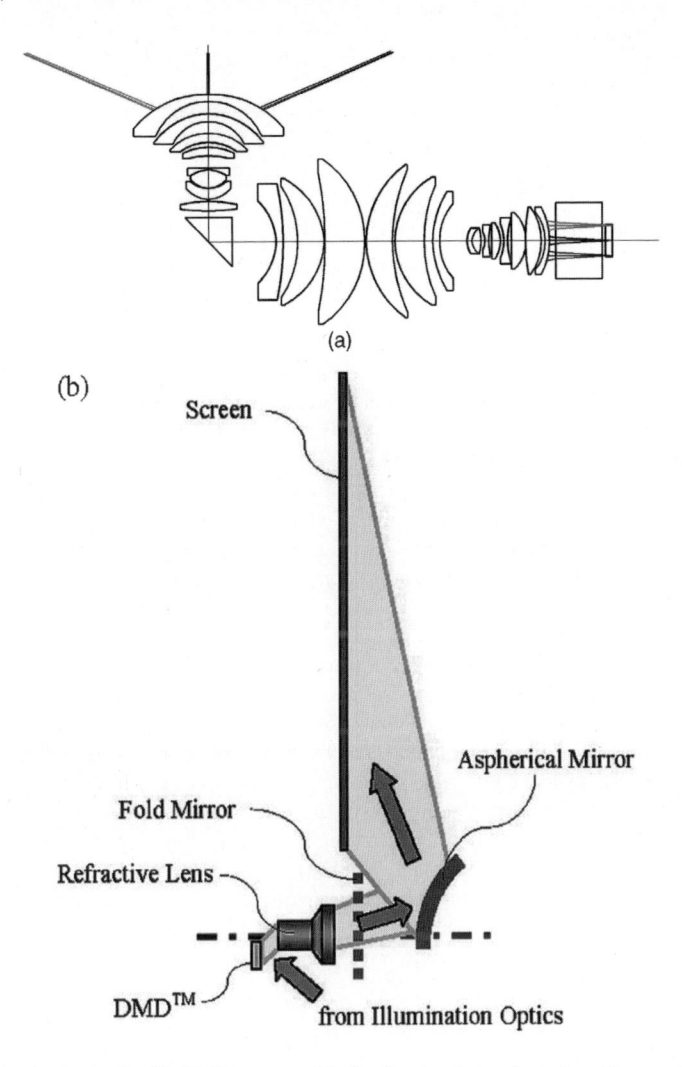

Figure 7.3 Projection lenses for thin RPTV systems (a) all-refractive design (reproduced by permission of © 2005 the Society for Information Display (Gohman 2005)); (b) refractive/reflective design with single aspherical mirror (reproduced by permission of © 2006 the Society for Information Display (Kuwata et al. 2006)).

---

[2] None of these three projector technologies is currently in production.

aspherical mirror to widen the projection angle, as shown in Figure 7.3(b) (Kuwata *et al.* 2006). Some refractive/reflective designs use more than a single aspherical mirror. In fact, it is possible to design all-reflective 'lenses' (Ogawa *et al.* 2004) with four mirrors to project the image on the screen. Due to the high cost of the all-refractive design, which includes 25 optical elements while a typical RPTV lens has 8–12, it is likely that future thin RPTV systems will include at least one aspherical mirror to expand the projection angle.

7.1.3 Zoom Lenses, Focal Length and Throw Ratio

One feature desired by projector manufacturer marketing departments that makes designing a projection lens more difficult is making the lens a zoom or vari-focus lens. Both of these lens types allow a variable image size to be projected on a screen. This allows the projector to be placed in a convenient position, and the image size then adjusted so the image exactly fills the screen. While design of zoom or vari-focus lenses is more difficult, and manufacturing is more expensive because of the extra elements, design of these lenses is a well-known art. It is covered in the standard texts on lens design mentioned at the beginning of this section and is supported by all professional lens design programs.

Zoom and vari-focus lenses differ in that, in a zoom lens, the image remains in focus as the image size is changed. This is normally accomplished by a variable cam mechanism designed into the lens that moves two or more elements or groups to automatically adjust focus as the lens is zoomed. This cam mechanism is missing from a vari-focus lens: after the image size is changed, the operator must manually refocus the image. The zoom lens is more expensive due to the extra mechanical and/or optical elements. With a zoom or vari-focus lens, projector users have a much wider choice of where they can set up the projector relative to the screen than they would have with a fixed focal length lens, or even with a series of interchangeable fixed focal length lenses.

Lenses for projection and other applications are normally designated by their focal lengths. For instance, a zoom lens for a 35 mm camera might have a range in focal lengths from 35 mm to 85 mm. This would mean a variation from a moderate wide-angle lens to a moderate telephoto lens. Designating a lens by its focal length is only meaningful if the size of the object is known. For a 35 mm camera, the size of the film is 24 × 36 mm, or 43.3 mm diagonal, and most regular users of cameras learn intuitively what focal length corresponds to what magnification of the image on the film.

In projectors, the focal length of the lens is a less useful way of designating the lens because there is a wide variety of light valve or microdisplay sizes in use, from a lower limit of less than about 13.7 mm (for VGA DMD devices) to an upper limit of about 250 mm diagonal for some matrix type single light-valve projectors made with panels from laptop computers. Microdisplays as small as 4.6 mm diagonal have been shown in prototype pico-projectors, although no projection systems with microdisplays this small have reached production. A more useful way to designate a projection lens is by the ratio between the distance to the screen and the width of the image on the screen. This ratio is sometimes called the 'throw ratio'. This terminology is not standardized, the same thing can also be called 'projection ratio', 'projection factor', 'lens factor' or some other term involving 'throw', 'projection', 'lens' or 'ratio'. Unfortunately, these same terms are sometimes used in slightly different combinations to designate a zoom lens: it is necessary to understand exactly what is being specified when reading projector marketing literature.

Projector manufacturers often designate how large a picture the projector is capable of making, for example, 'screen sizes from 20″ to 300″ (measured diagonally)'.[3] This sort of information is useful as a general guide, especially when combined with an ANSI lumen specification. However, since it says nothing about the distance to the screen, it says nothing about the projection lens.

---

[3] From Mitsubishi G1A LCD Video/Data Projector literature.

Examples of various throw-ratios are shown in Figure 7.4. A lens with a throw-ratio of 1:1 would be a very wide-angle lens that would produce an image whose width is the same as the distance from the lens to the screen. A 1:1 lens would normally be used in a rear projection application. This could either be a dedicated rear projector design or a front projector used with a separate rear projection screen. A 7:1 lens would have a long throw distance between the lens and the screen. This would allow a projector to be set up in the back of a large auditorium behind the entire audience and fill a screen large enough to be seen by everyone. For some projectors intended for use in large venues, lenses with 10:1 or higher throw-ratios are available. Projection lenses with throw ratios of about 3:1 are the most widely useful. They are especially suitable for small conference rooms where the projector will be placed on the conference table and used to project on a screen at one end of the room.

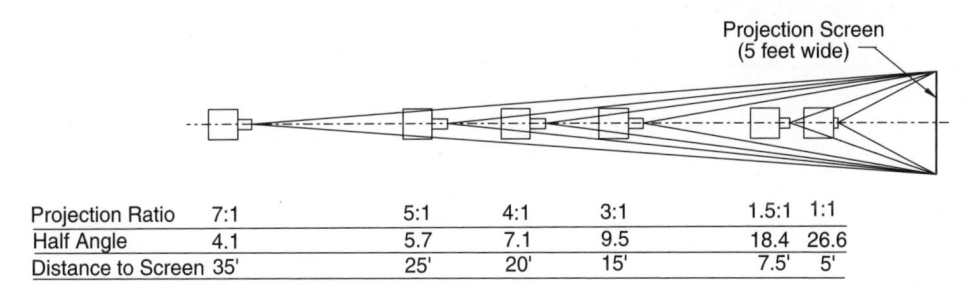

Projection Ratio	7:1	5:1	4:1	3:1	1.5:1	1:1
Half Angle	4.1	5.7	7.1	9.5	18.4	26.6
Distance to Screen	35'	25'	20'	15'	7.5'	5'

Figure 7.4 Throw ratios for projection lenses.

With a throw-ratio type of designation, an end-user can more easily get a feeling for how a particular projection lens would be used than with a focal length designation. Throw-ratio calculators are commonly available online, often at a manufacturers website. For example, Projector Central (http://www.projectorcentral.com) has an online calculator that will give the relationship between projection distance, image size and image brightness for virtually any projector commercially available in the US.

7.1.4 Projection Lens Offset

Another projection lens feature desired by marketing departments that makes design of a lens more difficult is the ability to offset the lens in order to project an image off the optical axis without distortion. This is shown in Figure 7.5 for an image above the level of the projector.

In Figure 7.5(a), the front end of the projector has been tilted up in order to project onto a screen above the projector. This causes two problems. First, the focus of the image at the top and the bottom of the screen will be different, and it may not be possible to get both areas in focus at the same time. Secondly, and more importantly, trapezoidal distortion will be induced in the image, and the top of the image will be wider than the bottom of the image.

In Figure 7.5(b), a projector has been placed on a table horizontally and the lens has been offset in order to project the image above the centerline. In this case, there is no trapezoidal distortion from the lens offset, and the image will remain rectangular on the screen. In addition, the image is in focus at all points, even though the top of the image is further from the projection lens than the bottom. Currently, the maximum practical amount of offset that can be designed into a projection lens is about 12°, as shown in Figure 7.5(b). This is sufficient offset to allow a lens with a throw ratio of 1.8:1 to produce a 4:3 image at the screen where the bottom of the image is at the level of the projector. This, for example, would allow a projector to sit flat on a conference room table and produce an image above the table and readily visible to all viewers.

138   PROJECTION LENSES AND SCREENS

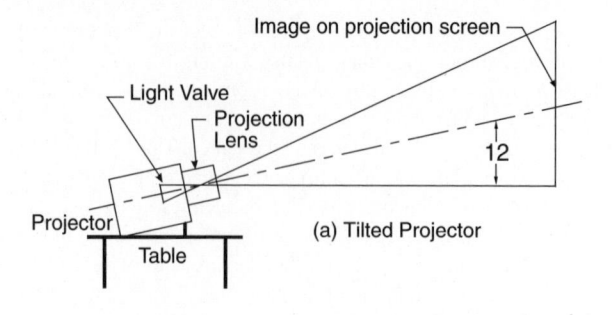

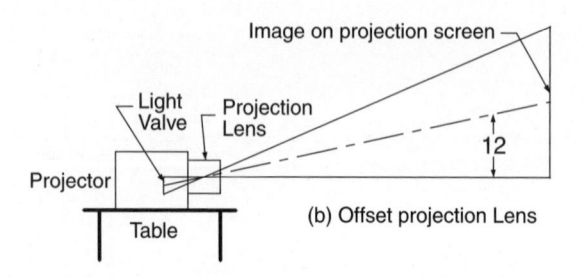

Figure 7.5   Application of an offset projection lens: (a) tilted projector; (b) offset projection lens.

This feature is particularly desirable, although difficult to implement, in zoom lenses. If the bottom of the image starts level with the projector, as the lens is zoomed the bottom of the image does not move. As the image gets larger, the bottom remains at the same place on the screen while the top rises and the sides move out. This makes it particularly easy to set up a portable projector in a variety of conference rooms. It also means that as the image size increases there is no problem with the table top obscuring the lower edge of the image.

When a projection lens is designed to be used with offset, the field of the projection lens must be larger than it would be used if there were no offset. This effect is shown in Figure 7.6. In this example, a lens with a throw ratio of 1.8:1 is offset 12°, enough so the bottom of the image is level with the projector.

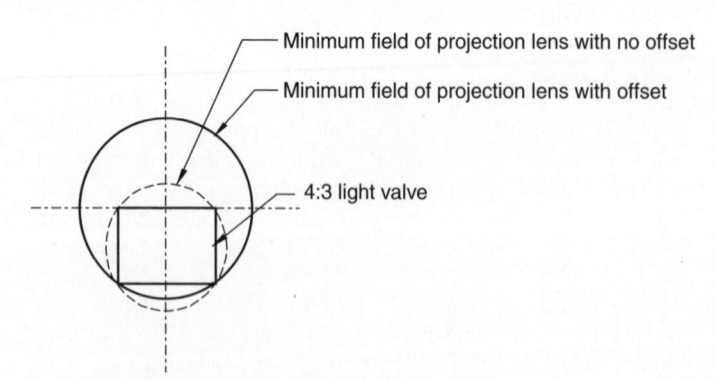

Figure 7.6   Field utilization with projection lens offset.

The fields of most projection lenses are circular, due to the circular nature of common lens elements. Obviously, a 4:3 rectangle can only use a portion of this circular field. The minimum acceptable field diameter is equal to the diagonal of the microdisplay. This occurs when the lens is centered on the microdisplay and there is no offset. When there is an offset, as shown in Figure 7.5(b), the field still must cover the entire microdisplay. In this case, the field increases in diameter by a factor of $\sqrt{2}$ and the area of the field doubles, as shown in Figure 7.6. This increased field size can significantly increase the cost and complexity of the lens.

Figure 7.5(b) shows how the microdisplay, the principal plane of the projection lens and the image on the screen remain parallel to each other as the lens and the image are offset. This is called the Scheimpflug condition.

Note in Figure 7.6 only the bottom half of the field is used to project above the axis. If the lens mount is designed to allow full x-y adjustment of the projection lens, it is not only possible to place the projector below the center of the screen, but it is possible to place the projector above or to one side of the screen without inducing trapezoidal distortion. In this way, a projector can be ceiling mounted and projecting down 12°, or mounted in a corner of the room and projecting to the left or right side by 12°. In this case, the top half or the right or left side of the lens field will be used.

Figure 7.7 shows the relationship between the illumination path (not to scale), the microdisplay, the projection lens and the image on the screen. Figure 7.7(a) shows the normal lens offset technique. In this case, the illumination path axis is perpendicular to the microdisplay and remains telecentric. The illumination path is f/2.4 in this example, with a cone of ±12°. The axis of this cone is parallel to the axis of the projection lens, so the lens also sees a cone angle of ±12°. Therefore, the lens f/# is still f/2.4.

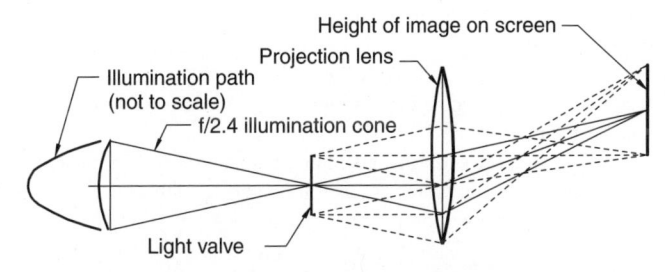

Figure 7.7 Illumination paths with projection lens offset: (a) lens offset without illumination path offset; (b) lens and illumination path offset.

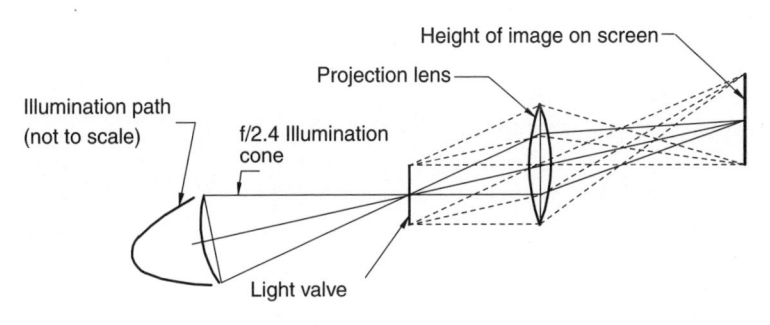

In Figure 7.7(b), the situation is changed. The illumination path is tilted relative to the microdisplay by the same amount as the lens offset. One thing this accomplishes is that the field increase requirement shown in Figure 7.6 is no longer required: the field needs to cover only the microdisplay and no additional area. On the other hand, the f/# of the lens must be reduced. While the cone angle of the light relative

140 PROJECTION LENSES AND SCREENS

to the illumination system axis is still ±12°, relative to the microdisplay or projection lens the angles go from 0° to 24° degrees in this example. Off-axis light rays of 24° require a f/1.23 lens for all to be transmitted. In this case it is the pupil that is underutilized, not the field, as shown in Figure 7.8.

Portion of pupil used with 12 degree offset and f/2.4 illumination at the LCD.

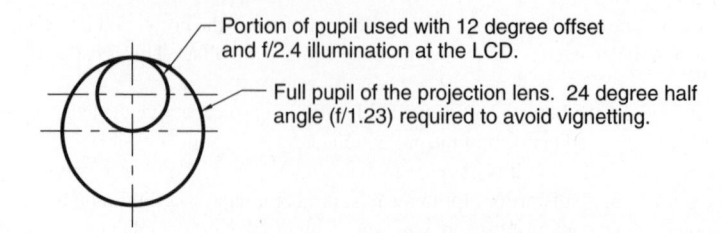

Full pupil of the projection lens. 24 degree half angle (f/1.23) required to avoid vignetting.

Figure 7.8 Pupil utilization with a tilted illumination path.

The design shown in Figure 7.7(a) is far more common than the arrangement in Figure 7.7(b). There are several reasons for this. The main reason is that in Figure 7.7(a) only the projection lens needs to be moved on an x-y stage if the designer desires variable offset. In Figure 7.7(b) it would be necessary to move both the lens and the illumination path to achieve variable offset. Moving the illumination path would involve not only a translation, but a tilt change as well.

Another reason for preferring the arrangement in Figure 7.7(a) over 7.7(b) is contrast of most microdisplays is a function of the angle the light passes through the microdisplay. In Figure 7.7(a), the maximum angle is 12° but in Figure 7.7(b) the maximum angle is 24°. Therefore, for most microdisplay and light-valve technologies, the arrangement in Figure 7.7(a) would show better contrast.

If a microdisplay requires a tilted illumination path, the arrangement in Figure 7.7(b) might be preferred. For example, the DMD requires 20° off-axis illumination and this arrangement may be preferred. Some LCD designs also perform better with off-axis illumination and this technique might be preferred.

If the lens has any inherent distortion such at pincushion or barrel distortion, this will affect the image whether the lens is offset or not. The effect will be more severe with lens offset, however, because higher field angles or lower f/# lenses are used. Also, with enough offset to allow a projector to sit flat on a table and not be obstructed by the edge of the table, the lower edge of the image goes roughly through the center of the lens. The top of the image goes through the edge of the lens and sees a different distortion. Therefore, the top and bottom edges of the image may not be parallel, with the bottom edge straight and the top edge curved.

In Figure 7.9 these various distortions are shown for a 4:3 image, when there is sufficient offset to allow the image to be level with the tabletop on which the projector is sitting. Clearly, the rectangular image with no distortion is preferred if it can be achieved in a cost effective lens design.

——— No Distortion
— — Pincushion Distortion
·········· Barrel Distortion
- - - - - Trapazoidal Distortion

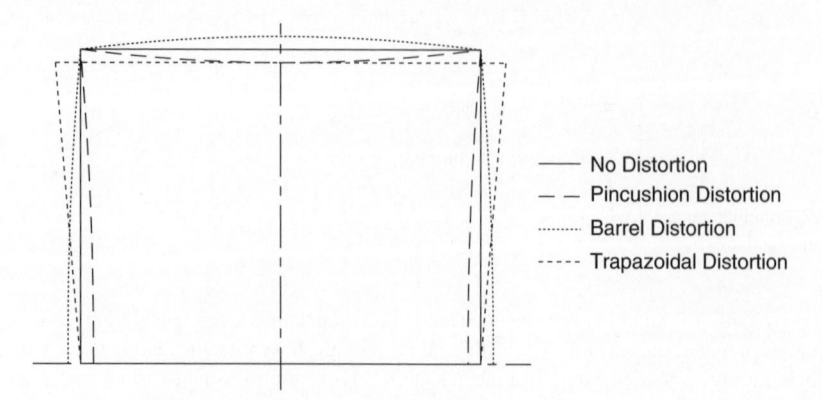

Figure 7.9 Distortions with image offset from the horizontal.

7.1.5 Matching the Projection Lens to the Illumination Optical Path

In a simple projection system, the illumination at the projected image at the field angle θ varies as $\cos^4(\theta)$. This law is derived, for example, in Boyd (1983). Another derivation using a different approach involving reference spheres is given by Koch (1992). The Boyd approach helps to elucidate the basic physical and geometrical principles of radiometry while the Koch approach can be applied directly to other luminance and throughput problems associated with projectors.

The $\cos^4$ law is illustrated in Figure 7.10. This law applies most directly to CRT projection sets with flat CRT faceplates. The $\cos^4$ law is not absolute. Manipulation of the pupil position and shape during lens design can reduce the center to edge fall-off relative to the $\cos^4$ law. A curved faceplate (Wolf 1937; Malang 1989) or an IARC CRT (Vriens 1988; Clarke 1988b) design can also modify the light falloff predicted by the law. With realistic, manufacturable designs for CRT projection lenses, however, the $\cos^4$ law is a moderately good predictor of the center to edge fall-off in luminance of a CRT projection system.

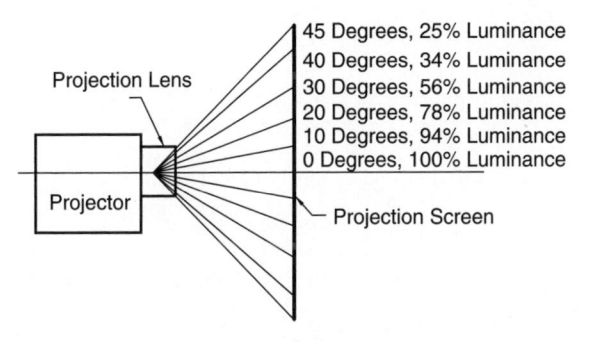

Figure 7.10 $\cos^4$ law.

For a given lens throw ratio and image aspect ratio, the field angle to the corner of the screen can be calculated. Knowing the field angle, the luminance at the corner can be estimated from the $\cos^4$ law. The results of this calculation are shown in Table 7.1 for 4:3 screens. As can be seen, for high throw ratios (long focal length lenses), 3.0:1 and higher, the luminance fall off from the center to the corner of the screen due to the $\cos^4$ law is not significant. For low throw ratios, this effect can be the major factor reducing the corner luminance compared to the center.

Table 7.1 Relative luminance at corner of screen due to $\cos^4$ law.

Throw ratio	Angle (to corner)	$\cos^4 \theta$
0.63	45.0	25 %
1.00	32.0	52 %
1.50	22.6	73 %
3.00	11.8	92 %
5.00	7.1	97 %
7.00	5.1	98 %
10.00	3.6	99 %

The $\cos^4$ law applies primarily to projectors with Lambertian or near-Lambertian sources, i.e. CRT projectors. The law does not necessarily apply to microdisplay and light-valve projectors. It has been shown (Shimizu 1995) that if the f/# of the projection lens is lower than the f/# of the illumination path, the $\cos^4$ law no longer applies and the relative luminance of the corner can be significantly higher than predicted. In this case, the uniformity of the illumination of the microdisplay is the governing factor in the uniformity of the screen. This uniformity is in turn governed by the design of the integrator or other illumination optics. When designing a microdisplay projection system with a low throw ratio lens (short focal length), it is especially important to keep this effect in mind.

The design of the projection lens must be matched to the design of the illumination path for another reason as well. When optimizing the design of the projection lens, the normal merit function involves either MTF (see Chapter 13) directly, or factors that contribute to the MTF, such as spot size, aberrations and color correction. The typical lens optimization program assumes uniform illumination over the pupil. Unfortunately, this is rarely the case in a projection system, and it can significantly affect the MTF of the optimized lens (Rykowski and Forkner 1995). Conversely, programs that do a good job of analyzing lumen throughput and uniformity rarely have satisfactory optimization tools to design a projection lens.

Rykowski recommends a five-step process when designing an illumination system for a microdisplay or light-valve projector and a matched projection lens. The design process would involve two software packages, one with good lens design optimization facilities and the other with good analysis facilities for non-uniform light distribution.

1. Do an initial design of the illumination system with the optimization program. This design must take into account the desired f/# of the projection lens, the target lumen throughput and the uniformity at the microdisplay.

2. Do an initial design of the projection optics with the optimization program. This design must have the correct f/#, field angles, MTF, etc. This stage of the design would be done with a nominal illumination system, not the system from step 1.

3. Analyze the projection optics from step 2 with the real illumination system from step 1.

4. Re-optimize the projection lens design with the data from step 3.

5. Re-optimize the illumination system design.

While the re-optimization of the illumination system should include only small changes that will have little effect on the projection lens MTF and other properties, a prudent person would add a sixth step: re-analyze the complete system to ensure the illumination path and projection path work together as expected.

7.2 Projection Screens

While projection screens are not a part of the projector light engine, the projector designer should have a basic understanding of projection screens to understand how the images generated in the projector are seen by the audience. Projection screens have been poorly documented in the literature, perhaps even less than other elements of projection system engineering. An old but still useful review of projection screens is given in Vlahos (1968).

There are two basic modes of operation of projection screens, front projection and rear projection, as shown in Figure 7.11. In front projection, the screen reflects the light and the projector and the audience are on the same side of the projection screen. In rear projection, the screen transmits the light, and the projector and audience are on opposite sides of the screen. Generally, the contrast of an image shown on a rear-projection screen is higher than the same image from the same projector on a front projection screen as shown in Figure 7.12 (Brubaker 2007). This figure should be considered an example, since there are many types of front and rear projection screens and each screen design will have its own effect on contrast. While this effect is most noticeable in the presence of ambient light, it can also occur in an otherwise dark room when light from the image itself reflects off the walls of the room and back onto the front projection screen. Front projection screen designers have worked to reduce this contrast differential between front and rear projection, as will be discussed in Section 7.2.4.1. While these efforts have improved the situation, the differential remains and if contrast with ambient light is an issue, rear projection typically provides higher performance than front projection.

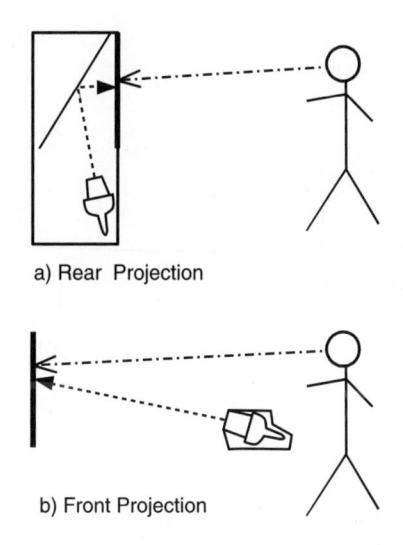

Figure 7.11 Front vs rear projection: (a) rear projection; (b) front projection.

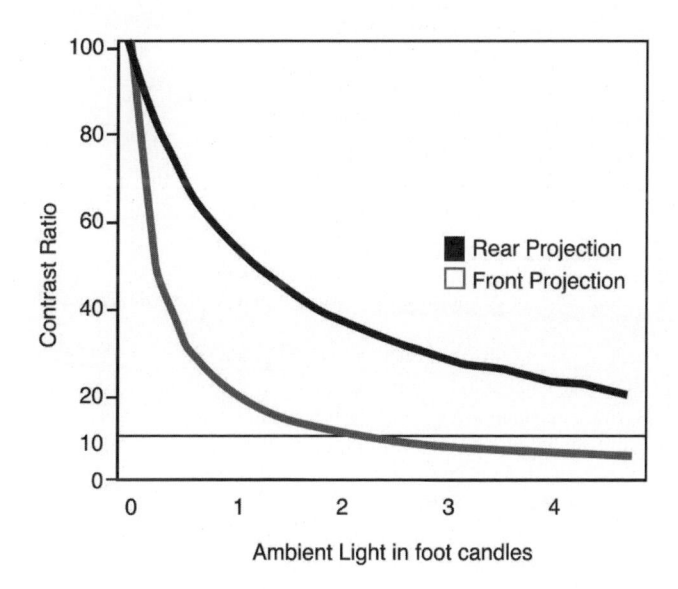

Figure 7.12 Contrast differential between front and rear projection in the presence of ambient light (reproduced by permission of © 2007 Da-Lite screen co., Inc. (Brubaker 2007)).

Sometimes the same projector is used in both front and rear applications. When the projector is used in rear projection, the image, as seen by the viewer, is reversed left to right compared to front projection. Most modern projectors can correct this by electronically reversing the image left to right prior to projection.

144   PROJECTION LENSES AND SCREENS

If not, it is necessary to use a turning mirror, as shown in Figure 7.13 to correct the image. This turning mirror may also allow the rear projection booth to be more compact.

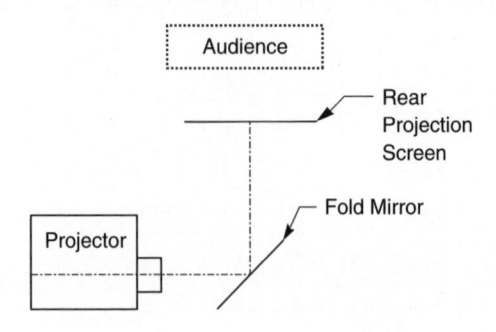

Figure 7.13 Fold mirror for rear projection.

7.2.1 Projection Screen Gain

Projection screens, as passive optical elements, don't have gain in the sense of increasing the total amount of light. They have gain in the sense that they direct some of the light that would normally go into angles where there is unlikely to be an audience into angles where the audience is likely to be. This increases the amount of light seen by the audience, increasing the perceived image brightness. A perfectly diffusing, 100 % reflecting screen would have a gain of 1.0. Such objects exist, or at least ones with 99.9 % reflectivity (Springsteen *et al.* 1994), and are used as standards when measuring screen gain (SMPTE 2003). The light distribution from a perfect gain 1.0 screen is called a Lambertian distribution.

A Lambertian light distribution is by definition one whose luminance or radiance is independent of viewing angle (Boyd 1983) (see also Appendix 1). Figure 7.14 illustrates the luminance distributions of several screens, including the luminance distribution for a Lambertian screen shown as a solid black line. This screen would have a gain of 1.0. However, a projection screen of this type with a perfectly diffusing surface would waste a major portion of the light from the projector by directing it toward the ceiling, the floor, and at extreme horizontal angles where there are unlikely to be viewers.

If the screen optical properties are controlled so the screen does not fully diffuse the light from the projector, the screen will have gain. For example, if a screen only diffuses the collimated light from the projector into a $\pm 60°$ horizontal angle and a $\pm 10°$ vertical angle, the light that a perfect diffuser would send outside these limits now will now remain in this smaller angular space. A screen with this range of angles might have a gain of about 5–6. Since these angles are typical of the angles subtended by the audience, the audience would see more light and a 5–6 x brighter image than it would with a fully diffusing screen. Viewers outside the intended $\pm 60°$ horizontal and $\pm 10°$ vertical viewing zone would see a very dim image, if they could see anything at all.

The dotted line of Figure 7.14 shows an ideal screen with a gain of 1.5. For this ideal screen, the luminance within a circularly symmetric cone with a half angle of about 55° would be independent of angle. Outside this cone, there is no light at all. For a viewer anywhere in the $\pm 55°$ viewing space, the image would appear the same brightness, and 1.5 x as bright as it would be if a fully diffusing screen were used. Outside the viewing space, the image would be invisible.

The third curve in Figure 7.14 shows a more realistic screen with a gain of 1.5. This curve is shown with a dash-dot pattern. For this screen, the gain is 1.5 on the centerline, but the light falls off with angle.

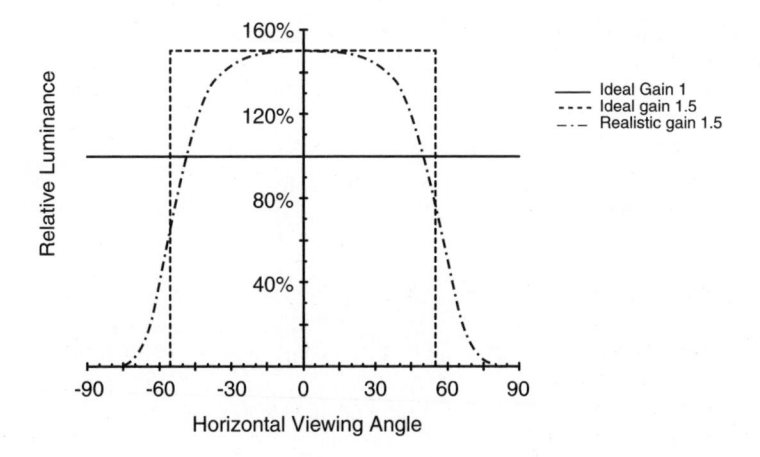

Figure 7.14 Light distribution with angle.

For viewers off axis, the image would not appear as bright as it would for viewers on axis. At 45° off axis, the image would be about as bright as with a gain 1.0 screen, and at 75° off axis the image would be lost completely.

This redistribution of light by the screen can be expressed mathematically by the equation:

$$\iint L(\theta, \phi) d\Omega dA \leq \Phi \tag{7.1}$$

where:
- $L(\theta, \phi, x, y)$ is the luminance distribution in audience space, where θ and ϕ are the azimuthal and polar angles and x and y are positions on the screen,
- $d\Omega$ is an element of solid angle,
- dA is an element of the screen area, and
- Φ is the total lumen output produced by the light engine.

In a front projection screen with no absorption and no transmission, the equal sign would be used in Equation (7.1). In a rear projection screen with some backscatter off the screen or contrast-enhancing absorption, the 'less than' symbol would be appropriate.

This equation simply repeats the statement that the screen is a passive element that does not have 'gain' in that it does not add light to the system. The light coming out of the screen into audience space is always less than or equal to the lumens going into the screen from the light engine.

Equation (7.1) is often used in a form that has already been averaged over area and angle for quick calculations of perceived luminance. This averaged form is:

$$L_V = \frac{G\Phi}{\pi A} \tag{7.2}$$

where:
- L_V is the average luminance for an observer centered in the audience space,
- G is the screen gain compared to a diffusing screen,
- Φ is the total lumen output of the projector, and
- A is the area of the screen.

146 PROJECTION LENSES AND SCREENS

Note that gain is normally defined in terms of the center of the audience space. This may be normal to the screen, but for consumer CRT rear projection applications, it is often slightly below the centerline. This is done to maximize the luminance for a seated viewer. This downward direction is done by adjusting the center of the Fresnel lens relative to the center of the screen, in the same way offsetting the projection lens in Section 7.1.4 directed light above the centerline. Fresnel lenses are discussed in Section 7.2.3.1.

Since L_V, Φ and A are all relatively easy to measure, Equation (7.2) can be inverted and used to calculate the gain of a projection screen

7.2.2 Multiple Projectors and Screen Gain

If multiple projectors are used in a projection system, only low gain screens can be used. This situation occurs in front projection where two or more projectors are projected on the same screen and edge matched to produce an extra-wide image. It also occurs in videowall systems where the images from multiple small rear projection units are tiled together to give a large image. This situation is shown in Figure 7.15.

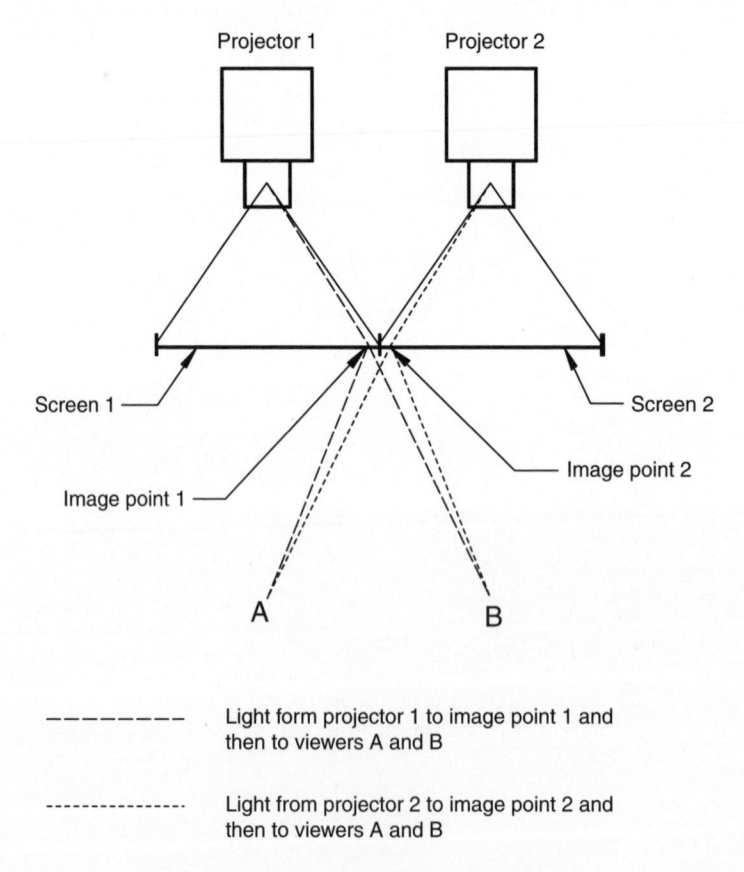

Figure 7.15 Edge matched projectors and screen gain.

Projector 1 is displaying its image on screen 1 and projector 2 is displaying on screen 2. This situation is the same as a videowall, with only two projectors shown. Two observers, A and B, are standing in front of the display. In the typical high gain screen, the gain is maximized for an observer on axis. Since observer A is on axis for projector 1, he will see a relatively high gain and high brightness for image point 1. Since A is off axis for projector 2, he will be in a low gain angular position for image point 2 and will see this point as relatively dim. Therefore, there will be a strong step in the luminance across the seam between screens, with the portion of the image displayed on screen 1 very much brighter than the portion displayed on screen 2.

Observer B will also see a strong step in the luminance, but it will be in the opposite direction. He is in a high gain region of projector 2 and will see image point 2 as quite bright. However, he is in a low gain position for projector 1 and will see image point 1 as relatively dim.

The only way to avoid this problem is to use a low gain screen, preferably a Lambertian screen, when edge matching images. Due to the losses in a projection screen, a Lambertian screen typically has a gain, as measured by Equation (7.2), of less than 1.0. A typical gain value for a projection screen with a Lambertian-enough angular distribution to be used with edge-matched projectors is on the order of 0.7.

7.2.3 Rear Projection Screens

In a rear projection system, the light engine produces an image that is focused on the rear of the projection screen. The rays from the exit pupil of the projection lens to each point on the screen define a conical bundle with a convergence angle θ. For a microdisplay projector, θ is about 0.25–0.5° degrees, while for a CRT projector it is about 2.75°. In addition, the ray bundles going to different points on the screen are not parallel to each other. With no Fresnel lens or screen, one could only see an image while one's eye was within a ray bundle. Since the ray bundles are not parallel, it would be impossible to have one's eye within the ray bundle for more than a small area of the screen.

The purpose of the Fresnel lens is to make the ray bundles converge to the point where the viewer is, so it would be possible to see the entire image at once. The purpose of the rear projection screen is to broaden the ray bundles out to wide enough angles so the image is visible to a large audience, not just to a single observer in a fixed position.

Rear projection screens intended for three CRT consumer systems are not appropriate either for microdisplay or for high resolution applications (Goldenberg et al. 1997). On the other hand, screens intended for microdisplay rear projection are not suitable for CRT projection systems. Therefore, this section will be broken into two sub-sections: one dealing with rear projection CRT systems and the other dealing with microdisplay rear projection. Since a Fresnel lens is required for either type of rear projection system, there will also be a separate sub-section on the Fresnel lens.

7.2.3.1 Fresnel lens

In the absence of additional optical elements in the light path, an observer looking at the screen center would be outside the angles that define the edge illumination. This would make for very severe uniformity problems, as discussed below. The principal optical element for correcting this is a field lens. Figure 7.16 illustrates the situations without and with this field lens.

Figure 7.16(a) shows a rear projection system with a screen that produces a viewer space of ±45°. On its own, the screen does not change the direction of the central ray of each ray bundle. Therefore, rays from both edges of the screen mostly miss observer A and the edges would appear dark. Observer B, on the other hand, sees the light from the near edge of the screen, and some of the light from the center but essentially none of the light from the far edge. Addition of a field lens in Figure 7.16(b) redirects the rays so they are slightly converging. Observer A is now in the central region of all light coming from both the central part of the screen as well as both edges. Observer B is now in the audience space and can see the center and both edges, although they are not as bright or as uniform as seen by Observer A.

148 PROJECTION LENSES AND SCREENS

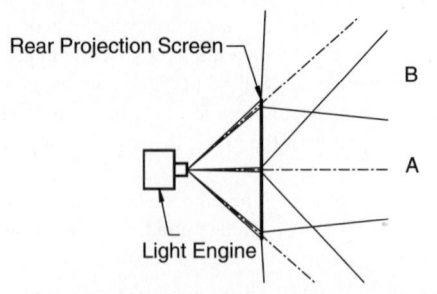

(a) Rear Projection System without a Field Lens

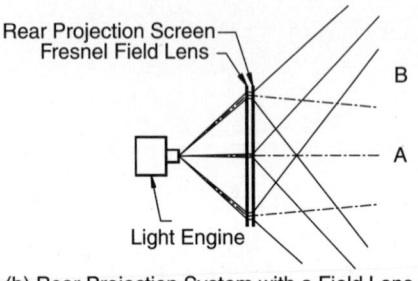

(b) Rear Projection System with a Field Lens

Figure 7.16 Field lens in a rear projection system: (a) without field lens; (b) with field lens.

Since the field lens must be the same size as the projection screen, it is impossible to use a conventional lens. A thin Fresnel lens is used instead (Strong and Hayward 1940; Milster 1995). The design of a Fresnel lens is illustrated in Figure 7.17. The contour of each small section of the lens is the same as the contour of the normal lens, with the parallel plate sections shown by the dotted lines removed. Since only parallel sections are removed, the first order properties such as focal length and f/# are the same for two lenses. In a Fresnel lens that is part of a projection screen, the individual segments are very small, and are normally made as flat segments rather than sections of a very large sphere.

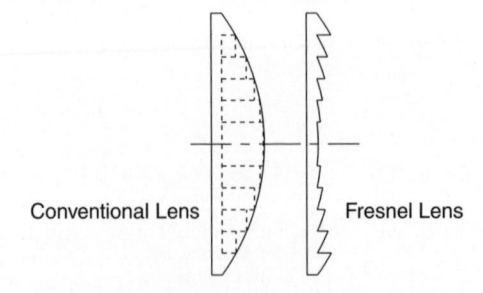

Figure 7.17 Fresnel lens.

Additionally, the focal length of the Fresnel is chosen to provide a conjugate distance compatible with the application. The conjugate distance is the position where all the rays at the center of each ray bundle converge to a point. In a consumer CRT projector, the conjugate distance is designed to be about three meters from the screen, corresponding to the typical seating position in a living room. This requires a Fresnel lens focal length that is slightly shorter than the distance from the screen to the projection lens. The centroid of the illumination is at angle slightly below the horizontal to maximize the system brightness for a seated viewer. This involves shifting the center of the Fresnel lens down slightly, so it is below the center of the screen.

For datagraphic displays, the focal length of the Fresnel lens is typically chosen to give a nearly infinite conjugate distance. This requires a focal length that is equal to the distance from the projection lens to the screen. This is done to provide the best possible image over the largest audience space, since the locations of viewers of datagraphic projectors are not as predictable as the locations of consumers watching TV.

There are two possible orientations for the facets of the Fresnel lens, as shown in Figure 7.18. Figure 7.18(a) shows a small section of the lens near the edge of the screen, with the light engine to the left and the screen and the audience to the right. This orientation is the preferred one for two reasons. First, no light is lost because all light hits the tilted surface and not the vertical step of the Fresnel lens. Secondly, the angle of incidence at the tilted segment is reduced in Figure 7.18(a) as compared to Figure 7.18(b), reducing the Fresnel reflection losses at the interface.

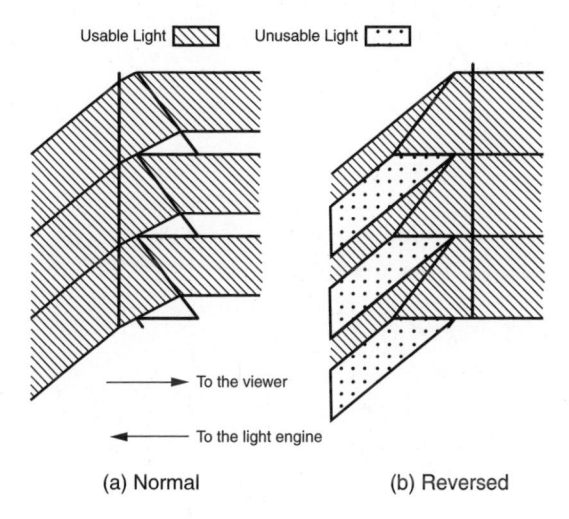

Figure 7.18 Fresnel lens orientation: (a) normal, (b) reversed.

Due to the large size and complex surface structure of a Fresnel lens, they are rarely antireflection coated in projection systems. The reflections reduce the throughput, particularly at the corners of the screen. Since the angles of incidence to a point on the screen are different for the blue and red tubes in a three-lens CRT system, the differing Fresnel reflections can affect the color balance of the corners relative to the center and the left side relative to the right side. Both of these effects can be reduced by using a high index plastic for the Fresnel lens (Lindegaard 1996).

The reason the orientation shown in Figure 7.18(b) would ever be considered is that it allows the Fresnel lens to be molded on the back of the screen itself. The orientation shown in Figure 7.18(a) requires the Fresnel lens to be a separate element than the screen. This creates several problems. First, the cost of a two-piece screen is significantly higher than the cost of a one-piece screen. Secondly, the air gap between the Fresnel lens and the screen is poorly controlled. In practical terms, there is no way to bond the screen to the Fresnel lens in the viewing area without these bonds becoming visible. As the screen or Fresnel lens bends due to temperature or humidity changes, the separation between the two can increase, especially in the center. Moreover, in areas where the two pieces are in contact, they can rub against each other and cause visible damage. This damage is especially likely to occur while the set is being shipped. Ghosting is also a problem in Fresnel lenses for large field angles (Davis *et al.* 2001).

7.2.3.2 Fresnel lenses for thin RPTV systems

In order to work as a field lens and collimate the light from the projection light engine, the Fresnel lens must have a focal length approximately equal to the distance from the light engine to the screen. When the angle of incidence at the Fresnel lens at the corner of the screen become greater than about 45°, even the arrangement shown in Figure 7.18(a) does not work very well. In the past, this was one factor limiting the minimum depth of a RPTV set. With competition from LCD and plasma TVs, the RPTV manufacturers have been forced to develop new Fresnel lens designs that work at angles greater than 45° off axis.

Figure 7.19 shows one design for a Fresnel lens operating at very high field angles. (Shikama *et al.* 2002) Near the center of the screen, the Fresnel lens has a design like the one shown in Figure 7.18(b). Far from the center of the screen, the system has pure TIR blades. In a TIR blade, the light enters the screen material through a window that is at near-normal incidence to the incident light. The light then strikes the back surface of the TIR blade at an angle greater than the TIR angle for the screen material. The blade angles are arranged so the light exits the front of the screen at normal incidence to the front surface. There is a transition region where the refractive blades gradually get smaller and the TIR blades gradually get larger. If there were a sharp demarcation between a refractive region and a TIR region, this would be visible in the image.

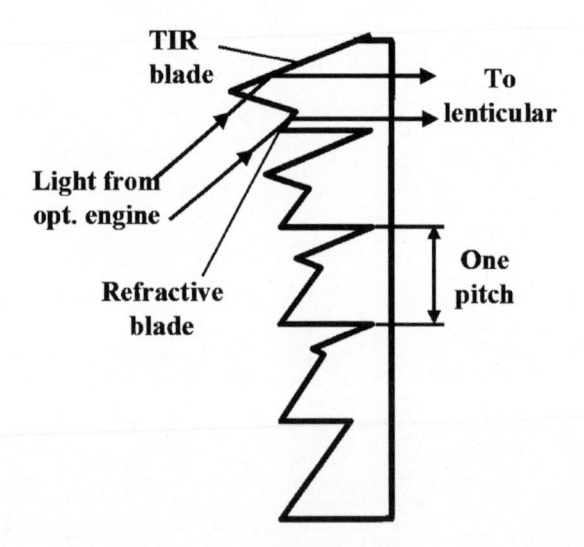

Figure 7.19 Fresnel lens for a thin rear projection TV (reproduced by permission of © 2002 the Society for Information Display (Shikama *et al.* 2002)).

Figure 7.20 shows a rear projection Fresnel lens intended for very thin RPTV systems, with a depth as low as perhaps 4.5" (Huang 2007). In this design, there is no fold mirror and the projected image strikes the screen at an extreme oblique angle. There are only TIR blades in this design, there are no refractive blades. As shown in the figure, the TIR surfaces are slightly curved, focusing the light falling on the blades onto an aperture in the screen structure. The apertures are separated from each other by an opaque black matrix. This matrix absorbs ambient light to increase the contrast of the system. Alternatively to curving the TIR portion of the blade, the input facet can be curved to perform the focusing function.

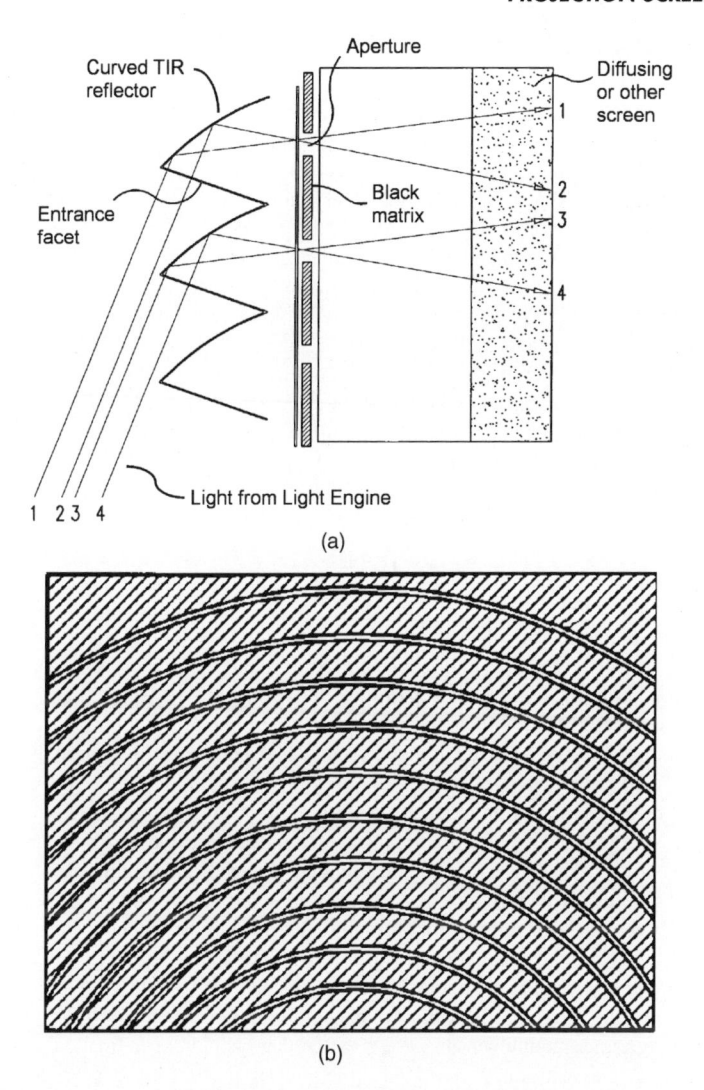

Figure 7.20 Integrated rear projection Fresnel and screen intended for very thin RPTV sets: (a) cross-section of the Fresnel lens, aperture structure and diffusing screen; (b) screen structures are in concentric rings centered below the lower screen edge (Huang 2007; United States Patent and Trademark Office).

Note that in this screen design, the blades and apertures are not straight lines, they are segments of concentric circles whose center is well off the bottom of the screen, as shown in Figure 7.20(b). Of course, in a real screen, there would be many more apertures and they would be small enough to be invisible at normal viewing distance.

For most rear projection screens and Fresnel lenses, the resolution of the system is limited by the size of the structures. In this design screen this is not necessarily true because of the focusing nature of the TIR blades. As can be seen in Figure 7.20(a), the light rays 1 through 4, corresponding perhaps to four different pixels on the microdisplay, remain ordered at the screen in the same way they were in the incoming image. Therefore the Fresnel structure does not limit the resolution although possibly the screen structure will.

152 PROJECTION LENSES AND SCREENS

7.2.3.3 Rear projection CRT screens

One of the defining features of rear projection CRT systems is that only one CRT (usually the green) is on axis and the other two are to the left and right of the optical centerline. This arrangement is shown in Figure 7.21. For a high-gain screen, where the direction of the light rays that emerge from the screen remain correlated with the direction before the screen, this poses a color uniformity problem, even on-axis. This effect is shown in Figure 7.22(a). Note that the central peaks of the red, green and blue light do not coincide. Therefore, members of the audience will see different red/green and green/blue luminance ratios than intended, and a single viewer will see different ratios depending on where he looks on the screen. This will be perceived as a color shift. A color correcting rear projection screen will bring the three peaks into approximate coincidence, greatly reducing the color shift. The red, green and blue distributions with a color correcting screen are shown in Figure 7.22(b).

A quantitative evaluation of this color shift by Bradley *et al.* (1985) defined a color error parameter δ. Bradley said that when δ was less than 0.5, the color uniformity on the screen was generally considered acceptable by observers for use in consumer CRT RPTV systems. There are several screen architectures that can achieve this low δ value. The most commonly used design was the double lenticular screen. Screens based on TIR structures could also achieve the required color correction. Both designs will be discussed in the following sections.

(a) Double lenticular screens

The cross-section of a double lenticular screen is shown in Figure 7.23 (Glenn 1970). This screen consists of two arrays of vertically oriented cylindrical lenses, on the front and rear of the screen. The lens on the rear of the screen focuses the light from each of the three CRTs onto the lens on the front of the screen. The red, green and blue beams are focused onto different portions of the front lens; each encounters a different slope of that lens. The refracted directions of the three colors are different so that all are directed in the forward direction.

The front and rear lenses working together disperse the light over a wide horizontal angle. Since a double lenticular screen is made with cylindrical lenses, it only affects the horizontal angles of the light. With no diffusion, the vertical cone angles would remain unchanged, and would be about 2.75° for a CRT projector, to give a vertical viewing angle also of 2.75°. This vertical viewing angel is not acceptable so there is normally some diffusion in the form of small scattering particles added to the bulk of the screen. This diffusion will widen the vertical angle to cover the desired audience space. The diffusion will also smooth out any irregularities of the horizontal brightness versus angle curve.

Figure 7.24 shows the color uniformity of a typical double lenticular screen. Note that this screen does not in fact meet the target requirement that $\delta_{max} \leq 0.5$. One would expect that color shifts would be noticeable at large angles with this screen. Since the largest color errors occur at large angles which

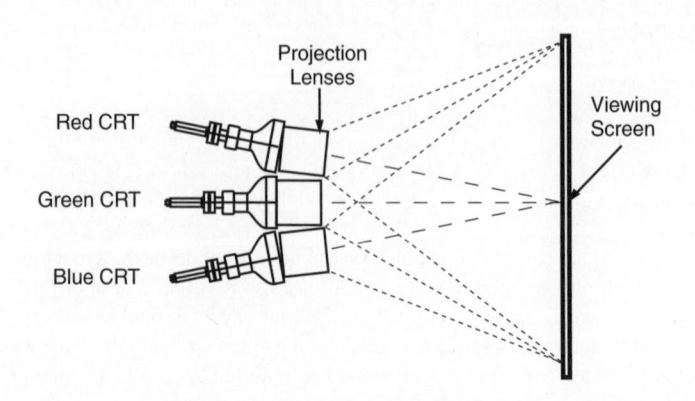

Figure 7.21 Rear projection CRT system.

PROJECTION SCREENS 153

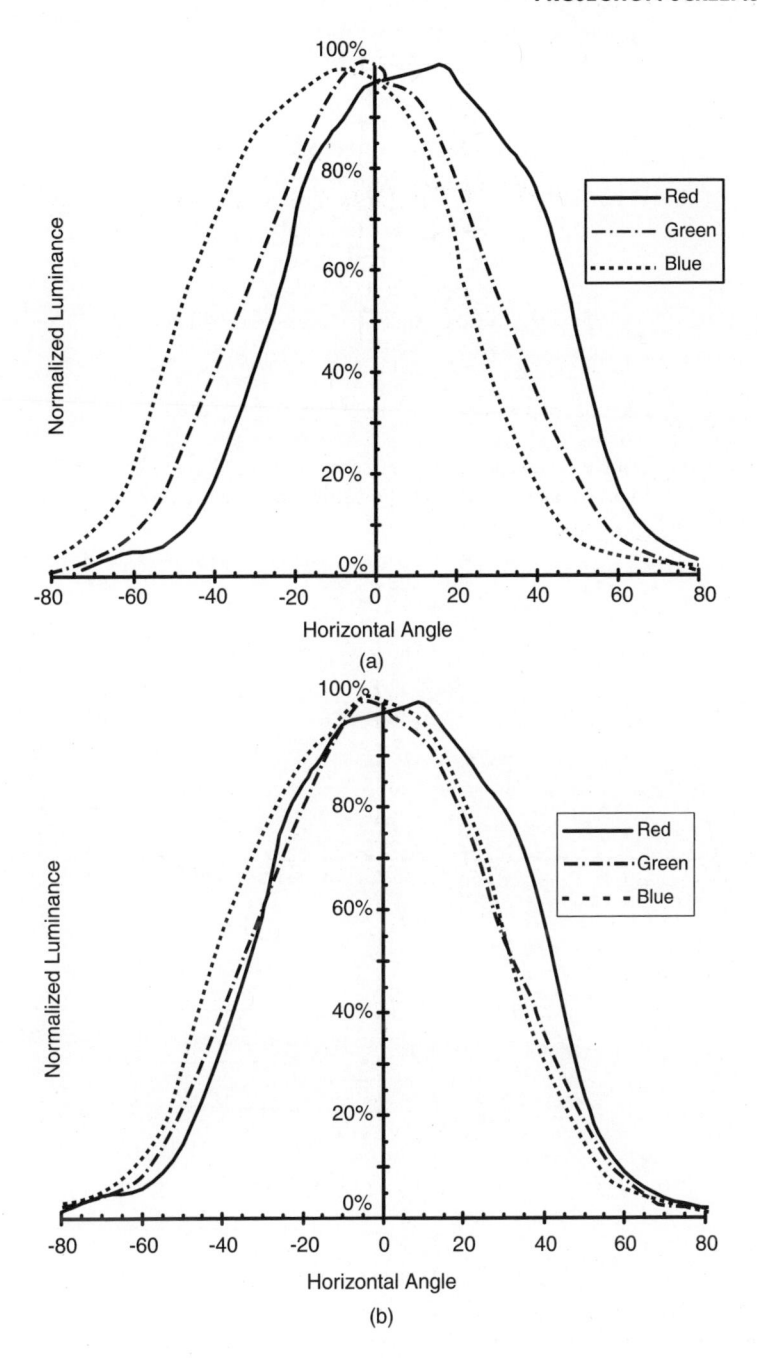

Figure 7.22 Angular light distribution in a rear projection CRT system: (a) without a color correcting screen; (b) with a color correcting screen.

154 PROJECTION LENSES AND SCREENS

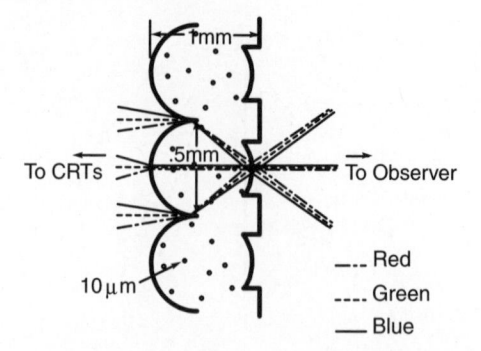

Figure 7.23 Cross-section of double lenticular screen.

correspond to the edge of the audience space, these errors were deemed commercially acceptable. Figure 7.25 shows the horizontal and vertical light distribution produced by a typical consumer CRT rear-projection double lenticular screen.

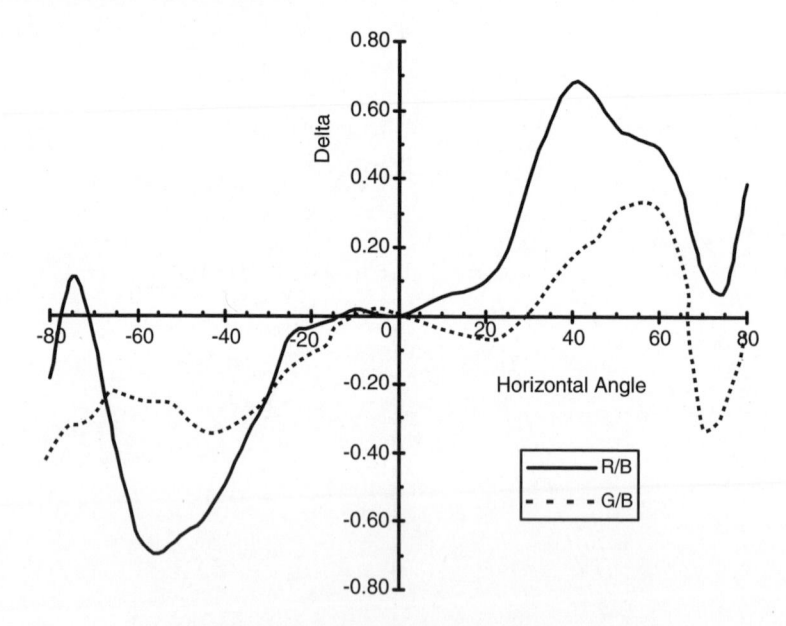

Figure 7.24 Color uniformity of a double lenticular screen.

In Figure 7.23 it can be seen that there are flat surfaces on the viewer side of the screen separating adjacent lenticular pairs. Since no light passes through these flat surfaces, they can be blackened to absorb ambient light and increase the contrast of the projected image.

One of the problems with double lenticular screens is that the finest pitch practical for high volume manufacturing is about 0.5 mm. The pitch is limited by two problems:

1. The lenses on the front and rear surfaces must be in registration with each other. For small pitches, this becomes more difficult.

2. For a screen material with a given index of refraction, the thickness of the screen is proportional to the pitch. Therefore smaller pitch screens are thinner and, if they are too thin, structural integrity is a problem.

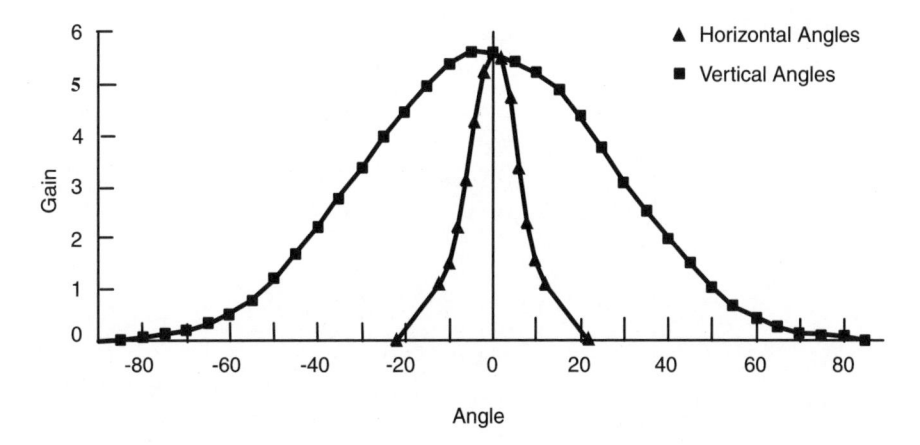

Figure 7.25 Horizontal and vertical light distribution with a consumer double lenticular screen.

The pitch of a screen is important because the pitch determines the resolution limit of a screen. For a 0.5 mm pitch screen, no portion of the image smaller than 0.5 mm can be resolved. In a 52″ diagonal 4:3 rear projection system the screen width will be 1056 mm and there will be approximately 2113 lenticular structures across the screen. Obviously, this will not limit the resolution of a NTSC video projector, with perhaps 400 lines of resolution in the horizontal direction. However, with a 1920×1080 HDTV image, resolution would be restricted and Moiré and aliasing can become problems. Moiré and aliasing are discussed in Chapter 14.

While satisfactory double lenticular screens are made for standard definition consumer CRT projectors, these are unlikely to be adequate for HDTV CRT projectors and certainly are not satisfactory for high-resolution data/graphics CRT projectors. These types of projectors will require higher resolution screen technologies.

(b) TIR screens

Another type of color correcting screen for rear projection CRT sets is based on total internal reflection (TIR) (Bradley *et al.* 1985; Bradley 1986). The basic structure of a TIR screen is shown in Figure 7.26. Light enters the screen from the light engine and fresnel lens that would be below this figure. Light that enters near the centerline of each volcano-shaped structure is refracted into a variety of angles by the tip as shown in Figure 7.26(a). Most of the light encounters the sides of the structure, as shown in Figure 7.26(b). Since the angle the light makes with the wall is larger than the critical angle, as given by Equation (6.12), this light TIRs. This light is further dispersed by refraction at the tip. Off axis rays, such as the ones coming from the side tubes in a CRT projector, are also dispersed over a wide angle, as shown in Figure 7.26(c).

The TIR structure serves as an angular integrator, averaging over angle rather than over area. The light emerges from the tip of the TIR structure with angles randomized over the audience space. This randomization of angles gives a TIR screen fairly good color correction. Like a double lenticular screen, a TIR screen does nothing to increase vertical angles. Therefore, bulk or surface diffusion is added to the screen to increase the vertical viewing angle.

Note that in the areas of this screen where TIR occurs, no light passes through the screen to the viewer. These areas can therefore be blackened in order to increase the absorption of ambient light and increase contrast. Various ways of blackening the screen are shown in Figure 7.27. The first valley between lenticules (a) is shown unblackened. The second valley (b) is shown filled solidly with a black absorbing material. This technique is the easiest to apply. However, in the areas coated with the black material,

156 PROJECTION LENSES AND SCREENS

(a) Central Rays (b) Off Center Rays (c) Off Axis Rays

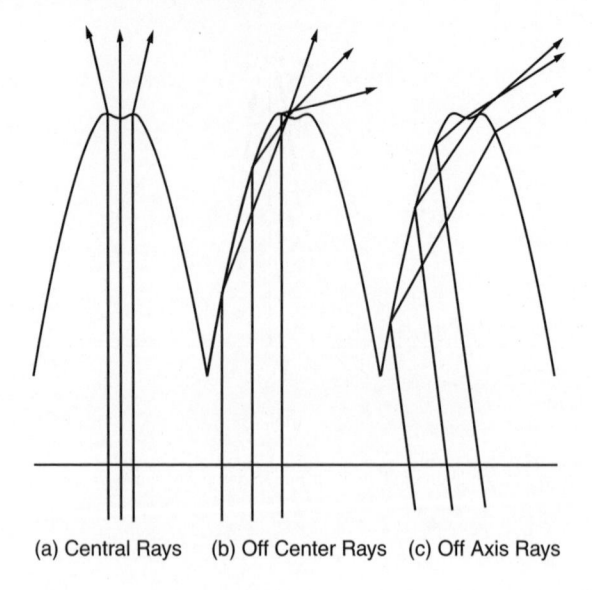

Figure 7.26 Structure of a TIR screen: (a) central rays; (b) off-center rays; (c) off-axis rays.

a. Unblackened

b. Filled Grove

c. One Large Bead

d. Many Small Beads

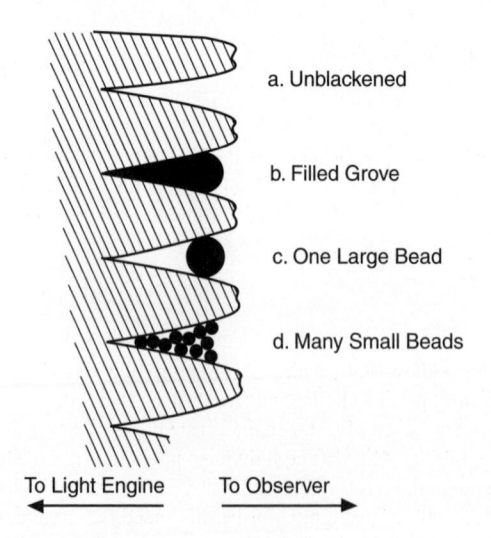

←To Light Engine   To Observer→

Figure 7.27 Bead blackening of a TIR screen: (a) unblackened, (b) filled grove, (c) one large bead, (d) many small beads.

optical contact between the black material and the screen would occur. Because of this optical contact, the conditions for total internal reflection may not be satisfied and the screen may no longer work as expected.

There are two possible solutions to this problem. First, a mirror coating could be applied to the screen under the black coating. The reflection would still occur as desired, only it would be based on conventional mirror reflection rather than TIR. A second solution is shown in Figure 7.27(c). Here the blackening is done by a long, thin bead or thread of black material deposited in the groove between TIR structures (Nakanishi and Inoue 1990).

From the front, this appears to completely fill the valley. Therefore, it effectively absorbs the ambient light that falls on the valley region. However, this bead only touches the screen at two lines along the length of the TIR structures. This minimizes the area where the TIR conditions are destroyed and has little effect on the efficiency of the screen. An alternative version of this technique is shown in Figure 7.27(d) where many small particles are used, instead of one large filament (Stanton 1986).

There are difficulties in manufacturing TIR screens. These include getting the correct contour in the TIR structure and the cost associated with blackening it. As a result of these difficulties, combined with the fact that double lenticular screens are able to achieve acceptably low color shifts, the TIR screen has not had a significant effect on the standard definition TV market.

Several advantages associated with the TIR screen may lead it to play a role in higher resolution systems. First, all the structure of a TIR screen is on the front surface. Therefore, it is unnecessary to register front and rear structures to each other, and this limit on screen pitch in double lenticular screens does not apply. TIR screens can be made with pitches as low as 0.25 mm. Similarly, the thickness of the screen plays no role in its function. Therefore, the base screen may be as thick as necessary to ensure structural integrity. Finally, since there is no structure on the rear surface, a Fresnel lens for a thin RPTV system such as the one shown in Figure 7.19 could be placed there. Due to the angular integration effect of the TIR screen, perhaps a simpler Fresnel lens could be used in this configuration. In this configuration, there would be no need to align the structures on the front and rear surfaces.

7.2.3.4 Microdisplay and light-valve rear-projection systems

There are several major differences between microdisplay projectors and CRT projectors that argue against using the same rear projection screen designs for both light engine technologies.

1. Microdisplay projectors can have higher resolutions than CRT projectors. Therefore, a finer pitch is required if the screen has any structure.

2. Most microdisplay projectors use single projection lenses. Therefore, they do not require the screen to provide color correction for off-axis colors.

3. Most microdisplay projectors have much smaller diameter projection lenses than CRT projectors. A consequence of this is a phenomenon called speckle that significantly affects the appearance of an image on the screen.

The difference in resolution has already been discussed in terms of the pitch of a double lenticular screen.

The single-lens design common to microdisplay projectors allows the use of single lenticular screens, illustrated in Figure 7.28. TIR and single lenticular elements can also be included in a single screen, as described by Noda (1989). The rear surface could also be used for horizontal lenticular elements, eliminating the need for diffusion to control vertical viewing angle. The single lenticular screen, like the TIR screen, only has structures on one surface. This simplifies manufacturing of the screen and lowers the achievable limit on screen pitch and resolution. On the other hand, the entire front surface of the screen is active area. Blackening a section of the screen to improve contrast would cover a portion of the screen that transmits light and would reduce the screen throughput. Since each portion of a lenticular sends light in a different direction, it would also affect the luminance vs angle properties of the screen. A dye can also be added to the screen material to improve contrast, at the expense of lumen output. The

158   PROJECTION LENSES AND SCREENS

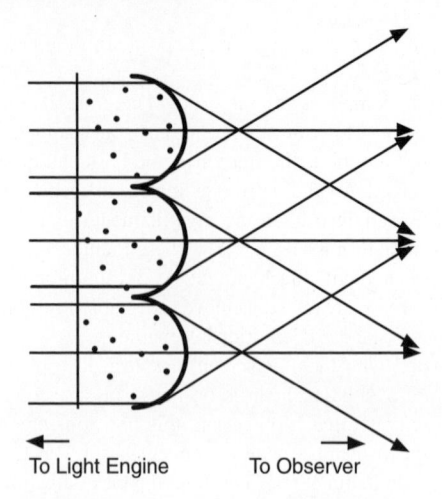

Figure 7.28   Single lenticular rear projection screen.

contrast vs output tradeoff must be done for each individual projection system, taking into account the number of lumens produced by the light engine, the source material expected to be shown, the size of the screen and expected ambient lighting conditions.

The single lenticular design can be reversed (Ebina 2002) as shown in Figure 7.29. The cylindrical lenses are in this case on the back side of the screen, facing the Fresnel. For the highest efficiency, these lenses should not have circular profiles but aspherical cross sections. The focal length of the lenses and the thickness of the screen are chosen so the lenses focus the light on or near the front surface of the screen. Because of this focusing effect, the light only passes through a small portion of the front surface and it is possible to blacken the other areas of the screen. This reduces the reflectivity of the screen and enhances the system contrast. This type of screen can be made with a very fine pitch, with a pitch as

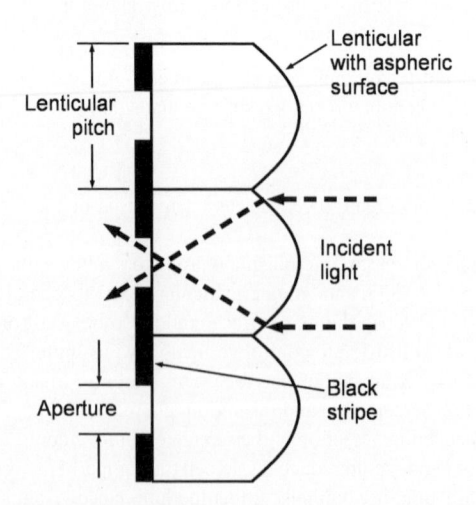

Figure 7.29   Contrast enhancing single lenticular rear projection screen (reproduced by permission of © 2005 the Society for Information Display (Iwata 2005)).

small as 64 μm reported by Iwata *et al.* (2005). Other designs of single lenticular, high contrast screens for RPTV are also available, for example, see (Oda *et al.* 2000).

It is also possible to make screens based on small beads embedded in a black matrix (Wolfe *et al.* 2003), as shown in Figure 7.30. These screens typically have very high ambient light rejection because almost the entire front surface is the black matrix material. The viewing angle of a screen of this type is typically circularly symmetric but can be adjusted in the screen design and construction over a range from a gain of 1.2 (or less) up to a gain of 4.8 (or more). Screens of this type are normally too expensive for a mass-market RPTV, but they are commonly used in installations where image quality and light-rejecting ability are more important than cost.

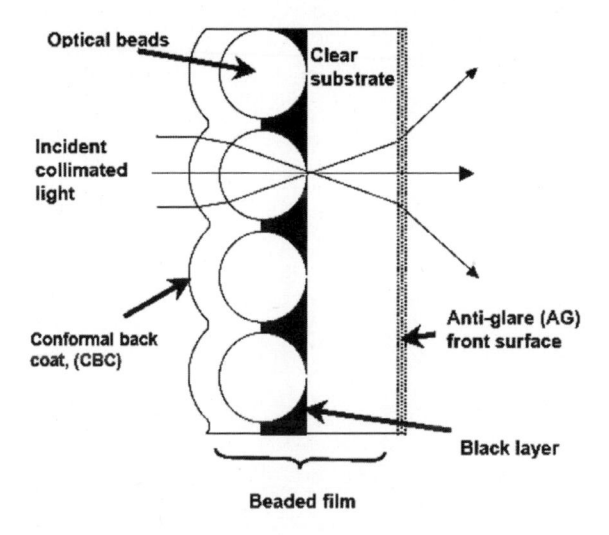

Figure 7.30 High-contrast beaded rear projection screen (reproduced by permission of © 2003 the Society for Information Display (Wolfe *et al.* 2003)).

7.2.4 Front Projection Screens

Front projection screens are not part of the projector and therefore it is difficult to predict the system performance with these screens. In some cases, the same portable projector will be used with many different screens. In the related case, a front projection screen may be used with many different projectors, sometimes in the same presentation. For example, a microdisplay-based projector may be used to give a main presentation supplemented by film clips and slides. In this case it is impossible to tie the design of the front projection screen too closely to the projector design. This is unlike the rear projection case, where a screen and a projector are normally used together permanently so the screen can be optimized to make the projected image look good.

The gains for the more common front projection screens are typically low, in the range of 1.1–2.8. These correspond to viewing angles from ±50° for a gain 1.1 screen to ±30° for a gain of 2.8 (Da-Lite 2007). A procedure for measuring screen gain for front projection screens is given in SMPTE (2003). With low-gain screens, the requirements for color correction and field (Fresnel) lenses are generally minor. Therefore, these screens can be used with three-lens CRT projectors as well as with single-lens microdisplay, light-valve and film projectors with few problems.

Higher gain front projection screens are available but are used only in limited applications. High-gain front projection screens are available with gains of 8 or more. Screens with this high a gain typically have a nearly specular component, with enough diffusion to give a usable viewing angle. It is also possible to design a high gain front projection screen similar to the TIR rear projection screen (Bradley 1990). With high gain, a field lens at the screen is required, as described in Section 7.2.3.1. The simplest way of achieving this is to curve the entire screen. In this way the screen itself acts as the field mirror, redirecting the light toward the audience.

7.2.4.1 Light rejecting front projection screens

Front projection screens can have serious contrast problems. Since they are designed to reflect light, they can reflect not only the light of the projected image but the ambient light in the room as well. The sequential contrast of a modern projector, when measured carefully in a dark room as described in Section 13.3.2, can range from 500:1 upward to 5000:1 or more.[4] On the other hand, the contrast in a well-lit room can be as low as 2:1, and 20:1 would be considered good. In this case, the contrast of the projector is almost irrelevant because the projector lumen output, the screen and the ambient light control the perceived contrast. This effect is explained in detail in Section 13.3.2.

Ambient light rejecting screens can significantly improve the situation. There are three major techniques to improve high ambient light contrast of a front projection system:

1. angular selective screens
2. polarization selective screens
3. wavelength selective screens.

The simplest contrast-enhancing screen is just a black screen. A black screen will reflect very little of the ambient light, but it also will reflect very little of the light from the projector. The simple solution to this is to just get a brighter projector. In the current market, with 30,000 lumen projectors available, this can be a viable solution for some applications.

More sophisticated solutions depend on the relative angles of the screen to the projector, the audience and the ambient light source. Ambient light often comes from above, for example from the ceiling lights in a conference room. The audience and the projector are often at the same level, with the projector on the conference room table and the audience sitting around the table. If the screen has gain with a relatively small vertical viewing angle and a wide horizontal viewing angle, all audience members will be able to see the image. The ambient light, however, is reflected downwards by the screen and misses the audience. Typically, the screen would be darkened, so the apparent gain on-axis is in the range of 0.8 to 1.1. The viewing angle of this type of screen is typically limited to about 40–50°. The audience would see a gain of about 1.0 for the projector light but a gain of 0.3 or 0.5 for the ambient light, giving a 2 x or 3 x improvement in the high ambient light contrast. Lower cost front projection screens where the gain is circularly symmetric will more typically improve the high ambient contrast by 1.5–2 x. The gain vs angle properties of a screen of this type is shown in Figure 7.31. If the ambient light source were outside the 30° viewing cone of this screen and the viewer and projector were within the 15–20° high-gain cone of the screen, the ambient light contrast would improve by approximately 2:1. The exact improvement would

---

[4] The maximum contrast of a commercial projector that the authors are aware of is 500 000:1 for the SEOS Zorro projector shown in Figure 10.8.

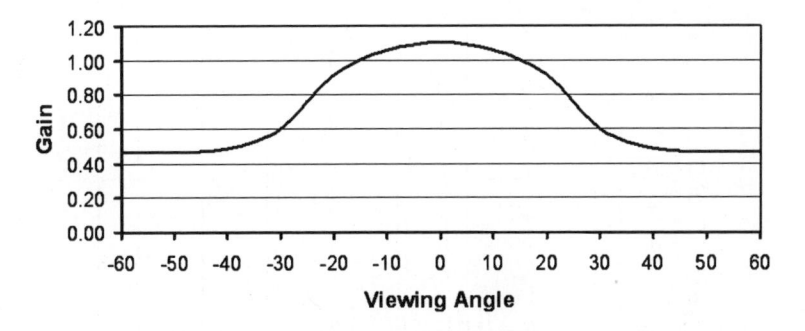

Figure 7.31 Circularly-symmetric ambient light rejecting front projection screen.

depend on the location of the projector, viewer and ambient sources in the room and the reflectivity of other objects in the room, especially the wall in a 30° cone directly behind the projector.

Since most ambient light comes from above in normal settings, it is possible to tailor the gain of light rejecting screens so they are not circularly-symmetric like the one shown in Figure 7.31. In this type of design, light sources above the level of the screen encounter very low gain and reflect little light into the audience space. Light sources on the horizontal centerline of the screen (i.e. the projector) encounter high gain and wide viewing angles. While screens of this type are beginning to appear on the market as of this time, published information on the details of the technology are non-existent. The Dai Nippon Supernova screen is said to include seven distinct layers, including a reflection layer, a black color layer, an optical lens film and a hard surface coating (DNP 2007).

If the output from the projector is polarized, a screen that reflects one polarization and absorbs the other polarization can increase the contrast ratio by a factor of 2 x. This occurs because ambient light is typically unpolarized so half of the ambient light is absorbed by the screen while nearly all of the light from the projector is reflected. See (Harada *et al.* 2002) for an example of a screen of this type.

A 'silver screen' that preserves polarization is required when polarization-dependent 3D images are projected. These screens typically have narrower viewing angles of around 30° and higher on-axis gains of about 2.3–2.5. As such, they can be quite good at rejecting ambient light, as long as the source is far enough off axis.

The contrast-enhancing ability of wavelength selective screens is based on the fact that the output of a projector is not full-spectrum light. Typically, the output spectrum is just red, green and blue light, with little or no overlap between the red, green and blue bands. In the extreme case of a laser projector, the output spectrum is just three wavelengths, one for the red, one for the green and one for the blue.

On the other hand, the ambient light is typically broad-spectrum and includes light at wavelengths between the red, green and blue used by the projector. If a screen can be designed and built that reflects the red, green and blue but absorbs the wavelengths in between, it can significantly attenuate the ambient light without greatly reducing the projector light, enhancing the contrast. The reflectivity characteristics of a screen intended to do this (Shimoda and Kawashima 2007) is shown in Figure 7.32.

This screen cannot be used with certain types of projectors that are becoming increasingly common. These are projectors that have more than three primary colors. The most common projector in this class uses red, green and blue but adds a white primary as well. These RGBW projectors are normally only used for business applications. The white primary will have broad spectral range and some of the white will be absorbed by the screen. This will not only reduce the brightness of the system but it may upset the color balance and the gray scale of the projector. Another design that is becoming more common for home theater projector is one that adds one or two additional primary colors, most commonly yellow and cyan at about 590 and 500 nm respectively. These colors would be strongly absorbed by the screen compared to the red, green and blue primaries, seriously affecting the color balance and system brightness.

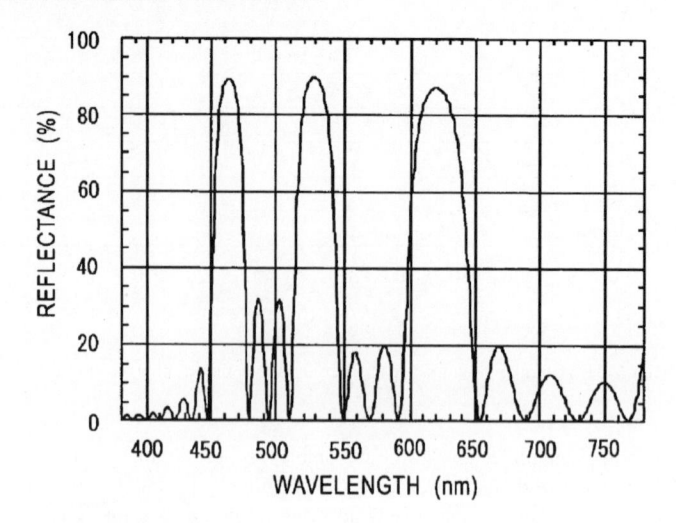

Figure 7.32 Spectral characteristics of the Sony 'ChromaVue' contrast enhancing screen (Shimoda 2007; United States Patent and Trademark Office).

7.3 Speckle in Projected Images

Speckle is a phenomenon associated with the coherence properties of light. Coherence is normally associated with lasers, but it is also associated with any light source that has a finite sized pupil, i.e. all real light sources. It was never observed in CRT projection system, or at least never an issue in the design of these projectors. Presumably this was because of the very large pupil of the typical CRT projector gave its output very low coherence.

Speckle first became a problem in electronic projection systems with the introduction of microdisplays, light valves and light amplifiers with much smaller pupils than the ones used in CRT systems. It was a special problem in rear projection systems where the typical screen design at the time would enhance speckle.

With the introduction of lasers with their much more coherent light, speckle has become an issue in front projection as well. In laser-based systems, it is normally necessary to reduce speckle through a light engine design effort: screen design can help but is not enough to eliminate speckle.

7.3.1 Speckle in Rear Projection Systems

Speckle is a serious problem for rear projection microdisplay systems. Much of the discussion of speckle in this section follows (Goldenberg *et al.* 1997). Speckle gives the entire image a grainy appearance. Microfilm viewers are an excellent place to see speckle, for those unfamiliar with the phenomenon. Figure 7.33 shows the intensity profile across one line of a display driven by a uniform video signal. The noise in the trace is caused by the optically generated speckle. Without speckle, this profile would be a straight, horizontal line.

The ratio of the standard deviation to the average in a trace like this is used as a measure of speckle and is called speckle contrast. Correlation between speckle contrast and human perception of speckle can be significantly increased if the a low pass filter is used on the data, with an upper frequency limit of about 60 cycles per degree is applied before the standard deviation is calculated (Watson and Ahumada 2005). More sophisticated measures of speckle and other spatial artifacts in an image will be discussed in Section 13.2.3. For the data shown in Figure 7.33 this speckle contrast is 9 %. Normally a speckle

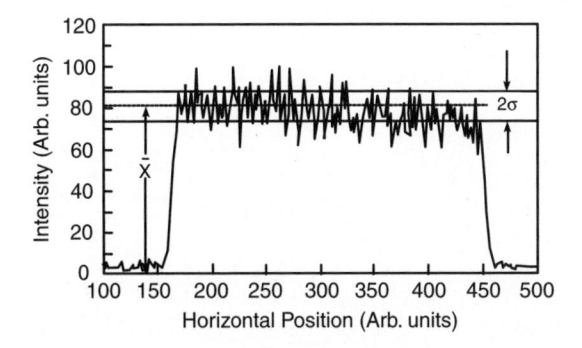

Figure 7.33 Line profile of a uniform light with speckle.

contrast greater than 2 % is visible and speckle contrast greater than 4 % is nearly always objectionable. Microdisplay projectors with a high-resolution diffusing screen can have speckle contrasts as high as 13 % or more. Speckle in a system with laser illumination can be even higher and will be discussed in the next section. Speckle significantly reduces the apparent resolution of the system, since the speckle appears as high spatial frequency noise that masks the high spatial frequency information content (Kozma and Christensen 1976).

Speckle is a phenomenon associated with coherent light, and is most commonly associated with laser illumination. However, light coming from a lens with a finite pupil diameter (i.e. all real lenses) is partially coherent. The coherence length is a function of the diameter of the projection lens and the distance to the screen (Born and Wolf 1980) and is given by:

$$\rho = 0.610 \frac{\bar{\lambda}}{\sin(\theta')} \tag{7.3}$$

where:
- ρ is the coherence length of the light at the screen,
- θ' is the half angle subtended by the exit pupil of the projection lens at a point on the screen and
- $\bar{\lambda}$ is the average wavelength of the light.

The smaller the pupil angle of the lens is, the longer is the coherence length. There is a direct relationship between coherence length and speckle: for a given screen, the longer the coherence length, the worse the appearance of speckle. For example, in a typical 52-inch projector with 7-inch CRTs, the coherence length would be about 7 μm. This short coherence length for CRT projectors would not result in a visible speckle problem. For a microdisplay projector with 35 mm diagonal microdisplays and an f/3.0 lens, the coherence length at the screen would be about 75 μm. In this case, speckle can have a major influence on the appearance and resolution of the image.

Most projection screens require diffusers. These can either be bulk diffusers with multiple small particles to scatter the light or surface diffusers embossed on one or both surfaces of the screen. This diffusion works with the lenticular or TIR structures to spread the light through the audience space. These diffusers, regardless of their exact technology, act as multiple scattering centers to scatter the light in the desired directions. If there are several scattering centers within one coherence length of each other, the coherence of the scatterings will result in interference patterns which are visible as speckle.

One cure for speckle has been known for many years (Leifer et al. 1961). It involves using two diffusers in close proximity to each other, and moving one diffuser slowly relative to the other. This technique works by time averaging the speckle over the integration period of the eye. If we take the required travel to be one coherence length (75 μm) and the integration time required to eliminate speckle to be a field time of 15 ms, the required velocity would be:

$$V_{\min} = \frac{75\,\mu m}{15\,ms} = 5\frac{mm}{s}$$

This rough estimate is on the high side: typically velocities in the range of 1–3^{mm}/$_s$ are enough to reduce speckle to unnoticeable levels. This technique is explained in more detail theoretically by Lowenthal and Joyeux (1971). Experimental studies of the problem are given by Okamoto and Asakura (1990) and Aizu and Asakura (1988), among others. (Shin *et al.* 2006) gives an example of the use of this system in a high-brightness laser-based RPTV. The system had a 720 p DLP and a 70″ screen with a total lumen output of 600 lumens. With no speckle reduction efforts, the system had a speckle contrast of 6.5 %. With a rotating diffuser in the illumination path, the speckle contrast became 3.1 %. With both the rotating diffuser and the moving diffuser at the screen, the speckle contrast was reduced to 1.6 %. According to Wang *et al.* (1998), any speckle contrast less than 4 % is invisible to the human eye. Wang based this limit on experiments using a rotating diffractive optical element (DOE) in the path of laser beam before the modulator. Five revolutions per second were found to be sufficient to eliminate speckle.

Note the discrepancy between the acceptable levels of speckle contrast discussed by Goldenberg (1997) and Wang (1998). Since there is no industry-wide acceptance level for speckle contrast, the value of speckle contrast must be determined by the user for each application. In addition, it is recommended the user use one of the more sophisticated measures of speckle in his evaluation, as will be discussed in Section 13.2.3.

Another method to reduce speckle is to have a diffuser that has a finite thickness, preferably greater than several times the coherence length. This thick diffuser can be either a single element with bulk diffusion spread through the volume, or two separate diffusers with a non-diffusing spacer. One way to achieve two separated surface diffusers is to emboss them on the front and back surfaces of a thin sheet. This thick diffuser approach does not eliminate speckle as completely as the moving diffuser method, but it can reduce it to acceptable levels. Speckle from thick diffusers and laser light has been evaluated theoretically by Shirley and George (1989). The authors are not aware of a definitive study of speckle with thick diffusers applicable to projection screens.

Unfortunately, this thick diffuser approach can significantly effect resolution in the screen, as shown in Figure 7.34. In Figure 7.34(a), a single pixel from the projector (not shown) is imaged onto a single thin diffuser. The diffuser increases the angle of the cone of light, but all rays still appear to come from a small region on the screen. This screen would have high resolution, if the speckle could be removed. In Figures 7.34(b,c), the light rays from a single pixel no longer appear to the viewer to be coming from as small a region as before. These screens would have less speckle, but the resolution would be degraded by the spreading out of the pixel image.

For the system requirements given by Goldenberg *et al.* (1997), it was possible to find an acceptable compromise between the speckle and the resolution reduction from two diffusers. Depending on the resolution and gain requirements, a satisfactory compromise may not be possible for other systems.

The state of the art in high-resolution rear projection screens to be used with lamp-based microdisplay projectors is that a suitable screen for almost any current application is available, at a price, from a custom screen vendor. The cost of the custom screen, however, can be more than the cost of the projection system.

7.3.2 Speckle with Laser Illumination

Projectors with laser illumination can have much longer coherence lengths than the coherence induced in systems with HID lamps by the lens pupil. Therefore, the speckle issues associated with laser projectors are much more serious than the problems associated with microdisplay-based projectors. The incremental speckle improvements from screen design changes discussed in the previous section are no longer sufficient to provide a satisfactory image. The moving screen approach can, however, provide a

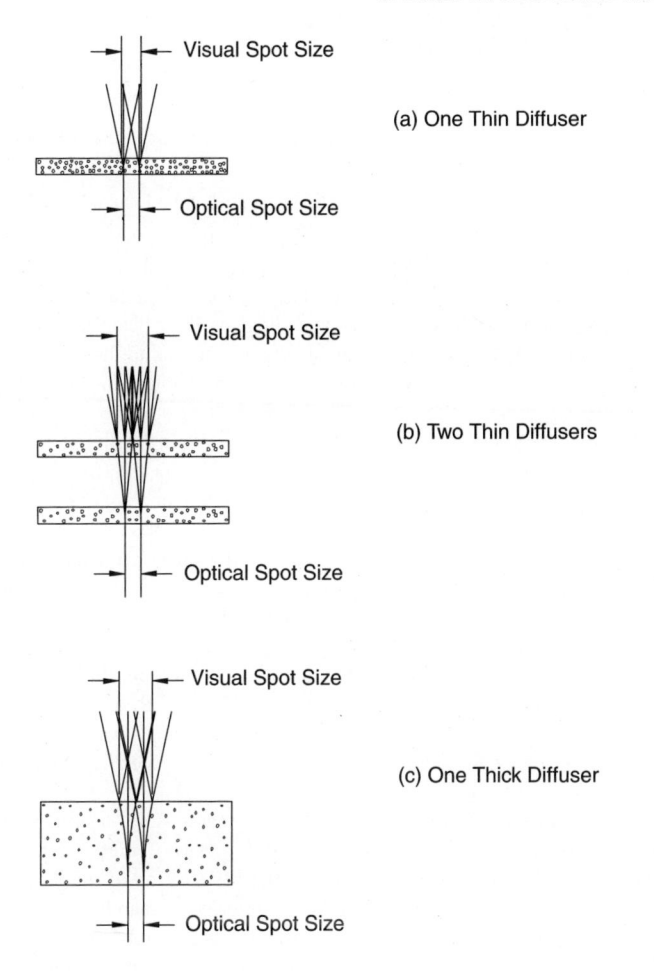

Figure 7.34 Effect of a finite thickness diffuser on resolution: (a) one thin diffuser; (b) two thin diffusers; (c) one thick diffuser.

satisfactory image. This is not necessarily desirable or cost-effective in a rear-projection system and it may not be possible at all in a front projector.

The best way to control speckle in a laser-illuminated projector is to reduce the coherence of the illumination in a laser illuminated projector. Iwai and Asakura (1996) suggest four methods to do this:

1. control of spatial coherence
2. control of temporal coherence
3. temporal averaging based on time-varying spatial sampling
4. spatial integration at the detector.

The moving diffuser approach already discussed is an example of using the third approach to speckle reduction, temporal averaging based on spatial sampling.

Screen design with a thick diffuser is an example of the fourth category, spatial integration. This method has already been discussed in section 7.3.1. Screen design alone is normally does not reduce speckle enough to make the image acceptable, except in the case of a moving diffuser in the screen.

When multiple lasers are used as in the laser bar system to be discussed in Section 8.3.2, the separate emitters are typically incoherent with respect to each other. This is a type of the first approach to speckle control, the control of spatial coherence. In this case, the speckle is reduced by a factor of $1/\sqrt{N}$, where N is the number of emitters. For a 24 emitter Novalux laser, for example, the spatial coherence and speckle is reduced to 20 % of the level that would be expected with a single emitter. This coherence reduction, coupled with a speckle reducing screen design, may be enough to produce a satisfactory image.

Spatial coherence reduction can also involve expanding the laser beam to a finite size and passing it through a retardation plate with a number of areas with different retardances. The beam then passes through an integrator so it can uniformly illuminate the microdisplay. In this way, the overall beam coherence is reduced.

Temporal coherence can be controlled either in the screen itself or in the illumination path. As described above, Shin (2006) actually used both techniques to produce a speckle-free image. Trisnadi (2002) describes a system of speckle reduction for a GLV laser projector using a vibrating binary phase grating in the imaging path of the projector. Unfortunately, this approach requires a relay lens to image the display on the phase-grating and then a second lens to reimage the display on the projection screen. The added cost of the relay lens makes this approach impractical for mass-market projectors. Temporal coherence can also be controlled by a moving diffuser in the illumination path. This can reduce the temporal coherence and therefore the speckle enough to produce a satisfactory image. Switchable Bragg gratings have also been used to reduce the speckle through reduced temporal coherence (Smith and Popovich 2000).

Another approach to reducing temporal coherence is discussed by Roddy and Markis (2003). In this design, the laser itself, by vibrating the output mirror with a RF drive, is wavelength-modulated enough to reduce speckle. Each wavelength generated produces a different speckle pattern, and these patterns average to a uniform image over the integration time of the eye.

Speckle reduction is a topic of active projection system research, since lasers are expected to play an increasingly important role in projection systems in the future. A very brief search of issued US patents produced 25 related to speckle reduction issued since the first edition of this book. In one published example of speckle reduction, Serafimovich *et al.* (2005) describes research going on at Samsung on the topic. Unfortunately, as this book goes to press, very little of this research has been published because of the commercial importance of anti-speckle intellectual property (IP). Speckle reduction can be thought of as a solved problem because speckle-free images generated by laser projectors have been publicly demonstrated at venues such as SID Symposium. Unfortunately, it is normally not revealed how the speckle was reduced, although in one case the authors were told that the speckle reduction element added less than $1 to the cost of the system.

References

Aizu, Y. and Asakura, T. (1988) Velocity dependence of dynamic speckles produced by a moving diffuser with background noise. *Optical Communications* **68**: 329–33.
Betensky, E. (1993) Postmodern lens design. *Optical Engineering* **32**: 1750–6.
Born, M. and Wolf, E. (1980) *Principles of Optics*, 6th edn. Oxford: Pergamon Press.
Boyd, R.W. (1983) *Radiometry and the Detection of Optical Radiation*. New York: John Wiley & Sons, Inc..
Bradley, H. (1986) *Rear Projection Screen*. US Patent 4,573,764.
Bradley, H. (1990) *Lenticular Arrays for Front Projection Screens and Contrast Improving Method and Device*. US Patent 4,964,695.
Bradley, R.J., Goldenberg, J.F. and McKechnie, T.S. (1985) Ultra-wide viewing angle rear projection television screen. *IEEE Transactions on Consumer Electronics* **CE-31**(3): 185–93.
Brubaker, B. (2007) Brightness vs contrast. *Angles of Reflection*, June 2007, newsletter. Da-Lite Screen Co., 4 pp.

REFERENCES

Chen, L.-H., Horng, W.-C., Chen, R.-H. and Lin, C.-C. (1995) Design and characterization of an aspherical zoom lens for projection TV. In M.H. Wu (ed.), *Proc. SPIE: Projection Displays* **2407** (April): 80–8.

Clarke, J. (1988a) Current trends in optics for projection TV. *Optical Engineering* **27**: 16–22.

Clarke, J. (1988b) Optical aspects of the interference filter projection CRT. *SID Intl. Symp. Digest of Technical Papers*, 214–17.

Da-Lite (2007) www.da-lite.com

DNP (2007) http://www.supernovascreen.com/

Davis, A., Bush, R.C., Harvey, J.C. and Foley, M.F. (2001) Fresnel lenses in rear projection displays. *SID Intl. Symp. Digest of Technical Papers*, Paper P-95, 934–7.

Ebina, K. (2002) Optical system architecture for rear projection screen. *SID Intl. Symp. Digest of Technical Papers*, Paper 51.3, 1342–5.

Fischer, R.F. and Tadic, B. (2007) *Optical System Design*, 2nd edn. New York: McGraw-Hill.

Glenn, Jr, W.E. (1970) *Composite Back Projection Screen*, US 3,523,717, issued 11 August 1970.

Gohman, J., Peterson, M. and Engle, S. (2005) Slim rear projection. *SID Intl. Symp. Digest of Technical Papers*, Paper 70.4, 1922–5.

Goldenberg, J., Huang, Q. and Shimizu, J.A. (1997) Rear projection screens. In M. Wu (ed.), *Proc. SPIE: Projection Displays III* **3013**.

Harada, T., Takamatsu, Y., Ishikawa, M., Bruzzone, C.L., Moshrefzadeh, R.S. et al. (2002) *Reflective Projection Screen and Projection System*, US Patent 6,381,068, issued 30 April 2002, 21 pp assigned to 3M.

Huang, Y. (2007) *Total Internal Reflection Fresnel-lens and Devices*, US Patent 7,230,758, issued 12 June 2007.

Iwai, T. and Asakura, T. (1996) Speckle reduction in coherent information processing. *Proceedings of the IEEE* **84**(5) May: 765–81.

Iwata, S. Kagotani, A., Igarashi, Y., Takahashi, S. and Abe, T. (2005) A super fine-pitch screen for rear projection TV, Paper 70.2, *SID Intl. Symp. Digest of Technical Papers*, 1914–17.

Kingslake, R. (1983) *Optical System Design*. New York: Academic Press.

Koch, D.G. (1992) Simplified irradiance/illuminance calculations in optical systems. In H. Zügge (ed.), *Proc. SPIE: Lens and Optical Systems Design* **1780**.

Kozma, A. and Christensen, C. (1976) Effects of speckle on resolution. *J. of the Optical Soc. of America* **66**: 1257–60.

Kuwata, M., Sasagawa, T., Kojima, K., Aizawa, J., Miyata, A. et al. (2006) Projection optical system for a compact rear projector. *J. SID* **14**(2): 199–206.

Leifer, I., Spencer, C.J.D., Welford, W.T. and Richmond, C.N. (1961) Grainless screens for projection microscopy. *J. of the Optical Soc. of America* **510**.

Lindegaard, S. (1996) Rear-projection Fresnel screens with short focal length. *SID Intl. Symp. Digest of Technical Papers*, 514–17.

Lowenthal, S. and Joyeux, D. (1971) Speckle removal by a slowly moving diffuser associated with a motionless diffuser. *J. of the Optical Soc. of America* **61**.

Malang, A.W. (1989) High brightness projection video display with concave phosphor surface. *Proc. SPIE* **1081**: 101–6.

Milster, T.D. (1995) Miniature and micro-optics. In M. Bass (ed.), *Handbook of Optics*, 2nd edn. Vol. 2, 7.18–7.27.

Moskovich, J. (1997) *Telecentric Lens Systems for Forming an Image of and Object Composed of Pixels* US Patent 5,625,495, issued 29 April 1997.

Moskovich, J. (2003) *High Performance Projection Television Lens Systems*, US Patent 6,509,937, issued 21 January 2003.

Nakanishi, Y. and Inoue, M. (1990) *Rear Projection Screen*. US Patent 4,927,233.

Noda, H., Muraji, T., Gohara, Y., Miki, Y., Tsuda, K. et al. (1989) High definition liquid crystal projection TV. Japan Display '89 (IDRC) Proceedings published by the SID, pp 256–259, 1989.

Oda, K., Sekiguchi, H. and Gotoh, M. (2000) Ultra high contrast screen. *SID Intl. Symp. Digest of Technical Papers*, Paper 15.2, 198–201.

Ogawa, J., Agata, K., Sakamoto, M., Urano, K. and Matsumoto, T. (2004) Super-short focus front projector with aspheric mirror projection optical system. *SID Intl. Symp. Digest of Technical Papers*, Paper 12.2, 170–3.

Okamoto, T. and Asakura, T. (1990) Velocity dependence of Image speckles produced by a moving diffuser under dynamic speckle illumination. *Optics Communications* **77**.

Ramachandran, G. and Prior, G. (2007) Cost-effective ultrathin RPTVs can reverse falling market share relative to FPDs. *SID Intl. Symp. Digest of Technical Papers*, Paper 9.6L, 113–16.

Roddy, J.E. and Markis, W.R. (2003) *Speckle Suppressed Laser Projection System with Partial Beam Reflection*, .US Patent 6,625,381, issued 23 September 2003.

Rykowski, R. and Forkner, J.F. (1995) Matching illumination system with projection optics. In M.H. Wu (ed.), *Proc. SPIE: Projection Displays*, **2407**: 48–52.

Serafimovich, P.G., Cheong, B.H., Ahn, P.S., Shin, J.K. and Kim, S.H. (2005) DMD illumination using diffractive optical elements. *SID Intl. Symp. Digest of Technical Papers*, Paper P-158, 902–5.

Shikama, S., Suzuki, H. and Teramoto, K. (2002) Optical system of ultra-thin rear projector equipped with refractive-reflective projection optics. *SID Intl. Symp. Digest of Technical Papers*, Paper 46.2, 1250–3.

Shimizu, J. (1995) Brightness distributions in projection displays. *SID Intl. Symp. Digest of Technical Papers*, 66–9.

Shimoda, K. and Kawashima, T. (2007) *Projection Screen and Method for Manufacturing the Same*, US Patent 7,251,074, issued 31 July 2007.

Shin, S.C., Yoo, S.S., Lee, S.Y., Park, C.-Y., Park, S.-Y. et al. (2006) Removal of hot spot speckle on rear projection screen using the rotating screen system. *Journal of Display Technology* **2**(1), March: 79–84.

Shirley, L.G. and George, N. (1989) Speckle from a cascade of two thin diffusers. *J. of the Optical Soc. of America* **6**: 765–81.

Smith, R. and Popovich, M. (2000) Electrically switchable Bragg grating technology for projection displays. *Proceedings of the 7th IDW*. ITE & SID, 1065–8.

Smith, W.J. (1990) *Modern Optical Engineering: The Design of Optical Systems*, 2nd edn. New York: McGraw-Hill, Inc.

Smith, W.J. (1992) *Modern Lens Design: A Resource Manual*. New York: McGraw-Hill Inc.

SMPTE (2003) ANSI/SMPTE 196M-2003, SMPTE Standard for Motion-Picture Film – Indoor Theater and Review Room Projection – Screen Luminance and Viewing Conditions.

Springsteen, A.W., Leland, J. and Ricker, T. (1994) A Guide to Reflectance Materials and Coatings. North Sutton, NH: Labsphere Inc.

Stanton, D.A. (1986) *Blackened Optical Transmission System*, US Patent 4,605,283, issued 12 August 1986.

Strong, J.D. and Hayward, R. (1940) *Transparent Projection Screen*, US Patent 2,200,646, issued 14 May 1940.

Trisnadi, J.I. (2002) Speckle contrast reduction in laser projection displays. In M.H. Wu (ed.), *Proc. SPIE: Projection Displays VIII* **4657** (April): 131–7.

Vlahos, P. (1868) Film-based projection systems. In H.R. Luxenberg and R.L. Kuehn (eds), *Display System Engineering*, McGraw-Hill, Inter-University Electronics Series, V. 5, 277–318.

Vriens, L. (1988) Interference CRT. *SID Intl. Symp. Digest of Technical Papers*, 217–21.

Wang, L., Tschudi, T., Halldórsson, T. and Pétursson, P.R. (1998) Speckle reduction in laser projection systems by diffractive optical elements, *Appl. Opt.* **37**(10): 1770–5.

Watanabe, T., Kabuto, N., Hirata, K. and Aoki, K. (2000) Recent progress in CRT projection display, *SID Intl. Symp. Digest of Technical Papers*, Paper 21.2, 306–9.

Watson, A.B. and Ahumada, Jr., A.J. (2005) A standard model for foveal detection of spatial contrast. *Journal of Vision* **5**(9): 6, 717–40, http://journalofvision.org/5/9/6/

Wolf, M. (1937) The enlarged projection of television pictures. *Philips Technical Review* **2**.

Wolfe, C.R, Esguerra, M., Lewis, K., Kinosian, K., Knutson, B. et al. (2003) Engineering Blackscreen™ for rear-projection television applications. *Proc IDMC*. SID, pp. 275–8.

8

Light Sources for Light-valve and Microdisplay Projection Systems

In this chapter, the properties of light sources and the way these properties affect projector performance will be discussed. The system performance of microdisplay and light-valve projectors is very much dependent on the characteristics of the projection lamp or other light source and its associated optics and power supply. The intention of this chapter is to provide the basic information about lamp properties that affect the design of a projector. This will allow the projector designer to make intelligent inquiries of the light source vendor.

The two major lamp types used for light-valve and microdisplay projection systems are tungsten-halogen lamps and high intensity discharge (HID) lamps, which include xenon lamps, metal-halide lamps, and ultra-high pressure mercury lamps. The UHP™ from Philips is the best known example of the latter lamp type. For all discharge lamp types, the part of the lamp assembly which generates the light is called the burner. Lamps of moderate power for projection applications are most often built with the burner permanently attached to a reflector. Higher power lamps (>1000 W) more commonly use a separate burner and reflector. Tungsten-halogen lamps are generally only used in low performance projectors.

Solid state light sources, including both LEDs and lasers, are beginning to appear in projection displays. These systems are not using just a new kind of light source; a completely different projection design is required. These design changes include factors such as the power supply, thermal management, light collection techniques, color management, microdisplay design, projection lens and safety features. These factors will all be discussed, either in this chapter or elsewhere in the book. Sections 8.3 and 8.4 of this chapter will focus primarily on LEDs and lasers as light sources and their optical properties. Section 8.9.5 will focus on the special light collection requirements for LEDs. Chapters 9 and 10 will discuss the new projection architectures enabled by lasers and LEDs respectively. Both lasers and LEDs can also be used in the same projector designs as lamps.

Projection Displays, Second Edition M. Brennesholtz, E. Stupp
© 2008 John Wiley & Sons, Ltd

8.1 Lamp Parameters

A number of lamp or other light source characteristics can have significant effects on the projector characteristics. Among these are:

1. lamp power,
2. lamp efficacy,
3. light emitting area size, shape and emergent solid angle,
4. lamp spectrum and
5. the presence of spectral emission lines.

Other lamp factors, such as lamp life and lamp warm-up, can affect the applications to which a specific lamp is applied and the way the projector including the lamp is marketed. Consumer applications are very demanding on these lamp qualities. Many short-arc metal-halide lamps have lifetimes only of a few hundred hours, which would likely preclude their use in consumer products. Lamps with long warm-up characteristics would also not be interesting for consumer projectors. For institutional applications, on the other hand, lamp life can be less important than some of the other parameters affecting performance. For virtually all applications, safety and environmental issues also determine whether specific lamps can be used.

Lamps specifically designed for projection displays appeared on the market from many of the major lamp manufacturers (Schnedler 1995; Kawai 1995; Higashi 1995a; Stewart 1997) when microdisplay-based projectors began to appear in the 1995 time frame. Most of these lamps shared several common characteristics, including short arc length, longer life and integrated reflector. The importance of each of these factors will be discussed in following sections. Much of the lamp R&D since then has been aimed at providing these lamps at lower cost, higher power, shorter arc and/or longer life.

8.2 Types of Projection Lamps

Most projection systems use arc lamps, also known as high intensity discharge (HID) lamps. A few low end and ultra-portable projectors have used tungsten-halogen lamps. A few LED and laser illumination systems are on the market and their numbers are expected to grow rapidly in the future.

A HID bulb consists of a sealed envelope containing the fill material. At operating temperature the fill is vaporized, although at room temperature the fill may be a gas such as xenon, a liquid such as mercury or a solid such as a metal-halide salt. Normally, two electrodes are used to provide power to the arc. There have been electrodeless lamps built in the past that were powered by microwaves (Dolan *et al.* 1992; Turner *et al.* 1995), but these types of lamps are only now beginning to reach commercial application in projection systems, as will be discussed in Section 8.2.5. The main promise of electrodeless lamps is very long life, since there is no electrode to be damaged, or to be sputtered and darken the inside of the envelope.

HID lamps can be driven with either AC or DC waveforms; the type of waveform is usually not optional but, rather, is specified by the manufacturer of the lamp. DC arcs tend to be more stable and may have longer life (Higashi 1995b), although one of the longest lived lamps available for projection applications is an AC lamp.

All the lamp spectra in this chapter are normalized so $Y = 1$. This normalization technique allows the easy comparison of the shapes of spectra without concern for their amplitudes. With this normalization, the spectrum, when convolved with the $\overline{Y}$ curve, would contain 683 lumens when the vertical axis of each graph is interpreted in units of W/nm. Another normalization technique that is useful is to normalize the spectrum so it contains a defined amount of energy over the visible range, such as one watt over the range of 460–680 nm. With another commonly encountered normalization technique, the peak value in the visible is set to 100 %. This last technique, while easy to implement, can lead to misleading impressions when comparing spectra of different lamps.

8.2.1 Xenon Lamps

Xenon lamps are HID lamps where the only fill material is the noble gas xenon. Xenon has several advantages over competing lamps. For one thing, the colorimetry of xenon lamps is very good (Yeralan 2005). As can be seen in Figure 8.1 the spectrum is nearly flat across the visible range. Another advantage of xenon lamps is that they are 'instant on' with no associated warm up period. They also pose no environmental hazard, since xenon is an inert gas. The final advantage of xenon lamps is they can be built in very high power versions for use in very high brightness projectors. 7 kW lamps are the norm in film and digital cinema applications and higher power lamps have been built for special applications.

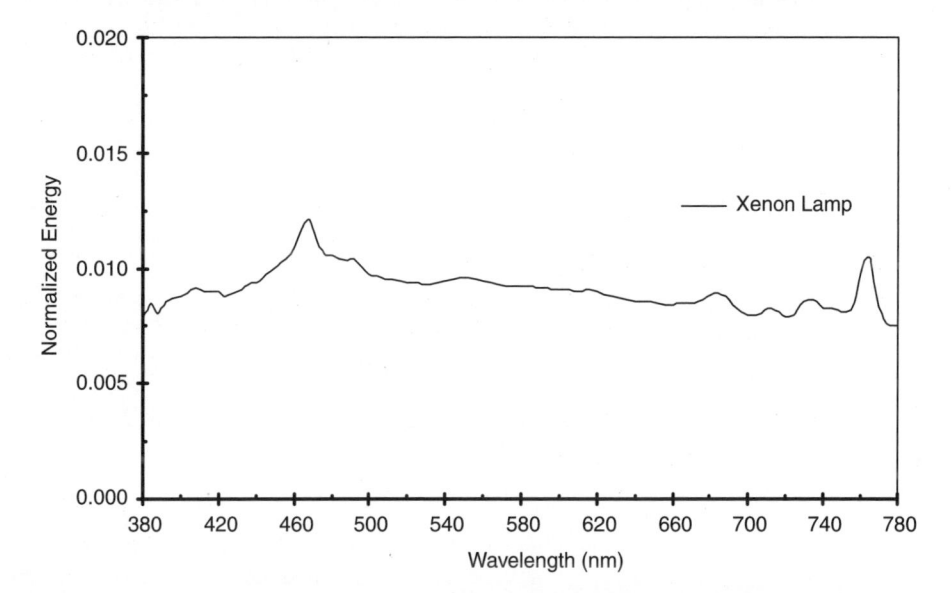

Figure 8.1 Spectrum of a xenon lamp. This lamp has a correlated color temperature of 6008 K and CIE x = 0.322 and CIE y = 0.327.

On the other hand, xenon lamps can pose a significant safety hazard. When cold, the typical pressure inside a lamp is about 10–13 atmospheres. When operating, the pressure rises to 50–59 atmospheres. Higher pressure lamps with cold pressures as high as 19 atmospheres and hot pressures of 76 atmospheres have been built (Smolka et al. 1995) and provide greater efficacy with a smaller arc size. The high internal pressure of a xenon lamps leads to a lamp explosion hazard.

The likelihood of a lamp explosion is the greatest when the lamp is operating. Not only is the internal pressure higher, but the high temperature can produce extra strain in the quartz envelope. During operation, however, the lamp is operated inside a lamp housing within the projector that presumably has been specifically designed to contain all the lamp shrapnel in the case of an explosion. The greatest danger to personnel is when lamps are being changed. In this case, the lamp may be outside it's protective shipping container and, if the technician is not wearing proper safety equipment, serious injury can result if the lamp explodes. It is *never* acceptable to ignore a lamp manufacturers safety instructions with a xenon lamp.

8.2.2 Metal-halide Types

Metal-halide lamps are HID lamps in which the fill material consists primarily of mercury, with a doping of a halide salt of the desired metal. There are a number of metal atoms that, in the free, gaseous state, have excellent spectral emission properties. Unfortunately, most of these metals in the pure state have a low vapor pressure and few atoms would be in the free, gaseous state at temperatures achievable in a HID lamp. Most free atoms would diffuse to the relatively cool quartz envelope, condense or react with the lamp envelope and be unavailable to the arc. This problem is solved by the use of a halide salt of the desired metal in the lamp.

Iodine is the halogen that is most commonly used. These halide salts have much higher vapor pressures at the envelope temperatures of a lamp than the pure metal and they are in the vapor state during normal operation. Gaseous metal-halide molecules can then diffuse or convect into the still higher temperature region of the arc. Here the molecules are dissociated, leaving the metal atoms free to be excited by the arc and emit their characteristic spectra. When the metal atoms eventually diffuse or move by convection out of the arc region, they form new metal-halide molecules with the free and very reactive halide atoms. The salts maintain a solid/gaseous equilibrium and repeat the cycle.

Over 50 different metals have been successfully used as the fill material in metal-halide lamps (Waymouth 1971). To get the desired optical properties, two or more metals can be used in combination. Dysprosium (Dy) and neodymium (Nd) salts are common ingredients in lamps intended for projection systems. Cesium iodide is often added to rare-earth metal halide lamps such as these in order to broaden and stabilize the arc. Lamps with these metals have high efficacy and fair colorimetry. Other fill materials have been reported in lamps intended for projection displays, including gadolinium (Gd) and lutetium (Lu) mixtures (Takeuchi *et al.* 1995).

Spectra of two different metal-halide lamps are shown in Figure 8.2. Even though mercury is the main ingredient in metal-halide lamps, the emission spectrum is dominated by the metal atoms. This occurs because the metal atoms commonly used in lamps generally have lower electron energy levels than the mercury atoms have. The energy from the arc goes into exciting these lower energy levels, rendering the energy unavailable to the mercury atoms.

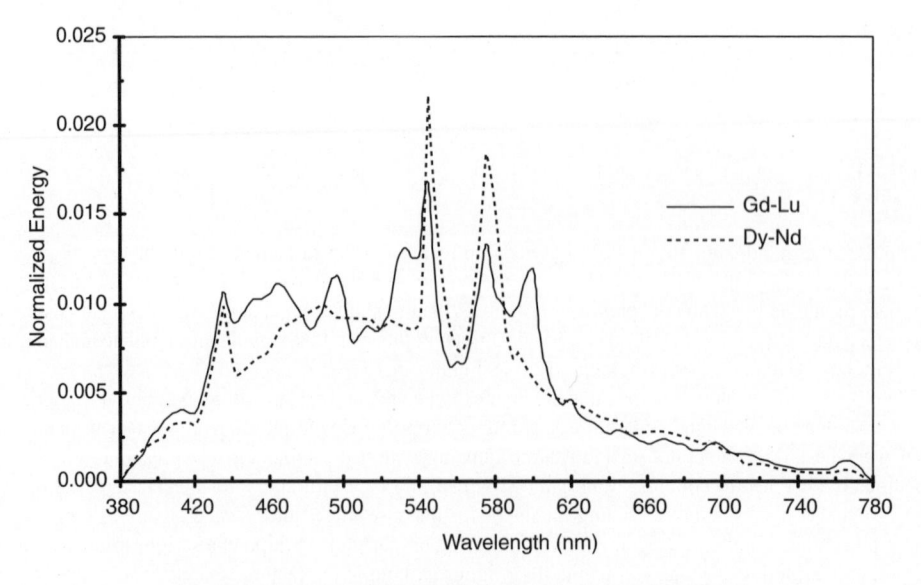

Figure 8.2 Spectra of metal-halide lamps (data courtesy of Matsushita).

TYPES OF PROJECTION LAMPS 173

Metal halide lamps can have good colorimetry and very high efficacy. Up to 100 lumens per watt is possible in a lamp optimized for projection applications. On the other hand, there are a number of problems associated with metal-halide lamps. Historically, metal halide lamps had relatively large arcs. This problem has largely been overcome in lamps designed specifically for projection applications. Metal halide lamps with arcs shorter than 1.5 mm are now available to the projector designer.

Another problem with metal halide lamps is the long warm-up time required to get full brightness and the proper color. This occurs because it takes time for the quartz envelope to heat up enough to fully vaporize the metal halide salts and establish the metal-halide cycle. Since the different metal salts in a lamp vaporize at different rates, there can be major color shifts before the lamp finally stabilizes. In older lamps, and lamps not specifically designed for projection, warm up times of 5–10 minutes are not uncommon. This problem has been greatly reduced in projection lamps, with warm-up times on the order of 1 minute now available.

Another problem with metal halide lamps is that different colors can be emitted from separate regions of the arc. This occurs because in a mixture of atomic types, the different metal atoms have different atomic weights, vapor pressures and diffusion rates. It is not uncommon for the area farther from the center of the arc to be redder than the regions nearer the center of the arc. The flux throughput of the colors in the outermost regions of the arc can be attenuated more by limiting apertures in the projector than those in the more central region, which can impact the color rendition by the projector. The implications of this change in color with aperture size will be discussed in more detail in terms of étendue in Chapter 11.

Due to the very high temperature gradients in a metal halide lamp, thermal convection within the lamp plays a strong role in the operation of these lamps. In horizontal operation, the arc is visibly bowed up by convection. Because of this, the lamps will have different optical properties and light distributions depending on the tilt angle of the lamp. This effect is shown in Figure 8.3. These two spectra are measured on the same metal halide lamp, except that for one spectrum the lamp was horizontal and the other spectrum was measured when the lamp was tilted 45°. Note the significant shift in the amount of red available from the lamp in the two different orientations. Most of this shift is due to the change in convection pattern. Different colored regions of the arc are brought into the center of the reflector, changing the spectral distribution of the emissions.

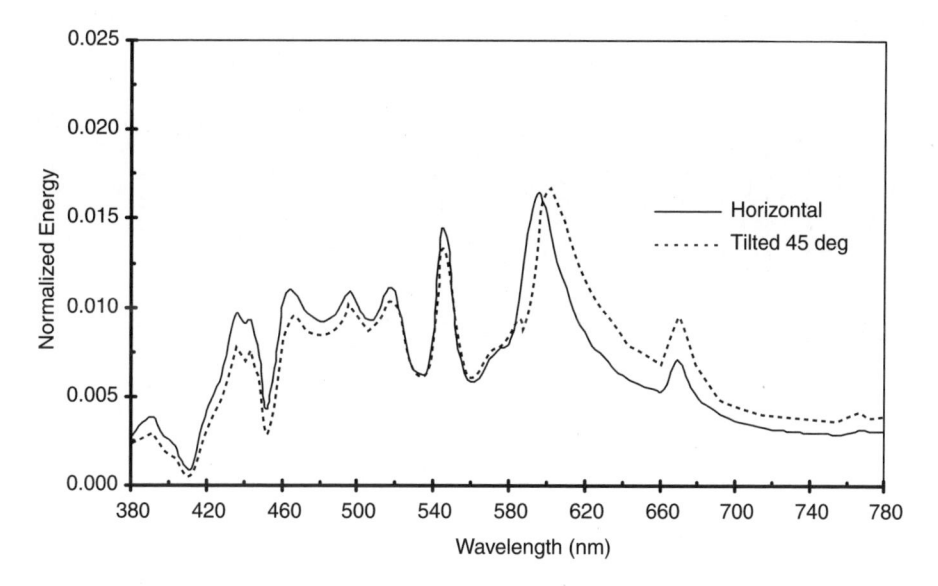

Figure 8.3 Effect of tilt on the spectrum of a metal halide lamp.

Due to the presence of two or more atomic species emitting light in a metal halide lamp, these lamps can change color over the life of the lamp. This occurs if the two metals react with the envelope at different rates into compounds with lower vapor pressures than the original metal halides. While the end of life is conventionally taken as the 50 % light output point, if there is a major color shift, the lamp may be unusable in a projector even while it is still emitting more than 50 % of its initial lumens.

In addition to color shifts with operation, devitrification of the burner envelope by the metal halides can occur. The envelope becomes translucent which causes an apparent increase in the arc size and can reduce the system throughput (see Chapter 11). Additionally, erosion of the electrodes during operation, which increases the arc gap, can also decrease the throughput. The usable life of short arc metal halide lamps, the type most interesting for projector application, is frequently only a few hundred hours. Some lamps made especially for projection applications, however, are now achieving satisfactory operation for more than one thousand hours.

While most metal halide lamps operate with AC power, lamps have been developed to employ DC drive (Higashi 1995; Higashi and Arimoto 1995). These lamps are claimed to have up to 4000 hour life.

8.2.3 The UHP Lamp

Since the introduction of the UHP[TM,1] by Philips in 1995, the ultra-high pressure short-arc mercury (Hg) lamp has come to dominate projection systems. Metal-halide lamps are rarely used in projection and xenon lamps are used only in high-end systems. Ultra-high pressure Hg lamps are now offered by several other companies in addition to Philips. In most cases, this HID lamp is offered as an integrated lamp/collector system with a power supply optimized to maximize the lamp life. Ultra-high pressure mercury lamps are available in versions from about 50 W up to about 900 W. These lamps provide a short arc (1.0 to 1.3 mm) and lamp lives of up to 8000 hours or more for the lower-power versions and as much as several thousand hours for the higher powered lamps (lamp life is discussed further in section 8.8). A typical emission spectrum of two of the lower power lamps is illustrated in Figure 8.4.

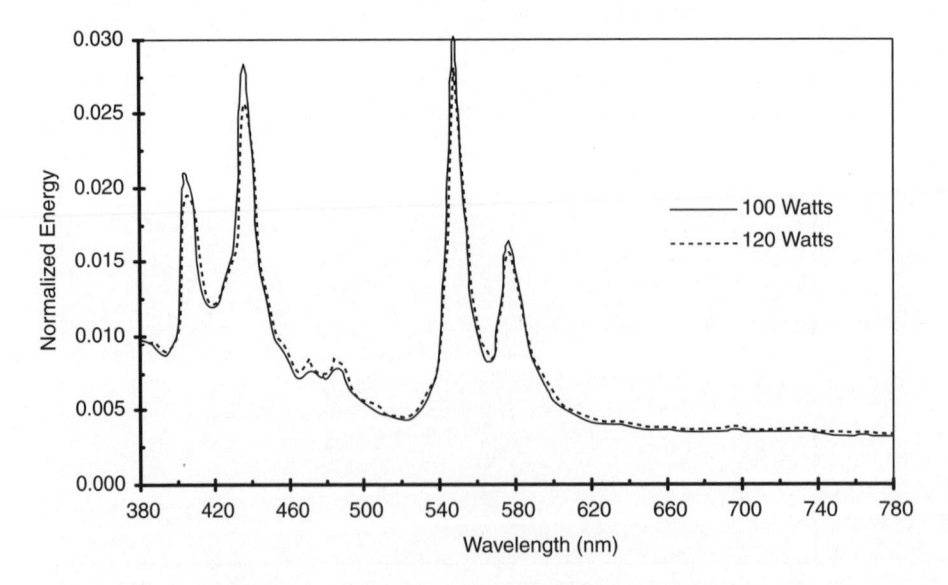

Figure 8.4 Emission spectra of 100 W and 120 W UHP lamps.

---

[1] [TM] Philips Lighting.

The spectrum of the UHP-type lamp presents several problems to the projector designer. First, while there are blue and green emission lines at 436 and 548 nm respectively, there is no red emission line. This leads to low red output compared to the blue and green energies and difficulty in achieving a good white balance. Secondly, there is an emission line in the deep blue at 407 nm. This emission line contributes little or nothing to either the color or brightness of the system, but the relatively high-energy 407 nm photons can cause rapid aging in organic components in the projector. This includes not only the obvious components such as the LC materials and polarizers in LC designs but also the less obvious organics such as optical cements or the oil used to prevent mirror sticking in DLP microdisplays. Therefore this line must be filtered out by a long-pass filter, such as the L-42 filter shown in Figure 6.13, without reducing the brightness of the 436 nm line, a key emission line in the blue output of the lamp.

Finally, the yellow line at 575 nm must also be filtered out in three-color projectors (red, green and blue). If the yellow is included in the green channel, the green becomes too yellow. If the yellow is included with the red, the red becomes much brighter, but it will be too orange to produce satisfactory television images. Due to the relatively narrow wavelength gap between green and red, the filters to remove the yellow line must be made to a relatively high tolerance. These colorimetry problems with the UHP-type lamp reduce its effective efficacy from 60 lumen/watt to 45 lumen per watt or less, depending on the colorimetry requirements for the system.

The properties of the original UHP lamp are summarized in Table 8.1. The very short arc gap combined with a parabolic reflector results in a narrow beam bundle. This is important for maximizing the system throughput, particularly for systems designed with smaller microdisplays. This lamp has a long usable life, maintaining at least 75% of the bundle characteristics for >8000 hours. Another version of this lamp, introduced in 1995, was rated at 120 W with an increase in light output that was approximately linear with the power.

Table 8.1 Properties of the UHP lamp.

Arc gap	1.3 mm
Lamp wattage	100 W
Luminous flux	6000 lm
Color temperature T_C	8500 K
Full width beam half angle	2°
Color rendering index R_a	60
Average life	4000 to 8000 hours
Bundle maintenance	75 % after 8000 hours

Source: Schnedler and Wijngaarde (1995).

Since the original introduction of the 100 W and 120 W UHP lamps, Philips has extended the technology to both higher and lower power levels. Lamps similar to the UHP have also appeared on the market from both European and Asian manufacturers. UHP-type lamps are available in powers from 100 W up to about 400 W. For a complete summary of all UHP properties and a good review of all improvements made in the UHP between its introduction and 2005, see Derra et al. (2005).

For low-brightness projectors Philips has introduced a 50 W lamp fundamentally similar to the UHP with the trade name Ujoy (Moench et al. 2007). This lamp is capable of supplying about 2000 lumens to a high-étendue system. Its normal application, however, would be in a low-brightness (perhaps 200 lumens) 'personal projector' with a small microdisplay that has a low étendue. In a system with an étendue of 5 mm^2 steradian, the lamp can deliver about 1100 lumens to the microdisplay. If the optical system is 20 % efficient, this will mean 220 lumens at the projection screen.

The spectrum of the Ujoy lamp is shown in Figure 8.5 compared to a standard UHP lamp. As discussed earlier, both spectra have been normalized so Y = 1 to facilitate comparison of the spectra. As can be seen, the emission lines of the Ujoy lamp are higher than they are for the UHP lamp, and the broad-band radiation, including the red light, is lower. This lack of red, even compared to a standard UHP, will make it difficult for a projector with a Ujoy lamp to have good colorimetry. The differences in Figure 8.5 in the emission lines at 407 nm are more the effect of the reflector coating design than the fundamental difference between the Ujoy and UHP spectra.

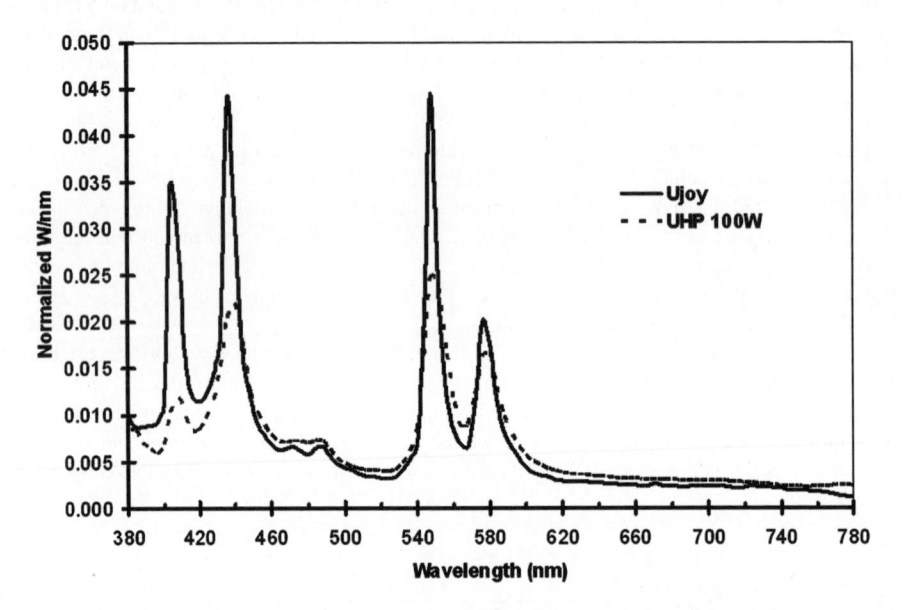

Figure 8.5 Spectrum of Philips Ujoy lamp compared to 100 W UHP (data courtesy of H. Moench at Philips Research, Aachen).

The designer of a high-brightness projector, greater than about 3000 lumens, who wants to use UHP-type technology has two choices. First, it is possible to use multiple UHP-type lamps. For example, the Christie LX1500 projector uses four 330W NSH[2] lamps to produce a projector lumen output of 15 000 lumens at the screen. The second choice would be to use the Philips CPL lamp, with powers up to 900 W. While this lamp has been described in the literature (Pollmann-Retsch *et al.* 2006) it has not yet been introduced as a product. The spectrum of the CPL as a function of power is shown in Figure 8.6.

Note that in Figure 8.6 as the lamp power increases, not only does the total light output increase, but the shape of the spectrum changes. This occurs because as the power increases the internal pressure in the lamp increases. High pressure increases the broad-band emission of the lamp at the expense of the line emission from the elemental mercury. In an extreme case, this can convert an emission line into an absorption line, as can be seen in the blue emission line at about 436 nm. For this line, at all power levels at or above 554 W the line is visible as an absorption line. This switch at high powers from line to broadband emission will improve the lamp colorimetry and allow the lamp to compete with xenon lamps in high-end applications.

The performance of the CPL lamp is limited by the properties of the quartz envelope. At 1400 K (1127 °C) quartz begins to soften and recrystallize rapidly. Sapphire (single crystal alumina or SCA) can tolerate higher temperatures and can be used for higher performance lamps. (Eastlund and Levis 2003) Unfortunately, the cost of sapphire and the difficulty in working the material have prevented lamps

---

[2] NSH is the type designation for ultra-high pressure mercury lamps made by Ushio.

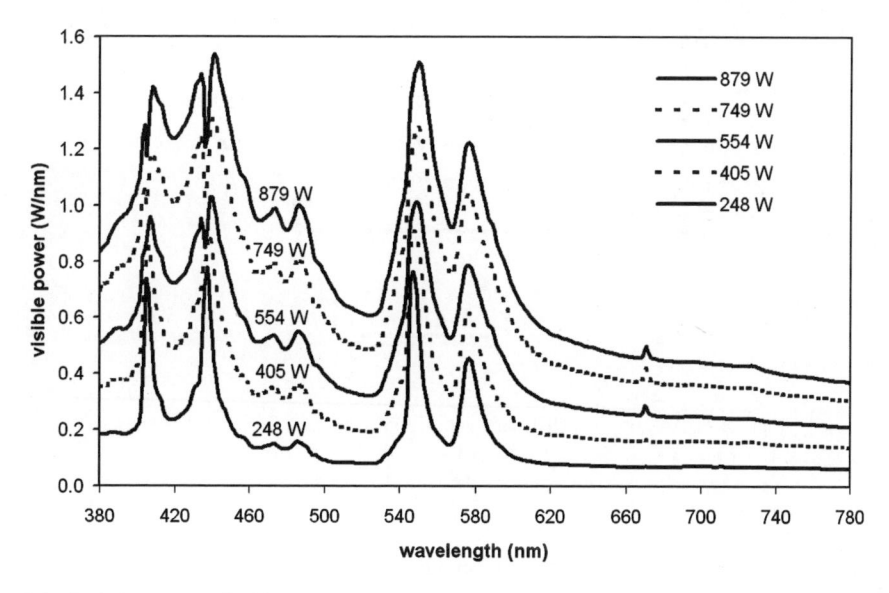

Figure 8.6 Emission spectra of Philips CPL Lamps at various powers (data courtesy of J. Pollmann-Retsch at Philips Research, Aachen).

with this material from entering the commercial marketplace. Lamps with polycrystalline alumina (PCA) envelopes are made commercially, however. Since PCA is an optically diffusing material, these lamps are not suited for use in étendue limited applications such as projection systems.

8.2.3.1 Temporal properties of UHP lamps

AC-driven ultra-high pressure mercury lamps such as the Philips UHP can suffer from a problem commonly called 'arc jump'. Normally, the arc is attached to two points, one on each of the electrodes. In operation, the arc can suddenly jump from one set of attachment points to another set. Typically the motion of the arc is in the order of 100–200 μm. Smaller motions can also occur, but they don't normally cause problems visible at the projection screen. Since the arc is nominally at the focus of the elliptical or parabolic reflector, this small motion relative to the focal point can change the collection efficiency, with a resultant change of the luminance at the screen. The luminance change may only be a few percent but the suddenness of the change makes it visible to the human eye. Even worse, the arc can alternate between two sets of attachment points at jump frequencies of 0.01–10 Hz. This creates a very disturbing sensation of flicker to the viewer.

Philips has solved this problem by adding a stabilization pulse to the trailing edge of each waveform, as shown in Figure 8.7 (Moench *et al.* 2000). This pulse has the tendency to modify the electrode shape until there is only a single attachment point on each electrode. In three-panel projectors, this stabilization pulse must be synchronized with the video rate but otherwise rarely causes problems. In DLP projectors where different time slices represent different bit planes of the video, the pulse must be used very carefully, or else the light generated during the pulse could be discarded completely.

Other solutions to the arc jump problem have been developed by other manufacturers of UHP-type lamps. Sugaya (2005) describes a method where the light output is constant during the stabilization period, eliminating the need for synchronization between the stabilization pulse and the video.

The light output of the UHP responds to drive power changes very quickly, on the order of 50 μS, with the instantaneous light output approximately linear with the instantaneous power. For proper lifetime,

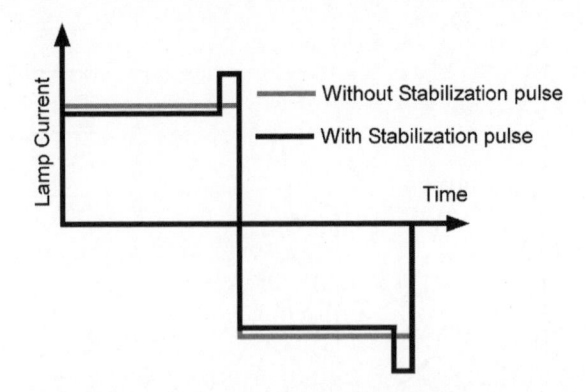

Figure 8.7 AC drive of a UHP lamp with stabilization pulse to reduce arc jumps.

the UHP lamp should be driven at the correct average power. Over milliseconds-long periods of time, the lamp can be driven safely at higher or lower powers, such as during the stabilization pulse. In an extreme example, UHP lamps were operated with the lamp running at about 3 x the average power with a 33 % duty cycle to give the correct average power (Stanton *et al.* 1996).

Field sequential projection systems can take advantage of this ability to efficiently provide a good color balance. The lamp would be run at higher powers during the red field where there is a shortage of light in the UHP and at lower powers during the green field where there is an excess of light. Over a complete frame time, the average power is the nominal lamp power. This allows a projector with a UHP lamp to achieve the desired color balance without discarding green light to match the red levels.

This ability was first demonstrated in LCoS projectors with rotating drums (Brennesholtz 2002; Brennesholtz *et al.* 2005). Philips later developed a ballast optimized to match the DLP in single-panel operation (Moench *et al.* 2006). In addition to driving the lamp at higher power when colors were displayed that required extra output to achieve color balance, the ballast could reduce the power while specific green bit planes were displayed by the DMD. This allowed the DLP system to achieve a better gray scale in scenes with low brightness. This is needed because the DMD can not switch fast enough to adequately display the low-order video bits in a single-panel system. One previous solution was to introduce dithering, an image processing solution that tended to introduce image artifacts and noise in dark scenes. Another solution was to include a dark-green segment into the color wheel (Hewlett *et al.* 2005). While the dark green filter was in the optical path, the low order bits could be displayed for a longer time to produce the desired light at the screen. This had the same effect as reduced power from the lamp during the desired bit planes but was much less efficient.

8.2.4 Tungsten-halogen Lamps

Tungsten-halogen lamps are incandescent lamps with halogen atoms incorporated in the gaseous fill surrounding the filament. At the operating temperature of any incandescent lamp, tungsten sublimes slowly from the filament. In normal incandescent lamps this blackens the lamp, reduces its output, and shortens the lamp life due to filament thinning and burnout. In tungsten-halogen lamps, the tungsten is transported away from the hot quartz walls back to the tungsten filament. This occurs by the formation of tungsten halides at the wall, diffusion of these molecules to the filament, and dissociation of the molecules at the high temperature of the filament, leaving metallic tungsten.

Because of the halogen transport cycle, these lamps can operate at higher filament temperatures than conventional incandescent lamps, which gives the tungsten-halogen lamps higher efficacies and color temperatures, along with better lifetimes. Figure 8.8 shows the spectra of a tungsten-halogen lamp run at both 3350 K and 2986 K and a conventional household 100 W tungsten lamp operated at 2888 K. The bulb at 3350 K provides significantly more blue light than either of the other bulbs, but the lamp is still blue deficient. This lack of blue and the relatively short life of the tungsten-halogen bulb run at 3350 K, as compared with HID lamps, explains why these lamps are not used in high performance projection systems. They are sometimes designed into very low end projectors where the low cost of both the lamp and its power supply aid in keeping the overall system cost low.

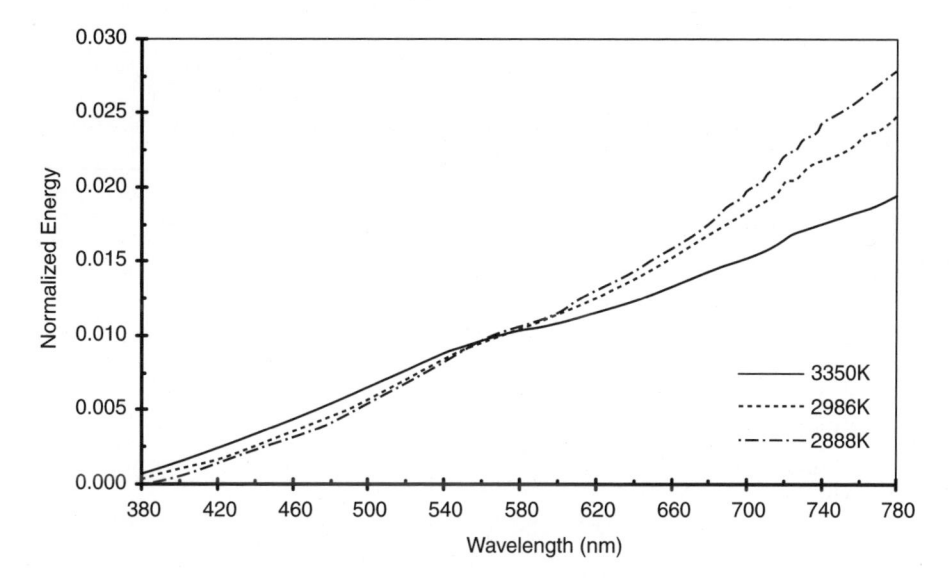

Figure 8.8 Normalized spectra of incandescent lamps.

8.2.5 Electrodeless Lamps

Lamps with no electrodes and powered by microwaves instead are just beginning to appear in commercial use in projection systems (Kirkpatrick *et al.* 2003; Pothoven and Pothoven 2003; Espiau *et al.* 2004; Espiau and Chang 2005; Joshi 2007; McGettigan 2007). In these lamp designs, a microwave source, much like the source in a microwave oven, drives a resonant cavity. Inside this resonant cavity, at a node of maximum intensity, is a sealed bulb containing a noble gas, mercury and the metal halide materials. Early electrodeless lamps used sulfur as a fill material (Turner 1995) but the fill material currently in use in projection systems has not been identified by the lamp vendors.

Since there are no electrodes to interfere with the chemical cycles inside the lamp, there is a wide variety of fill materials possible in microwave lamps. The materials have been chosen, however, to give as xenon-like a spectrum as possible in order to improve system colorimetry over the colors possible with an ultra-high-pressure mercury lamp.

Layout of a typical microwave lamp in its resonant cavity is shown in Figure 8.9. The puck is made of a dielectric material and then coated with a highly conductive coating. The shape of the puck is designed resonate with a node in the center, where the bulb is located. This provides the maximum field intensity at the lamp. The lower half of the bulb is mirrored so all of the light comes out the top of the lamp.

The spectra of the Luxim LiFi 4000+ with three different fill mixtures are shown in Figure 8.10. These three spectra were normalized to produce the same Y value. All of these fill mixtures were tailored to meet different rear projection TV needs. Note that the first and second mixtures are very similar except the second mixture produces more red in the spectrum beyond 600 nm than the first sample. This shows

180   LIGHT SOURCES FOR LIGHT-VALVE AND MICRODISPLAY PROJECTION SYSTEMS

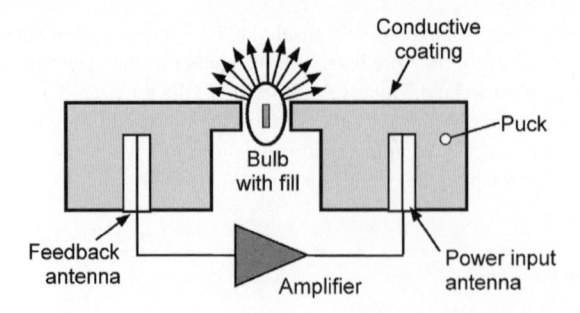

Figure 8.9 Electrodeless lamp in a resonant microwave cavity

up not only in the spectrum but in the color temperature of the two lamps, which were 7550 K and 7355 K, respectively, for the first and second mixtures. The third mixture has a very different spectrum with prominent emission lines in the green and blue. This lamp is intended for RPTV applications where very high color temperatures are desired by the designer. These are just three samples of possible spectra. Since the lamp is a metal-halide lamp the spectrum can be tailored to suit the application over a fairly wide range.

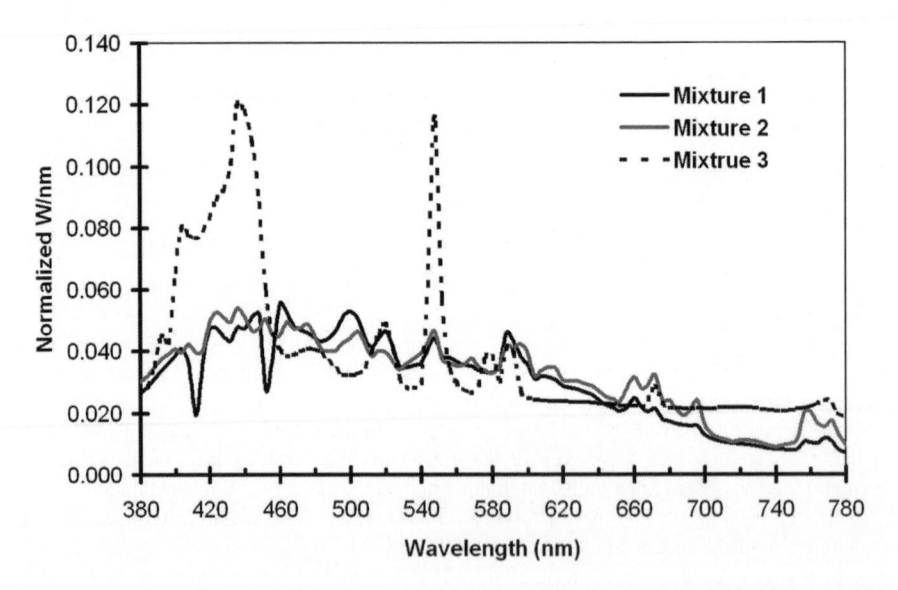

Figure 8.10 Spectra of three Luxim LiFi 4000+ lamps with different fill mixtures (data courtesy of T. McGettigan at Luxim Corporation).

8.3 Lasers as Projection Light Sources

Solid state light sources include both solid-state lasers and light emitting diodes (LEDs). Gas lasers are also available but will not be considered in this section because they are not considered suitable for mass-production projection applications. Solid state lasers will be discussed in this section, LEDs as light sources will be discussed in Section 8.4.

Lasers were proposed as light sources for projection use since just after the laser was invented (TI 1966). Prototype laser projectors have been built, but laser projectors did not enter production until recently, mainly because of the very high cost of the lasers compared to the lamps. Interest in designing lasers specifically for display applications began in earnest in the 1990s (Fink *et al.* 1996; Nakamura *et al.* 1997).

Currently there are at least two high-end laser projectors in production from Jenoptik (Hollemann 2000, 2001) and Evans & Sutherland (E&S) (Bloom and Tanner 2007). The lasers used in these projectors are custom designed and are extremely expensive, in part because their production volumes represent perhaps 20–50 total units of the two designs per year.

Most of the currently available visible-light lasers are targeted at the scientific, industrial and medical markets and are far too expensive for use as light sources in displays. Currently there are a number of companies working to develop low-cost lasers for mass-market projection displays. Projectors containing these lasers are likely to enter the market in 2008 or 2009, so it is important to understand how these lasers operate and the advantages and disadvantages of lasers compared to lamps.

This section will not give an in-depth discussion of solid-state laser technology; it is far beyond the scope of this book. The standard reference for those who need that level of detail is the text *Solid-State Laser Engineering* (Koechner 2006). The purpose of this section is to elucidate the factors in a laser that are most important to one who wants to use a laser in a projector.

8.3.1 Choice of Laser Wavelengths

In a HID lamp, the spectral output of the lamp is determined by the chemical composition of the lamp fill material and the projector designer has little chance to change that spectrum. With laser illumination, however, the source is monochromatic and, within broad limits, lasers of any desired wavelength can be designed.

Therefore, it is up to the projector designer to specify the exact wavelengths to be used for his projection display. The decision on what wavelengths to be used will be based on several factors, including:

1. commercially available laser wavelengths, including both availability and cost,

2. target color gamut of the display, and

3. special features to be accommodated, such as multiple primary colors or Infitec 3D technology.

The authors have encountered commercially available solid-state lasers with all of the wavelengths shown in Table 8.2 while investigating lasers for projection systems. In addition, some laser manufacturers claim they can tailor a laser to any desired wavelength with an accuracy of about 1 nm.

Table 8.2 Partial list of available laser wavelengths.

Color	Wavelengths available (nm)
Deep blue	405, 415 (Sometimes called 'royal blue')
Blue	430, 440, 445, 447, 460, 473
Cyan	485, 488, 490, 491, 495, 501, 507, 517
Green	520, 523, 527, 532, 542
Yellow	555, 561, 575, 584, 588, 593, 606
Red	611, 615, 620, 625, 628, 630, 633, 635, 638, 640, 642, 645, 650, 654, 660, 670, 671

The largest color gamut currently specified for any display system is the Digital Cinema Initiative (DCI) color gamut (DCI 2007), although special purpose projectors may have a need for a wider color gamut.

182   LIGHT SOURCES FOR LIGHT-VALVE AND MICRODISPLAY PROJECTION SYSTEMS

This DCI color gamut is shown in Figure 8.11, along with a nominal laser color gamut that completely covers the DCI gamut. For reference, a color gamut for cinema film is included. Note this film gamut has a very different shape than the gamut of an electronic projector. This is due to the fact that film produces colors with a subtractive process and electronic projectors produce colors with an additive technique. This film gamut is representative and the exact gamut for film varies with the film type and processing[3].

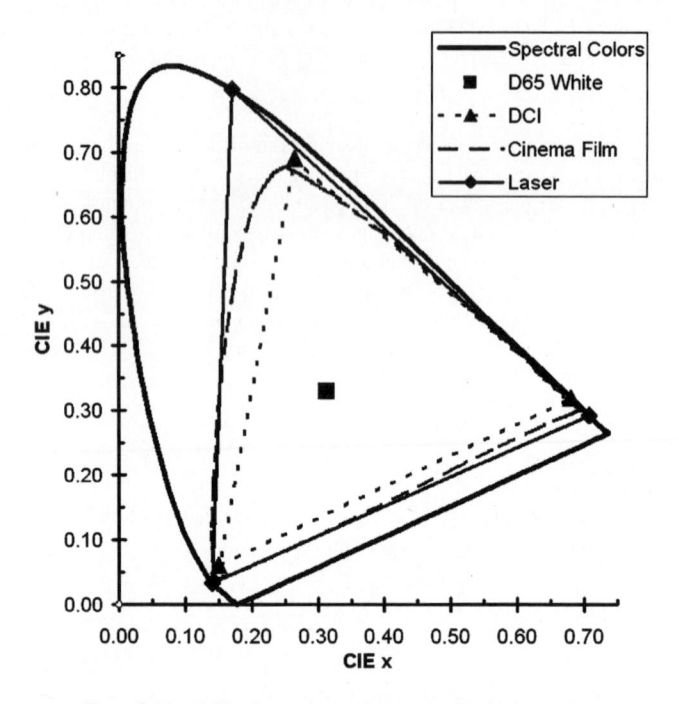

Figure 8.11   DCI color gamut and corresponding laser gamut.

Table 8.3 shows the minimum and maximum wavelengths usable if the projector designer desires to cover the entire DCI color gamut. If a green or blue wavelength is chosen outside the minimum to maximum range, or a red wavelength shorter than the minimum shown, the projector will not be able to display all colors required of a digital cinema projector. If a red wavelength is chosen above the 650 nm shown as a maximum wavelength, the projector will still show all DCI colors but excessive laser power will be required. Longer red wavelengths than 650 nm add little to the color gamut and there would be no reason to use these longer wavelengths in a projector.

Table 8.3   Laser wavelengths to completely cover the DCI color gamut.

Color	Minimum	Nominal	Maximum	Jenoptik	E&S	Novalux
Red	616	630	650	628	631	625
Green	523	532	545	532	532	532
Blue	455	462	468	446	448	465

---

[3] The data for this film gamut were supplied to the authors by Shmuel Roth of Genoa Color Technologies.

Table 8.3 also shows the wavelengths specified for the Jenoptik, E&S and Novalux[4] NECSEL lasers. Note the Jenoptik and E&S blue wavelengths are shorter than the minimum wavelengths for the DCI color gamut. While this short-wavelength blue makes it possible to display very 'blue' colors, it makes it impossible for the display to show some of the important colors in the cyan (blue-green) region.

Section 13.3.3.4 discusses other color gamuts used by electronic projectors, including the gamuts for high-definition and standard-definition TV. These other color gamuts are generally smaller than the DCI color gamut and any set of lasers that can make all DCI colors can also make all standard- or high-definition colors as well.

8.3.2 Laser Designs Suitable for Projection Applications

This section will discuss laser designs that are likely to be used in projection displays in the near future. Currently only solid-state lasers are under consideration as light sources for projection displays, so other laser designs including gas lasers will not be discussed.

Solid state lasers can be characterized by two major design features:

1. Layout, design and materials of the laser, including the method to pump the laser gain medium and
2. How the laser achieves the target wavelength and power.

Several layouts and designs of solid state lasers appear to be suitable for lasers for display applications, including:

1. Edge-emitting laser diodes and laser bars,
2. Vertical cavity lasers,
3. Optically pumped semiconductor (OPS) lasers,
4. Area emitting lasers,
5. Fiber lasers,
6. Master oscillator/power amplifier (MOPA) designs.

Each of these laser designs will be discussed in Section 8.3.2.1.

The final output of the laser for projection displays must be visible light at the wavelengths discussed earlier. Several techniques can be used by solid-state lasers to generate the correct wavelength of visible light for laser projection systems:

1. Direct emission of the desired visible wavelength by the solid-state laser
2. Direct emission of an IR wavelength by the laser diode with non-linear frequency doubling into the desired visible wavelength.
3. More complex non-linear schemes based on optical parametric oscillator designs.

Each of these wavelength generation techniques will be discussed in Section 8.3.2.2.

---

[4] Novalux was acquired by Arasor in late 2007. It is expected use of the name Novalux will be discontinued but the laser name NECSEL™ will continue in use.

8.3.2.1 Laser architectures

Edge emitting laser diodes are more or less conventional semiconductor diode designs made out of special materials that allow them to emit light of the desired wavelengths. The laser channel is parallel to the surface of the semiconductor. It is only exposed when these diodes are diced from the wafer. This dicing is normally done along a crystal plane to provide an optically flat surface to form the mirrors at either end of the lasing channel.

Figure 8.12 shows a typical edge-emitting laser diode. The laser channel in the active layer is typically 1–3 μm thick. Current passes through the active area over a band typically between 7–100 μm wide. This asymmetry in the laser channel leads to an asymmetry in the output laser beam, with the beam divergence in the vertical direction typically much greater than the divergence in the horizontal direction. Sometimes crossed cylindrical lenses are used to correct this asymmetry and bring the beam into collimation with roughly equal divergence angles. Nagahama and Mukai (2005) provide a description of a typical edge-emitting blue laser diode intended for use in display applications.

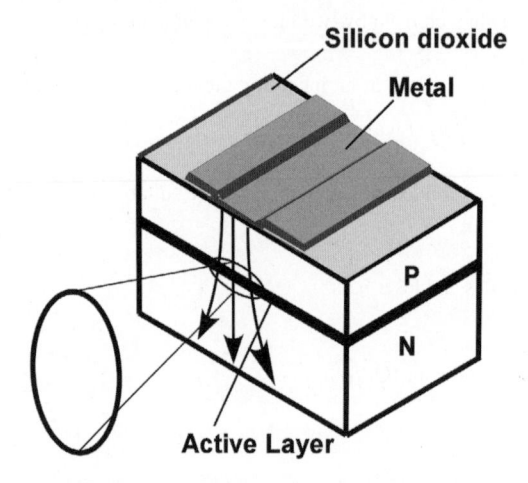

Figure 8.12 Typical edge-emitting laser diode.

Many individual die like the one shown in Figure 8.12 can be cut from a wafer. Sometimes a group of emitters are not separated from each other and this group of attached laser diodes is called a laser bar. To a projection engineer, the most important type of laser bar is a high-powered IR bar used to pump an optically-pumped semiconductor laser such as the one shown in Figure 8.14.

Vertical cavity surface emitting lasers (VCSEL) are another type of diode laser where the laser beam emerges from the top surface of the laser, not the edge of the diced chip (Mooradian 2001). In a pure VCSEL laser, the mirrors forming the laser resonator can be formed into the die itself. Alternatively, the lasers are designed with an external mirror as a vertical external cavity surface emitting laser or VECSEL. Novalux is committed to developing these lasers for projection applications and calls their design the Novalux extended cavity surface emitting laser or NECSEL™ (Niven and Mooradian 2006; Jansen *et al.* 2007; Mooradian *et al.* 2007). While these lasers are not available commercially as of this writing, they are expected in 2008 and 2009.

These lasers are based on GaInAs quantum well devices. The extended cavity allows for intracavity frequency doubling (see Section 8.3.2.2) with the nonlinear optical material, periodically polled lithium niobate, inside the laser cavity where the intensity is the highest to provide maximum conversion efficiency. These lasers can be run either continuous wave (CW) or in a pulsed mode with about 1 MHz pulse repetition frequency. Multiple emitters can be placed on a single substrate, the vertical cavity equivalent

LASERS AS PROJECTION LIGHT SOURCES 185

of a laser bar. Novalux plans on placing 24 emitters on a substrate in their lasers designed for high brightness projectors. This actually results in 48 emergent beams, since Novalux uses a mirror arrangement to recover the backward traveling frequency doubled beam, as shown in Figure 8.13

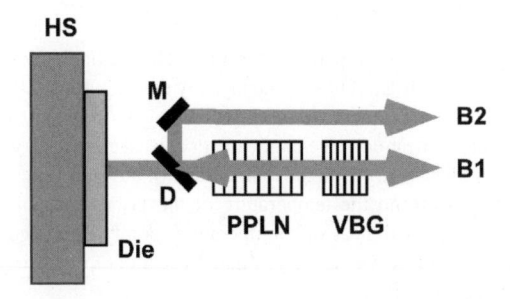

Figure 8.13 Novalux arrangement to recover the backward traveling visible light beam (reproduced by permission of © 2007 the Society for Information Display (Mooradian *et al.* 2007)).

where

    Die is the surface-emitting IR semiconductor diode laser,
    HS is the heat sink,
    D is a dichroic mirror that transmits the IR wavelength and reflects the visible,
    M is a mirror,
    PPLN is a periodically-polled lithium niobate crystal,
    VBG is a volume Bragg grating (Volodin *et al.* 2004) or a dichroic filter that reflects the IR and transmits the visible light,
    B1 is the visible beam created by the forward-going IR beam.
    B2 is the visible beam created by the IR beam reflected from the VBG,

The edge- and surface-emitting laser diodes already described pump the gain medium with the electrical current passing through the diode structure. It is also possible to pump the gain medium with optical power (Chilla *et al.* 2005), as shown in Figure 8.14.

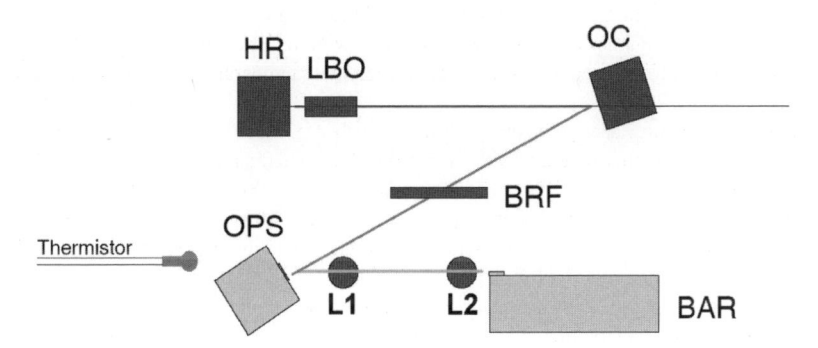

Figure 8.14 Optically pumped semiconductor laser (reproduced by permission of © 2005 SPIE (Chilla *et al.* 2005)).

In this figure,

    BAR is a edge emitting bar laser diode, normally at 808 nm,
    L1, L2 are optics to focus the output of the BAR on the OPS,
    OPS is the optically pumped semiconductor laser gain medium,

BRF is a birefringent filter,
OC is the output coupler for the emergent visible beam,
LBO is a Lithium Triborate frequency doubling crystal,
HR is a high reflectivity mirror.

This type of design is typically very temperature sensitive. This particular design included thermoelectric coolers (TEC) for both the BAR and OPS. Typically, any design with non-linear frequency doubling would include a heater to keep the frequency-doubling material at a constant and elevated temperature.

All lasers that directly emit the desired wavelength are, or at least can be, very temperature sensitive. In particular, the laser wavelength can vary with the temperature of the laser junction. Junction temperature varies with both the laser power and the temperature of the laser heat sink. Figure 8.15 shows the variation of emission wavelength with heat sink temperature for a typical edge-emitting red diode laser run at constant power. In addition to wavelength, other laser diode properties vary with temperature, including threshold, efficiency and lifetime.

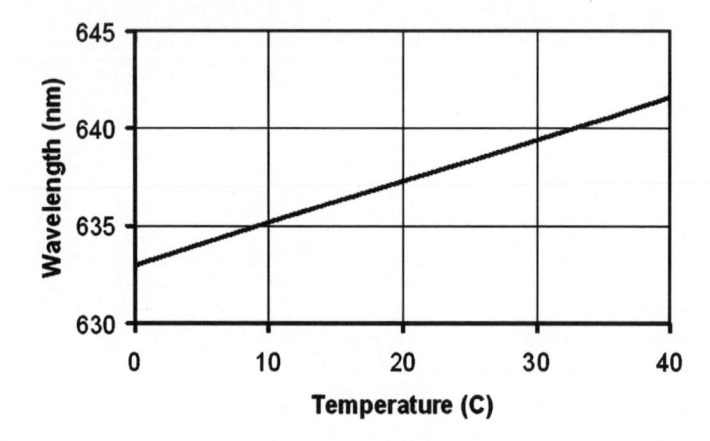

Figure 8.15 Wavelength variation with heat-sink temperature for a red edge-emitting laser.

Lasers with the design illustrated in Figure 8.13 show little or no variation in wavelength with temperature because the wavelength is controlled primarily by the volume Bragg grating rather than the physics of the laser junction itself. Efficiency can still vary strongly with temperature, especially the temperature of the PPLN (Liu *et al.* 2006).

Area emitting lasers are lasers that emit a broad area beam, much larger than the beams of laser diodes. The only laser of this type currently under development for display applications is the CRT-based, electron-beam pumped laser described in Bondarev *et al.* (2005). Electron beam pumped lasers have a long research history, especially in the former USSR and Russia. For example, Bogdankevich (1994) cites 200 papers dating back to 1951. With the introduction of low-cost electrically pumped visible light lasers, the future of these lasers for display applications is in doubt.

Fiber lasers are diode pumped and are capable of operating at very high powers, with up to 20 kW CW reported for IR fiber lasers. Commercial lasers that are currently available produce IR light, although visible light lasers are possible and have been reported in the scientific literature. For lower power fiber lasers, the diode is coupled into one end of the fiber. For higher power fiber lasers, the pump light can be coupled into the fiber through the cladding or through multiple pigtails. One of the reasons such high powers can be achieved is that the light from multiple pumping diodes can feed into a single lasing fiber. Lasing fibers for IR output are doped with ytterbium, thulium, erbium and other rare-earth metals to produce the gain medium. The IR output can be converted to visible light using one of the techniques discussed in Section 8.3.2.2. New fiber materials and/or dopants would need to be developed for fiber lasers to directly generate visible light. (Mizushima *et al.* 2006) describes the use of a fiber laser to generate light for the green channel of a rear projection TV (RPTV).

Master oscillator/power amplifier designs (MOPA) use a relatively low power laser to generate light of the correct wavelength. This light is then amplified by one or more external gain stages. Multiple series and parallel gain stages can be used to generate extremely high power outputs for industrial and scientific applications. A laser of this type designed for a projection display application is described by Moulton *et al.* (2002).

MOPA designs can also be integrated into a single substrate, as shown in Figure 8.16 (Osowski *et al.* 2007). In this laser, the initial single mode beam is generated in the narrow-waveguide distributed-feedback (DFB) section, which is driven by 700 mA. When the tapered amplifier section is driven with 5 A, this beam is amplified to produce >3 W of output of 1040 nm light.

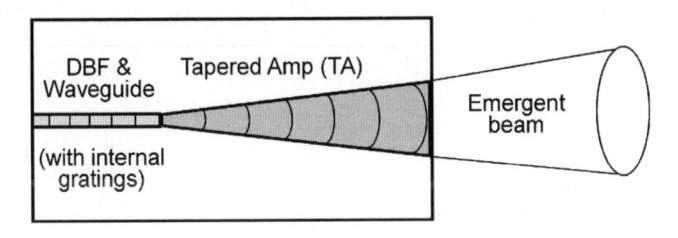

Figure 8.16 Master oscillator/power amplifier (MOPA) laser integrated on a single substrate (reproduced by permission of © 2007 SPIE (Osowski *et al.* 2007)).

8.3.2.2 Laser wavelength generation

While laser diodes that use direct emission have been demonstrated for most laser wavelengths, these laser diodes often have lifetime, power output, manufacturing or other problems, especially in the blue and green regions. Red laser diodes in the visible range typically use an InGaAlP material and can produce relatively high powers. Unfortunately, it is difficult to get direct emission red laser diodes with wavelengths much shorter than 638 nm; 650 nm is perhaps the optimum wavelength generated with this material. GaN can be used to make blue lasers, with an optimum wavelength of about 405 nm, although the material can be pushed to 440 nm or beyond for visible light. Green emission with the II-VI semiconductor material ZnCdSe/ZnSSe is possible, but development of this material is not as far along as the development of the more common III-V semiconductor materials.

Therefore, one of the non-linear approaches is commonly used for green, and sometimes for blue and red. For example, the lasers used in the Jenoptik and E&S projectors use non-linear conversion for all three colors. Novalux is developing low-cost lasers where all three colors are generated by non-linear conversion (Niven and Mooradian 2006).

Non-linear wavelength conversion works by combining two low energy photons into a single high energy photon or splitting a single high energy photon into two low energy photons. The energy of a photon is given by the equation:

$$E = h\nu = h\frac{c}{\lambda}. \tag{8.1}$$

where:
- E is the energy of a photon,
- h is Plank's constant,
- ν is the frequency of the photon,
- c is the speed of light in a vacuum, and
- λ is the wavelength of the photon.

This wavelength conversion process conserves the energy of the photon, so the frequencies and wavelengths of the photons involved are related by:

$$v_1 = v_2 + v_3$$
$$\frac{1}{\lambda_1} = \frac{1}{\lambda_2} + \frac{1}{\lambda_3}. \tag{8.2}$$

where:

- v_1, λ_1 correspond to the high-energy, short wavelength photon,
- v_2, λ_2 correspond to the first low-energy, long wavelength photon,
- v_3, λ_3 correspond to the second low-energy, long wavelength photon.

The most common case of non-linear wavelength conversion is when $\lambda_2 = \lambda_3$ and $\lambda_1 = \lambda_2/2$. This case is commonly called 'frequency doubling' and it will turn an infra-red (IR) laser beam into a beam of visible light. Since IR lasers are commonly available at a variety of wavelengths, relatively high powers and relatively low cost, frequency doubling allows the designers of visible light lasers to draw on the extensive infrastructure used to manufacture IR lasers to produce visible light lasers.

These photon conversion processes all require non-linear optical materials. Non-linear in this case means the index of refraction of the material depends on the electric field amplitude of the light wave passing through it. These strong electric fields correspond to very bright light beams. Non-linear optical materials used in lasers include lithium niobate ($LiNbO_3$), lithium triborate (LBO or LiB_3O_5) and potassium niobate ($KNbO_3$).

One common material for frequency doubling is periodically polled lithium niobate, commonly abbreviated PPLN and pronounced 'pip-lin'. Lithium niobate is a ferroelectric material that retains an applied DC electric field. Periodically polled material has a strong electric field applied during the manufacturing process to produce spatially varying 'up' and 'down' electric fields (Byer et al. 1991). In virtually any optical material, including the non-linear optical materials used in lasers, the index of refraction varies with the wavelength of light. Without periodic polling, the long-wavelength input laser beam and the short-wavelength output beam would see different indices of refraction. While the output beam would have exactly double the frequency of the input beam, it would not have exactly half of the wavelength inside the non-linear crystal. Therefore, the input and output beams would rapidly fall out of phase with each other and the conversion efficiency from long wavelength to short wavelength would be low. Periodic polling the material creates a virtual index of refraction that is equal for the two wavelengths. They remain in phase throughout the crystal and wavelength conversion efficiency can be very high.

The photon combination and splitting processes can be combined into a single system. For example, Figure 8.17 shows a RGB laser from Q-Peak where the system uses a single 1047 nm laser and then multiple non-linear optical processes to derive the red, green and blue wavelengths (Snell et al. 2000).

In this system there is only one laser, a diode-pumped Nd:YLF laser producing 1047 nm IR light; 808 nm edge-emitting IR laser diodes were used to pump the Nd:YLF lasing medium. The IR output from this laser is frequency doubled in a second-harmonic generation (SHG) stage to produce 523 nm green light. This green light in turn pumps an optical parametric oscillator (OPO) which generates a signal beam of 896 nm and an idler beam of 1256 nm. Each of these wavelengths in turn is frequency doubled in its own SHG stage to produce blue and red respectively. The OPO was not 100 % efficient at converting green photons into the 896 nm signal and 1256 nm idler photons. This remaining green light was used as the green output of the system. By correctly tuning the stages, the red, green and blue outputs could be designed to match the desired D65 white point.

More complex arrangements are possible, including ones that use sum frequency mixing. Figure 8.18 shows one of two sum frequency mixing stages of the Jenoptik laser.

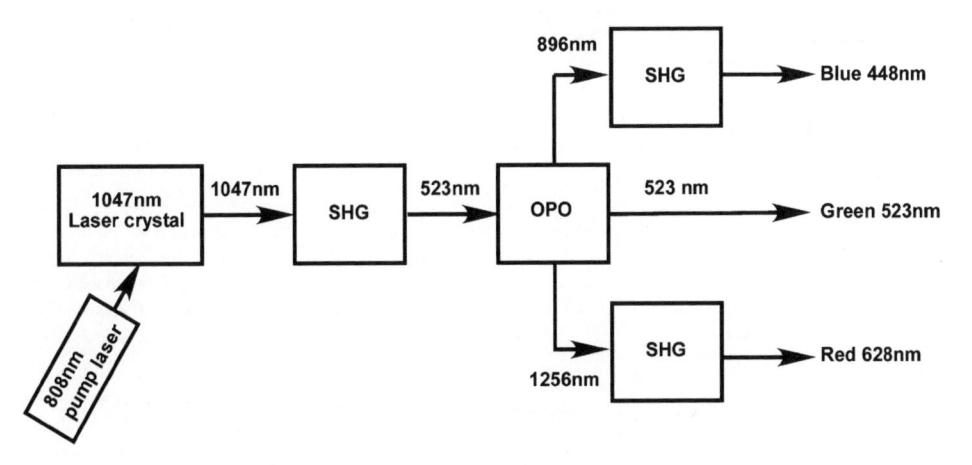

Figure 8.17 Red, green and blue wavelengths derived from a single IR laser.

Figure 8.18 Sum frequency mixing stage of the Jenoptik laser.

In this system, the 1064 nm was derived directly from a diode pumped Nd:YVO$_4$ laser and the 1536 nm beam was the signal beam from a OPO that used the 1064 nm as input. The 3.5 μm idler beam generated in this OPO system was discarded.

8.3.3 Laser Safety

Laser safety can be a significant issue in the design of a laser-based projector. A laser projector must be safe by three different criteria to be commercially viable:

1. The system must satisfy the applicable safety regulations.
2. The system must be perceived as safe by the end user and consumer.
3. The system must, in fact, be safe and unlikely to cause injury to the manufacturing personnel, maintenance personnel, end user or consumer.

Multiple classes of lasers and systems containing lasers are defined by the regulators for safety applications. Table 8.4 shows the most important classes in terms of laser projector design. The class definitions are generally uniform from country to country, but the regulations regarding lasers in a given class can vary greatly in different countries, or even different regions is a single country. Therefore, it is important for a projector manufacturer using laser light sources to consult a laser safety expert familiar with the laws of every country where a projector will be marketed.

Laser safety is not an add-on to be done at the end of the projector design cycle, it is a critical factor in the design of a projector. A laser safety expert must be included in any design team working on laser

Table 8.4 Laser safety classes.

Class 1	A Class 1 laser is considered safe based upon current medical knowledge. This class includes all lasers or laser systems which cannot emit levels of optical radiation above the exposure limits for the eye under any exposure conditions inherent in the design of the laser product. There may be a more hazardous laser embedded in the enclosure of a Class 1 product, but no harmful radiation can escape the enclosure.
Class 2	A Class 2 laser or laser system must emit a visible laser beam. Because of its brightness, Class 2 laser light will be too dazzling to stare into for extended periods. Momentary viewing is not considered hazardous since the upper radiant power limit on this type of device is less than the MPE (Maximum Permissible Exposure) for momentary exposure of 0.25 second or less. Intentional extended viewing, however, is considered hazardous.
Class 3	A Class 3 laser or laser system can emit any wavelength, but it cannot produce a diffuse (not mirror-like) reflection hazard unless focused or viewed for extended periods at close range. It is also not considered a fire hazard or serious skin hazard. Any continuous wave (CW) laser that is not Class 1 or Class 2 is a Class 3 device if its output power is 0.5 W or less. Since the output beam of such a laser is definitely hazardous for intrabeam viewing, control measures center on eliminating this possibility.
Class 3R	These less hazardous Class 3 lasers may receive a type-certificate and individual installations need not have a permit. Any projector intended to sell in even moderate volume must, in practical terms, rate as Class 3R or below.
Class 3B	These more hazardous Class 3 lasers have very stringent safety regulations. Every individual installation of a Class 3B laser system must be permitted (in the US).
Class 4	A Class 4 laser or laser system is any that exceeds the output limits (Accessible Emission Limits, AELs) of a Class 3 device. As would be expected, these lasers may be either a fire or skin hazard or a diffuse reflection hazard. Very stringent control measures are required for a Class 4 laser or laser system.

projection systems. A poorly designed, from the safety standpoint, projector with 200 mW output (about 66 lumens) can rate as class 3B, preventing its commercial sale in any significant volume. On the other hand, a properly designed high-output projector can rate as class 3R, allowing its sale with a minimum of regulatory overhead.

8.4 Light Emitting Diodes as Projection Light Sources

Light emitting diodes (LEDs) are currently (2007–08) used in a few commercially available projection systems, primarily 'pocket projectors' and RPTV systems. See Chapter 2 for a brief overview of the various market segments in the projection business. It is believed that LEDs will be limited in projection systems to maximum brightness of about 1000 lumens, or perhaps 2000. Projectors with higher light outputs than this are expected to use either lasers or lamps as the light source, not LEDs.

This section cannot give a complete overview of all major LED technology issues. For more details on LED technology, see one of the many texts on the technology, such as Schubert (2003), large parts of which are available online at www.lightemittingdiodes.org.

The optical properties of an LED can be difficult to measure (Nägele and Distl 1999) and the specifications of an LED can often be difficult to understand in terms of how they apply to projection systems. This comes from a number of factors including:

1. Some LEDs have a strong variation of the light output as a function of angle,

2. Variation in color and light output of a LED with temperature,

3. Variation in color and light output of a LED with drive level,
4. Variation of LED properties from piece to piece in a single production lot.

LED light output in the commercial specifications of LED are often given in candela or millicandela. Candela is a measure of the intensity of a light source in a particular direction. While this measure is sometimes a useful specification if the LED is to be used as an indicator light, projection engineers are most commonly concerned in the total output of a LED integrated over all angles or a range of angles. This output is measured in lumens, which is integrated over area and angle, or luminance, which is integrated only over angle and is lumens per unit area.

The color of an LED is sometimes specified by a dominant wavelength. If a straight line is drawn from the color point corresponding to a reference white spectrum through the color point of the LED to the boundary of the color diagram, the wavelength corresponding to this intersection is the dominant wavelength of the LED. The equal energy spectrum (sometimes referred to as Illuminant E) with x = 0.333 and y = 0.333 is most commonly taken as the reference point. Roughly speaking, the dominant wavelength of a LED corresponds to the wavelength of a laser that is closest in color to the LED. See Section 8.4.2 for more details on the colorimetry of LEDs.

Most projection designs involving LEDs are severely constrained by étendue considerations (Joachim and Alexander 2007). In general, this limits the maximum size of a LED die that can be used with a given microdisplay and f/#. For example, the largest die that can be used efficiently with a 0.55″ DLP imager in an f/2.4 projection system is 4 mm^2 (2 mm on a side). This produces an illumination at the DLP and ultimately at the projection screen that is proportional to the luminance (lumens/mm^2) of the die. In order to increase the luminance at the screen, it is necessary to increase the luminance of the die. The 1000–2000 lumen output discussed above is simply a reflection of the maximum realistic power density and luminance of a LED. It also takes into account the maximum realistic size and étendue of a microdisplay. Simply increasing the die area to increase the total lumen output of the die has little or no effect on the screen luminance. This étendue issue is identical to the étendue issue that drove lamp-based projectors to use the shortest possible arc with the highest possible are luminance.

LEDs, like the surface and edge emitting lasers described in Section 8.3.2.1, are semiconductors produced on wafer processing lines and then diced to the correct size. The materials used in LEDs are similar to the materials used in laser diodes. In contrast to a semiconductor laser, there is no organization to the emitted photons when the electron-hole pair recombine in a LED: they are typically emitted at any angle and polarization and over a relatively broad wavelength range, at least compared to a laser wavelength range.

High-brightness LEDs (HB-LEDs) suitable for projection applications are an offshoot of HB-LEDs developed for other applications. For example, one of the original and largest markets for HB-LEDs is backlight illumination for the LCD in a cell phone or other portable device. The high energy efficiency of these HB-LEDs makes them attractive in portable devices where conservation of the battery power source is important.

HB-LEDs are about two years ahead of lasers, in terms of their development as projection light sources. This is due in large measure to the investments made in HB-LEDs for other applications such as cell phones have driven the entire technology. Lasers with the power levels and the wavelengths for display applications have few applications outside the display field. Therefore, the investments in lasers for displays have been lower, leading to slower but steady progress.

8.4.1 Performance Improvements in LEDs for Projection

According to Haitz's law, the light output from a LED package has doubled every 18–24 months since the 1970s. This 'law' is sometimes referred to as the LED analog of Moore's law for semiconductor transistor counts. This analogy breaks down for several reasons. First, Haitz's law is sometimes stated in terms of light output per package. Extremely large packages have been built containing a large number of die and with very high total light output. Haitz's law, when stated in terms of light output per package, clearly does not apply to these packages, nor does it apply to cases where unusually large die are used. From a projection engineer's point of view, the correct way to state the law is in terms of an area-based measure such as luminance (candela/m^2) or luminous emittance (lumen/m^2). Second, LEDs are much closer than transistors to the theoretical limits imposed by physics and not too many more doublings will be possible before further progress is blocked. Finally, HB-LED technology is really two technologies. The first is AlInGaP technology used for red LEDs since the late 1960s, where the luminance has been increasing by a factor of 10x every decade. The second is InGaN technology that is used for green, blue and white LEDs and where the luminance has been increasing by 10x every 5 years since they were introduced in the early 1990s (Harbers *et al.* 2007).

Regardless of how it is measured or what it is called, there have been great strides in LED light output and further improvements are highly likely. These improvements come in four general categories:

1. Epitaxial wafer improvements that provide for higher internal quantum efficiency;

2. Chip design improvements that allow for higher extraction efficiency of photons from the high-index semiconductor into the air;

3. Optical improvements that allow more of the light extracted from the die to be focused on the projection microdisplay; and

4. Packaging improvements that will allow the die to be run at higher power densities (W/mm^2).

Even with the limitations set by physics, improvements in all these factors are possible. Some factors are already close to their limits, however. For example, die designs allowing 80 % extraction efficiency are well known and are implemented in various versions by all LED manufacturers (Bierhuizen *et al.* 2007). Internal quantum efficiency (IQE) values of ultra-violet emitting LEDs have been reported to be as high as 70 %, although the IQE value for green LEDs is more typically 30 %. Red and blue efficiencies are between these two values. Typically the green LED is the limiting factor in the light output of a projector with LED illumination. One reason this occurs is because the green wavelengths fall at the limit of what the InGaN technology can achieve. A breakthrough such as the development of a new material technology will be needed to allow green efficacy to match blue or red. Much of the extraction efficiency and IQE improvements in the coming years are likely to come from introducing technologies already demonstrated in research labs into industrial production. These two factors are close enough to their theoretical limits that even one additional doubling from these factors is unlikely.

Optical collection efficiency is a well understood factor. Much of what was learned in light collection for projection from HID lamps can be or already has been applied to LEDs. Recycling optics seem to be particularly suited for LEDs, as will be discussed in Section 8.9.5.

Packaging for improved thermal properties is likely to be a fruitful field of development for LEDs for years to come. For étendue reasons, it is necessary for a LED to emit as much light as possible from as small an area as possible. If efficiency improvements are blocked, it is necessary to increase the power density to increase the total light output. Package design to optimize thermal performance will be discussed in Section 8.4.3.

8.4.2 Color with LEDs

LEDs can produce colored light through two basic mechanisms:

1. Direct emission of the desired wavelengths

2. The LED can emit UV or blue light which can be converted to the desired wavelength with a phosphor or fluorescent material.

This second approach is the only way to produce white LEDs. LEDs have a finite spectral bandwidth that is much wider than a laser bandwidth, but it is not wide enough to cover red, green and blue from a single device. The most common way to make a white LED is to use a blue LED and a yellow phosphor, most commonly a YAG:Ce phosphor. Some of the blue light is absorbed by and excites the yellow phosphor. The remaining blue light mixes with the emitted yellow light to produce a white output. By adjusting the yellow phosphor used and its density, LEDs with a variety of color temperatures can be made. These white LEDs are among the most efficient light sources known, with some low power laboratory devices claiming up to 300 lumen/watt, which represents a near-perfect conversion of electrical power into white light.

More realistically, 130 lumens per watt have been achieved in low-power production devices. Production high-brightness white LEDs made with this technology more typically have efficacies in the range of 60–100 lumens/watt. Unfortunately, these white LEDs made with the blue/yellow system are not normally suited for use in projectors because of their extremely poor colorimetry. While the blue is satisfactory, dividing the yellow into red and green components for a full color display produces both poor color saturation and poor efficiency. These LEDs are used in systems like cell phone displays where power consumption is paramount and poor colorimetry can be tolerated.

Other phosphors besides YAG:Ce can be used to produce white LEDs. In particular, a mixture of green and red phosphors with a blue LED can produce a much better white color with a much higher color rendering index than the blue/yellow combination. Still better colorimetry can be done if a UV LED is used in combination with red, green and blue phosphors. Unfortunately, these two approaches have lower efficacy than the blue/yellow approach.

As noted in Section 8.4.1, the efficacy of green-emitting InGaN LEDs is relatively low and likely to remain low. Green LEDs with a higher efficacy can actually be made using a InGaN UV LED and a green phosphor. Unfortunately the available green phosphors have too broad a bandwidth to make a good green color point. To improve the color so that it would make a good green primary it is necessary to use narrow band filters and the efficacy advantage of the phosphor approach is lost.

The more common case in projection applications is to use separate red, green and blue LEDs. The spectra of typical red, green and blue LEDs for a projection application are shown in Figure 8.19, with dominant wavelengths of 616.5, 538.9 and 460.5, respectively. The amplitudes of the red, green and blue

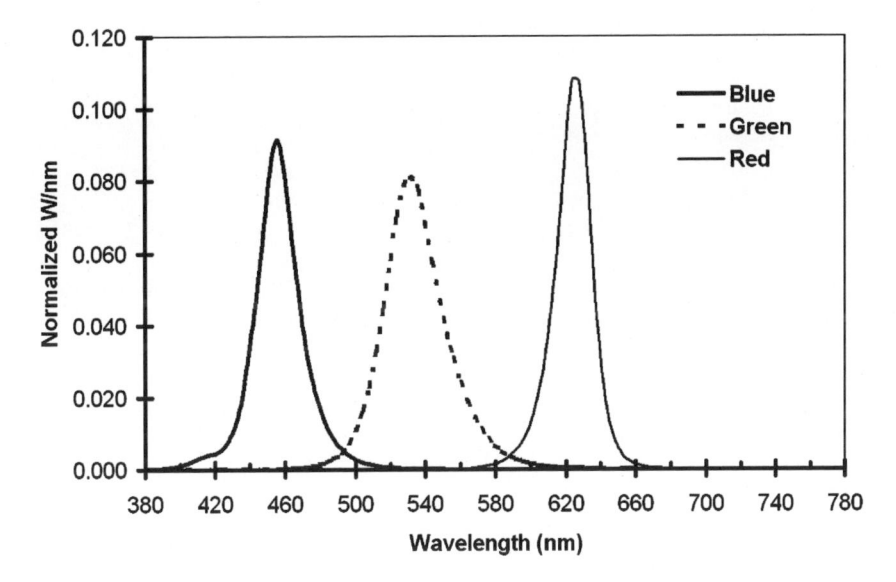

Figure 8.19 Spectral characteristics of red, green and blue LEDs intended for use in a projection application (data courtesy of Osram Optical Semiconductors).

194 LIGHT SOURCES FOR LIGHT-VALVE AND MICRODISPLAY PROJECTION SYSTEMS

spectra have been adjusted so the white spectrum produced by the three is color-balanced to a D65 white point, the normal target for video applications. The spectra were normalized so the white output had $Y = 1$.

Figure 8.20 shows the red, green and blue primary colors for the spectra shown in Figure 8.19. Also marked are the color points for lasers that have the same dominant wavelength as the LEDs. Note that the red and blue LEDs and lasers have almost the same color, but there is a significant difference between the green LED and the green laser. This points out one of the limitations in specifying dominant wavelength. The specified DCI color gamut is also plotted as a comparison.

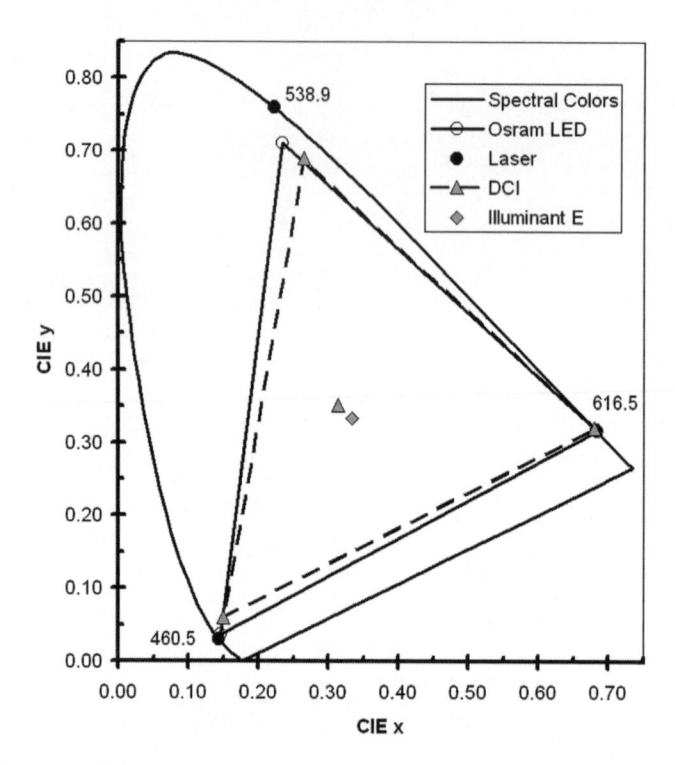

Figure 8.20 Color gamut of LEDs and Lasers with the Same Dominant Wavelength (Data courtesy of Osram Optical Semiconductors).

8.4.3 Thermal Issues with LEDs

In general, the goal of a HID lamp designer is to get the highest possible arc temperature compatible with the electrode and bulb materials. For example, the CPL lamp envelope and cooling system is carefully designed to maintain the quartz envelope below 1400 K (1127 °C), the point where quartz begins to soften and recrystallize. In general, with a HID lamp the higher the temperature of the arc, the higher the luminance of the arc and the efficacy of the system. In many ways, this simplifies the thermal design of the system because it is relatively easy to cool the high temperature arc with ambient air.

Unfortunately, this is not true for a LED. All properties of an LED deteriorate at higher operating temperatures. These properties include:

1. Efficacy of the LED (lumens/watt) is reduced;
2. Lifetime of the LED is reduced;

3. Color and wavelength of the LED changes, changing the color of the final image;
4. Forward voltage of the diode junction decreases, requiring higher current to maintain power.

Properties in LED data sheets are often specified at 25 °C junction temperature, an unreasonably low temperature for any projection design. Different LEDs from different manufacturers have different specifications on the maximum operating junction temperature, but typically this specified to be at or below 125 °C.

The étendue limitations associated with LEDs require the highest possible power densities, typically several watts/mm^2, to allow projectors with reasonable light output levels. This requires all LEDs for projection applications be packaged in special packages with high thermal conductivity to remove the heat from the die without exceeding the specified junction temperature of the die (Kückmann 2006).

Even if the maximum temperature is not exceeded, there is a strong drive to keep the LED temperature as low as possible. Figure 8.21 shows the typical variation of the normalized output of a LED with the temperature of the thermal pad on the LED package, which is normally several degrees lower than the junction temperature. As can be seen, the variation is different for the different colors. Blue and green are similar, since these LEDs are typically made of similar materials. Red, however, is made from a different material and is significantly different than blue or green. The large variation in red can affect the color balance of the system. As the system warms up, if all three LEDs increase in temperature at the same rate, the color balance will shift to higher color temperatures. Feed back to the red, green and blue LED currents is probably necessary to maintain the color balance within acceptable limits.

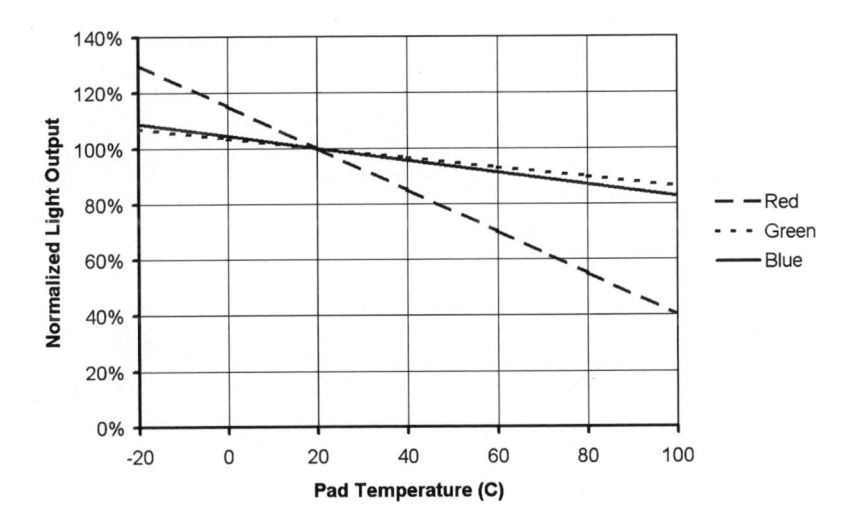

Figure 8.21 Light output variation of typical red, green and blue LEDs with temperature.

8.4.4 LED Drive Issues

While lamp drive circuits are often supplied by the lamp vendor, LED drive circuits are normally supplied by the user of the LED. These in turn may be supplied by third-party drive suppliers. LEDs are normally nearly constant voltage devices characterized by their forward voltage which is typically in the range of 2–2.5 V for red and 3.5–4.5 V for blue or green, as shown in Figure 8.22. While this data applies specifically to the Luminus Devices PT120 chipset, it is typical of all AlInGaP and InGaN LEDs. Because of the large current change with a small voltage change, the drive circuit needs to drive a given current. High-brightness LEDs are driven at currents in the range of 350 mA to 1000 mA/mm^2 or more, for powers in the range 1–4 Watts/mm^2. The name ultra-high brightness (UHB-LEDs) is sometimes used for LEDs with >1 A/mm^2. It is anticipated that eventually UHB-LEDs will reach power densities of 10 W/mm^2 or higher in projection applications, corresponding to even higher current densities than the ones shown in Figure 8.22.

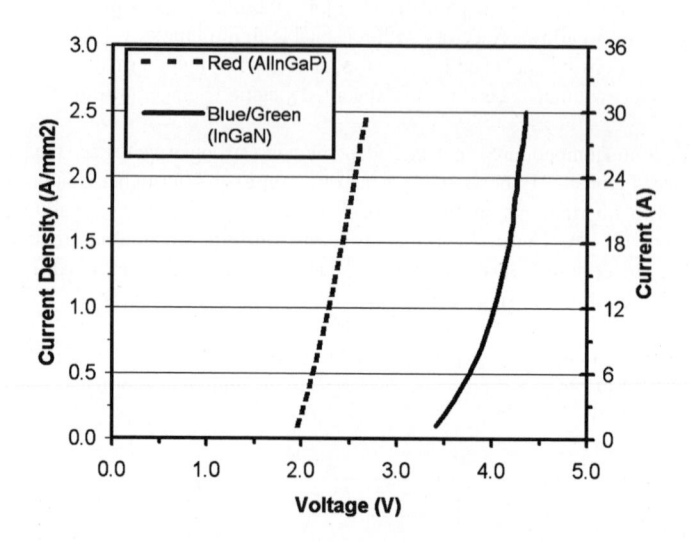

Figure 8.22 Typical forward voltages of HB-LEDs and UHB-LEDs (data courtesy of Christian Hoepfner at Luminus Devices).

LEDs respond very quickly to changes in drive and response speeds of 30 MHz and higher are normal. Much higher speeds are in fact possible but they are rarely needed by a projector designer. This fast response is of particular importance in the color sequential systems to be discussed in more detail in Section 10.4.2. Like the UHP lamp, they can be overdriven as long as the thermal requirements are met. Unlike the UHP, the light output of a LED is not linear with current. Example brightness vs drive curves for high-brightness LEDs intended for use in projection systems are shown in Figure 8.23 (Harbers et al. 2003). InGaN is typically used for blue, cyan and green LEDs while AlInGaP is commonly used for yellow, orange and red LEDs. These curves were measured with 1 mS pulses and a 25 % duty cycle.

Note that these curves are not linear even though the LED junction temperature was kept constant. For example, doubling the current from the nominal produces only about 1.8 x the light output in the InGaN system and about 1.65 x the light output in the AlInGaP system This non-linearity of light output with drive is called 'droop' and is a normal feature of all high-current LEDs, regardless of technology. In addition to droop, driving the LED at higher current and power is likely to increase the junction temperature. This increased temperature will reduce the efficiency of the LED and further reduce the light output.

The dominant wavelength of a LED can also change with drive level (Luminus 2007). Figure 8.24(a), (b) and (c) show the change of dominant wavelength with drive for red, green and blue LEDs respectively. These curves represent the PT54 projection chipset from Luminus Devices. Note the curves are different

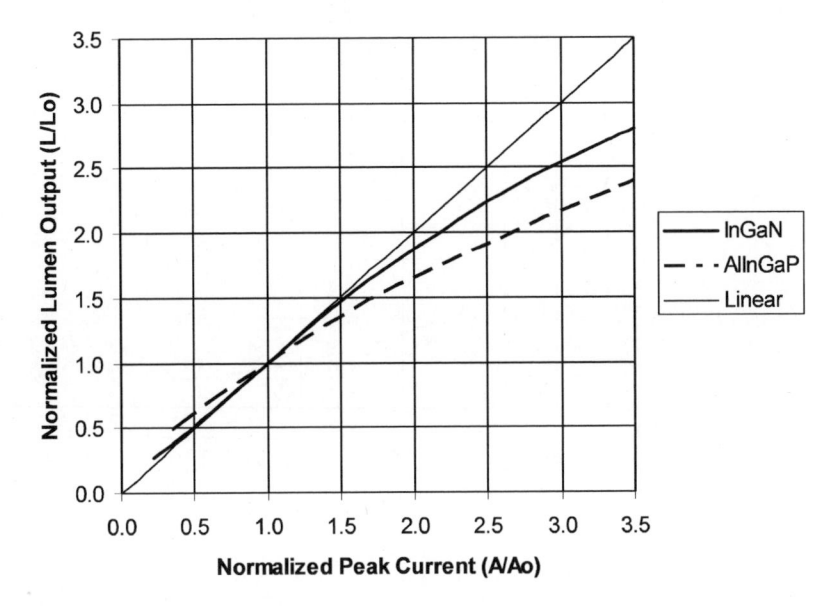

Figure 8.23 Normalized lumen output curves for two LED materials as a function of drive current.

for continuous wave (CW) drive, such as would be used in a three microdisplay system, and in a single-panel color sequential projector where a pulsed drive is more likely to be used. These LEDs are designed to be used at very high currents to produce sufficient light output for a projection application. They can operate at 36A total, or $3A/mm^2$.

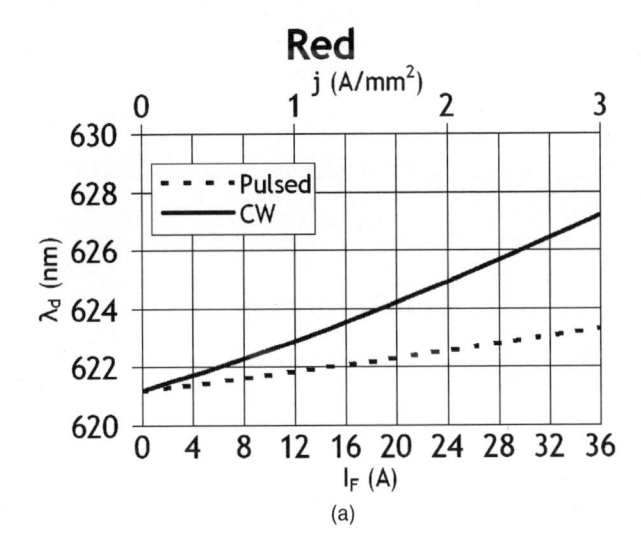

Figure 8.24 Variation in dominant wavelength with drive in LEDs for a projection application (data courtesy of Christian Hoepfner at Luminus Devices): (a) red, (b) green, (c) blue.

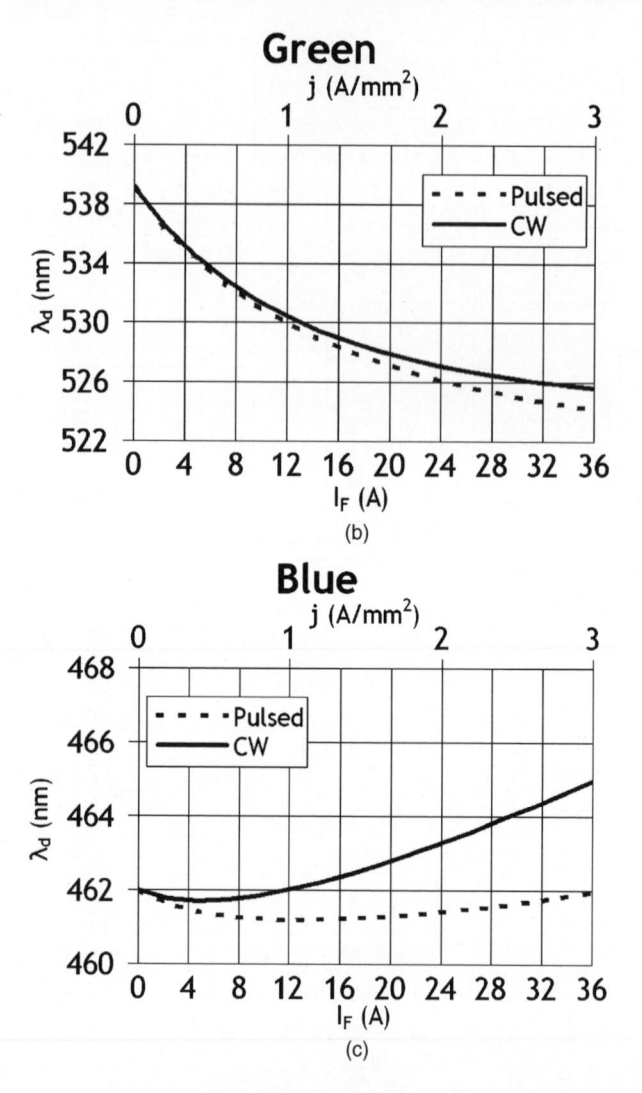

Figure 8.24 (Continued).

8.5 Efficacy and Lumen Output

Lamp, laser or LED efficacy is the amount of visible light output from a light source per unit electrical power input, measured in lumens/watt. Efficacy is a photometrically weighted measure. The efficacy of lamps that have been used in commercial projectors varies widely, from a low of about 20 lm/W for tungsten-halogen or low power xenon lamps to a high of about 100 lm/W for high efficiency metal-halide lamps. The typical xenon lamp for projection is about 25 lumens/watt and UHP-type lamps are typically about 60 lumens/watt.

Superficially, it would be thought that the higher the efficacy the better, but, as is frequently the case, there are tradeoffs to be made. High efficacy lamps often have very large arcs, poor colorimetry, emission lines at undesirable wavelengths and/or long warm-up times. Xenon arcs, on the other hand, have low efficacy but have excellent colorimetry and can have very short arcs. Xenon lamps can be designed to

EFFICACY AND LUMEN OUTPUT   199

operate at very high power, making them desirable for applications in which total input power is only of secondary importance.

The total lumen output of a lamp is the product of the electrical power into the lamp multiplied by the lamp efficacy. For example, if a lamp has an efficacy of 60 lm/W and is excited with 100 W of electrical power, the total lumen output is 6000 lm. Many lamps have a design power level at which they should be operated in order to maintain the appropriate temperatures within the lamp. For instance, both tungsten-halogen and metal-halide lamps must be operated at or near design power levels for the halogen cycle to operate correctly. Other lamps, especially xenon lamps, can be operated at any power up to the maximum design power.

Operation of any lamp at a power significantly above its design power will dramatically reduce lamp life, sometimes to zero. With some lamp types, operation above the design power level can also be a safety hazard.

Lamp efficacy or total lumen output from the light source is typically measured in an integrating sphere, with no collection optics, as shown in Figure 8.25. Integrating spheres must be significantly larger than the lamp, and lamp manufacturers often use integrating spheres 2 m (> 6 feet) or larger in diameter. The value measured by this technique is obviously not affected by any property of the reflector, since there is no reflector in the system.

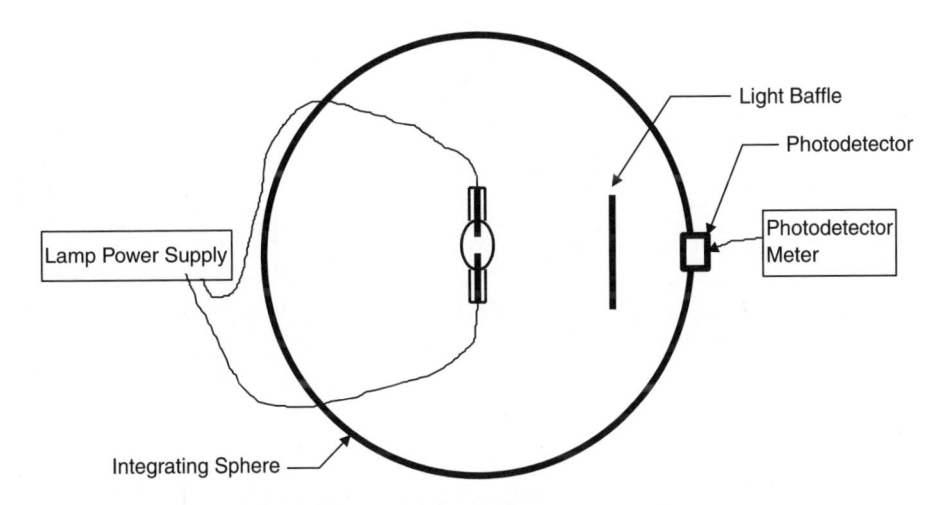

Figure 8.25 Lamp measurement in an integrating sphere.

Another technique, shown in Figure 8.26, measures the total light collected by the reflector from a lamp. In many ways, output measured this way will be more useful to the projector designer than true lumen output as measured in Figure 8.25. In Figure 8.26, the lamp and reflector are arranged so all the light collected in the beam enters the entrance port of a calibrated integrating sphere. Most light that is not collected into the beam never enters into the sphere and does not contribute to the measured value.

A third technique, shown in Figure 8.27, provides the most useful information to the projector designer. Here the lamp and reflector are focused into an aperture in the integrating sphere that has the same étendue as the projector in which the lamp will be used. With this technique, only light from the lamp/reflector combination that can actually be used in a projector will be measured. Another advantage of this technique for the projector designer is that it requires a smaller integrating sphere than the methods shown in Figures 8.25 or 8.26. Typically, a 0.25–0.5 m ($\approx$ 10–20 inch) diameter sphere would be sufficient to use this technique, with the requirement that the diameter of the entrance aperture be small compared with the sphere diameter. This technique will be expanded upon in Chapter 11.

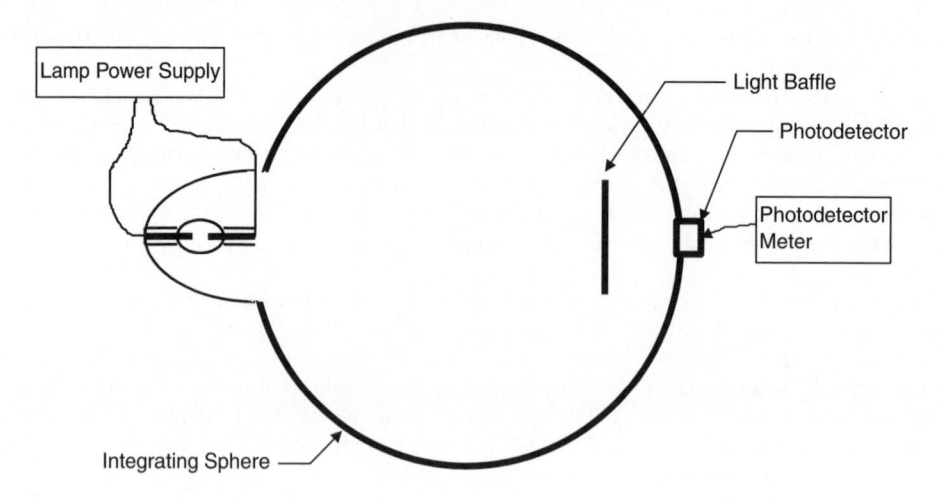

Figure 8.26 Lamp and reflector measurement with an integrating sphere.

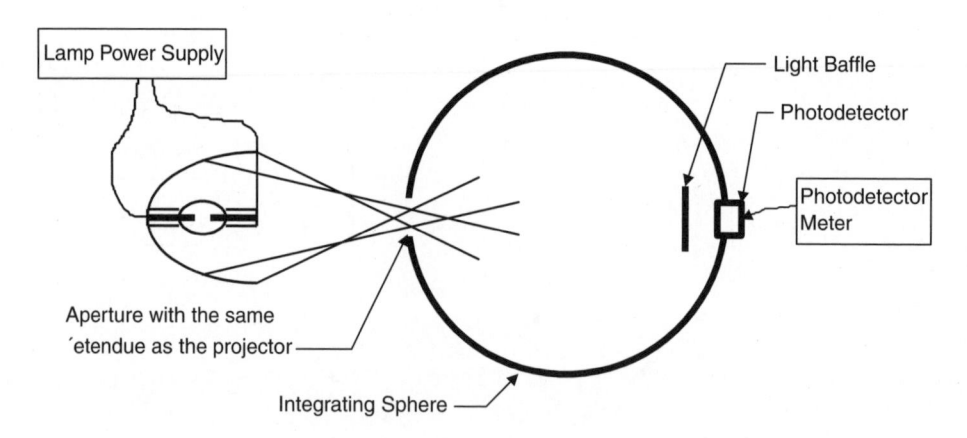

Figure 8.27 Lamp and reflector measurement with an integrating sphere and aperture.

8.6 Spectral Characteristics of Lamps

In order to produce useful data for the projector designer, the integrating sphere and photodetector combinations in Figures 8.25, 8.26 and 8.27 must be corrected so that their spectral sensitivity is the same as the human eye response curve, the $\overline{Y}$ curve. Any of these methods can produce a complete emission spectrum of the lamp by replacing the detector with a fiber-optic probe that leads to the entrance slit of a spectroradiometer.

8.6.1 Lamp Spectral Emission Lines

Some lamps have significant power in emission lines at wavelengths which are difficult to use in a color projector. The mercury emission lines at 577 and 579 nm and the sodium doublet at 589 nm are particularly bad in this respect. While mercury cannot be avoided in most HID lamps, sodium is never added to a lamp intended for projection applications.

The light from these emission lines is yellow. When the light from the lamp is divided into red, green and blue beams, it is difficult to include these bright yellow emission lines in either the red or the green channels. If the dichroic filters include the yellow emission lines into the green channel, the green can become a yellow-green. If the lines are included into the red channel, the red may become an orange-red. Therefore, if yellow emission lines are bright, it may be necessary to reject the light from the lines entirely.

A yellow emission line causes another problem. It is difficult and expensive to make dichroic filters with cutoff values controlled more accurately than ±1 % of the cutoff wavelength. A typical dichroic filter to divide red from green will have a cutoff wavelength of about 585 ± 6 nm. Filters with cutoff wavelengths from the low end of the distribution (~579 nm) will include a yellow line with the red, turning it orange, but making the green color a very good green. Filters near the upper end of the distribution (~591 nm) will produce a good red, but a very yellowish green. Therefore, different projectors of the same design will have very different colorimetry.

The problem can also occur with strong spectral lines in the blue/green transition region, 490–510 nm. Since there is more overlap in human visual response between the blue and green than there is between the green and red, lines in this region are rarely as serious a problem as are the yellow sodium and mercury lines.

8.7 Light Distribution from a HID Lamp

Typically a lamp vendor will supply light distribution information on his lamps. Most commonly, a plan view of a lamp and an angular distribution polar plot will be given. For example, Figure 8.28 shows the light distribution in an Osram HBO 1000W/CL lamp (Osram 1996). This data illustrates the typical light distribution information a projector designer can expect from the lamp vendor.

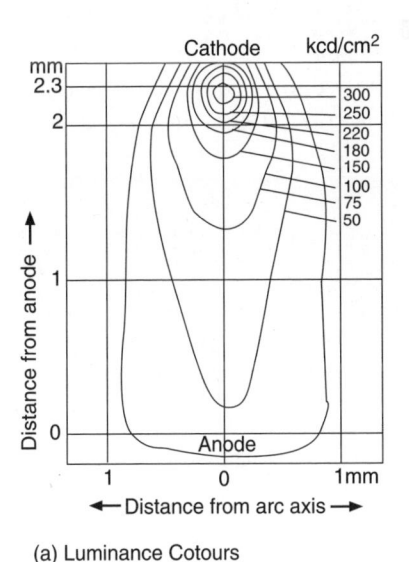

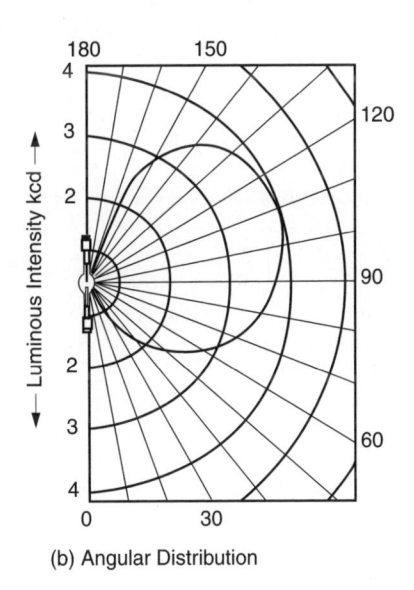

Figure 8.28 Typical light distribution data (Osram HBO 1000 W/CL lamp): (a) luminance contours; (b) angular distribution (reproduced courtesy of Osram Sylvania, Inc.).

202 LIGHT SOURCES FOR LIGHT-VALVE AND MICRODISPLAY PROJECTION SYSTEMS

Another way the distribution data can be presented is in the form of angular deviations from nominally collimated light, as shown in Figure 8.29. This figure shows the much narrower light distribution from the short arc UHP lamp compared with the wider distribution from a conventional 200 W metal halide lamp with a 4.5 mm arc gap.

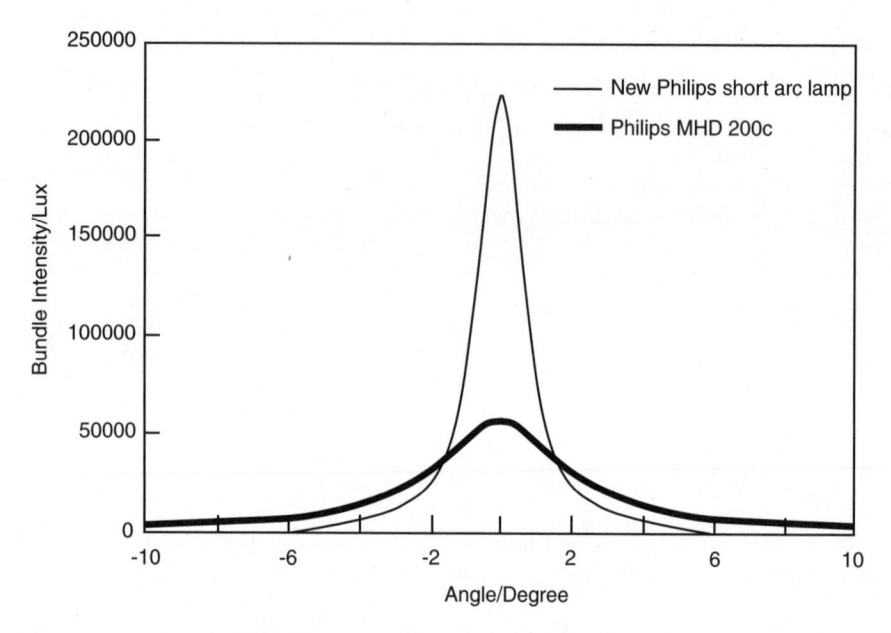

Figure 8.29 Distribution of light emission from 4.5 mm arc gap, 200 W metal halide lamp and from the 100 W UHP lamp (reproduced from Schnedler and Wijngaarde 1995; with permission of SID).

Distribution data, as shown in either Figure 8.28 or Figure 8.29, can be used to construct lamp models that will allow the projector designer to estimate collection efficiency and projection system throughput. This subject will be covered in more detail in Chapter 11.

8.8 Lamp Life

Lamp life is an issue of vital concern to a projector designer. For other optical components, the normal goal of a projector designer is that the component will last for the life of the projector and never need replacing. For the current generation of HID lamps this may not be a realistic goal. Microwave-powered electrodeless lamps, LEDs and lasers all have very long lives and can be designed to last the life of the projector. Projector life, other than lamp life, is likely to be in excess of 10 000 hours and 20 000 hours is a typical consumer electronics specification. Therefore, the projector designer must take lamp life into account, and make provisions for changing the lamp when it fails.

Projectors intended for professional use may use shorter-lived lamps than projectors intended for consumer use. Currently, a 1000 hour life is considered the minimum target in a projector intended for professional use. A target for consumer projectors would be more like 20 000 hours life, i.e. no lamp change would be required during the life of the projector.

8.8.1 Lamp Servicing

In a few projector designs, the lamp can only be changed by a service technician. More commonly, lamps are intended to be changed by the end user. The projector designer must take the following considerations into account when designing a system with a replaceable lamp:

- Safe handling of both the old and the new lamp;
- Prevention of damage to the new lamp by the person changing the lamp;
- Environmental considerations in the disposal of the old lamp;
- Accurate alignment between the new lamp and the projector optical path;
- Reliable and safe high voltage power connections between the lamp and the power supply;
- Interlocks to prevent lamp changing while the projector power is on.

8.8.2 Failure Mechanisms

There are a number of different mechanisms involved in lamp life times. Some of the major mechanisms that limit the lifetime of a lamp in a projector are:

- Loss of an active ingredient of the lamp, as a result of chemical or physical binding of this species at some inappropriate place inside the bulb.
- Depositing of electrode material on the inside of the lamp envelope as a result of sputtering or sublimation of the electrodes. This leads to a darkening of the envelope. While the halide transport cycle helps to minimize this in tungsten-halogen lamps or HID lamps with tungsten electrodes, the cycle is more or less effective in different parts of the lamp due to significant temperature variations of the lamp envelope.
- The removal of material from the electrodes, which may lead to an increase in the arc gap. This may in turn lead to a change in the power or efficacy of the lamp, affecting the light output into an integrating sphere. This increase in the arc gap leads to an increase in lamp étendue (see Chapter 11) and possibly reduced light throughput in the system.
- Clouding of the quartz envelope due to crystallization or etching increases the scattering of the light, also resulting in a larger apparent arc size.
- Catastrophic failure of the lamp envelope.
- Inability of the power supply to produce enough voltage to ignite or run-up the lamp.

The increase in the arc gap and the clouding of the envelope do not normally significantly affect the lumens measured in an integrating sphere as in Figure 8.25. These factors can have significant effect on lumens as measured in Figure 8.27 and they will affect projector throughput. Obviously, more than one aging mechanism can be active in any given lamp.

8.8.2.1 Measurement of lamp life

Measuring lamp life requires care in the measurement set up, patience to allow the lamps to run for a sufficiently long time, typically 1,000 hours or more, and multiple samples of the lamp since there can

204   LIGHT SOURCES FOR LIGHT-VALVE AND MICRODISPLAY PROJECTION SYSTEMS

be significant variation in life properties from lamp to lamp within a single lamp type. Most projector manufacturers rely on the lamp vendor to provide the basic life data, and only run enough tests to confirm there is nothing in the projector design adversely affecting lamp life.

Since the starting pulses associated with igniting a HID lamp can have a significant affect on life, lamp life is normally measured with an on/off cycle. Two hours on/one hour off is one commonly used cycle. The off period must be long enough to ensure the lamp cools completely. With this cycle, a lamp would accumulate 16 on-hours a day, and 62 days would be required to validate a 1000 hour lifetime. To validate a 10 000 hour life for a consumer product would require almost two years. Unfortunately, there are no generally accepted accelerated life tests for lamps.

When life is specified for a lamp in hours, this most commonly indicates the number of hours until the light output is reduced to 50 % of the initial light output or time to either catastrophic or ignition failure, whichever comes first. Since there are aging mechanisms that decrease the usable light in a projector, but do not affect the total light output into an integrating sphere, light output over life for a projection lamp should be measured using the technique shown in Figure 8.27. When a projector designer is evaluating life data from a lamp vendor, it is essential that the techniques used by the vendor to measure life are understood.

When light output as a function of hours is plotted for projection lamps, there are several types of curves that are observed. Three typical curves are shown in Figure 8.30. Note that in all three curves, the initial light output is normalized to 100 %, the end of life is taken as 50 % of initial output and the life is 3500 hours. These are illustrative curves and do not represent the life for any specific lamp. In Figure 8.30(a), the lamp has a rapid initial decline, then is relatively stable for the remainder of its life. In Figure 8.30(b), the light output declines slowly and steadily to the end of life. Finally, in Figure 8.30(c), the lamp is relatively stable for most of its life, and life ends abruptly with total failure of the lamp. Note that a change in the definition of end-of-life will affect these three cases differently. If end-of-life for a

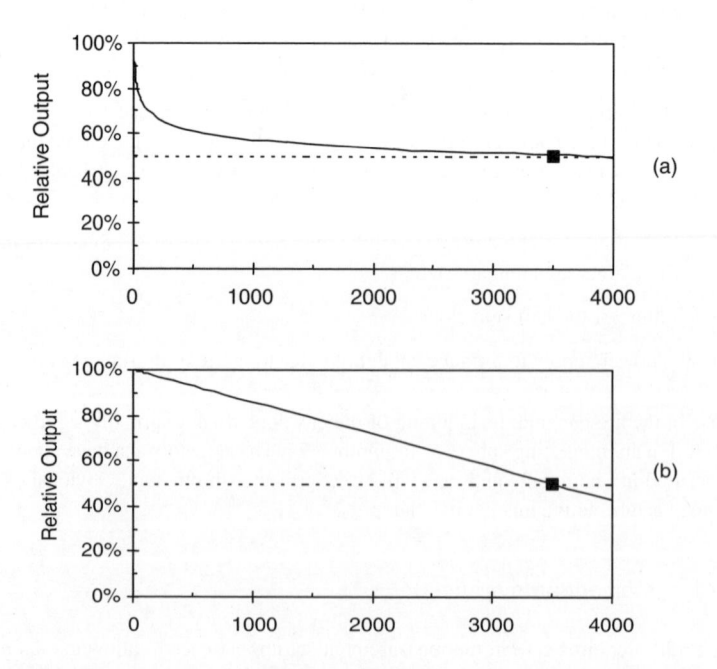

Figure 8.30   Typical lamp aging curves: (a) rapid initial decline, then relatively stable; (b) steady decline; (c) relatively stable, then catastrophic failure.

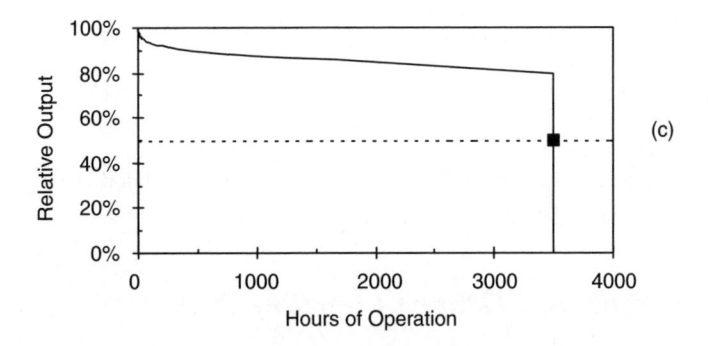

Figure 8.30 (Continued).

given application is 75 % of initial light output, the life in Figure 8.30(a) is cut to almost zero, the life in Figure 8.30(b) is cut in half and the life in Figure 8.30(c) is not affected at all.

The measurement method also affects the lamp lifetime. Measurement of the total light output, as shown in Figure 8.31(a), is nearly meaningless. The correct method is to measure the output of a lamp into a detector whose étendue matches the étendue of the target system, as shown in Figure 8.31(b). Each

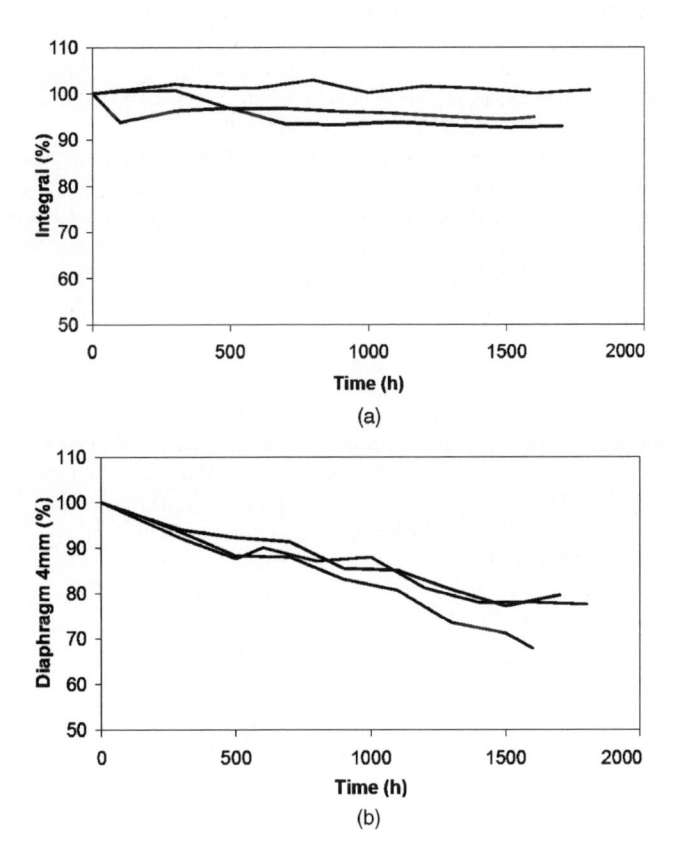

Figure 8.31 Lamp aging curves with different measurement methods (reproduced by permission of © 2007 SPIE (Moench *et al.* 2007)): (a) total lumen output; (b) lumen output into a 4 mm aperture.

figure represents the same three batches of Philips Ujoy lamps (Moench *et al.* 2007). In the first case, the lamps are measured at intervals in an apparatus similar to the one shown in Figure 8.25. As can be seen, the total light output of the lamps remains nearly constant. In Figure 8.31(b), the same lamps were measured using the technique shown in Figure 8.27, with a 4 mm aperture. This aperture size, combined with the convergence angle of the Ujoy lamp, corresponded to an étendue of 10 mm^2 steradian, which is a typical value for a small microdisplay. As can be seen, under these conditions, the light output of the lamps declined significantly over life.

8.9 Reflectors and Other Collection Systems

Light collection systems fall into two broad categories: imaging and non-imaging systems. An imaging system is one where an image of the arc, normally with major distortions and aberrations, is (or could be) produced somewhere in the optical path. With critical illumination, this image is at the microdisplay and with Köhler illumination, this image is at the entrance pupil of the projection lens. In a lenslet integrator such as the one shown in Figure 6.28, multiple images of the arc are produced at the second integrator plate which is in turn imaged onto the pupil of the projection lens. Therefore, this system can be said to be a Köhler system.

A non-imaging system does not produce an image of the arc or other light source. Instead, the emphasis is on the optical beam with its aspect ratio, area, and divergence. All collection systems in use in projectors today are of the imaging type, although non-imaging collection systems have been proposed (Stewart *et al.* 1998) for HID lamps. While non-imaging systems have some theoretical advantages, the practical difficulties in implementing them has kept them out of projection systems up to this point. Non-imaging systems are used commonly with LED light sources and will be covered in Section 8.9.5. Anyone interested in this type of collection system for a projection application is referred to the standard text on the subject (Welford and Winston 1989).

The following sections will describe some of the major types of imaging collection systems usable in projection systems.

8.9.1 Reflectors with Conic Sections

Most projectors use reflectors to collect the light from the lamp and form a beam that can be manipulated by the remainder of the illumination optics, modified by the microdisplay and projected on the screen. The most common types of reflectors are conic sections: ellipses and parabolas. An ellipse will bring light to a focus at a point and a parabola will generate a collimated beam of light. Both of these, shown in Figure 8.32, are based on a point light source and are degraded by the finite size of the lamp arc.

Reflectors have also been designed where the basic contour is parabolic or elliptical, but the reflector is subdivided into small segments (Chen *et al.* 1995). These individual segments can either be flat or have some curvature that is different from the overall conic section. Reflectors of this type can significantly increase the étendue of the light beam, compared to the étendue of a reflector that has the base conic section without segments. The advantage of this type of reflector is that it can perform optical integration. This may make it possible to eliminate the separate integrator in the optical path. This may in some designs increase total light throughput and decrease system cost and size.

8.9.2 Compound Reflectors

Compound reflectors, in which a sphere is mated to either a parabola or an ellipse, have also been used (True 1979). Figure 8.33 shows a compound elliptical and spherical reflector system similar to the one that was used for many years in the Talaria projector.

REFLECTORS AND OTHER COLLECTION SYSTEMS

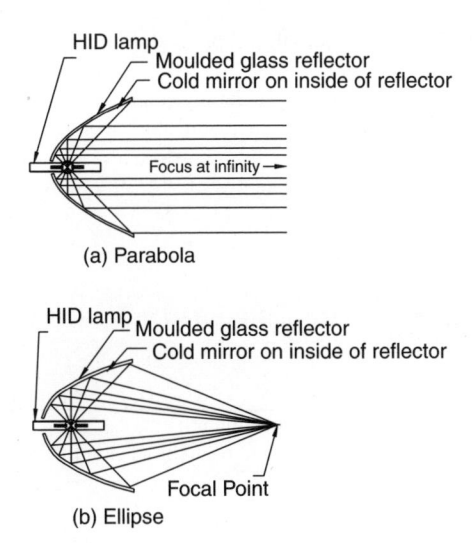

Figure 8.32 Parabolic and elliptical reflectors: (a) parabola, (b) ellipse.

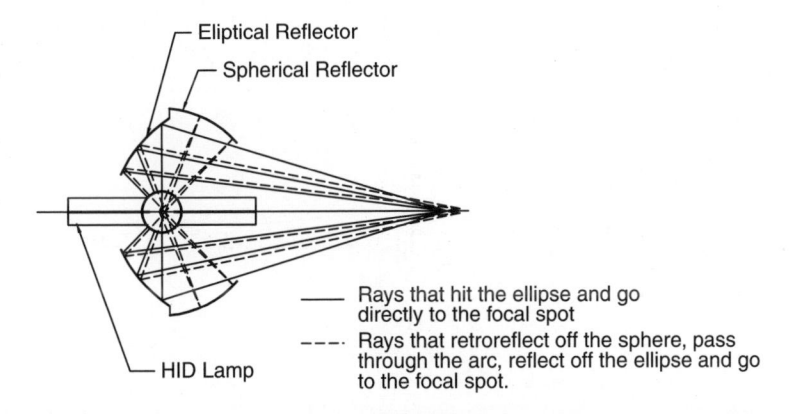

Figure 8.33 Compound reflector system.

In this system, any light that fell on the ellipse after being emitted from the lamp was focused to a point. Light that missed the ellipse would strike the sphere, whose center was also the center of the arc. Light would retro-reflect off the sphere, pass through the arc and (hopefully) reach the elliptical reflector. There it would be focused at about the same point where the light that struck the ellipse on the first pass would focus. The actual collection system used by the Talaria had a negative lens in the output beam of the collection system, not shown in Figure 8.33. The purpose of the negative lens was to increase the focal length of the reflector while maintaining a compact package. Similar negative lenses are sometimes seen coupled to an elliptical reflector in a modern LCD projector for the same reason: make the system more compact.

Due to the distortions introduced by the extra passes through lamp envelope, the absorption by the arc, and other factors, the focusing of the light from the sphere is never as good as the focusing of the light that comes directly from the ellipse. Nevertheless, reflectors of this type can collect significantly more (30 % or so) light than conventional single element reflectors. While light that strikes the sphere first is collected less efficiently than light that strikes the ellipse first, this is light that would otherwise be lost from the system completely.

The sphere in this design needs to be very carefully aligned to the center of the arc. If a true retroreflecting material is used (Wang *et al.* 2007), alignment is much less critical.

Since the light from the sphere passes through the arc before it reaches the ellipse, this configuration may affect the lamp. In particular, it may significantly affect lamp life by putting an additional thermal burden on the arc, the envelope or the electrodes. A projector designer should check with the lamp manufacturer before using this type of reflector in a system.

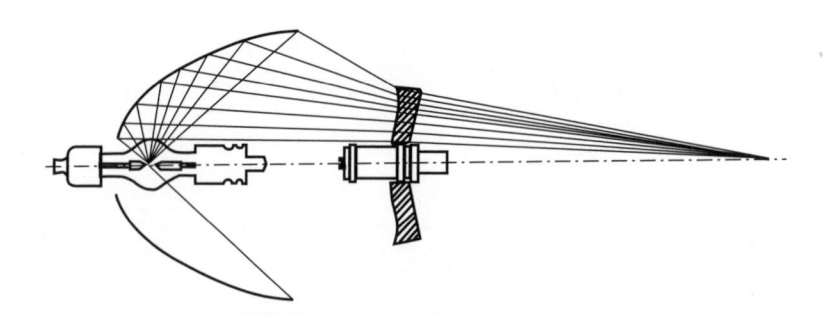

8.9.3 Constant Magnification Reflectors

Another type of reflector that can be used in a projection system is a constant magnification reflector (True 1987; Sekiguchi *et al.* 2004). A constant magnification reflector can be designed as a 'parabola-like' or 'ellipse-like' system, i.e. it can produce a collimated beam or a beam that comes to a focus. The goal of a constant magnification design is to correct the variation in lateral magnification inherent in reflectors with conic section cross-sections.

In order to do this, two optical elements with aspherical contours are required. The simplest arrangement is an aspherical reflector coupled to a corrector plate, as shown in Figure 8.34. Constant magnification systems only provide benefits for étendue limited projection systems: they do not increase the total amount of light collected. Étendue limited systems and the lumens vs étendue function will be discussed in detail in Chapter 11. This lateral magnification correction can also be built into the lenslet integrator, as was described in Chapter 7.

Figure 8.34 Constant magnification reflector system.

8.9.4 Refractive Collection Systems

A typical refractive collection system is shown in Figure 8.35. The condenser lens is normally an aspherical lens. While Figure 8.35 shows the lens as a doublet, it obviously can be a singlet or have more than two elements. While the total collected light by a f/0.7 refractive condenser is significantly less than the light collected by a f/0.15 reflector, the aberrations are also less. For small microdisplays and étendue limited projection systems, the total collected light usable by the device is sometimes higher than the collected light from a reflector. Many refractive collection systems use a spherical mirror to increase the light collected, in much the same manner as was described in Section 8.7.2 for compound reflector systems.

Systems with two collecting lenses have been proposed and used in projection systems (Williams 1997; Stroomer and Timmers 1991). A system with two condensing lenses is shown in Figure 8.36. In

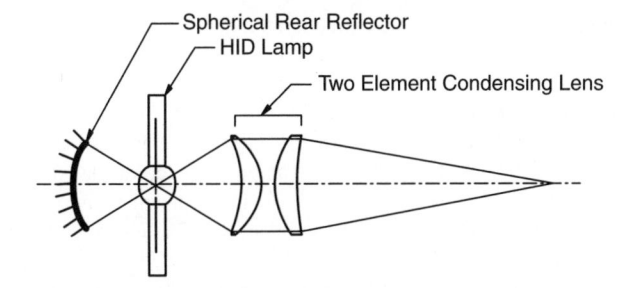

Figure 8.35 Refractive collection system.

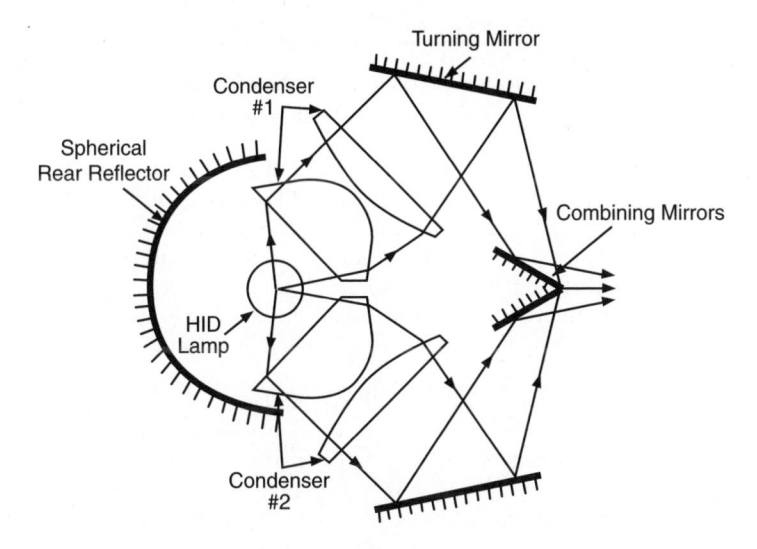

Figure 8.36 Double condensing lens refractive collection system.

this system, the two collection lenses are orthogonal to each other, and produce two images of the arc side by side. A single spherical mirror serves as a back reflector for both lenses.

8.9.5 Collection Systems for LEDs

LEDs can use the same sorts of collection systems as lamps, including conic section reflectors, compound reflectors, constant magnification reflectors, refractive collection systems and non-imaging collection systems. Obviously, any optical design that requires the light to pass through the emitting area, such as the compound reflector shown in Figure 8.33, cannot be used with LEDs since LEDs are opaque. Two types of systems seem to be more common in LED-based systems than in lamp based systems. First, non-imaging optics forms the basis of most collection systems described in the literature to date. Secondly, many LED collection systems include light recycling. Light recycling has been used in lamp-based systems, such as the one shown in Figure 6.59, but it has not commonly reached production.

The compound parabolic concentrator (or collector), commonly known as the CPC, forms the basis of many LED collection systems. The CPC can be hollow with a mirrored inside surface or it can be a solid optical glass or plastic, and depend on total internal reflection. TIR-based designs are preferred, both because of the low cost of injection molded plastic and because the plastic has a higher index than air and provides a better optical coupling and extraction efficiency when used with the high-index die. Other non-imaging designs are normally used, in part because a CPC typically has a relatively large size compared to the LED size.

Figure 8.37 shows a basic CPC plus a CPC with two types of recycling added. Figure 8.37(a) shows a basic CPC used with a LED. The CPC would normally have a circular cross section, but it can have a rectangular cross section to match the shape of the LED and/or the microdisplay. Three sample light rays are shown. Ray A is emitted from the LED at a position and angle such that it never touches the CPC and emerges directly from the open end of the CPC unmodified. Ray B is emitted at a position/angle combination that it reflects once off the CPC before exiting. Ray C is emitted at a position/angle combination so it reflects twice off the CPC before emerging. While LEDs in general can have very narrow emission angles or unusual angular distributions of the emergent light, LEDs used in projection applications normally have a Lambertian or near-Lambertian angular distribution. The CPC is designed to reflect rays with extreme emergent angles into the forward direction and only allows rays to emerge within a limited range of angles. Due to the limits imposed by étendue, the opening of the CPC must be larger than the LED area. Since the cone angle has been reduced, the area must be increased; an optical system such as a CPC can only increase or maintain étendue, never decrease it. A LED/CPC combination like this would normally produce unpolarized light.

Figure 8.37(b) shows a CPC with a mirrored aperture at the output plane of the CPC with two representative rays D and E. Ray D emerges from the LED at a position/angle combination that allows it to reflect once off the CPC and emerge from the CPC through the aperture. Ray E strikes the mirrored aperture and reflects back toward the LED, reflecting off the CPC on the way. Here it reflects off the LED back toward the aperture. In this case it passes through the aperture. If by chance it had struck the mirrored surface a second time, it would have been reflected back toward the LED and after reflecting there would have had a third chance. This design also produces unpolarized light.

Figure 8.37(c) has the same mirrored aperture as the system shown in Figure 8.37(b), but in this case the opening is covered by a reflective polarizer such as a wire grid polarizer. There may also be a quarter-wave plate installed in the system. In this case, if a ray strikes the mirrored aperture, it is recycled, exactly like ray E in Figure 8.37(b). The example ray F is emitted from the LED and reaches the opening in the aperture. The reflective polarizer then splits the ray into two rays, F_1 and F_2. F_1 has its electric field vector perpendicular to the wires in the wire grid polarizer so it passes through the aperture. Ray F_2 has its electric field vector parallel to the wires so it is reflected back toward the LED. It passes through the quarter wave plate twice, once before and once after reflecting off the LED. This rotates its polarization by 90° so when it reaches the aperture the second time its electric field vector is perpendicular to the wires and passes through. This design emits polarized light and could be used with a LCD or LCoS microdisplay. Recycling light can provide advantages in both luminance and polarization even when no CPC or other collection system is used (Beeson 2006).

A tapered light pipe as shown in Figure 8.38 can also serve as a collection and collimation optics (Teijdo et al. 2006). If the light pipe is long enough, it can also serve as an integrator to ensure spatial uniformity at the output. The tapered sides provide a collimation effect and if properly designed the system can provide a uniform cone angle and telecentric light at the output face, as well as spatial uniformity.

LEDs sometimes use collection systems that are a combination of an imaging and a non-imaging collection, as shown in Figure 8.39. In this simple projector, the LED would be a white LED and the LCD panel would have a color filter array. The central portion of the collector is an imaging type, although no actual image is produced. The outer portion is non-imaging and collects the oblique rays from the LED and sends them forward toward the edges of the panel (Keuper et al. 2003). The tapered light pipe design shown in Figure 8.38 can also be used in this type of simple projector, with the microdisplay placed at the output of the light pipe.

REFLECTORS AND OTHER COLLECTION SYSTEMS 211

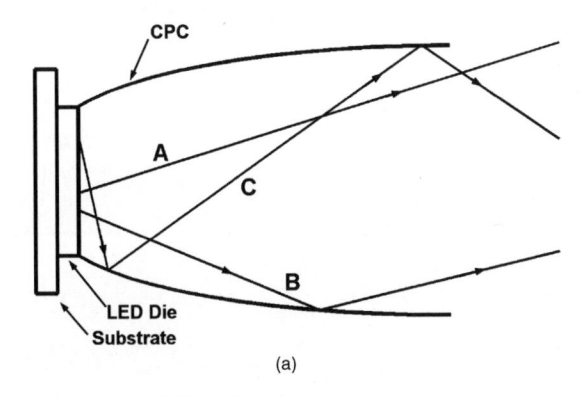

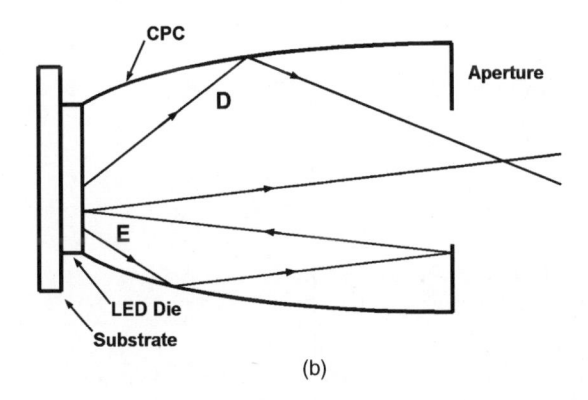

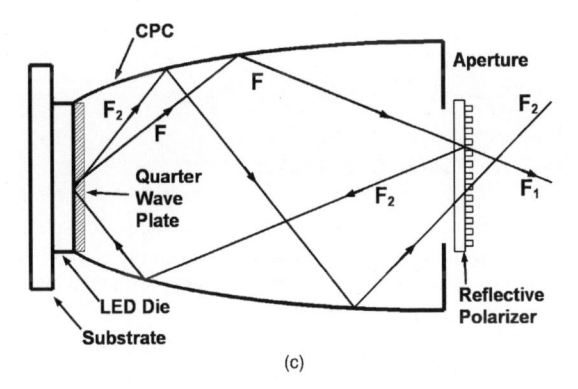

Figure 8.37 Compound Parabolic Concentrator designs for LED light collection: (a) CPC used alone; (b) CPC with a recycling aperture to produce unpolarized light; (c) CPC with a recycling aperture and polarizer to produce polarized light.

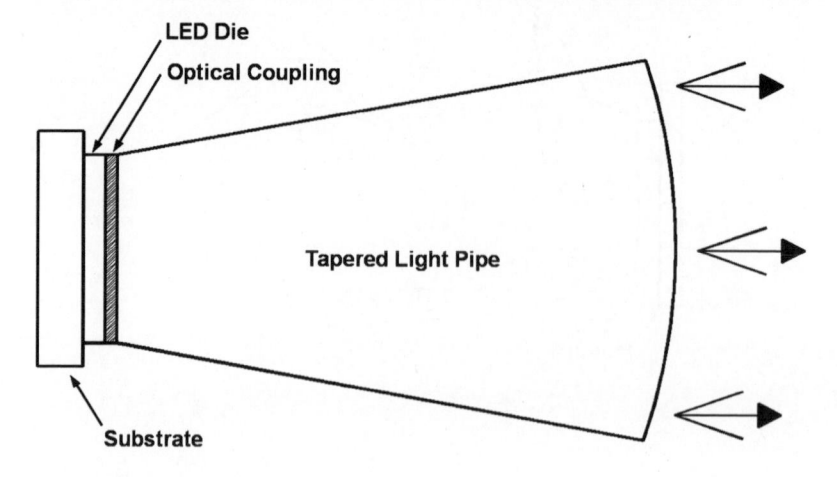

212   **LIGHT SOURCES FOR LIGHT-VALVE AND MICRODISPLAY PROJECTION SYSTEMS**

Figure 8.38 Tapered light pipe providing collection, collimation and integration functions.

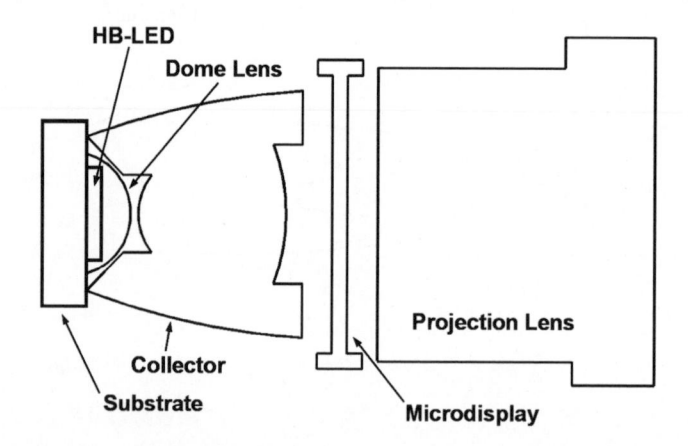

Figure 8.39 LED collection system combining imaging and non-imaging optics (reproduced by permission of © 2003 the Society for Information Display (Keuper et al. 2003)).

8.10 Lamp Ballasts and Ignitors

HID lamp power supplies are commonly called 'ballasts'. Ballasts are often sold by the lamp vendor as a mated set with the ignitor, lamp and reflector. By purchasing a complete set, the projector manufacturer can ensure he uses a supply that drives the lamp at the optimum power and with the optimum waveform. The lamp manufacturer benefits not only from the selling price of the ballast, but he also knows the lamp will be driven properly for optimum lamp life, minimizing warranty returns. If a projector manufacturer decides to obtain the ballast from a source other than the lamp manufacturer, it must be ensured that the ballast will provide the current, voltage, power, waveform and starting pulses to operate the lamp correctly. Historically magnetic ballasts were used for HID lamps, but currently virtually all projection systems use electronic power supplies.

The operating voltage of a HID lamp, as provided by a ballast, is typically in the range of 50–100 Volts. This voltage is all it takes to maintain an arc but is an insufficient voltage to initiate the electrical breakdown within the lamp that leads to the arc. The ignitor or starter circuit provides high voltage to the lamp in order to initiate this breakdown. The igniting voltages can be pulse trains with peak voltages as high as 25 kV. Ignitors are normally designed to be as close to the lamp as possible, with leads as short as possible. This both minimizes the amount of voltage required to start the lamp and minimizes the effect of the starting pulses on the other electronic circuitry in the projector (Coaton and Marsden 1997).[5]

Most HID lamps suffer from a 'hot restrike' problem. This means that if a lamp is operating and then is extinguished, it cannot be restarted immediately even with the high voltage pulses put out by the ignitors. This problem is especially common in mercury and metal halide lamps containing mercury. It is caused by the fact that the lamp remains hot after it is shut down. At this temperature, the vapor pressure of mercury is quite high, so the pressure inside the lamp is high. This high pressure suppresses the electrical breakdown that leads to ignition of the arc. Cool down times of 5 minutes or more can be required before the vapor pressure is low enough to allow the starter to reignite the lamp.

Lamp, ballast and starter combinations intended for projector applications are normally optimized to minimize or eliminate the hot restrike problem. The projector designer should inquire of the lamp maker about hot restrike prior to committing a projector design to use a particular lamp and ballast combination. Issues related to ballasts, starters and hot restrike are covered in standard lamp texts such as Waymouth (1978).

Some lamp types must operate at the nominal average power level, but may be pulsed to higher peak powers with a reduced duty cycle, keeping the average power constant. Obviously, a special ballast is required to operate a lamp in this mode. This may be an advantage in certain projector architectures (Stanton *et al.* 1996), where the higher peak brightness and shorter duty cycle can be utilized. Reduced duty cycle operation will normally also affect lamp efficacy, spectrum and life. The manufacturer of a lamp should be consulted before attempting to drive a lamp in this manner.

References

Beeson, K., Zimmerman, S., Livesay, W., Ross, R., Livesay, C. *et al.* (2006) LED-based light-recycling light sources for projection displays. *SID Intl. Symp. Digest of Technical Papers*, Paper 61.5, 1823–6.

Bierhuizen, M. Krames, S., Harbers, G. and Weijers, G. (2007) Performance and trends of high power Light Emitting Diodes. In I. T. Ferguson *et al.* (eds), *SPIE Proceedings Vol. 6669, Seventh International Conference on Solid State Lighting*.

Bloom, D.M. and Tanner, A.H. (2007) Twenty megapixel MEMS-based laser projector. *SID Intl. Symp. Digest of Technical Papers*, Paper 3.3, 8–11.

Bogdankevich, O.V. (1994) Electron-beam-pumped semiconductor lasers. *Quantum Electronics* **24**(12): 1031–53.

Bondarev, V.Y., Kozlovsky, V.I., Sannikov, D.A., Kuznetsov, P.I., Jitov, V.A. *et al.* (2005) Electron-beam pumped green VCSEL on MOVPE-grown ZnCdSe/ZnSSe MQW structure. *13th Int. Symp.: Nanostructures: Physics and Technology.* St Petersburg, Russia, 20–25 June 2005, 2 pp.

Brennesholtz, M.S. (2002) Color-sequential LCoS projector with a rotating drum. *SID Intl. Symp. Digest of Technical Papers*, Paper 51.4, 1346–9.

Brennesholtz, M.S., McClain, S.C., Roth, S. and Malka, D. (2005) A single panel LCOS engine with a rotating drum and a wide color gamut. *SID Intl. Symp. Digest of Technical Papers*, Paper 64.3.

Byer, R.L., Fejer, M.M. and Lim, E.J. (1991) *Nonlinear Optical Radiation Generator and Method of Controlling Regions of Ferroelectric Polarization Domains in Solid State Bodies*, US 5,036,220 issued 30 July 1991.

Chen, K.-V., Chen, S.-M. and Hao, Z.-D. (1995) Analytical light source illuminance in liquid crystal projection TV. In M.H. Wu (ed.), *Proc. SPIE: Projection Displays* **2407**: 53–6.

Chilla, J.L.A., Zhou, H., Weiss, E., Caprara, A.L., Shou, Q. *et al.* (2005) Blue and green optically-pumped semiconductor lasers for display. In M.H. Wu (ed.), *Proc. SPIE: Projection Displays* **5740**: 41–7.

Christie Digital Systems (2007) *LX1500 Product Brochure*, September.

---

[5] Most projector engineers have destroyed at least one electronic instrument in their careers by forgetting to disconnect it from a lamp under test during the ignition cycle.

Coaton, J.R. and Marsden, A.M. (eds) (1997) *Lamps and Lighting*, 4th edn. London: Hodder Headline Group and New York: Wiley & Sons, Inc.

Derra, G., Moench, H., Fischer, E., Giese, H., Hechtfischer, U. et al. (2005) UHP lamp systems for projection applications. *J. Phys. D: Appl. Phys.* **38:** 2995–3010, 16 pp.

Digital Cinema Initiatives, LLC, (2007) *Digital Cinema System Specification, Version 1.1*, 12 April 12 2007, 158 pp. This specification can be downloaded at no cost from http://www.dcimovies.com/

Dolan, T., Ury, M.G and Wood, C.H. (1992) A novel high efficacy microwave powered light source. In L. Bartha and F.J. Kedves (eds), *6th International Symposium on the Science & Technology of Light Sources (LS:6)* Budapest, 30 Aug–3 Sep.

Eastlund, B.J. and Levis, M.E. (2003) *Sapphire High Intensity Discharge Projector Lamp*, US Patent 6,661,174, issued 9 Dec 2003.

Espiau, F. M., Joshi, C.J. and Chang, Y. (2004) *Plasma Lamp with Dielectric Waveguide*, US Patent 6,737,809, issued 18 May 2004.

Espiau, F.M. and Chang, Y. (2005) *Microwave Energized Plasma Lamp with Solid Dielectric Waveguide*, US Patent 6,922,021, issued 26 July 2005.

Fink, C.G., Bergstedt, R., Flint, G., Hargis, D. and Peppler, P. (1996) Microlaser-based projection display for simulation. *Interservice/Industry Training Systems and Education Conference*, Orlando, FL

Harbers, G., Keuper, M. and Paolini, S. (2003) Performance of High Power LED Illuminators in Color Sequential Projection Displays. *Proceedings of the 10th International Display Workshops, IDW '03*, 4 Dec 2003, Fukuoka, Japan: ITE & SID, Paper LAD3-4, 1585–8.

Harbers, G., Bierhuizen, S.J. and Krames, M.R. (2007) Performance of high power light emitting diodes in display illumination applications, *Journal of Display Technology* **3**(2), June: 98–109.

Hewlett, G.J., Penn, S.M., Pettitt, G.S. and Segler, Jr., D.F. (2005) *Neutral Density Color Filters*, US Patent 6,879,451 issued 12 April 2005.

Higashi, T. (1995) Life performance improvement of the short arc metal halide lamp by DC operation. In R. Itatani and S. Kamiya (eds), *Proceedings of the 7th International Symposium on the Science & Technology of Light Sources*, Paper 32:P. Tokyo: The Illuminating Engineering Institute of Japan.

Higashi, T. and Arimoto, T. (1995) Long life DC metal-halide lamps for LCD projectors. *SID Intl. Symp. Digest of Technical Papers*, Paper 11.2.

Hollemann, G., Braun, B., Dorsch, F., Hennig, P., Heist, P. et al. (2000) RGB lasers for laser projection displays. In M.H. Wu (ed.), *Proc. SPIE: Projection Displays* **3954**, April, 140–51.

Hollemann, G., Braun, B., Heist, P., Symanowski, J., Krause, U. et al. (2001) High-power laser projection displays. In M.H. Wu (ed.), *Proc. SPIE: Projection Displays* **4294**, March, 36–46.

Jansen, M., Carey, G.P., Carico, R., Dato, R., Earman, A.M. et al. (2007) Visible laser sources for projection displays. In M.H. Wu and H.Y. Lin (eds), *Proceedings of SPIE: Projection Displays* XII, **6489**.

Joachim, R. and Alexander, W. (2007) Requirements on LEDs for advanced optical systems. In K.P. Streubel and H. Jeon (eds), *Light-Emitting Diodes: Research, Manufacturing, and Applications XI, Proc. of SPIE* **6486**, 11 pp.

Joshi, C.J. (2007) Development of Long Life, Full Spectrum Light Source for Projection Display. *SID Intl. Symp. Digest of Technical Papers*, Paper 14.4, 3 pp.

Kawai, K. and Matsumoto, M. (1995) Short arc metal halide lamp suitable for projector application. *SPIE Proceedings: Projection Displays* **2407**.

Keuper, M.H., Paolini, S., Harbers, G. and Tsang, P. (2003) Ultra-compact LED based image projector for portable applications. *SID Intl. Symp. Digest of Technical Papers*, Paper P-126, 713–15.

Kirkpatrick, D.A., Dolan, J.T., MacLennan, D.A., Turner, B.P. and Simpson, J.E. (2003) *High Brightness Microwave Lamp*, US Patent 6,617,806, issued 9 September 2003.

Koechner, W. (2006) *Solid-State Laser Engineering*, 6th edn. Springer Series in Optical Sciences, 747 pp.

Kückmann, O. (2006) High power LED arrays: Special requirements on packaging technology. In K.P. Streubel, H.W. Yao and E. Fred Schubert (eds), *Light-Emitting Diodes: Research, Manufacturing and Applications X, Proc. of SPIE* **6134**.

Liu, X., Hu, M.H., Hughes, L.C. and Zah, C.-E. (2006) Wavelength matching and tuning in green laser packaging using second harmonic generation. *IEEE 2006 Electronic Components and Technology Conference*, 1064–9.

Luminus (2007) *PT54 - Preliminary Datasheet – Revision 02*, Luminus Devices, 7 pp.

McGettigan, T. (2007) New generation LIFI light sources. *Projection Summit 2007*, Paper SS-1-2. Norwalk CT: Insight Media.

REFERENCES

Mizushima, T., Furuya, H., Mizuuchi, K., Yokoyama, T., Morikawa, A. et al. (2006) Laser projection display with low electric consumption and wide color gamut by using efficient green SHG laser and new illumination optics. *SID Intl. Symp. Digest of Technical Papers*, Paper L-9, 1681–4.

Moench, H., Derra, G., Fischer, E. and Riederer, X. (2000) Arc stabilization for short arc projection lamps. *SID Intl. Symp. Digest of Technical Papers*, Paper 9.1, 84–7.

Moench, H., Deppe, S., Pekarski, P. and Munters, T. (2006) An ultra-high-pressure (UHP) lamp drive system for time-sequential displays. *SID Intl. Symp. Digest of Technical Papers*, Paper 55.3, 1720–3.

Moench, H., Mackens, U., Pekarski, P., Ritz, A., S'Heeren, G. et al. (2007) Personal projection with Ujoy technology. In M.H. Wu and H.Y. Lin (eds), *Proc. of SPIE: Projection Displays XII* **6489**.

Mooradian, A., Carey, G., Carico, R., Dato, R., Dudley, J. et al. (2007) Surface-emitting-diode lasers and laser arrays for displays. *Journal of the SID* **15**(10): 805–9.

Mooradian, V. (2001) *High Power Laser Devices*, US Patent 6,243,407 issued 5 June 2001.

Moulton, P.F., Snell, K.J., Lee, D., Wall, K.F. and Bergstedt, R. (2002) High-power RGB laser source for displays. *The Image Society Image 2002 Conference*, Scottsdale, AZ.

Nagahama, S. and Mukai, T. (2005) High-power InGaN Blue – laser diodes for displays. *SID Intl. Symp. Digest of Technical Papers*, Paper 52.3, 1605–7.

Nägele, T. and Distl, R. (1999) *Handbook of LED Metrology*. Instrument Systems GmbH, version 1.1, 40 pp.

Nakamura, S., Senoh, M., Nagahama, S., Iwasa, N., Yamada, T. et al. (1997) Room-temperature continuous-wave operation of InGaN multi-quantum-well-structure laser diodes with a long lifetime. *Appl. Phys. Lett.* **70**: 868–70.

Niven, G. and Mooradian, A. (2006) Low cost lasers and laser arrays for projection displays. *SID Intl. Symp. Digest of Technical Papers*, Paper 67.1, 1904–7.

Osowski, M. L., Hu, W., Lammert, R. M., Liu, T., Ma, Y. et al. (2007) High brightness semiconductor lasers. In M.S. Zediker (ed.), *High-Power Diode Laser Technology and Applications V, Proc. of the SPIE* **6456**, 7 pp.

Osram (1996) *Lighting Program Photo Optics '96/97*. Danvers, MA: Osram Sylvania, August.

Pollmann-Retsch, J., Giese, H., Mackens, U. and Moench, H. (2006) Compact power lights for high-end projection applications. *SID Intl. Symp. Digest of Technical Papers*, Paper 55.2, 1716–19.

Pothoven, F.R. and Pothoven, T.A. (2003) *Method for Manufacturing an Electrodeless Lamp*, US Patent 6,666,739, issued 23 December 2003.

Schnedler, E. and Wijngaarde, H. v. (1995) Ultrahigh-intensity short-arc long-life lamp system. *SID Intl. Symp. Digest of Technical Papers*, Paper 11.1, 131–4.

Schubert, E.F. (2003) *Light-Emitting Diodes*. Cambridge: Cambridge University Press.

Sekiguchi, A., Sasagawa, T. and Goto, Y. (2004) Étendue-density homogenization lens optimized for high-pressure mercury lamps. *Journal of the SID* **12**(1): 105–11.

Smolka, E., Schnabl, A. and Schilling, F. (1995) Doped and undoped xenon short arc lamps with extremely high spectral radiance. In R. Itatani and S. Kamiya (eds), *Proceedings of the 7th International Symposium on the Science & Technology of Light Sources*, Paper 53:P. Tokyo: The Illuminating Engineering Institute of Japan.

Snell, K.J., Lee, D., Pati, B. and Moulton, P.F. (2000) RGB optical parametric oscillator source for compact laser projection displays, In M.H. Wu (ed.), *SPIE Proceedings: Projection Displays* **3954**, 25 April, 158–62.

Stanton, D.A., Shimizu, J.A. and Dean, J.E. (1996) Three lamp single light valve projector. *SID Intl. Symp. Digest of Technical Papers*, Paper 36.3, 839–42.

Stewart, C.N., Rutan, D.M. and Savage, D.J. (1997) High efficiency metal halide lighting systems for compact LCD projectors. In M.H. Wu (ed.), *SPIE Proceedings: Projection Displays III* **3013**: 72–9.

Stewart, C.N., Rutan, D.M., Jacobson, B.A. and Gengelbach, R.D. (1998) Metal halide lighting systems and optics for high efficiency compact LCD projectors. In M. Wu (ed.), *SPIE Proceedings: Projection Displays IV* **3296**, January.

Stroomer, M.V.C. and Timmers, W.A.C. (1991) *Illumination System*. US Patent 5,046,837, issued 10 September 1991.

Sugaya, K., Horikawa, Y., Suzuki, Y., Arimoto, T., Tsukamoto, T. et al. (2005) A new driving scheme for AC-type ultra-high-pressure mercury lamp with highly stabilized light output. *SID Intl. Symp. Digest of Technical Papers*, Paper 52.1, 1596–9.

Takeuchi, N., Kitahara, Y., Wakamiya, M., Omura, H. and Tabata, M. (1995) Short arc metal halide lamp with new ingredients for LCD projector. In R. Itatani and S. Kamiya (eds), *Proceedings of the 7th International Symposium on the Science & Technology of Light Sources*, Paper 28:L. Tokyo: The Illuminating Engineering Institute of Japan.

Teijdo, J.M., Ludley, F., Ripoll, O., Ueda, M., Oshima, Y. et al. (2006) Compact three panel LED projector engine for portable applications. *SID Intl. Symp. Digest of Technical Papers*, Paper 73.2, 2011–14.

TI (1966) Texas Instrument Bulletin No. DLA 1324, *Experimental Laser Display for Large Screen Presentation*, January.

True, T. (1979) Recent advances in high-brightness and high-resolution color light-valve projectors. *SID Intl. Symp. Digest of Technical Papers*, Paper 5.1.

True, T. (1987) *Constant Magnification Light Collection System*. US Patent 4,642,740, issued 10 February 1987.

Turner, B.P., Ury, M.G., MacLennan, D.A. and Leng, Y. (1995) Progress in sulfur lamp technology. In R. Itatani and S. Kamiya (eds), *Proceedings of the 7th International Symposium on the Science & Technology of Light Sources*, Paper 35:P. Tokyo: The Illuminating Engineering Institute of Japan.

Volodin, B.L., Dolgy, S.V., Melnik, E.D., Downs, E., Shaw, J. *et al.* (2004) Wavelength stabilization and spectrum narrowing of high-power multimode laser diodes and arrays by use of volume Bragg gratings. *Optics Letters* **29**(16), 15 August: 1891–3.

Wang, Y.-J., Li, K. and Inatsugu, S. (2007) New retroreflector technology for light-collecting systems. *Optical Engineering* **46**(8), August.

Waymouth, J.F. (1978) *Electric Discharge Lamps*. Cambridge: The MIT Press.

Welford, W.T. and Winston, R. (1989) *High Collection Nonimaging Optics*. Burlington, MA: Academic Press, Inc.

Williams, S.A. (1997) History of Eidophor projection in North America. In M.H. Wu (ed.), *SPIE Proceedings: Projection Displays III*, **3013**.

Yeralan, S., Doughty, D., Blondia, R. and Hamburger, R. (2005) Advantages of using high-pressure, short arc xenon lamps for display systems. In M.H. Wu (ed.), *Proceedings of SPIE: Projection Displays XI* **5740**: 27–36.

9
Scanned Projection Systems

The systems discussed in this chapter are two classes of projectors where the image is created one pixel at a time: CRT projectors and scanned laser projectors. CRT-based projection systems are the original projection technology and are rapidly losing market share to newer technologies, although they are still used in a few specialized applications. Scanned laser systems were proposed soon after the laser was invented but they are only now reaching commercial production. The high cost of lasers prevented their widespread use in displays, but the ongoing laser developments described in Chapter 8 are likely to bring them into the mass market in the near future.

Other raster-scanned projection systems have been used in the past to generate images. These systems include the mechanically scanned Scophony projector, the oil-film based Eidophor and Talaria projectors and the liquid crystal-based ILA systems. Since these systems are no longer in production and the technologies are not likely to return to production they are not covered in this book. For a review of the technology of the Eidophor, Talaria and ILA systems, see the first edition of this book. They are also covered, along with the Scophony projector, in Brennesholtz (2007).

9.1 CRT Projectors

In all projection systems, the subsystem which converts the electrical signal into the optical image is called the light engine. The light engines of most CRT projectors are based on three CRTs and three lenses, one for each color. The tubes are optically coupled to the lenses through a cooling liquid between the tube face and the rearmost optical surface of the lens. Three-CRT/one-lens systems have been made but they are infrequently used in current projection systems. Some high luminance systems for professional applications have used more than one tube of each color, leading to six- or nine-tube systems.

Full color tubes, such as the shadow mask or tension mask tubes used in direct-view displays, are not used in projection due to the inefficiencies of the electron beam utilization. The masks, essential in direct-view tubes for separating the three beams for color purity, typically intercept 75–80 % of the

218 SCANNED PROJECTION SYSTEMS

energy. As a result, it would be difficult to generate adequate luminance in the projected image with a full spectrum tube of the types used for direct-view applications.

Many of the system issues are common to three-lens and one-lens systems. In the following sections, the more common CRT projection architectures and system issues will be discussed.

9.1.1 Three-lens CRT Projectors

Three-lens CRT projectors are made for both front screen, i.e. where the viewer and projector are on the same side of the screen, and rear screen installations. The configuration for the light engine, shown in Figure 9.1, is the same for both screen types. Figure 9.2 is a photograph of the light engine from a consumer rear-screen CRT projector.

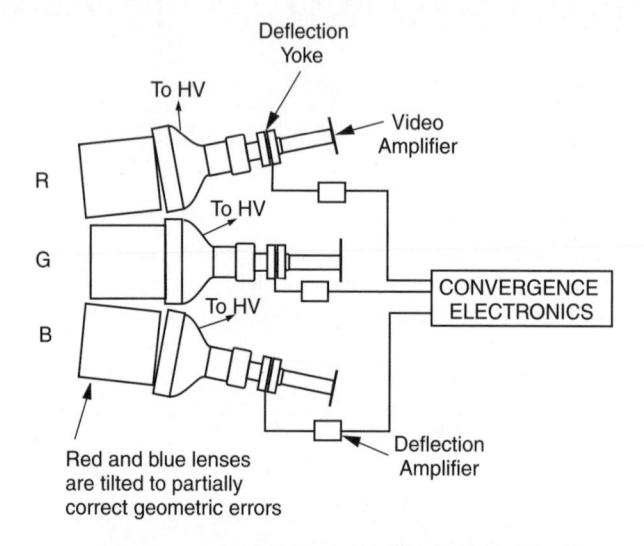

Figure 9.1 Schematic of three-lens CRT light engine.

Figure 9.2 Photograph of light engine from contemporary consumer rear screen CRT projector (courtesy Philips Consumer Electronics Corp.).

For three-lens systems, the components and subsystems of the light engine are given in Table 9.1.

Table 9.1 Components of a CRT light engine.

CRTs (R,G,B)	3
Video amplifiers	3
Lenses	3
Focus and deflection coils	3
Deflection amplifiers	3
Convergence electronics	1
HV power supply	1
Mechanical assembly	1

The optical coupling of the tube to the lens is typically designed with cooling liquid between the tube faceplate and the first optical surface of the lens. The cooling liquid may have added dyes, different for each tube color, to improve the spectral purity of the light from the tube. Additionally, the tube faceplate is frequently curved to improve the light collection by the lens and simplify the lens design. The speed of CRT projection lenses (higher speed = lower f/#) are often in the range of f/1.

Rear screen systems are mounted in a cabinet with an optical path that is folded to minimize cabinet depth, as illustrated in Figure 9.3. Alternatively, the screen can be flush mounted in a wall with the projection system behind the wall. The distance from the fixed focal length lenses to the screen is about 750 mm. As will be discussed below, the non-symmetric angles of illumination of the screen by the tubes and non-linearities in the deflections produced by the tube yokes result in the raster on the screen being distorted. The location of a picture element is non-linearly dependent on the pixel position. The deflection circuits must correct for these errors.

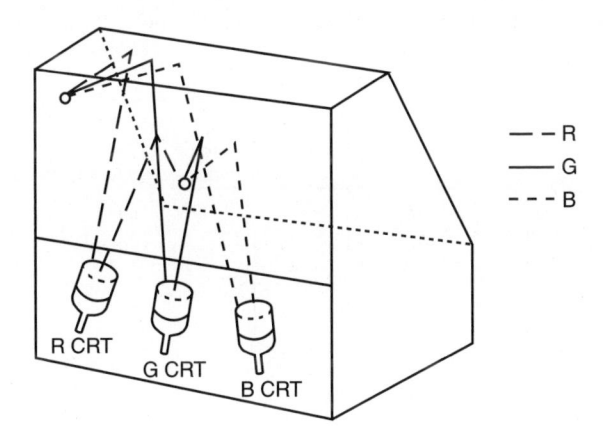

Figure 9.3 Configuration of optical paths in a rear-screen CRT projector.

For consumer television and other space sensitive applications, the optical path can be folded into a cabinet depth that is typically less than 600 mm with a 1370 mm (54 inch) diagonal, 4:3 aspect-ratio screen. A 54" diagonal rear projection system typically uses three 7-inch projection tubes, one on-axis and two off-axis, offset by about 10°. Rear projection systems for professional use and larger consumer systems may use 9-inch CRTs.

Front-screen applications frequently require longer focal length lenses. The trend in CRT projection lens design for rear screen application is to make shorter focal length lenses to achieve decreased optical throw distances, and hence, shallower cabinets. Shorter focal length lenses result in increased angles between the tubes. This requires increased error-correcting signals to the raster deflection to correct for the additional distortions this introduces. Some correction is derived from having the lens and tube axes at an angle to each other.

Additionally, higher angles of incidence at the corners of the Fresnel lens in the screen result in luminance non-uniformities. The acceptance of the incident light by the Fresnel lens is angle-of-incidence sensitive. Reflections of the incident light from high Fresnel facet angles result in reduced luminance of the projector in the corners relative to the image center. Fresnel lenses with reduced facet angles are being developed using higher index plastics. The optical losses in the corners of CRT projectors currently limit the depth of consumer rear projection systems. Rear projection screens are discussed in greater detail in Chapter 7.

Three-tube/three-lens rear projection systems are sometimes used in a matrix combination to form a videowall. Each three-tube/three-lens light engine constitutes a module. Combinations from 1×2 projector modules up to n × m modules are used to form very large images. The electronics and software controlling these systems permit each module to show its own image or to have one image displayed across all, as if they were a single display. Significant signal processing is required when an image covers more than one module to minimize luminance and chrominance step errors across the seams between modules. The modules are essentially the same as the professional rear projectors, usually without the fold mirror.

9.1.2 One-lens CRT Projectors

While the vast majority of CRT projection systems are based on three-lens architectures, there have been front-screen systems which employ only one lens (Malang 1989). These typically have been developed for professional front-screen application. A typical arrangement for these systems is illustrated in Figure 9.4. The outputs of the three tubes are combined using crossed dichroic filters. With the single-lens systems, the requisite liquid coupling is between the tube face and the dichroic recombiner. This liquid also acts as a coolant.

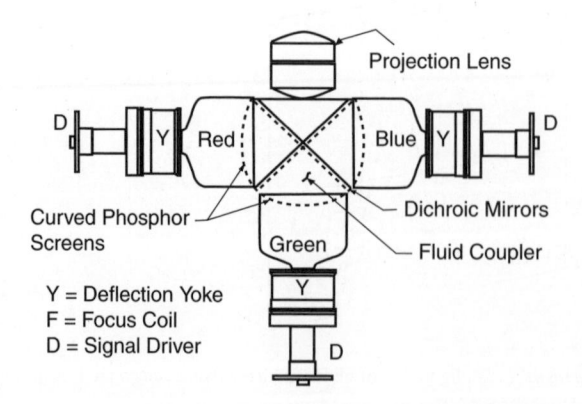

Figure 9.4 Typical architecture of 3-CRT/1-Lens projection system (from Malang 1989; with permission of SPIE).

The trapezoidal distortions, discussed below, resulting from off-axis illumination are not present in single-lens systems and convergence is much less problematic. Moreover, to a first approximation, convergence need not be readjusted when the screen/lens distance is changed. Mechanical convergence is factory set internally and usually need not be subsequently readjusted. Corrections for other raster errors, discussed below, are essentially the same in one-lens systems as in three-lens systems.

The interposition of the dichroic recombiner between the tubes and the lens makes the back focal distance larger and, as a result, the lens is more complex. It becomes difficult to accommodate this distance in the design and maintain the low f/# and the cost competitiveness of the lens. Therefore, these designs are only used where cost competitiveness is not an issue such as in simulation systems.

Additionally, a low f/# lens results in a large variation in the angles the light path makes with the dichroics. The higher angles result in a shift in the cutoff wavelengths of the dichroics, resulting in a shift in the color coordinates of the image with position. To minimize this, a higher f/# lenses may be required, resulting in reduced image luminance.

9.1.3 Convergence of CRT Projection Systems

It is desirable to converge the images from the three tubes to within about one-half pixel (corresponding to about 0.22 % of the picture width for NTSC, even smaller percentages for higher resolution systems) over the full image area. Consumer CRT systems rarely achieve this goal, especially in the corners and edges of the screen. It is common to find some region of the screen where the convergence is off by as much as two pixels. The three tubes in a three lens system are normally arranged in a single line, most often with the green tube on-axis and the red and blue tube on each side. If a rectangular image is generated on each of the CRT faces, the projected images from the off-axis tubes will appear trapezoidal, as illustrated in Figure 9.5.

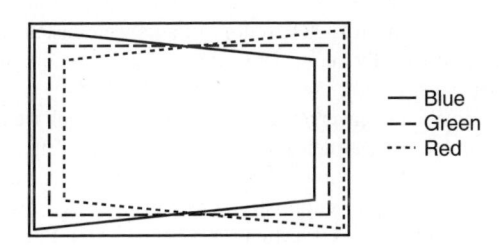

Figure 9.5 Keystone rasters generated by red and blue tubes.

Without considering the non-linearities in the electron-beam deflection system or the optics, the on-axis tube, typically the green CRT, will generate a rectangular image. Each of the off-axis tubes, however, will generate trapezoidal images due to the oblique projection in the horizontal direction. However, one of the trapezoids opens to the left and one to the right. If the light path is not normal to the screen in the vertical direction, a vertical trapezoidal distortion will also be present. Trapezoidal distortions can be partially corrected optically by having the center line of the off-axis tubes rotated with respect to the lens center line (Hockenbrock and Rowe 1982).

Single-lens CRT projectors do not suffer from convergence errors due to trapezoidal distortion. There are, however, many other sources of misconvergence common to all CRT projectors. These other errors, generated optically, electrically, and/or magnetically further distort the rasters. These distortions must be corrected to achieve convergence. Figure 9.6 illustrates some of these errors (George 1995). These include DC offset of the images (centers of images not coincident), image rotations, size differences, bowing of the raster, and pincushion distortion due to the flatness of the faceplate (which is accentuated with curved faceplates). Non-linearities and distortions in the deflection system and optics can add higher order errors.

222 SCANNED PROJECTION SYSTEMS

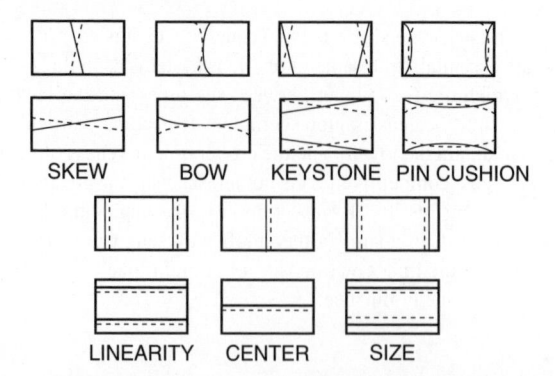

Figure 9.6 Errors in the projected image of rectangular rasters generated by a CRT projector. The solid and dashed lines indicate each of the distortions can be in either direction relative to the undistorted raster. (From George 1995 (© 1995, IEEE; reproduced with permission.)

Three-lens front-screen systems have light engines similar in architecture to those used in rear systems but the convergence issues can be quite different. Whereas the rear systems have fixed lens to screen distances, the front systems are sometimes used with variable distances to the screen. This leads to the requirement for reconvergence of the system each time the system is moved. User adjustments of the angles between the off-axis and on-axis tubes are provided for coarse convergence. The system generates a test pattern for electronic adjustment of the convergence coefficients. In some systems, this is done manually by a skilled operator adjusting about 30 or more trimming potentiometers. This tedious process of reconvergence has typically meant that CRT front projectors were installed in a fixed location and never moved. The process is frequently automated in more sophisticated systems using an auxiliary optical system and photodetectors.

Correction of the geometrical and convergence errors can be accomplished by applying correction waveforms to the horizontal and vertical deflection circuits for each of the tubes. Each tube has its own set of horizontal and vertical correction waveforms. The lower-order correction waveforms for convergence and geometry errors in projection CRT systems can be described by a truncated power series expansions in x and y for the x-axis corrections (Washburn 1995). Equations (9.1) and (9.2) give the x-axis deflection waveform $f_x(x,y)$ with up to third order corrections. For many projection system applications, it has been found that correcting to third order is sufficient to get satisfactory convergence.

$$f_x(x, y) = C_x + A_x x + \Delta X(x, y)$$
$$f_y(x, y) = C_y + A_y y + \Delta Y(x, y)$$
(9.1)

where the geometric distortion $\Delta X(x,y)$ is

$$\Delta X(x, y) = K_x y + L_x(x^2 - 1) + B_x(y^2 - 1) + T_x xy + R_x x(x^2 - 1) + S_x y(y^2 - 1)$$
$$+ P_x xy^2 + E_x x^2 y + \text{higher order terms}$$
(9.2)

The geometric distortion $\Delta Y(x,y)$ is obtained by interchanging x and y in Equation (9.2) and using coefficients appropriate for this direction.

The deflection and correction coefficients below define the function of each term:

x,y are the horizontal and vertical positions on the screen of the tube,
C is the DC offset correction for centering,
A is the linear displacement, corresponding to the deflection input,

K is the skew correction,
L is the differential non-linearity in x-direction,
B is the bowing in the y-direction,
T is the trapezoidal distortion,
R is the x-deflection nonlinearity, expanding at the outside, compressing toward the center,
S is the 'S'-shaped deflection nonlinearity, and
P, E are pincushion related terms.

The convergence electronics generates a signal for each of the variables in Equation (9.2), and fractions of these are selected, dependent on the magnitudes of the coefficients. These are summed and are added to the linear ramp signals and DC offsets to provide the total deflection signal to the deflection coils, given by Equation (9.1). Practical implementation of the convergence corrections has been done with analog, digital (Buttar *et al.* 1992; Ohsawa *et al.* 1993) or hybrid circuits. Most three-tube projectors now use digital convergence methods (Watanabe *et al.* 2000).

One design for a digital convergence system that depends on photosensors around the periphery of the screen is shown in Figure 9.7. This system works because a CRT rear projector typically overscans the image so it more than fills the projection screen. While some of the image is lost to the overscan, this process gives the image sharp, square edges. Note that the eight sensors shown in Figure 9.7 are not sufficient to correct all the types of geometric errors shown in Figure 9.6. These geometric errors tend to be stable, however, and do not need readjusting in the field if they are corrected properly in the factory. Convergence of the three images can drift in the field. One source of convergence drift is changes in the Earth's magnetic field due to the differences between the manufacturing location and the location where the set is used. Eight sensors are sufficient to correct this type of convergence error and other types commonly encountered in the field. While Figure 9.7 shows only a single CRT for clarity, the concept of convergence is, of course, meaningless except in the presence of images generated by multiple tubes.

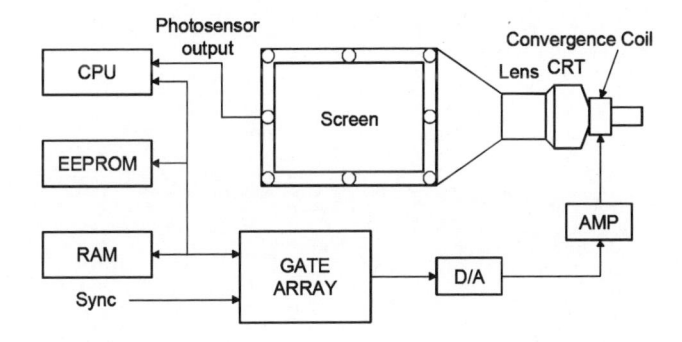

Figure 9.7 Digital Convergence system using sensors outside the image area (reproduced by permission of © 2000 the Society for Information Display (Watanabe 2000)).

9.1.4 Lumen Output of CRT Projectors

The illuminance of the screen produced by the three CRTs is given by Kingslake (1983):

$$E_V = \sum_i \frac{L_{V_i}(\lambda) T(\lambda)_i}{4(f/\#)^2 (1 + m^2)} \qquad (9.3)$$

224 SCANNED PROJECTION SYSTEMS

where
- E_V is the screen illuminance,
- i is red, green, or blue,
- $L_{V_i}(\lambda)$ is the luminance of the R, G, or B tube,
- $T(\lambda)_i$ are the corresponding spectral transmissions in each of the color bands,
- f/# is the f-number of the projection lens(es), and
- m is the magnification.

The screen illuminance is the sum of the illuminance produced by the three tubes. The total luminous flux, Φ_V, is the product of E_V and the illuminated area. If CRTs with flat faceplates are employed, the radiation from the CRT is nearly Lambertian with the centroids of the distributions from various parts of the tube face being all perpendicular to the face. If curved faceplate tubes are used, as is common in consumer projectors, the radiation pattern is still Lambertian but the centroids of the radiation from different parts of the tube face are directed toward the tube axis. This leads to reduced vignetting of the off-axis radiation (Figure 3.7).

The luminous flux that is delivered by a CRT projector is dependent on the maximum power and power density that can be handled by the CRT faceplate. This in turn is dependent on the tube construction, the raster size, the efficacy of the faceplate cooling, and the high voltage power supply. Other factors affecting the flux are the f/# of the projection lens and the spectral filtering implemented. This latter subject is discussed below.

The projection lenses typically are very fast, having f/#s typically between 0.9 and 1.1. These are large diameter lenses, to match the tube diagonal of 7–9″. The lenses for standard consumer CRT projectors typically use all plastic elements, but professional and some high-end consumer lenses use a combination of glass and plastic elements (Ohsawa *et al.* 1993). One reason even high-end lenses use plastic elements is because aspheric surfaces can be made relatively inexpensively using molding techniques.

A consumer projector with 7-inch tubes will produce about 150 lm when all three tubes are driven with 100 % signals over the full raster area. This luminance is adequate for screen diagonals up to about 60 inches, with screen gains up to about six. Systems with 9-inch tubes have significantly higher outputs and are suitable for screen sizes of about 80 inches. For even larger size screens, six or even more tubes are used, either in a single projector or by stacking two, or more, three-tube projectors, to increase the illuminance of the screen. Convergence in a system with six or more tubes is a not insignificant task.

The screen luminance produced is related to the screen gain by

$$L_V(screen) = G E_V \qquad (9.4)$$

where
- L_V is the output luminance,
- E_V is given by Equation (9.3), and
- G is the screen gain.

Screen gain, discussed in Chapter 7, is the ratio of the maximum system luminance with the screen relative to that obtained with a Lambertian screen, i.e. a small area on the screen appears equally bright in all directions. The gain includes the transmission or reflection losses of the rear or front screen, respectively. With a screen gain of 6, the full area luminance of a 52-inch consumer rear projector would be about 340 cd/m², when viewed normal to the screen.

The above assumes uniform illumination of the screen. In practice, there is significant fall-off in the illuminance of the screen, particularly when high-speed lenses are used. The illumination of the screen decreases as $\cos^4 \theta$, where θ is the half-angle at the center of the exit pupil that the screen subtends (Smith 1990).

CRT projectors are frequently designed so that the peak luminance in a local area can increase significantly above the normal maximum when the average scene luminance is low. Figure 9.8 illustrates how the relative luminance could vary as a function of the emitting area. The illustration shows a four times increase in local area luminance when the emitting area is 10 % of the total area. Some CRT projectors will have an increase in local luminance as much as eight times. This characteristic, sometimes called 'punch', gives CRT projectors wider dynamic range and higher contrast under low average luminance conditions.

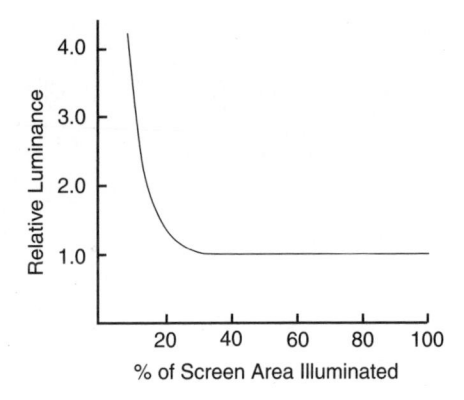

Figure 9.8 Luminance of a projection CRT as a function of emitting area relative to a 100 % white field.

Video has a typical average video voltage of 29 % (Doyle 1990) which produces an average brightness in a $\gamma = 2.2$ system of 9 %. Since the normal limitation on CRTs is average power, not peak power, this allows video highlights to be shown with much higher currents and therefore much higher luminance. Professional video signals often have much higher average video signals. For example, reverse video (black characters on a white background) has a very high average video signal. Therefore, professional CRT projectors do not benefit as much from this peak luminance ability as do consumer CRT projectors.

9.2 Scanned Laser Projectors

There are several fundamental optical architectures for laser projectors:

1. Systems that scan the laser beam on two axes and modulate the laser light at the video rate;
2. Systems that use a linear array to modulate the laser light and use a single mirror to scan this line image in the other axis to generate the full image;
3. Systems that use lasers to illuminate a conventional microdisplay.

The first two types of systems with two-axis or single-axis scanning normally can only use laser illumination and cannot use lamps or LEDs as light sources. These two types of scanned systems will be covered in this section. The third type of laser projector, where the laser is used as a light source for a more or less conventional microdisplay projection system will be discussed in Chapter 10 along with lamp and LED-based microdisplay projectors.

226 SCANNED PROJECTION SYSTEMS

In addition to these three laser architectures, it is possible, at least in theory, to produce a pixelated or non-pixelated, area-emitting laser and then project the image produced by the laser. This can be done with either a continuous lasing medium where the pixels are addressed by an electron beam or by a two-dimensional array of vertical cavity, surface emitting lasers (VCSELs), or by some other laser design that produces a one- or two-dimensional array. While there have been research efforts along these lines (Zory 1993; Chua *et al.* 1998; or Bondarev *et al.* 2005), none of the designs appears to be near production and they will not be discussed further.

9.2.1 Raster Scan Patterns

Generally, CRT projectors and other projectors with raster scanned electron beams use linear ramps on both the x and y axes. These linear ramps are corrected for geometric errors in the system but normally the corrections are small compared to the main linear ramps.

Scanned laser projectors, on the other hand, sometimes use sinusoidal drive on one or both axes. This sinusoidal scans are normally associated with the resonant frequency of some scanning component such as a single- or bi-axial MEMS mirrors discussed in Section 5.3. Figure 9.9(a) shows a conventional raster scan, with a linear ramp on both axes. In this type of scan, there is a 1:1 correspondence between time

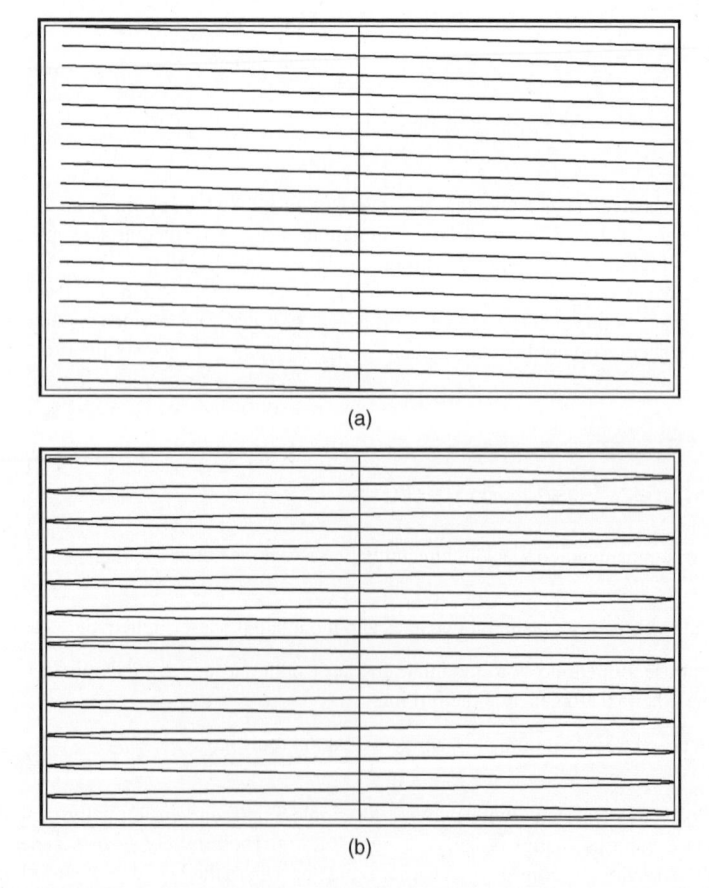

Figure 9.9 Patterns formed in scanned projectors: (a) linear ramp on both axes; (b) linear ramp vertically and sinusoidal scan horizontally; (c) sinusoidal scan on both axes.

(c)

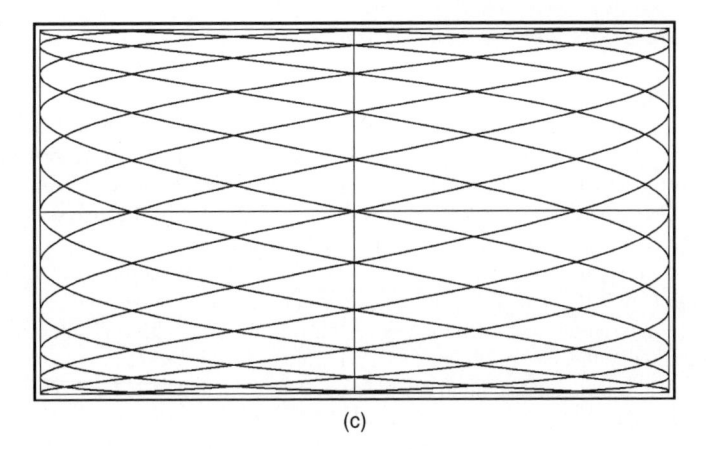

Figure 9.9 (Continued).

and the pixel to be displayed. Also, all pixels in this system would be shown with the same brightness. Therefore, this is the simplest system in terms of the electronics to drive the modulator, whether the modulator is an electron gun in a CRT, a laser that can be modulated directly or an acousto-optic modulator for lasers or other light sources that cannot be modulated at the video rate. Note that this raster has a noticeable skew to it. This is an issue only in the figure with 19 raster lines. In a real display with perhaps 1080 lines, it is not an issue.

One problem with the raster in Figure 9.9(a) is the horizontal retrace must be very fast. For a nearly-massless electron beam in a CRT, this is not much of an issue. The issue in a CRT is the inductance of the yoke scanning the beam and most CRT projector designs use this inductance to generate the high voltage for the CRT anode. In a MEMS moving mirror system, it is not practical to have this nearly instantaneous retrace. Figure 9.9(b) shows a raster where the horizontal scan is done by a resonant mirror moving in a sinusoidal pattern. While this simplifies issues for the mirror, it complicates issues for the electronics, since there is no longer a 1:1 correspondence between time and pixel location. This occurs because the scanned beam moves fastest near the center of the image and slows down near the edges, coming momentarily to a complete stop at the very edge. Therefore, the clock rate of pixels shifted into the modulator must be variable from the center to the edge. The brightness of the pixel can also vary since near the edges where the beam dwells longer, the inherent brightness is high. When this type of scan is used, often only the central region of it is used to minimize these problems. This causes problems of its own. For example, if only the central half of the scan is used, the laser can only run at a 50 % duty cycle, significantly reducing the lumen output of the projector unless a higher power and more expensive laser is used. In addition, the clock rate for the video data must be increased by 2 x in order to include all the pixels in the allocated time. In some designs, only the forward (e.g. left-to-right) scan is used and not the retrace scan (e.g. right-to-left). This further reduces the duty cycle, further increasing the clock rate and the power required from the lasers when they are on.

The system shown in Figure 9.9(b) still had a linear ramp on the vertical axis. In some biaxial scanning mirror systems, the vertical axis is also sinusoidal, creating a scan like the one shown in Figure 9.9(c) (James et al. 2007). This scan is commonly called a Lissajou scan, or just a Lissajou figure. This scan also has pixel placement and pixel brightness problems like the scan shown in Figure 9.9(b), but they are even more severe. In addition, it has a more subtle problem of variable refresh rate (Brown et al. 2002). If the horizontal and vertical frequencies are chosen to ensure a 25 Hz refresh rate for every pixel, there are regions of the image that are refreshed at 25 Hz but other regions of the image are refreshed at higher speeds, up to 60 Hz. This can partially compensate for the luminance uniformity problem, but it requires some pixels to be written to the laser beam several times in each raster cycle. One of the effects of this is to increase the required clock rate.

9.2.2 Laser Projectors with Two-axis Scanning

The light from the CRT projection systems described so far is incoherent and uncollimated. An alternative approach uses collimated laser output. Several approaches to laser projection are being developed, including mechanically scanned systems (Fink *et al.* 1996) that can use either macroscopic mirror systems or the MEMS mirrors discussed in Section 5.3, acousto-optical scanning (Yariv and Yeh 1984; Gurevich 2003), and laser CRTs (Nasibov *et al.* 1992). Other phenomena, such as diffractive scanning (Bos *et al.* 1995) and computed holograms (Banks *et al.* 2006) can be applied.

One design for a mechanically scanned two axis systems employs two orthogonal scanning mirrors, as illustrated in Figure 9.10. One of the mirrors deflects at the frame rate and the other deflects at the line rate to provide a raster. It is common for the field-rate mirror to be to be a galvanometer mirror. This mirror structure allows either a resonant or a linear scan, although a linear scan would be more common. The scan of one field is from top to bottom and the next scans from bottom to top. This type of scanning can produce flicker due to half field-rate frequency components near the top and bottom margins of the display. Scanning in the horizontal direction is accomplished with a multi-faceted polygonal mirror rotating at very high speeds. These high-speed rotating drum mirrors have been used for many years and were originally developed for the Scophony projector in the 1930s (Okolicsanyi 1937; Brennesholtz 2007). This faceted mirror drum produces the instant retrace needed to provide a raster scan such as the one shown in Figure 9.9(a). Note Figure 9.10 includes a projection lens. Depending on the range of deflection angles produced by the two scanning mirrors, this may not be needed. If the scanning system produces very small deflection angles, this lens needs to be a non-imaging lens that increases these angles. The galvanometer mirror in Figure 9.10 can be replaced by a second, low speed drum for the vertical scanning. The Jenoptik laser projector (Hollemann *et al.* 2001) uses the scan system shown in Figure 9.10 with a galvanometer mirror for vertical scan and a drum for horizontal scan.

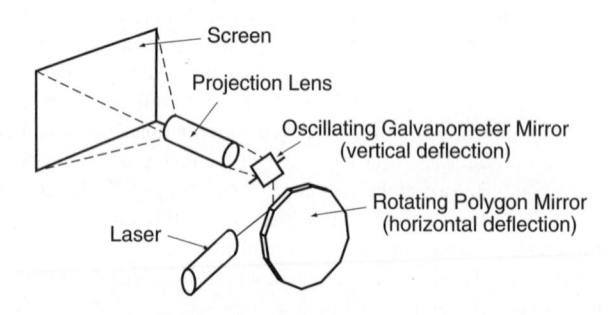

Figure 9.10 Mechanically scanned laser projector.

Due to the complexity, cost and size of the mechanical scanner shown in Figure 9.10, laser projector designs using the MEMS scanners discussed in Section 5.3 have been proposed. Systems with a bi-axial MEMS scanner can be extremely compact, and it has been proposed they can be integrated into cell phones. These bi-axial scanners have already been used in retinal scanning displays, which are miniature projectors that use the retina of the eye as the projection screen (Sprague 2005). The same scanner, when used with higher power lasers in the 50 mW to 200 mW range, can be used as projectors as well.

These MEMS devices typically produce a scan pattern such as the one shown in Figure 9.9(b) (Sprague *et al.* 2005) or Figure 9.9(c) (James *et al.* 2007). Due to the complexity of the electronics needed for a Lissajou scan, it is unlikely they can be fully integrated into a cell phone. The system described by James *et al.* (2007) took advantage of the very small size of the scanning head (8 mm diameter) for a medical application but the scanning head was tethered by fiber-optic and electronic cables to an external piece of electronic equipment.

One problem with a bi-axial scanning MEMS system is it produces a 'bow-tie' image distortion similar to the pincushion distortion shown in Figure 9.6. While this is fundamental in the optics of the single scanning mirror regardless of the scan type, with a Lissajou scan that needs major geometry correction anyway it is not so much of a problem. With a ramp on each axis, as shown in Figure 9.9(a) the ramp drive can be adjusted to correct for the problem. The only scan type where this effect causes noticeable problems is with the ramp/sinusoidal scan as shown in Figure 9.9(b). For a description of the geometric errors associated with various polar coordinate systems see Fiske (1998).

An alternative to a single scanning MEMS mirror is to use two different mirrors, one for horizontal and the other for vertical deflection (Stern *et al.* 2006). For very small projectors, typically the horizontal high speed scan would be a single axis MEMS device. The lower speed vertical scan mirror can be either a MEMS mirror or a conventional galvanometer mirror. This system with two different mirrors does not produce the bow-tie distortion of a single mirror.

9.2.3 Laser Projectors with a Single Scanning Axis

These projectors use a linear array such as the ones described in Section 5.2 to generate a line image. This image is then scanned by a moving mirror along the second axis to produce a full 2 dimensional image. A typical optical system to use one of these linear arrays is shown in Figure 9.11. The scanning mirror is on the 'slow axis' would be driven by a ramp signal at the video field rate. At these rates, a moving galvanometer mirror can be used to generate a ramp. Alternatively a rotating faceted drum can be used.

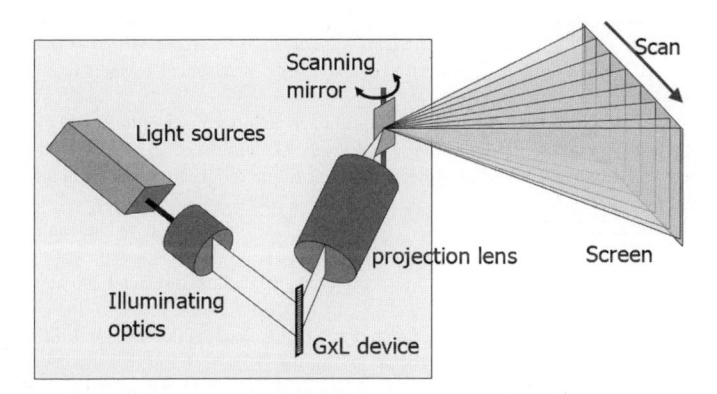

Figure 9.11 Optical system used in a Sony GxL projector for the Aichi World Expo (reproduced by permission of © 2006 the Society for Information Display (Eguchi 2006)).

Figure 9.11 shows essentially a single color channel of a projector. These systems can produce full color in a number of ways:

1. Color sequential operation;
2. Three channels identical to the one shown in Figure 9.11 with the images converged at the projection screen;
3. Three channels from the light source to the linear array and color combining dichroic filters between the linear array and the projection lens;
4. Three channels from the light source to the projection lens and color combining dichroic filters after the projection lens (Brazas 2004).

The color sequential option is not currently used for two reasons. First, laser projectors with linear arrays are very expensive and are only used in applications where image quality, not cost, is the paramount feature. Secondly, even if identical linear arrays are used for all three colors, the angles corresponding to the diffraction orders depend on wavelength. Therefore, the Schlieren optics needed for each color are different.

Note that the projection lens in Figure 9.11 is a true imaging lens, not a non-imaging lens as can be used in Figure 9.10. The lens must image the linear array onto the projection screen.

This linear array/scanning mirror arrangement can be used to produce some of the highest resolution images available today. For example, the Evans & Sutherland projector can produce a 4 K × 5 K pixel image. To the authors' knowledge, this is the highest resolution image available without tiling.

References

Banks, L., Birch, M., Krueerke, D., Buckley, E., Cable, A. et al. (2006) Real-time diffractive video projector employing ferroelectric LCOS SLM. *SID Intl. Symp. Digest of Technical Papers*, 2018–21.

Bondarev, V.Y., Kozlovsky, V.I., Sannikov, D.A., Kuznetsov, P.I., Jitov, V.A. et al. (2005) Electron-beam pumped green VCSEL on MOVPE-grown ZnCdSe/ZnSSe MQW structure. *13th Int. Symp. 'Nanostructures: Physics and Technology,'* St. Petersburg, Russia, 20–25 June 2005.

Bos, P.J., Chen, J., Doane, J.W., Smith, B., Holton, C. et al. (1995) An optically active diffractive device for a high-efficiency light valve. *SID Intl. Symp. Digest of Technical Papers*, 601–4.

Brazas, J.C. and Kowarz, M.W. (2004) High-resolution laser-projection display system using a grating electromechanical system (GEMS). *MOEMS Display and Imaging Systems II, SPIE* Proc. **5348**: 65–75.

Brennesholtz, M.S. (2007) The evolution of projection displays: from mechanical scanners to microdisplays. *J. SID* **15**(10): 759–74.

Brown, M., Freeman, M., Zobkiw, C. and Lewis, J. (2002) Image quality considerations in bi-sinusoidally scanned retinal scanning display systems. *SID Intl. Symp. Digest of Technical Papers*, 1320–3

Buttar, A., Jobling, D. and Rusznyak, A. (1992) A high performance digital convergence and focus system for projection TV. *IEEE Trans. Consumer Elect.* **38:** 734–40.

Chua, C.L., Thornton, R.L., Treat, D.W. and Donaldson, R.M. (1998) Independently addressable VCSEL arrays on 3mm pitch. *IEEE Photonics Technology Letters* **10**(7), July: 917–19.

Doyle, T. and Lammers, M.J.G. (1990) Statistics of television images. Philips Research Laboratories Technical Note 175/90.

Eguchi, N. (2006) GxL Laser Dream Theater at the Aichi Expo (equivalent to 2005-in TV). *Proc. 13th International Display Workshops*, 9–12.

Fink, L.G., Bergstedt, R., Flint, G., Hargis, D. and Peppler, P. (1996) Microlaser-based projection display for simulation. *Interservice/Industry Training Systems and Education Conference*, Orlando, FL.

Fiske, T.G., Silverstein, L.D. and Kelley, E. F. (1998) Viewing angle: a matter of perspective. *SID Intl. Symp. Digest of Technical Papers*, 937–9.

George, J.G. (1995) Minimum adjustment analog convergence system for curved faceplate projection tubes. *IEEE Trans. Consumer Elect.* **41**: 536–9.

Gurevich, B.S. (2003) Laser projection displays based on acoustooptic devices. *J. Opt. Technology* **70**: 500–3.

Hockenbrock, R. and Rowe, W. (1982) Self-convergent, 3-CRT projection TV system. *SID Intl. Symp. Digest of Technical Papers*, 108–9.

Hollemann, G., Braun, B., Heist, P., Symanowski, J., Krause, U. et al. (2001) High-power laser projection displays. In M.H. Wu (ed.), *Proceedings of SPIE: Projection Displays VII* **4294**: 36–46.

James, R., Gibson, G., Metting, F., Davis, W. and Drabe, C. (2007) Update on MEMS-based scanned beam imager. In D.L. Dickensheets, B.P. Gogoi, H. Schenk (eds), *Proc. of SPIE: MOEMS and Miniaturized Systems VI* **6466**.

Kingslake, R. (1983) *Optical System Design*. New York Academic Press, p. 131.

Malang, A.W. (1989) High brightness projection video display with concave phosphor surface. *Proc. SPIE*, **1081**: 101–6.

Nasibov, A.S., Kozlovsky, V.I., Reznikov, P.V., Skasyrsky. K.K. and Popov, Y.M. (1992) Full color TV projector based on A_2B_6 electron-beam pumped semiconductor lasers. *J. Crystal Growth* **117**: 1040–5.

Ohsawa, M., Onozawa, M., Hirata, K., Watanabe, T., Kimura, Y. *et al.* (1993) A 46-inch high-resolution rear-projection display. *J. of the SID* **1**: 23–34.

Okolicsanyi, F. (1937) The wave-slot, an optical television system, *Wireless Engineering* **14**: 526.

Smith, W.J. (1990) *Modern Optical Engineering: The Design of Optical Systems*, 2nd edn, New York: McGraw Hill, p. 145.

Sprague, R., Montague, T. and Brown, D. (2005) Bi-axial magnetic drive for scanned beam display mirrors. In H. Ürey and D.L. Dickensheets (eds), *Proc. of SPIE: MOEMS Display and Imaging Systems III* **5721**.

Stern, M., Yavid, D., Tan, C., Wood, F., Wittenberg, C. *et al.* (2006) Ultra-miniature projector: a high resolution, battery powered laser display. *SID Intl. Symp. Digest of Technical Papers*, Paper 73.3, 2015–17.

Watanabe, T., Kabuto, N., Hirata, K. and Aoki, K. (2000) Recent progress in CRT projection display. *SID Intl. Symp. Digest of Technical Papers*, 306–9.

Washburn, C.A. (1995) A magnetic deflection up-date: field equations, CRT geometry, the distortions and their corrections. *IEEE Trans. Consumer Elect.* **41**: 963–78.

Yariv, A. and Yeh, P. (1984) *Optical Waves in Crystals*. New York: John Wiley & Sons, Inc.

Zory, P.S. (ed.) *Quantum Well Lasers*. New York: Academic Press.

10
Microdisplay System Architectures

Light-valve or microdisplay projection systems use one or more microdisplays, light valves or light amplifiers to modulate the light generated by an external lamp. The microdisplay and light-valve system architectures described in this chapter can be built with either reflective or transmissive devices and are based on color-additive designs. Three-panel systems have also been designed based on color subtractive principles.

In CRT projectors, the light is generated as needed by the electron beam: when the video signal calls for black, no light is generated. In microdisplay projectors, the light for the maximum achievable white is normally generated continuously by the lamp or lamps. The microdisplay then modulates this light to produce the image with the desired spatial characteristics and gray scale. The 'blackest-black' is determined by how well the microdisplay can attenuate this maximum light. Color can be produced by using multiple microdisplays, with each providing the modulation for one or sometimes two colors. Alternatively, in some architectures with some types of microdisplays, full color can be produced from a single panel. There are advantages and limitations associated with each approach, as discussed below.

10.1 Microdisplay Systems

The dominant optics approach with microdisplay systems is the use of a single projection lens, but the three-lens approach has sometimes been used. Most microdisplay projection systems are color additive systems, like CRT projectors. In these systems, the light from the lamp is first separated into red, green and blue components. Each color is modulated separately and the images are recombined for the viewer.

In three-panel configurations, the light from the lamp is separated into three beams of red, green and blue light. These beams are then modulated by separate microdisplays and recombined to form the full color image. Convergence is one of the major issues with three-panel systems. The common three-panel architectures will be discussed in the next section.

Single-panel systems use only one microdisplay to modulate all three colors. There are a variety of approaches used for single-panel architectures. Among these are color-field sequential systems, color micro-filter designs, and angular color separation (ACS) systems. In color field-sequential systems, the pixels in the microdisplay are sequentially driven with information for each color field and illuminated with that color. This approach does not require more sub-pixels in the panel but does require higher field and data rates. Both color micro-filter and microlens panels require three times the number of subpixels but operate at the same frame rates as three-panel systems. These approaches will be discussed in more detail in Section 10.4.1.

Two-panel systems use one microdisplay to modulate one color and the other microdisplay to modulate the other two colors. This is normally only useful under certain special conditions:

1. The panel can modulate two colors satisfactorily, but not three, due to system or device limitations;
2. The projection lamp is deficient in one color, requiring a longer period of time than for the other two colors to get enough light of the deficient color to achieve white balance; or
3. The color sequential or other artifacts in a singe panel system are unacceptable for the application but the reduced artifacts in a two-panel system are acceptable.

The architectures of two-panel systems are based on similar projector architecture principles as those of three-panel systems. These are discussed in Section 10.4. Two examples of two-panel systems that have seen commercial production are the Talaria MLV projector from General Electric and the two DMD projector from Texas Instruments (Florence and Yoder 1996). The Talaria MLV projector (True 1987) is an example of a panel technology capable of modulating two colors easily. While a single light Talaria light valve was capable of modulating three colors, this mode of operation produced serious image artifacts. The two DMD projector used one DMD to modulate blue and green and the other DMD to modulate red. This was done because of the shortage of red in the chosen projection lamp.

Although major improvements in light utilization are being made, the system efficiency, lumens output divided by lamp lumens, of the best of the microdisplay projectors is less than about 10 %. Three-panel systems are normally more efficient than the single-panel systems. Most architectures for microdisplay projection can be built in either transmissive or reflective versions, depending on the technology of the LV. In general, reflective systems are more efficient and have higher throughput than transmissive systems. The throughput of microdisplay projectors will be discussed in detail in Chapters 11 and 12.

10.2 Three-panel Systems with Transmissive Microdisplays

The majority of microdisplay projectors commercially available have a three-panel, single-lens, one-lamp architecture based on color additive principals. Each of the microdisplays in these systems is dedicated to one color, red green, or blue. This allows for an optimum usage of the available light and gives the designer the widest possible choice of lamps. While the discussion below is mostly related to microdisplays employing TN-LC effects, much of the information is applicable to other light modulators as well. Differences, where they exist, will be discussed later in the chapter.

There are two main disadvantages of three-panel architectures: convergence and cost. The system designer tries to achieve convergence to better than one-half pixel over the full image to maintain the resolution capabilities of one of the panels. Convergence in these systems is achieved by six-axis adjustment $(x, y, z, \theta, \phi, \psi)$ of each of the panels. The optical path lengths between the panels and the lens must be well matched. As the resolution of a system increases, it becomes more difficult to achieve good convergence. Misconvergence can easily eliminate any resolution advantage of a system, especially near the edge of the screen where lack of convergence is normally more pronounced. The effect of misconvergence on the resolution performance of the system is discussed in Section 13.2.2.4.

The basic cost disadvantage of three-panel systems results from the requirement for three panels. The cost of the panels is a significant part of the total cost of the system. In addition, there are significant costs associated with the color recombination optics, convergence, more expensive projection lenses due to long backfocal requirements, and, for the equal path architecture discussed next, correction for astigmatism introduced by the tilted dichroics. For many applications, cost is an overriding factor. As a result, a variety of one-panel systems, some highly innovative, have been designed.

Two generic versions of three-panel architectures exist. These are the equal path system and the recombination cube system. Both architectures have been used in commercial projectors although the unequal path, recombination cube design currently dominates commercial production.

10.2.1 Three-panel Equal Path

Figure 10.1 shows the equal path configuration (McKechnie et al. 1989). Field and relay lenses are not shown in this figure for simplicity. The light from the illumination system is separated into three color bands using dichroic filters. Each of the color bands uses an identical microdisplay which modulates that color band. The modulated color bands are recombined using dichroic filters in a process which is the inverse of that in the illumination side of the system. The recombination must accurately overlay at the projection screen the illumination of each color subpixel onto that from the corresponding subpixels of the other colors. Sources of error in this process will be discussed below.

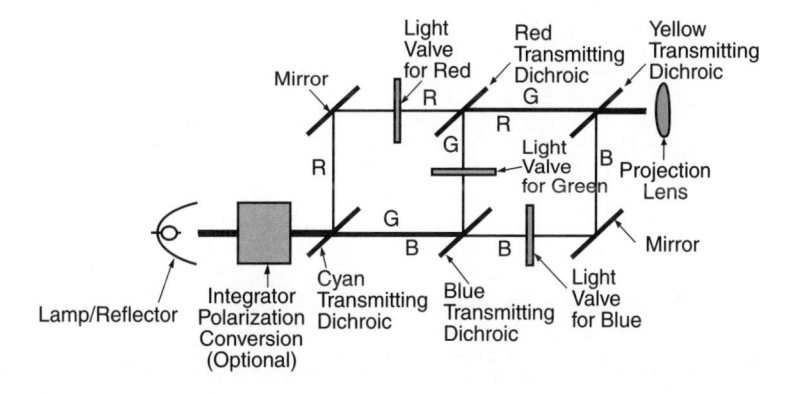

Figure 10.1 Architecture for equal path three-panel projectors with transmissive panels.

This architecture has a desirable symmetrical optical layout and uses, for the most part, less expensive components. The optical path lengths for all three color bands from the light collection optics to the panels are equal and the light paths from the panels to the projection lens also are equal for all three primary colors. Optics to correct for differences between the three paths are not required. The dichroic filters are very efficient, designed to transmit or reflect the limited wavelength bands of light with sharp cut-off and cut-on characteristics and negligible absorption.

The filter in the illumination path that reflects red transmits cyan for the following two channels and, in the recombination path, the filter that transmits yellow reflects blue. At least two of the microdisplays are mounted in six-axis adjustable mounts. Relatively simple, inexpensive mount structures have been developed for implementing these adjustments, which are done once in the factory. Obviously, the design of the mount must be shock resistant to maintain convergence.

236　MICRODISPLAY SYSTEM ARCHITECTURES

There are several disadvantages to this architecture. While the other components in these systems tend to be less expensive, the projection lens is more complex and therefore costly. This is due to the relatively large distance from the panel to the lens, requiring a projection lens with a large backfocal length. Lenses with this type of design tend to be more complex and expensive.

The dichroic filters used for recombination of the sub-images in the equal path system introduce aberrations and distortion in the image. Due to the large (45°) angle of incidence and finite thickness of the filters, coma is introduced into the transmitted beam. Since the optical paths for the three colors are not identical with respect to the number of filter plates traversed in the recombination path, these coma errors are not the same for all wavelengths. This gives rise to convergence errors.

Additionally, other distortions can be introduced into the reflected image if the filter is not perfectly flat. A curved filter plate acts as a lens for the image component reflected from its surface but not for the transmitted component, resulting in magnification variations. Filters are constructed with multiple thin films on transparent glass or quartz substrates. The thermal expansion coefficients of these films are not the same as that of the substrate, leading to a stressed structure. Curvature resulting from such stress is minimized with thicker substrates.

Figure 10.2 shows the calculated minimum pixel size to permit convergence to better than one half pixel due to the plate filter aberrations. The calculation assumes perfect mechanical convergence and allows the optical effects to produce a maximum of one-half pixel misconvergence. The effects are significant because thicker filter substrates are required to maintain filter flatness but thinner filters are required to reduce coma. A compensation method based on an additional plate in the optical path has been developed to partially correct these aberrations (Shimizu and Janssen 1996).

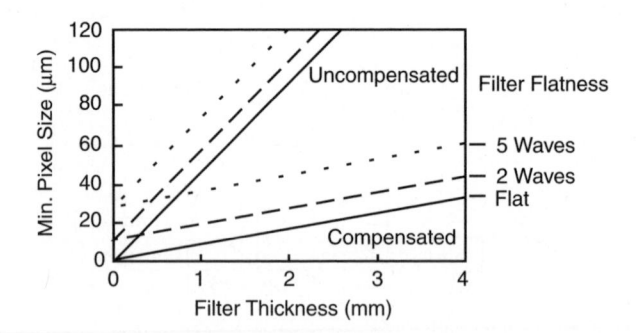

Figure 10.2　Minimum pixel size for which convergence can be maintained to better than half a pixel as a result of plate filter aberrations.

The image of the panel in the red light path in Figure 10.1 goes directly to the projection lens without being reflected in a mirror. The image in the green path undergoes one reflection and the blue image undergoes two reflections. Optical paths with an odd number of reflections are reversed, left-to-right, as compared with those having an even number of reflections. This is corrected electronically by reversing the scan direction for sampling the column information. The column addressing circuitry, whether integrated or in separate integrated circuits, typically has provision for reversing the scan direction from left-to-right to right-to-left.

The equal path architecture is susceptible to mechanical shock-induced misconvergence. The optical path lengths between the panels and the projection lens are long, which tend to magnify small displacements of the panels. Solutions for these problems often require more elaborate and expensive panel-mounting hardware to maintain convergence during shipment and subsequent handling of the projector.

10.2.2 Unequal Path Systems

The other architecture for three-panel, one-lamp, one-lens systems is illustrated in Figure 10.3. In this approach, crossed dichroic filters are used to recombine the light passing through the three microdisplays that are located at three faces of a cube (Shikama *et al.* 1993). The entrance pupil of the projection lens is on the fourth side of the cube. The crossed dichroic elements are most frequently in the form of an optical glass prism in which the dichroic filters are deposited on internal diagonal facets. The prism appears square when seen from above, and each side face has about the same aspect ratio as the microdisplay. This embedded dichroic design avoids all the problems of plate dichroics, as was discussed in Section 6.3.2.3.

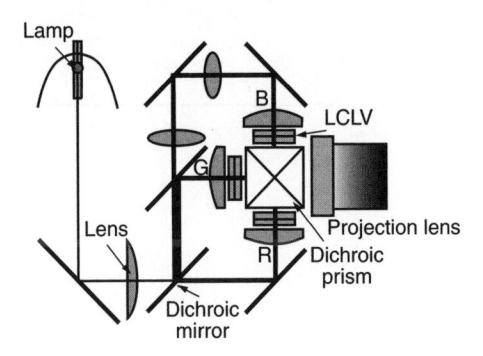

Figure 10.3 Architectures for recombination cube three-panel, one-lamp, one-lens projectors with transmissive panels.

Another advantage of this design is the shorter optical path between the panels and the lens, allowing for simpler projection lens designs and more compact systems, especially with rear projection systems. Additionally, it is common to mount the panels permanently on the cube directly. This provides a more rugged construction since all three panels are mounted to the same glass prism. With properly designed microdisplay mounts, convergence would not be likely to change unless there was a shock large enough to damage the combining prism itself. The microdisplay mounts must be adjustable during assembly to allow factory convergence before the positions of the panels are permanently fixed.

Previously, these combining dichroic prisms were expensive because of the glass quality required and the precision alignment. Manufacturing technology has advanced to the point where these prisms can be manufactured at relatively low cost and in high volume.

The length of at least one of the illumination paths differs from the others in this architecture. This gives rise to the name 'unequal path'. Additional relay and field lenses, as shown in Figure 10.3, are employed to compensate for this difference to produce the appropriate fill factor for the illuminating light. However, if the illumination is non-uniform, the three illumination paths may have different center to edge gradients, which can lead to color gradients in the final image.

A variation on the unequal path architecture is to use three transmissive panels and combine them with the same crossed dichroic prism. Instead of the lamp, splitting dichroics and relay lenses as shown in Figure 10.3, three separate light sources are used, each with their own rod integrator and sometimes polarization conversion system. (Willemsen *et al.* 2005; Teijdo *et al.* 2006) The light sources most commonly used in this design are red, green and blue LEDs. Of course, if lasers were used there would be no need of a polarization conversion system. A system of this type described by Teijdo *et al.* (2006) that includes tapered lightpipe integrators and polarization conversion is shown in Figure 10.4.

It is normal for the polarization of the green light in either Figure 10.3 or 10.4 to be rotated by 90° with respect to the polarization of the red or blue light. This allows the transmission band for the green light in the recombination cube to overlap with the reflection bands of the red and blue light. With a UHP-type lamp, it is especially important to allow the blue and green bands to overlap. This causes the output of the

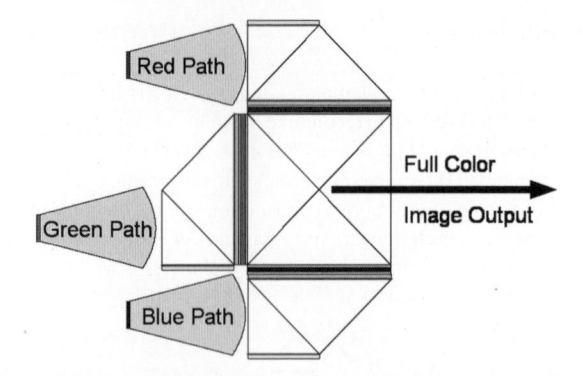

Figure 10.4 Recombination cube projector design with three illumination sources (reproduced by permission of © 2006 the Society for Information Display (Teijdo *et al.* 2006)).

projector to be polarized in one direction for green and the other for red and blue. If it is important for the polarization to be the same for all three colors, this can be done with a wavelength-sensitive polarization rotator such as the ones described in Section 6.3.4.

The same left-to-right reversal of the image on the center microdisplay compared to the images on the two side microdisplays occurs in the dichroic cube architecture as occurs in the equal path architecture. Here the center image is not reflected at all while each of the side images is reflected once. This is corrected in the same manner as for equal path systems.

The imaging path in a three-panel projector, either equal or unequal path design, is normally telecentric. This avoids the problem that would occur in a non-telecentric design where the left and right sides of the panel would see different angles of incidence at the dichroic filters. Sometimes, however, these designs, especially the equal-path version, are designed to be non-telecentric. This leads to an angle of incidence variation which causes a left to right color variation in the image. This can be partially compensated by a left-to-right gradation in the dichroic properties (Bowron *et al.* 1996). Due to the large size of the optics required for a telecentric equal-path design, or the high cost of dichroic filters with a left-to-right gradation in them, plus other advantages such as more compact form factor and more rugged convergence, the unequal path design has entirely supplanted the equal path design in commercial production.

The principal components of a typical light engine in three-panel, color additive systems are given in Table 10.1. Specific designs may add additional components, e.g., additional lenses to compensate for path differences in dichroic cross recombination systems.

Table 10.1 Components of light engine for three-panel projector.

Components	Equal path system	Unequal path system
Lamp/collector system	1	1
Lamp power supply	1	1
Integrator	1	1
Polarization recovery	1 (optional)	1 (optional)
Dichroic color separation filters	2	2
Path-correcting lenses	—	1 set
Polarizers	3	3
Microdisplays	3	3
6-axis convergence assembly	3	3
Analyzers	3	3
Field lenses for panel	3 (optional)	—
Dichroic recombination filters	2	dichroic cube
Fold mirrors	2	3
Projection lens	1	1

10.3 Three-panel LCoS Projector Architectures

The LCoS projectors to be discussed in Section 10.4.2 are based on a single LCoS panel and that panel needs to be fast enough to operate in a color sequential mode. Typically a twisted nematic (TN) effect would be used in one of these panels. In fact, the most common liquid crystal effect used in LCoS panels is vertically aligned nematic (VAN), which is not fast enough to operate color sequentially. These panels can only be used in three-panel architectures. There are two common LCoS architectures. The first is based on three polarizing beamsplitters and a dichroic combining prism much like the one used in an unequal path three transmissive panel projector. This class of projector will be discussed in Section 10.3.1. The other class of projectors is based on four cube beam splitters. These cube beam splitters can either be polarizing beam splitters or dichroic splitters, or there can be cubes of each type in the design. This class of projectors will be discussed in Section 10.3.2.

Other projector designs have been used for three-panel LCoS projectors in the past, including single PBS/Plumbicon prism (Doany *et al.* 1998) and single PBS/X-cube designs (Meuret and De Visschere 2003) and off-axis projection (Bone 2001). These designs could not normally meet the requirements imposed on LCoS projectors, especially in terms of contrast and uniformity, and will not be discussed further.

10.3.1 Three Polarizing Beamsplitters with a Dichroic Combiner

A three-panel LCoS projector can be built that is fundamentally similar in design to the three-panel transmissive LCD projector shown in Figure 10.3. The transmissive LCD microdisplays are replaced by reflective LCoS microdisplays mated to a PBS, as shown in Figure 10.5 (Bowron and Schmidt 1998).

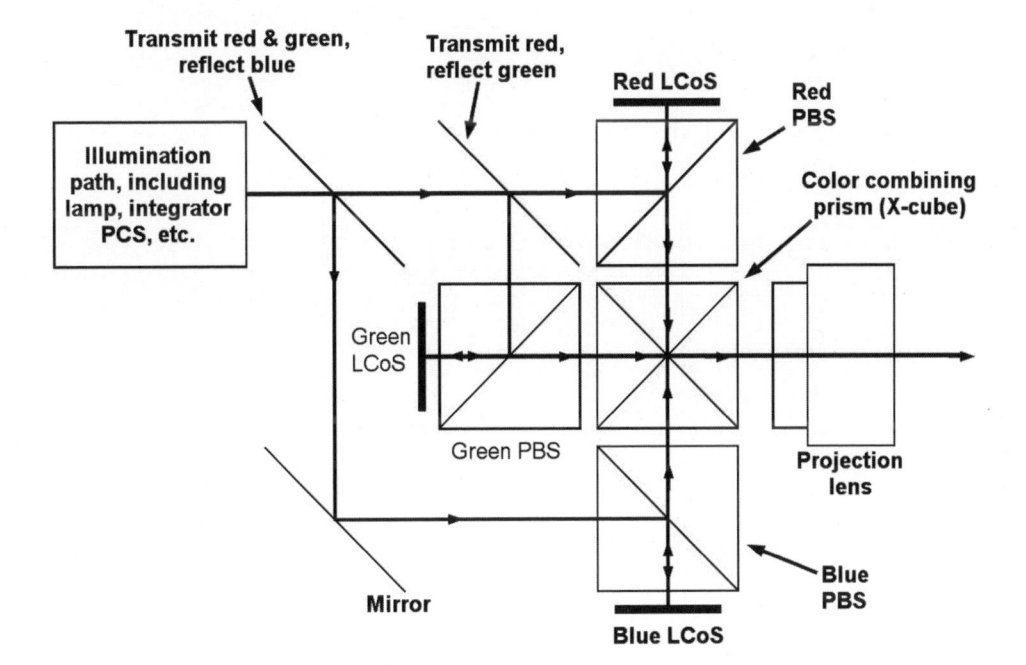

Figure 10.5 LCoS engine design with three MacNeille prism PBSs.

240   MICRODISPLAY SYSTEM ARCHITECTURES

MacNeille prisms can be used in this application despite their problems since each PBS is only used by a single wavelength band, red, green or blue. With these narrow wavelength bands, the PBS can be optimized for performance. This implies, of course, that each of the PBS prisms in Figure 10.5 has a different coating design, even if the glass parts of the prisms are identical. MacNeille prisms can be used so the imaging path is either in transmission, as shown in Figure 10.5, or in reflection. In reflection, each LCoS panel would be moved to the adjacent side of the prism where the incident light would transmit through the prism and the modulated light from the LCoS panel would reflect off the coating. This would also involve changing the PCS in the illumination path to produce p-polarized light at the tilted coatings and redesign the dichroic filters, both the flat plates in the illumination path and the embedded dichroics in the X-prism, to operate correctly in the new polarization of light.

MacNeille prisms are not the only PBSs that have been used in this design. Three PBS projectors using Moxtek wire grid polarizers (Pentico *et al.* 2003) and 3M Vikuiti (Bruzzone *et al.* 2003) PBSs have also been manufactured. Since the 3M PBS should only be used in transmission in the imaging path, the drawing of a Vikuiti optical core would look exactly like Figure 10.5. The Moxtek polarizer, however, should only be used in the imaging path in reflection. Therefore, the three LCoS panels should be moved to the other side of each PBS where the incident light would transmit through the PBS instead of reflecting off it.

10.3.2 Four-cube LCoS Architectures

Titling this section 'Four Cube' is perhaps a misnomer because, as can be seen in Figure 10.5, the three PBS architectures discussed in the previous section also involve four cubes. Since roughly the same number of components are used in the three PBS and four cube architectures, cost of the two can be comparable. The final decision on which architecture to use then rests on performance criteria and a detailed cost analysis.

The original four-cube architecture is based on four PBS prisms arranged in a square as shown in Figure 10.6 (Robinson *et al.* 2000, 2006).

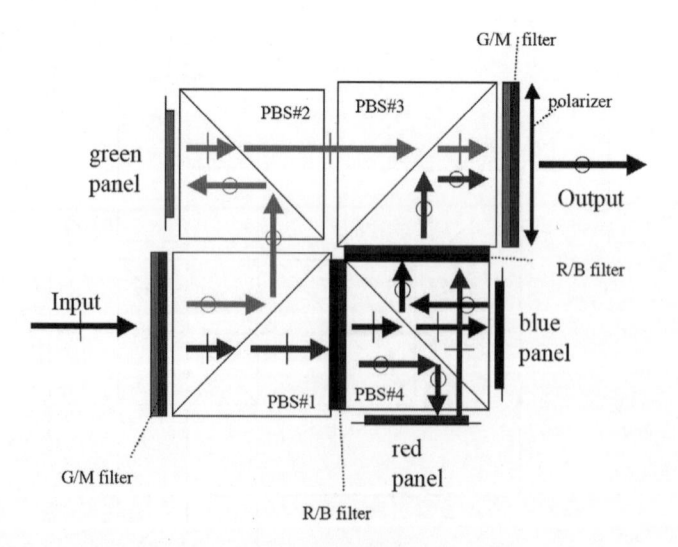

Figure 10.6 Four-cube LCoS architecture based on four PBS prisms (reproduced by permission of © 2000 the Society for Information Display (Robinson *et al.* 2000)).

In this projector architecture, the input light comes from a conventional lamp, integrator and PCS system, not shown in Figure 10.6. The light is horizontally polarized or p-polarized in relation to the tilted coatings on all four PBS prisms.

The light first encounters a green/magenta polarization rotating filter such as the ones described in Section 6.3.4. The polarization of the green light is rotated 90° by this filter and the polarization of the magenta (red plus blue) light is unchanged. When the light encounters the coating of PBS #1, the s-polarized green light is reflected and the p-polarized magenta light is transmitted. The green light then encounters the coating of PBS #2 and is reflected again where it encounters a LCoS panel. This panel modulates the polarization of the light according to the green video signal. For pixels where maximum green is called for by the video signal, the LCoS panel rotates the polarization by 90°. The light then passes through the coatings of both PBS #2 and PBS #3.

The magenta light that passed through the coating of PBS #1 encounters a red/blue polarization filter. This filter rotates the polarization of the red light 90° while leaving the polarization of the blue light unchanged. The s-polarized red light reflects off the coating of PBS #4 and encounters the red LCoS panel. If the video signal calls for maximum red, the polarization of the red is rotated another 90° so when it re-encounters the coating of PBS #4 it is transmitted.

Meanwhile, the blue light is still p-polarized and transmits through the coating of PBS #4 and encounters the blue panel. If the video calls for maximum blue, the polarization is rotated 90° and when the blue re-encounters the coating of PBS #4 it is s-polarized and reflects.

At this point, the red and blue light encounters another red/blue polarizing filter and the polarization of the red light is rotated one more time. When the red and blue encounter the coating of PBS #3, they are both s-polarized and both reflect. The green light transmitted through PBS #3 and the red and blue light reflected from PBS #3 encounter another green/magenta polarizing filter, rotating the polarization of the green light to match the polarization of the red and blue light. This white light finally encounters a clean-up polarizer to maximize system contrast. The light then enters the projection lens and the image is projected on the screen.

A lower-cost variation on this design replaces the first PBS with a dichroic filter, as shown in Figure 10.7 (Chen *et al.* 2006). The dichroic is designed to split either yellow (red plus green) from blue or red from cyan (green plus blue). The details of the wavelength sensitive polarizers in the remainder of the design depend on exactly how the dichroic splits the light, but the principle of operation is the same as the previous design.

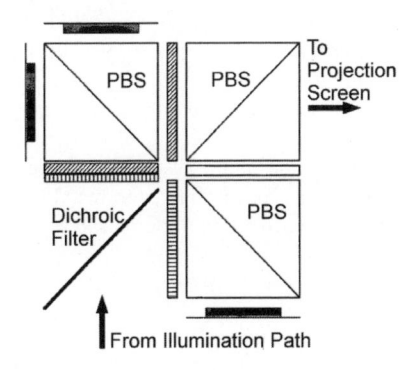

Figure 10.7 Four-cube LCoS architecture based on three PBS prisms and a dichroic filter (Chen 2006; United States Patent and Trademark Office).

10.3.2.1 Four-panel, high contrast LCoS architecture

A variation on the four-cube LCoS design is one that adds a fourth LCoS panel to enhance the system contrast. Figure 10.8 is a schematic of this design (Bridgwater *et al.* 2006).

The first four cubes, labeled 22, 24, 26 and 34 in Figure 10.8, form a conventional four-cube LCoS projector with three LCoS panels 28, 30 and 32. The full color image generated by this subsection of the projector is relayed by the lens 36 onto a fourth LCoS panel 40. This panel further modulates the

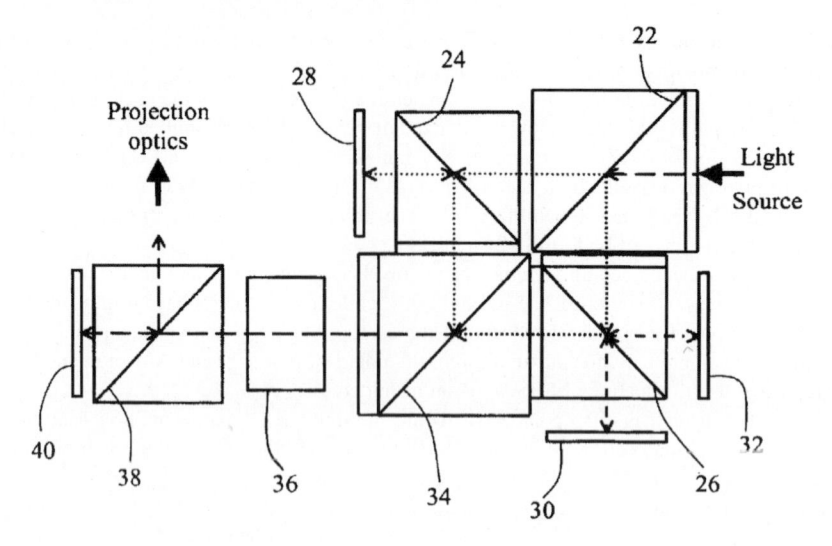

Figure 10.8 Four-panel LCoS projector with ultra-high contrast (Bridgwater 2006; United States Patent and Trademark Office).

light, especially in the dark regions, so the projector can produce contrast levels of 150,000:1 or more. This extremely high contrast is needed in some visualization and simulation applications that show predominantly dark scenes with a few bright highlights. Examples of applications where this contrast is needed include planetariums and aviation night vision simulators.

10.3.3 Three-panel, Three-lens Projectors

This type of projector is shown schematically in Figure 10.9 with transmissive microdisplays. The microdisplays, light valves or light amplifiers can be either transmissive or reflective devices. Projectors employing three lenses are not internally converged but rather are converged on the screen, as is done in CRT projectors. Because two of the three color channels are off-axis, the images these channels produce on the screen would show keystone distortions in the absence of compensation. This compensation can be difficult to achieve with pixelated displays having a fixed format. If a three-lens configuration is used with pixelated microdisplays, normally the two off-axis microdisplays would use offset projection lenses as shown in Figure 7.5 to avoid the keystone problem.

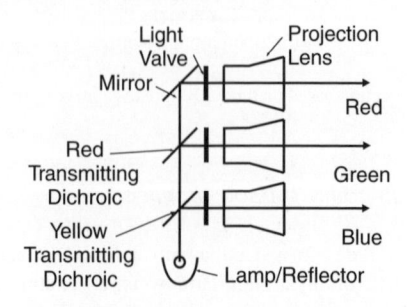

Figure 10.9 Schematic of 3-LV, three-lens projection system.

When the microdisplays, light valves or light amplifiers used in three-lens systems use electron-beam-addressing, the system can write distorted rasters on the outer devices. These rasters, when projected from off-axis positions, will be corrected to rectangular images with the correct aspect ratio and size to achieve convergence.

Light amplifiers are devices which are addressed with very low power optical images and produce the same image in the output, typically at very high lumen levels (Bleha 1995, 1997). Light amplifiers are reflective devices and the systems based on them share architectural features with others that are based on other reflective devices, such as crystalline silicon active-matrix microdisplays.

The writing source in a light-amplifier system is usually a high resolution, small diagonal CRT. A schematic of the optical path of a three-light-amplifier/three-projection-lens system is shown in Figure 10.10. Only one of the channels is shown fully. The illumination paths for the other two channels are indicated. The portions of these two channels not shown are essentially the same as the one shown.

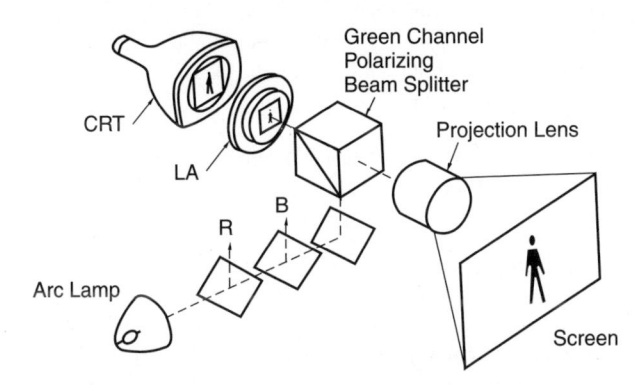

Figure 10.10 Three-lens light-amplifier projector. Parts of the optical paths for the red and blue channels are suppressed for simplicity but are essentially the same as the green channel illustrated (adapted from Bleha 1995; with permission of SID).

Writing with a raster image source produces many similarities between some light-amplifier systems and CRT projectors. Among these are variable raster sizes and shapes, and systems with one or three projection lenses. The highest output light-amplifier system used three lenses and delivered 12,000 lumen output from a 7000 watt xenon arc lamp. These projectors were used in the 1999 public demonstration of electronic cinema with the screening of *Star Wars Episode I* (Bleha and Sterling 1999).

Single-lens light-amplifier systems have been built in the past using combining dichroics before the projection lens (Ledebuhr 1986). Their performance was never satisfactory and they will not be discussed further.

10.4 Single-panel Projectors

Single-panel systems have several advantages. Their cost may be lower since only one microdisplay is needed, and microdisplay cost is often a significant fraction or the total cost of a projector. Depending on the architecture, the complexity of the projector optics in a single-panel projector may be reduced compared with three-panel systems. One of the biggest advantages of single-panel architectures is that they are autoconverged. This occurs because red, green and blue are all modulated by the same physical pixel.

On the other hand, the panel used in these systems is often larger and more complex than the individual panels in a three-panel system. In order to modulate all three colors with a single panel, some performance characteristic of the microdisplay must be significantly higher than the equivalent performance characteristic in a three-panel system. For instance, in projectors using subpixelation (three-color subpixels equal one white pixel), the active matrix has three times the number of independently addressable points in the array as compared to the pixel counts of the individual panels in three-panel projectors. This has implications for system resolution, as discussed in Section 10.4.1.3. Artifacts due to spatially separated color subpixels may also be observable, especially in lower resolution projectors.

For time-sequential systems, the response time of the panels must be more than three times faster than the response time in a three-panel projector. In addition to a high-speed panel, color sequential projectors require the video data formatting electronics to be faster and more complex. Many single-panel architectures have lower throughput than three-panel designs. The convergence and panel count advantages of time sequential systems are not completely free.

Single-panel subpixelated systems which use a polarization-dependent liquid crystal effect for operation have an additional advantage compared to three-panel systems. The aging of polarizers, LC material and LC alignment layer is due almost exclusively to the blue light with its high-energy photons (Yakovenko *et al.* 2004). In a three-panel system, the blue light falls on a single-panel and polarizer/analyzer combination. This panel is normally smaller than the panel used in a single-panel projector. Therefore, the power density for blue is higher and the aging takes place at a faster rate. This can lead to a differential aging in the blue panel compared to the red and green panels which in turn leads to a color shift in the image. While this was a major problem in early three-panel LCD projectors, more robust panel materials and improved blue filters have largely eliminated this as a problem in current LCD projectors.

10.4.1 Sub-pixelated Projectors

This class of single-panel projectors use three or more color subpixels on the panel to produce one 'white' pixel. These microdisplays employ color filters on each subpixel or geometrically separate the incoming color beams, as discussed in the following sections.

10.4.1.1 Microfilter projector

The simplest of the single-panel architectures employs a panel with color microfilters on subpixels (Takamatsu *et al.* 1992; Sperger *et al.* 1993). This panel is similar to that used for direct-view applications and, in some designs, the panel is in fact a direct-view panel. With this system, illustrated in Figure 10.11, no color separation is required before the panel and no color recombination is required afterwards. This approach removes many of the more costly elements of a three-panel projection system. The panel-to-lens distance can be set by the lens designer and the illumination path is much simplified.

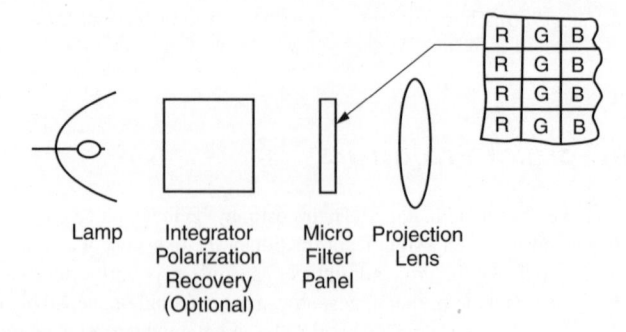

Figure 10.11 Architecture for single-panel system with color microfilter panel. The panel has the color filters for the subpixels arranged in vertical stripes.

The components of this light engine are given in Table 10.2. Given a source of microfilter panels, it is relatively straightforward to construct a video projector. Each of the addressable points in the array with its microfilter is called a subpixel. A full pixel consists of the subpixels which, when viewed together, produce white. For most systems, this consists of the three primary subpixels, R, G, and B. There have been several panel constructions in which four subpixels have been employed. These have included designs with a second copy of one subpixel color to achieve enhancement of one of the primaries.

Table 10.2 Components of light engine for one-panel projector with color microfilter panel.

Component	Number required
Lamp/collector system	1
Lamp power supply	1
Integrator	1 (optional)
Polarization recovery	1 (optional)
Field lens	1 (optional)
Polarizer	1
Microdisplay	1
Analyzer	1
Projection lens	1

Also, there have been four subpixel arrangements with one of the subpixels being white, i.e. not having a microfilter. The intent of this approach is to transmit more light when white images are presented. Unfortunately, this arrangement would also reduce the color saturation unless this extra pixel was completely turned off (black state) when a fully saturated color mixture was present in the image. If the white pixel is allowed to produce some of the luminance when saturated colors are shown, the image will have a washed out appearance, compared to an image made only from red, green and blue.

Note in Figure 10.11 that the red, green and blue sub-pixels are spatially separated from each other. When the image of these sub-pixels is projected on a screen, the magnification must be kept low enough so the eye will not resolve the three sub-pixels but will fuse them into a single full-color pixel. Diagonal lines in the image can exhibit 'jaggies', the stepwise approximation to the lines, of one or more colors.

The light throughput of all single-panel systems employing color microfilters is compromised by the optical losses in the filters. Under the most optimistic assumptions of 100 % transmission of each of the filters within its color band and no losses between color bands due to the shape of the filter transmission band, each filter would transmit about one-third of the light incident on it. This reduces overall system efficiency to one-third of what it might otherwise be. This issue also pertains to the color filter-wheel architecture discussed later in this section.

Historically, this system architecture was most often used in low luminance, lower cost applications. Because the architecture can be very inexpensive compared with other architectures, a significant number of color microfilter projectors have been sold. Typically, the panels in these projectors are quite large compared to the normal microdisplays without color filters. Panels up to about 3″ in diagonal are currently used in these projectors and panels up to 11″ diagonal have been used in the past.

10.4.1.2 Angular color separation projectors

An innovative alternative to using color microfilters in a single-panel system is based on microlenses (Hamada *et al.* 1994) to guide the three color bands to specific subpixels. The system is illustrated in Figure 10.12. Each microlens has a pitch equal to three of the subpixels in the array. Three dichroic mirrors, a holographic optical element (HOE) (Loiseaux *et al.* 1995) or a binary optical element (Gale and

246   MICRODISPLAY SYSTEM ARCHITECTURES

Swanson 1997), separate the incoming white light from the lamp into three angularly separated color beams. The microlens brings these beams to a focus on three adjacent subpixels, which modulate the transmitted light with the information appropriate to the incident color. Since each subpixel is illuminated by only one color of light, it does not need to have microfilters. Hence, this major source of efficiency loss which occurs in color microfilter panels is avoided. High brightness projectors based on this principal have been produced.

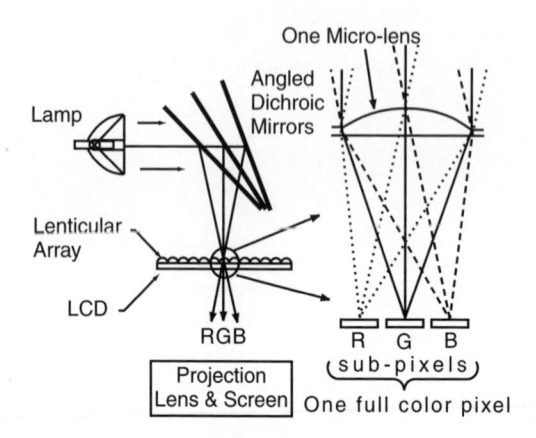

Figure 10.12   Principal of angular color separation for single-panel projector.

Obviously, there are complications associated with this type of system. The dichroic color separation filters must be very flat to avoid distortion of the light paths in the three channels, which could result in color crosstalk due to the light from one band falling on the subpixel associated with another band. The angles of illumination of these filters must also be restricted to avoid color shifts. The accurate manufacturing of the microlenses and their alignment to the subpixels is not easy. The presence of the off-axis beams due to the angled dichroics requires a projection lens with a low f/# in order to avoid vignetting of the off-axis colors. If there is any vignetting, it could lead to unacceptable color gradients. Angular color separation projectors also have the same resolution issues as color microfilter panels that were discussed above. Since three subpixels are required to make one pixel, the panels used for these systems are normally larger than those used in three-panel systems.

Due to the image quality problems related to projectors with angular color separation, the authors know of no projectors of this type currently in production. The cost savings associated with these designs have not materialized, at least to a high enough degree to justify the image problems.

10.4.1.3   Resolution of sub-pixelated projectors

A variety of color microfilter and angular color separation subpixel configurations have been proposed. Figure 10.13 illustrates three principal arrangements (Chen and Hasegawa 1992) for a panel with square subpixels in addition to a panel without subpixelation. Figure 10.13(a) shows the configuration without subpixelation. In Figure 10.13(b), the subpixel colors are arranged in vertical stripes. Figure 10.13(c) shows the colors arranged as diagonal stripes and Figure 10.13(d) shows a triangular arrangement. Also shown in this figure are the spatial frequencies displayable by each of these panels.

The highest spatial frequency in a given direction that a panel can show is equal to the reciprocal of twice the pixel pitch. This is the Nyquist frequency. It is the highest frequency for which one complete spatial frequency cycle can be displayed. Thus, in Figure 10.13(a), the highest horizontal or vertical frequency is displayed by two subpixels and, thus, the Nyquist frequency is $0.5/d$. This is the resolution

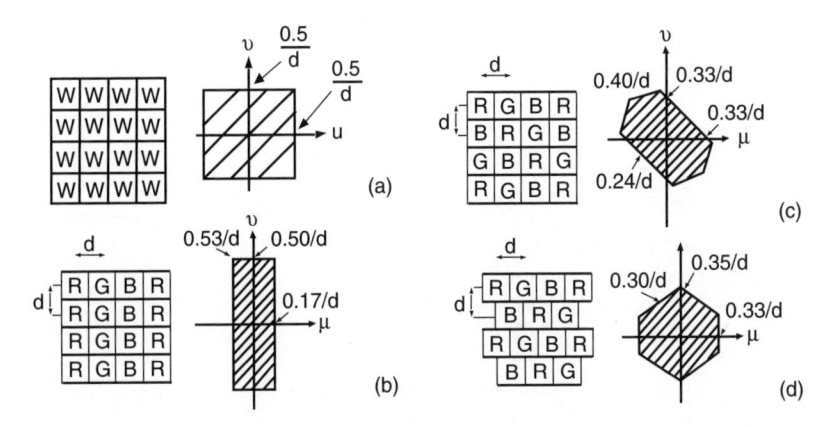

Figure 10.13 Subpixel arrangements and displayable spatial frequencies for microfilter panels with square subpixels with pitch d: (a) no subpixelation; (b) colors arranged in vertical stripes; (c) colors arranged in diagonal stripes; and (d) offset pixels with triangular color pattern (adapted from Chen and Hasegawa 1992; with permission of Elsevier Science).

limit of a three-panel system. In practice, it is often not possible to display frequencies satisfactorily all the way to the Nyquist limit. This is discussed further in Chap. 14.

With the subpixelation arranged with the colors in vertical stripes (Figure 10.3(b)), the white pixel pitch in the horizontal direction is equal to 3d and the Nyquist limit in this direction is 0.17/d. It is common to design the panels for this type of system with non-square subpixels, i.e. with the pitch in the vertical equal to 3d and the horizontal pitch equal to d. The white pixel would then be square with pitches equal to 3d in both directions. With this subpixel shape, the Nyquist frequency in both directions in then 0.17/d.

Compared with the resolution of a three-panel system, two observations can be made. First, three times the number of subpixels are required in this single-panel approach as in each panel of the three-panel system to achieve the same number of resolution elements. Secondly, for a given (sub)pixel pitch, the subpixelated panel would have larger diagonal, which usually translates to higher system size and cost. Alternatively, the pitch can be reduced to achieve the same diagonal. However, this has negative consequences on the aperture ratio and light throughput. These issues are discussed further in Section 11.4.1.

The diagonal arrangement of subpixels shown in Figure 10.13(c) alleviates some of the problems associated with the vertical configuration but introduces a new set. The Nyquist frequencies in both the horizontal and vertical directions are now equal. However, the first and third quadrant diagonal resolutions are now much lower than those in the second and fourth quadrants. This is usually not a significant deficiency for video applications since the human visual system is less sensitive to higher frequency information in the diagonal direction than in the horizontal or vertical directions. A more important deficiency of this configuration is its effect on the presentation of textual data, particularly when only one of the primaries is activated. The designer of the system with this subpixel configuration has a choice of non-straight edges for the text or edges consisting of a series of dots, rather than continuous lines.

The final pixel configuration shown in Figure 10.13 uses a triangular configuration of subpixels. This arrangement is the best of these for resolution, being nearly equal in all directions. However, it suffers from the same deficiencies in presenting datagraphics as the diagonally disposed color-microfilter architecture.

Other arrangements of pixels are possible, including arrangements where the pixels are not square. One such arrangement is shown in Figure 10.14 (Brown Elliott 2005). Note that this figure includes 8 pixels–each block actually represents two pixels. This arrangement uses one blue sub-pixel to go with two red and two green sub-pixels. This arrangement reduces the number of addressable pixels by 17 % while simultaneously equalizing and maximizing the MTF in the horizontal, vertical and diagonal directions. The elimination of the extra blue sub-pixel is possible because the blue channel carries almost none of the resolution information and is really only needed to give the correct white balance.

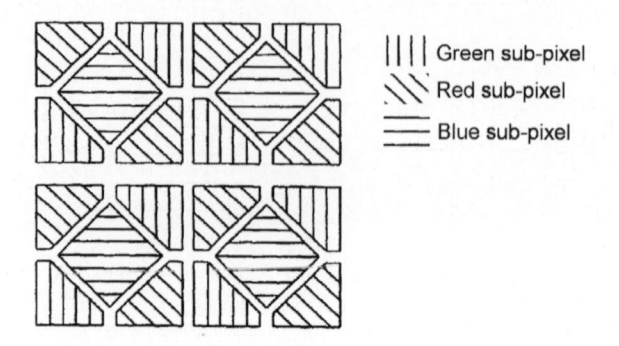

Figure 10.14 Sub-pixel arrangement with non-square pixels (Brown Elliott 2005; United States Patent and Trademark Office).

While the arrangement of pixels in Figure 10.14 has not proven practical in production, other pixel arrangements are possible, including ones with white sub-pixels. The white sub-pixels can significantly increase the resolution and the system luminance, at the expense of the luminance of saturated colors. A typical pixel arrangement for a micro-filter panel with a white sub-pixel is shown in Figure 10.15.

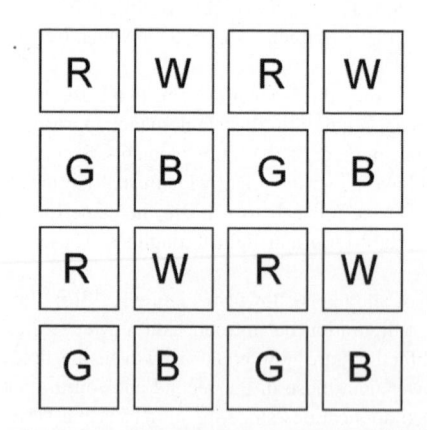

Figure 10.15 Portion of a micro-filter panel with red, green, blue and white sub-pixels.

In this arrangement, most of the resolution information is carried by the green and white sub-pixels. The red and blue sub-pixels are, of course, needed to provide a full color image. Since each pixel has one green and one white sub-pixel, the resolution of the system is significantly higher than a RGB system. In addition, the lumen throughput is much higher than a RGB system. The maximum lumen increase would be about 50 %, depending on exact filter design and other factors. Designing the system to use this maximum increase would normally be unacceptable from a colorimetry point of view, so a lumen increase target of 20–35 % would be more common. The same resolution improvement can be achieved with a RGGB pixel arrangement, but there would be no lumen throughput increase.

10.4.2 Color-field Sequential Systems

Color-field sequential systems, i.e. systems in which the three or more primary colors are presented in rapid succession, are an alternative to using subpixelation to achieve full color in single-panel projection systems.

The panels in these systems can have the same pixel layout as those used in three-panel systems. Each pixel displays in succession the information for R, G, and B synchronized with the color illumination. If the color-field time is shorter than the response time of the eye, color fusion will take place in the human visual system without flicker. Artifacts of color field sequential systems such as color breakup will be discussed in Section 14.2.3.

In a panel in which information for only one color is written to a pixel, as in color microfilter panels or panels in three-panel systems, the information from field-to-field is strongly correlated, i.e. the information on a pixel usually changes little field-to-field. Even if the display medium is slow, it typically does not have to undergo a major change in successive fields. It is only when there is rapid motion in the image that the 'slowness' of some electro-optical modulators becomes observable as image smear.

In a field sequential system, the information is not correlated between the sub-field images. For instance, with 60 Hz field rate, when a bright red object is displayed, the red sub-field is at maximum brightness while the previous blue and following green sub-fields are at minimum brightness. In this case, the optical modulator must switch from minimum to maximum and back to minimum all in 16.6 ms maximum, or 5.5 ms per color. To produce the most accurate color rendition and maximum lumen throughput in a color-field sequential display, the steady-state must occupy most of this 5.5 ms color period. This requires the switching of the light modulating medium from one color to the next be done in a small fraction of the color-field time, about 1 ms or less.

This can be a significant issue for systems that use panels based on TN LCDs. Virtually all microdisplay systems are progressively scanned, i.e., the frame time equals the field time. Within the 5.5 ms per color period, a change in the electrical input to the panel must result in a rapid change of the transmission of the microdisplay to produce the modulation appropriate to the incident light color.

The most common field-sequential projection systems are based on the DMD device from Texas Instruments. Other companies have tried to develop MEMS devices competitive with the DLP (Huibers 2006) but none is on the market as of this writing. Other MEMS systems such as the GLV or the GEMS devices are fast enough for field sequential operation but are not currently used in that mode. This occurs because any projector based on these MEMS systems are so expensive that image quality is of paramount importance and color-field sequential operation is not satisfactory. Field sequential projectors based on twisted nematic (TN) and ferroelectric (FLC) liquid crystal effects have also been built. In the past, color sequential CRT and oil-film light-valve systems have been built but none is in production now. All the microdisplay device types in current use were discussed in Chapters 4 and 5.

10.4.2.1 Addressing color-field sequential systems

There is a strong relationship between the data addressing technique used on a microdisplay and the types of optical architectures in which the microdisplay can be used (Brennesholtz 2003). Virtually all microdisplays, regardless of the active matrix or optical modulator technology, are addressed one row of pixels at a time, as shown in Figure 10.16. After a row is selected and the data is loaded into the pixels from the column electrodes, the row is deselected and another row is selected. Normally, but not always, the next row is adjacent to the current row. This leads to two different regions of the active matrix, one that has the new field data loaded into it and a second that still has the old field data.

Microdisplays have two basic types of response when addressed with this new video data:

1. The electro-optical modulator changes state as soon as the video data is loaded into the pixel; or

2. The electro-optical modulator does not change state when the data is loaded but requires a reset signal.

250 MICRODISPLAY SYSTEM ARCHITECTURES

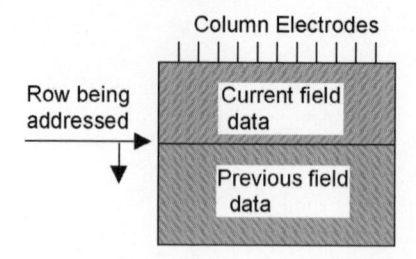

Figure 10.16 Panel with row at a time addressing of the active matrix.

Most LCD and LCoS systems fall into the first category and the DLP falls into the second category. In three-panel systems, the type of electro-optical response makes little difference. When the red panel in a three-panel system, for example, is addressed with updated video data, it is still red video data, even though different parts of the panel have data from the current or the previous field. Therefore, the entire panel can be illuminated at all times with red light.

In single-panel, color sequential systems, the situation changes. In a normal LC or LCoS system, the liquid crystal state changes as soon as the row is addressed. Since the new data corresponds to red video, for example, and the previous field data corresponds to green video, for example, the liquid crystal in different portions of the panel would correspond to red and green video content and the entire panel cannot be flooded with a single color of light. There are several optical architectures to address this problem that will discussed in Sections 10.4.2.4, 10.4.2.5 and 10.4.2.6.

In a system that requires a reset pulse to change the state of the electro-optical modulator, the situation is different, as shown in Figure 10.17. In this example, a DLP imager is being addressed one row at a time, as is normal. The data is loaded into the SRAM storage cell associated with each DLP pixel. At this point, the data does not affect the mirrors. Therefore, although the SRAM backplane contains data for two different video color fields as shown in Figure 10.17(a), the mirrors are all in the correct positions for the previous color field, as shown in Figure 10.17(b). After all rows have been addressed the entire SRAM array will contain the new color data, but the entire mirror array will positioned for the previous color data. At this instant, two things should happen simultaneously. First, the global reset pulse (or pulse train) should be applied to all pixels to switch the mirrors from the previous field position to the new field position. Secondly, the color of the illumination light should change from the color for the previous field data to the color for the new field data. This is relatively easy to do, as will be discussed in Sections 10.4.2.2 and 10.4.2.3.

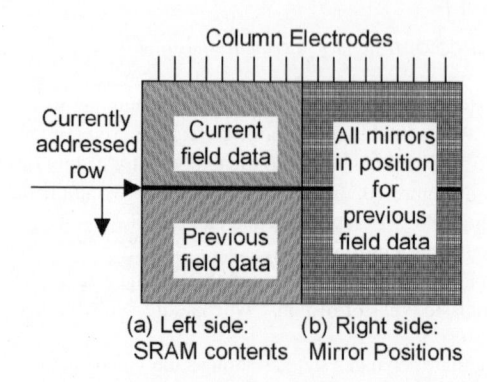

Figure 10.17 Addressing of a DLP imager or other microdisplay requiring a reset pulse.

Note that this system requires two memory cells per pixel. In the DLP, these memory cells are provided by the SRAM memory and the mirror position itself, which is a 1-bit memory cell.

LCoS backplanes have been designed to provide two memory cells per pixel so they can use reset addressing. Since there is no memory associated with the LC materials used in projection systems, the second memory must be designed into the silicon backplane. One such design is shown in Figure 10.18 (Huang *et al.* 2007). This figure with five transistors represents one pixel on the display. The first memory is labeled bit n_1 and consists of the left-hand transistor and its associated capacitor while the second memory for this pixel is bit n_2 and consists of another transistor and capacitor. When signal ϕ_1 activates its associated transistor, bit n_1 controls the liquid crystal. During this time, signal ϕ_2 de-activates its transistor and bit n_2 can load its data into the capacitor without affecting the liquid crystal state. After all data has been loaded into all the bit n_2 locations of active matrix array, ϕ_1 can de-activate and ϕ_2 can activate, loading all the new field data into the pixels in parallel. The cycle is actually more complex than this, since ϕ_1, ϕ_2 and ϕ_3 are cycled to both clear the pixel and allow the voltage on the counter-electrode to be changed.

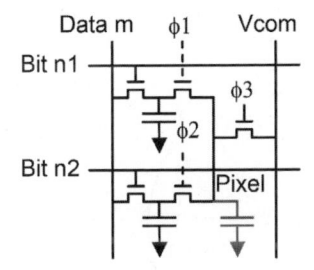

Figure 10.18 LCoS backplane design that allows a reset addressing scheme (reproduced by permission of © 2007 the Society for Information Display (Huang 2007)).

Other LCoS backplanes to accomplish global reset addressing are discussed in, for example, Borel *et al.* (2006) and Akimoto and Hashimoto (2000). Due to the high transistor count per pixel and the high-speed operation required, it is believed that global reset addressing cannot be implemented in transmissive microdisplays. If for no other reason, the high transistor and capacitor count would leave little space for the open aperture.

10.4.2.2 Color wheel and related systems

The most common and one of the simplest color-field sequential system is based on a color wheel, as illustrated in Figure 10.19. Figure 10.19(a) shows a color wheel used with a transmissive microdisplay and Figure 10.19(b) shows the more common case where the color wheel is used with a reflective microdisplay, in this case a DLP imager.

The transition times between colors are minimized by focusing the lamp onto the filter disc. For electro-optic media which have throughput during the color-field to color-field transition periods, it might be necessary to extinguish the lamp during these periods to avoid color errors. The illumination energy lost during the inactive periods may be recovered by pulsing the lamp with an instantaneous power greater than its rated power, while maintaining the average power constant (Stanton *et al.* 1996). This type of system has the same light utilization issues as does the color microfilter single-panel systems, viz., about two-thirds of the available light is rejected by the color filters. In this case, the light is lost because each color has approximately one-third temporal duty cycle. Nevertheless, satisfactory performance is achieved for many applications at a lower system cost.

252   MICRODISPLAY SYSTEM ARCHITECTURES

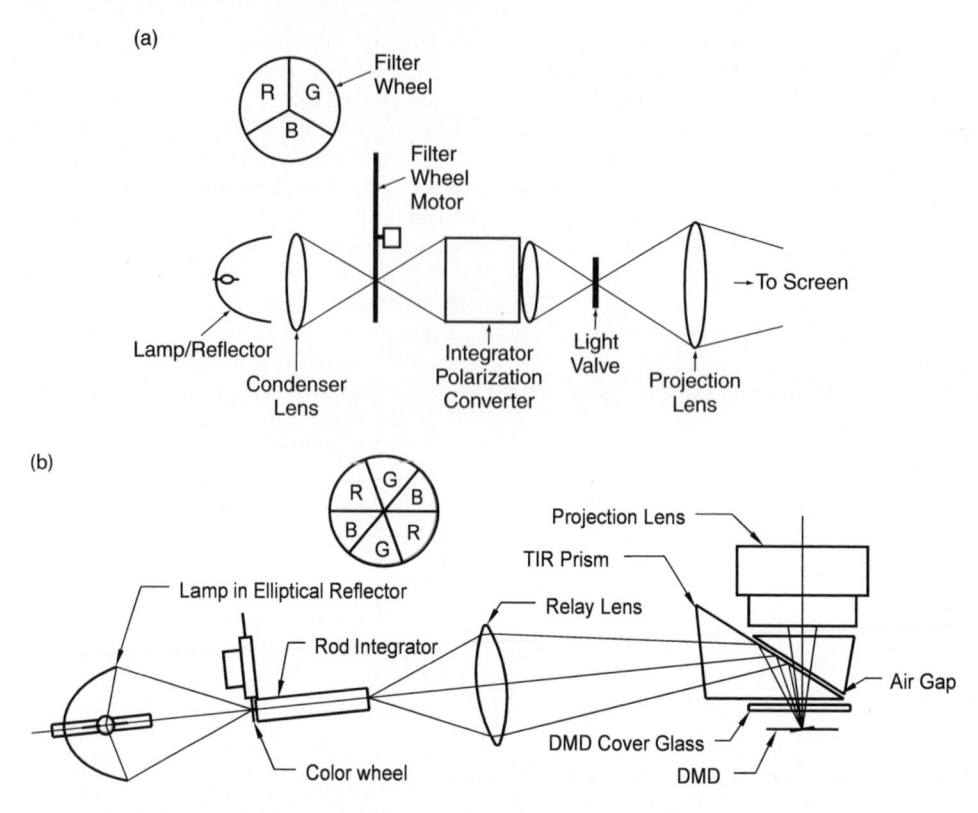

Figure 10.19 Schematic of color-wheel single-panel projector: (a) color wheel used with a transmissive microdisplay; (b) color wheel used with a DLP microdisplay.

In this system, as the spoke between adjacent colors on the color wheel passes through the light beam, the color changes very rapidly from one color to the next. Since the focused spot is in the pupil plane of the system, the color change is uniform over the entire panel. This means that only the global-reset addressing scheme as discussed in Section 10.4.2.1 can be used. If the electro-optical effect changes as soon as the row is addressed, there would be a mismatch between the data on the panel and the color of light illuminating the panel during a portion of the addressing cycle. Color-wheel based systems with row at a time addressing have been built, but their efficiency and colorimetry has not been good.

There is no limitation in a color wheel system on either the number of segments or the colors on the segments. While three segments are shown in Figure 10.19(a), it is common to have six segments representing two cycles of red, green and blue, as shown in Figure 10.19(b). This reduces the rotation speed of the color wheel by one-half.

If more than three colors are on the wheel, the system is often called a multiple primary color system. Systems with up to six primary colors (red, green, blue, cyan, yellow and magenta) have been commercialized with DLP technology (van Kessel 2006). It is also possible for the color wheel to have a white segment to increase the total lumen output of the system. This is the equivalent of the white sub-pixel discussed in Section 10.4.2.1. Due to the colorimetry problems associated with white segments, they are used only where lumen output is more important than colorimetry. These applications include conference room projectors and other business-oriented applications. Projectors designed to show primarily video material rarely have a white segment. See Section 13.3.3.5 for a discussion of this issue.

The color wheels illustrated in Figure 10.19 can be replaced by electrically tuned color filters (Schadt and Schmitt 1997; Wang 1997). These tunable filters are discussed in Section 6.3.4.

10.4.2.3 Three-light-source field sequential systems

An all-electronic equivalent to a rotating color wheel was demonstrated (Stanton *et al.* 1996). This non-mechanical execution of color-sequential operation used a single DMD, three UHP lamps and dichroic filters. The filters are chosen so the one reflects red light and transmits green light. The other filter reflects blue light and transmits green and red. Thus, each lamp could provide one band of colors as the input illumination of the DMD. While the lamps in Figure 10.20 are labeled 'red', 'green', and 'blue', they are in fact identical.

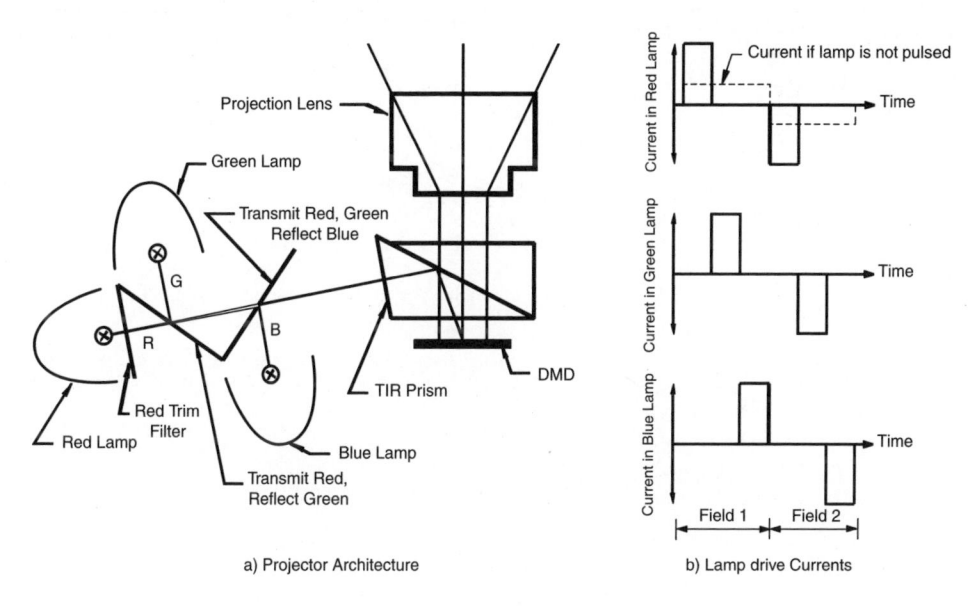

Figure 10.20 Color sequential system using three pulsed UHP lamps and one DMD.

The system achieved the color-sequential operation by pulsing each lamp in succession with an instantaneous power of 300 W, or three times larger than the lamp rating of 100 W. Each of the lamps was on for one-third of the field time (33 % duty cycle). The result is the average power for each lamp was equal its rated power. While the efficiency was not significantly different than the efficiency of a color-wheel system, the system output was three times greater than could be obtained with only one lamp and a color wheel.

Other light sources than the UHP are currently used in this architecture. The architecture is particularly suited for light sources such as LEDs and lasers where the three different light sources generate the desired color bands, rather than white light. In fact, with a narrow-band light source such as an LED or a laser, one is not limited to three light sources and projectors with four or more microdisplays can be built (Harbers *et al.* 2003). In these systems, the efficiency is nearly 3 × the efficiency of a color-wheel system when used with a white light source, since little or no light needs to be discarded. For example, Keuper *et al.* (2004) describes an illumination system of this type based on red, green and blue LEDs. While Keuper focuses on the illumination path, this basic design is used in DMD-based 'pocket projectors'. Figure 10.21 shows a photograph of a working pico-projector prototype with laser illumination from SCRAM technologies (Shanley 2007) that uses this architecture with a LCoS spatial

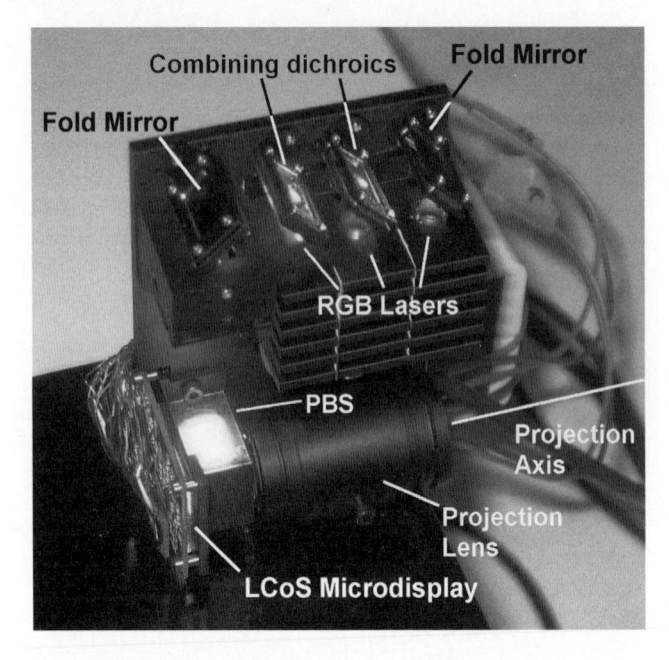

Figure 10.21 Laser pico-projector demonstration unit from SCRAM Technologies (photo by M.S. Brennesholtz).

light modulator. Since the lasers have polarized output, there is no throughput penalty associated with the polarization-dependent LCoS panel.

When lasers and LEDs are used in this design, the requirement that the average power be a constant for all three light sources does not apply. Therefore, the white balance can be achieved by either by:

1. Maintaining the 33 % duty cycle for each light source and varying the power to achieve the right white balance; or

2. Maintaining the nominal power for each laser or LED while varying the time allocated to each color.

Either of these approaches, or a combination of the two, will give the correct white point. This white point can be made programmable by the end user if desired.

Since lasers or LEDs normally give a larger color gamut than is needed for an application, the colors can be mixed slightly to give the exact desired color targets. For example, the red from a laser or LED is normally 'redder' than required. By leaving the blue and green laser or LED on at low power during the red field not only is the color primary brought to the target primary but the system lumen output is normally increased.

10.4.2.4 Address-and-flash systems

The 'address-and-flash' projector architecture (Koma *et al.* 2000) is an adaptation of the three light source field sequential system for displays, particularly liquid microdisplays, that lack global reset addressing. Physically, the system can look like the ones shown in Figure 10.20 or 10.21. In these systems, if a high-duty cycle is used for the three light sources, serious and unacceptable color artifacts can occur when a section of the microdisplay has the LC in the state for one color but the illumination light is for a different color. In this architecture, the light sources are turned off while the data are loaded into the memory and left off while the liquid crystal of the last row addressed settles into the desired state. This settling time is shown as T_{LC} in Figure 10.22.

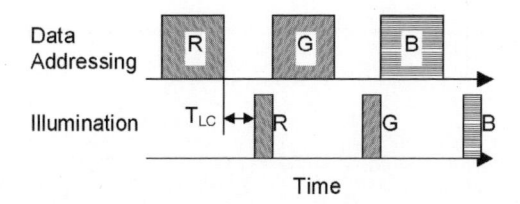

Figure 10.22 Timing diagram for an address and flash projector.

In this design, the light source 'On' time is typically a very low percentage of the total time. Most of the time is consumed with addressing the microdisplay and allowing the liquid crystal to settle into its final state. Therefore, this design is not normally suited for a high-output projection system. It is, however, often used in head mounted displays where throughput is not a major issue.

10.4.2.5 Rotating drum systems

If the rotating color wheel is not put at a pupil position but is instead imaged onto the microdisplay, the division between colors sweeps across the microdisplay as shown in Figure 10.23. This figure shows the image of the division (spoke) between the red and green dichroics at three slightly different color wheel rotation angles. Above the division, the microdisplay is illuminated by green light and below the division the illumination is red light. In Figure 10.23(a) the spoke is near the top of the panel and everything above the spoke is illuminated with red light and everything below is illuminated with green light. At this instant in time, the panel is addressed with red data at row n_1, the average position of the spoke. Note there are two regions of color error. In region A, the panel has red data but still has green light illuminating it. In region B, the panel has green data on it but is now illuminated with red light. In the center of the panel, Figure 10.23(b), the spoke is horizontal, data are addressed to row n_2 and there are no regions where there is a data and illumination mismatch. In Figure 10.23(c) near the bottom of the panel, the data is loaded into row n_3 and the color mismatch reappears, but is reversed left to right.

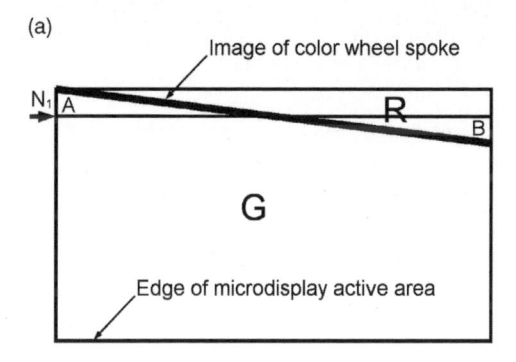

Figure 10.23 Scanning action produced by a color wheel on a microdisplay without global reset addressing: (a) spoke at top of panel; (b) spoke at center of panel; (c) spoke at bottom of panel.

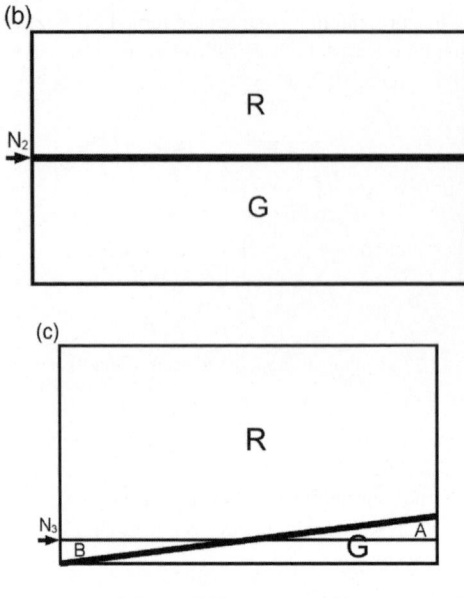

Figure 10.23 (Continued).

If the color wheel is large enough in diameter and the spoke width with no illumination is wide enough, the actual color errors produced by this system will not be large and may be acceptable, especially in low-cost systems. Large diameter color wheels are expensive; however, wide spokes waste light and most HDTV applications require very good colorimetry. Therefore, projector designers have searched for alternative geometries to avoid these color errors. One solution is to use a color drum instead of a color wheel.

If instead of a color wheel, the dichroic filters are arranged in a cylindrical shape around the rotation axis, it is possible to image the divisions between the dichroics as straight and horizontal lines on the microdisplay, avoiding the type A and type B errors shown in Figure 10.23 (Gleckman 2001; Brennesholtz 2002; Brennesholtz et al. 2005; Park et al. 2005).

While some drums have been made as true cylinders, it is simpler in production to assemble them from flat segments, as shown in Figure 10.24. Note that this drum not only has flat segments, but the segments are not all the same size. Since the drum was intended for use with a UHP lamp, the red segment is larger in order to provide the proper color gamut. Drum projectors, like color wheel projectors, are simple to design with more than three primary colors. For example, the system described in Brennesholtz et al. (2005) had five primary colors, with yellow and cyan added to the normal red, green and blue.

The optics to image the straight division between adjacent segments onto the panel can be quite complex and expensive compared to the optics of a color wheel. Therefore, drum projectors are not commonly used.

10.4.2.6 Scrolling color systems

A novel approach to single-panel projection has been demonstrated which does not incur the duty-cycle light losses of normal color-sequential operation or use subpixelation (Janssen et al. 1993). While this is a color sequential system, none of the light is rejected by color filters. Rather, three different parts of the panel are used simultaneously for the three color bands. The approach is illustrated in Figure 10.25, which shows the simplified optical path for this system. The light from the HID lamp is separated in the

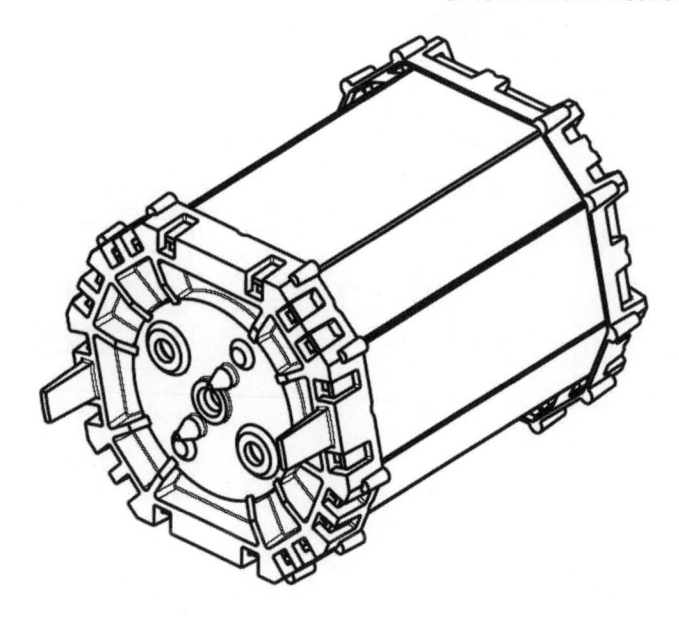

Figure 10.24 Color drum assembled from flat dichroic segments (reproduced by permission of Royal Philips Electronics).

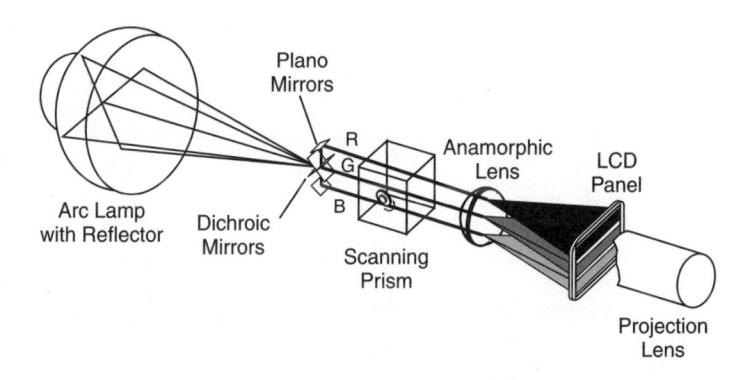

Figure 10.25 Optical system for a single-panel projector based on time-multiplexing of the panel (from Janssen 1993; with permission of the SID).

three color bands with dichroic filters. This light is formed into color bars with an aspect ratio of about 16:3 for a 16:9 display. The light passes through a rotating prism which scans these three beams across the microdisplay. The three beams are separated by narrow dark bands without illumination.

As the prism rotates 90°, the three color stripes scroll down the panel, as illustrated in Figure 10.26. The optical system is configured so that when a color leaves the bottom of the panel, an apex of the prism passes through the corresponding color stripe and causes that color to jump to the top of the panel. The cycle repeats for each of the colors. This action results in the name given to this architecture: scrolling color, single-panel. As in other color sequential systems, color fusion takes place in the human visual system if the scrolling rate is fast enough.

258   MICRODISPLAY SYSTEM ARCHITECTURES

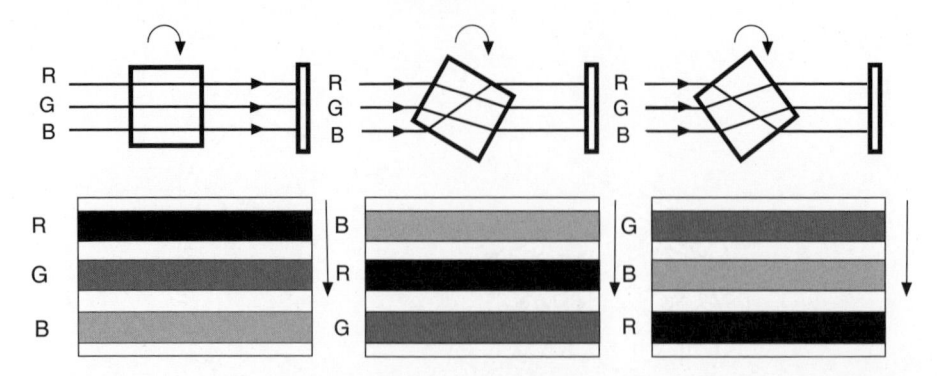

Figure 10.26 Scanning action produced by rotating prism (from Janssen 1993; with permission of the SID).

The single large prism as shown in Figures 10.25 and 10.26 proved both to be too expensive for production and to introduce too many aberrations in the off-axis colors, red and blue. This large prism was replaced by three small prisms, one to scan each color. This allowed on-axis operation for all colors (Shimizu 2001).

The electrical addressing of the panel is more complex than other sequential systems. Instead of writing each color field in succession, different parts of each field are written onto the panel at the same time. As an example, consider a panel having 480 total rows and the color bands and intervening dark bands each occupying 80 rows. At the instant of time shown in Figure 10.26(a), red band is shown to occupy rows 1–80, the green band rows 161–240, and the blue band rows 321–400. The other rows are not illuminated at this instant of time.

It is presumed for now that each of the illuminated rows has the appropriate signal level for the incident color. As the colors scroll, lines 1, 161 and 321 become no longer illuminated. In succession, line 1 would now receive the electrical signal for blue information, line 161 would receive the appropriate red signal, and line 321 would receive the green signal. This process now repeats in synchronism with the color scan for lines 2, 162, and 322, then 3, 163 and 323 and so forth. During the time it takes the dark interval to move 80 lines, the transmission of the LC switches from its previous color state to the new one. Each of the lines is then illuminated with the color appropriate to the information on that line. In this manner, all three colors are on the panel simultaneously. While this example assumes the color stripes are equal sizes, they need not be. For example, if a UHP lamp is used and the system is red-deficient, it is possible to make the red stripe larger than the blue and green to achieve the required white point. The relative addresses, of course, must be adjusted to match the stripe widths.

This architecture avoids one source of light loss that occurs with other color-field sequential systems by utilizing all three colors simultaneously, rather than discarding two at any given time. However, the effective area of the panel is reduced by the fraction of the total panel area occupied by one color band. Thus, if one of the color bands is one-sixth of the panel height, the effective area of the panel from an illumination point of view is one-sixth of its geometric area. This has implications for the light throughput of the system, discussed in Chapter 11. Typically, a scrolling color projector would need to use a $3 \times$ larger panel (by area) than a panel in a three-panel system in order to have a light output comparable to the output of the three-panel projector. This would lead to a diagonal that would typically be $1.7 \times$ the diagonal of the panel in a three-panel system.

When large panels are used and étendue is only a minor issue, scrolling color designs like this are more efficient than drum projectors (Janssen 2002). As panels become smaller, there is a crossover point and the drum architectures become more efficient than the scrolling color design. Since one of the major goals in projector design is to reduce the projector cost and one of the major tools in reducing cost is reducing panel size, this is a powerful argument in favor of color wheel or drum projectors.

10.5 Two-panel Systems

With the higher cost and complexity of three-panel architectures and the lower performance and/or increased complexity of one-panel approaches, compromises have been sought in which two panels are used. In these, it is most common to use one panel for one color band and to use the second for the other two. The choice of color usage is frequently based on the spectral content of the light from the lamp. If the light source is somewhat deficient in one of the color bands, that part of the spectrum would be directed to a dedicated microdisplay. The other two bands would share the second microdisplay, possibly by using a color wheel. The weaker color band illumination is used for twice the time of the other two. The effect of this is to improve the color balance as well as the system throughput.

An example of this architecture is shown in Figure 10.27 with two DMDs (Florence and Yoder 1996). The lamp in this system is an ultra-high pressure mercury lamp which is red deficient. A color wheel filters the lamp light alternately into yellow and magenta. Since yellow consists of the red plus green bands and magenta consists of the red plus blue bands, the output of the filter wheel is a continuous red superimposed with alternating green and blue. The color splitting prism directs the blue and green light to DMD-1 and the red light to DMD-2. The blue/green light path thus constitutes a color sequential system where each color field is now one-half of the frame time. The design of the optical path combines and converges the modulated light from the two DMDs for projection with a single lens.

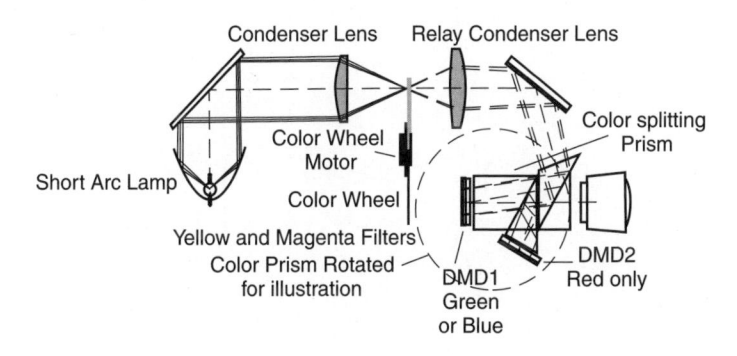

Figure 10.27 Two DMD system architecture (from Florence and Yoder 1996; with permission of SPIE).

Compared with a single DMD system where red light is passed only one-third of the time, the two DMD system yields about three times the amount of red light output. The blue and green outputs are also increased by about 50 % since each is now on for one-half of the time compared with one-third of the time for a single microdisplay approach.

The temporal response requirements for this two-band panel are significantly relaxed as compared with single-panel color-field sequential systems. A normal color sequential projector must run at no less than 360 Hz field rate, and preferably 540 Hz or more, to avoid color breakup in the image (see Section 14.2.3). Since red is on continuously, color breakup is greatly suppressed and the green/blue channel need only run at a field rate of about 240 Hz, or even less. The green channel would normally run at a 120 Hz field rate to ensure proper inversion of the LC, as described in Section 4.2.6.

10.6 Schlieren Optics-based Projectors

The Schlieren method (Hecht 1990) was originally developed for the study of defects in lenses. It permits small optical effects, such as localized index of refraction variations, to be visualized through their effects on the changes of the angles produced by the effects. Schlieren optical systems have been used in

260 MICRODISPLAY SYSTEM ARCHITECTURES

projectors in which the light-valve action is to produce an angular separation between the modulated and unmodulated beams. They are among the earliest optical configurations employed in electronic projection (Fischer 1940; Glenn 1958, 1970; Johannes 1989). While these oil-film based projectors have vanished from the projection scene, there is considerable current interest in Schlieren systems based on diffraction from MEMS and LC devices.

10.6.1 Dark Field and Bright Field Systems

Two types of projection Schlieren systems are used, viz., dark field and bright field optics. These are illustrated in Figures 10.28 and 10.30, respectively.

A. When the projection panel does not change the angle of light passing through it, the lamp is imaged on the output stop and no light passes through the lens to the projection screen.

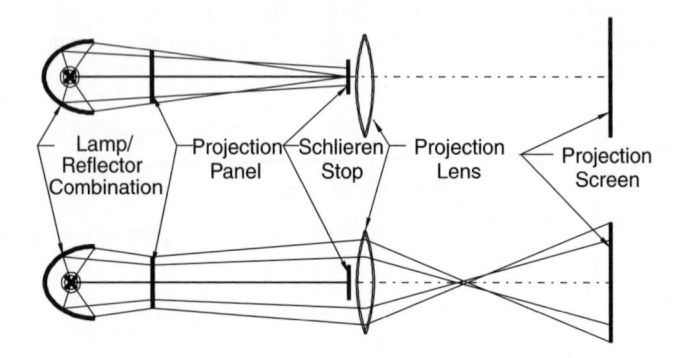

B. When the projection panel changes the angle of light passing through it, the lamp is no longer imaged on the schlieren stop and most of the light pass through the lens to the projection screen.

If not all of the light is modulated, any unmodulated light hits the stop and does not reach the projection screen.

The lamp may be imaged either by the reflector and condensor lens, (as shown) or by a separate schlieren lens.

Figure 10.28 Dark field Schlieren optical path.

The dark field system does not permit light into the projected beam in the absence of a perturbing element, such as a modulation on the light valve, because the lamp is imaged onto the output stop. When the projection panel changes the angle of light passing through it, the lamp is no longer completely imaged onto the stop. The fraction that misses the stop is imaged onto the screen. With defect-free optics, dark-field Schlieren systems are very effective in producing high contrast images. However, the requirement for defect-free optical materials and surfaces for dark field Schlieren translates to expensive, difficult-to-fabricate optical systems.

Two high output projection systems using dark field Schlieren optics were the Eidophor® (Johannes 1989) and Talaria® (Glenn 1958) systems. Both of the systems were based on diffractive effects in an oil film and were electron-beam addressed. Figure 10.29 shows one of the color light paths in the Eidophor system. The oil film is coated on the surface of a spherical mirror. An electron-beam scans a raster at constant intensity and variable speed, which is determined by a voltage with a variable amplitude and

SCHLIEREN OPTICS-BASED PROJECTORS   261

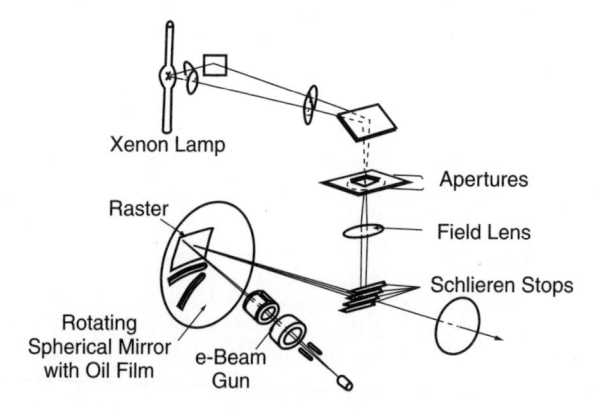

Figure 10.29 Schematic diagram of one channel in an Eidophor projector (from Williams 1997; with permission of SPIE).

constant frequency superimposed on the sweep voltage. The voltage induced on the oil film causes this film to deform. If the charge from the e-beam is deposited uniformly, the surface of the oil film is not deformed and the light reflected from the spherical mirror is intercepted by the Schlieren bars and a dark field is produced.

When signal is present, the amount of diffracted light passing the Schlieren bars depends on the depth of modulation, which in turn is dependent on the amplitude of the alternating scan frequency. Eidophor projectors use separate modulators for each color. Talaria projectors were similar in principle except they used transmissive oil films. Different versions of Talaria employed modulation techniques which achieved full color imaging with one, two, or three light valves.

A. When the projection panel does not change the angle of light passing through it, the lamp is imaged on the pupil of the projection lens and all light passes through the lens to the projection screen.

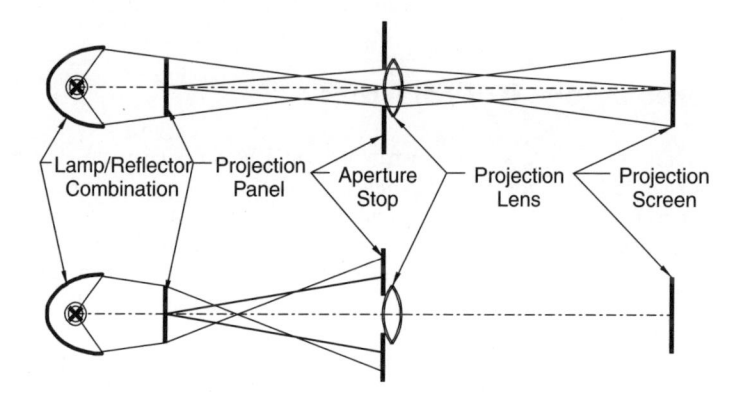

B. When the projection panel changes the angle of light passing through it, the lamp is no longer imaged on the pupil of the projection lens and little light passes through the lens to the projection screen.

Figure 10.30 Bright field Schlieren optical path.

Figure 10.30 illustrates the bright field Schlieren optical path. When the projection panel does not change the angle of the light passing through it, the lamp is imaged onto the pupil of the projection lens and all light passes through the lens to the projection screen. When the projection panel changes the angle of light passing through it, some of the light is intercepted by the aperture stop. These systems are most useful when the angular deviations are large and the beam can be completely modulated. If the modulation is less than 100 %, all the unmodulated light will reach the screen, spoiling the contrast.

10.6.2 Schlieren Light-valve Systems

There are several microdisplay and light-valve technologies developed after oil films which are used in Schlieren or Schlieren-like systems. Included among these are LC diffraction grating based microdisplays (Komanduri *et al.* 2007), polymer-dispersed liquid crystal (PDLC) microdisplays (Coates 1993), DMDs (Florence and Yoder 1996), and the deformable grating light valve (DGLV) (Apte *et al.* 1993, 1994).

Figure 10.31 illustrates the operation of a Schlieren light-valve projection system based on reflective PDLC valves. Schlieren projection systems can have any of the single-panel, two-panel or three-panel architectures discussed in this chapter. They can also have either one, two or three projection lenses.

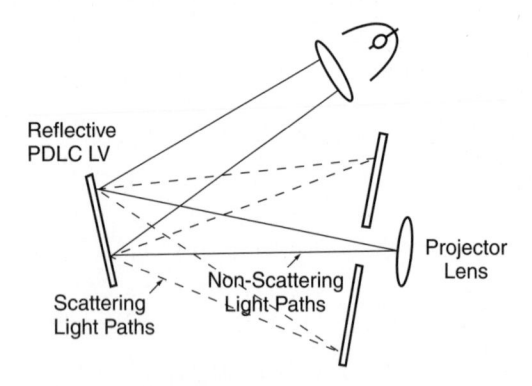

Figure 10.31 Schematic operation of one color channel in PDLC Schlieren-like projector.

10.7 Stereoscopic 3D Projectors

Projection of stereoscopic 3D images has been a part of the film industry since almost the start of movies with the earliest known 3D film, *L'Arrivee du Train*, shown in 1904. Stereoscopic 3D is difficult to do with film projection and the 3D fad in the 1950s ended in part because the projection technology at the time was not satisfactory for showing 3D movies in the mass-market. With the introduction of digital cinema, however, 3D movies can now be shown almost as easily as 2D movies (Lipton 2001). The Digital Cinema Initiative has amended the DCI specification to accommodate 3D movies (DCI 2007).

Stereoscopic 3D images are made when a slightly different image is presented to each eye. All practical stereoscopic projection technologies require the viewer to wear glasses of some type to separate the left eye and right eye images. When viewed without glasses, the two images overlap and present a confusing image to the viewer.

There are three common techniques to separate the left eye image from the right eye image:

1. Use orthogonal polarization states for the two images.

2. Use different wavelengths of light to project the two images and separate them with color filters.

3. Separate the two images in time in an eye sequential system.

One of the big problems with 3D applications is the low throughput, typically 14 %, of a 3D projector compared to the throughput of a similar projector showing 2D content. None of the three separation techniques currently has a major advantage over the other two in terms of throughput.

10.7.1 Separation by Polarization

When polarization is used to separate the two images, the orthogonal states can either be two perpendicular linear polarization states or left and right circularly polarized light. Typically, linear polarization states provide better left eye/right eye separation when the viewer is right on axis and does not tip his/her head. Circular polarization will provide a larger viewer space since it is more tolerant of off-axis viewing and head tipping in the audience.

This increased separation on-axis of linear polarization means that it is used in some visualization and simulation applications where the maximum possible separation is needed. For more general purpose professional applications circular polarization is used because of the larger audience space. While 3D cinema requires a large audience space, it normally uses linear polarization instead of circular polarization because of the low cost of the disposable glasses used with linear polarization systems. The polarization states are normally at $\pm 45°$ to the horizontal rather than horizontal and vertical.

The two different polarized images can be produced by a single projector, if the projector is fast enough to produce the required field rates. In this case, a polarization switch is used at the output lens of the projection system (Sharp and Robinson 2007). This technology is the basis of the RealD 3D digital cinema system, the most common system in use for 3D cinema as of this writing.

If the projector is not capable of the required frame rates, it is possible to stack two projectors and put orthogonal passive polarizers in front of each projection lens. This is the technology used in most of the 3D movies shown in the 1950s. Two side-by-side film projectors were synchronized and showed the two films with left and right eye images. Linear polarizers in front of each projection lens polarized the two images differently. The audience then wore linear polarized glasses to separate the two images.

With the low prices and high outputs of modern projectors, stacking a pair of projectors is a legitimate option for a low-cost stereoscopic 3D system. For example, the GeoWall Consortium (GeoWall 2008) provides guidance on construction of a 3D stereoscopic visualization system based on stacked 2D projectors. The consortium estimates that a complete 3D visualization system can be constructed for $6000–$8000.

When polarization is used to separate the image in a 3D system, the projection screen used must be a polarization-preserving screen. While the most commonly encountered screens involve scattering and are not polarization-preserving, both front and rear projection screens are available that do not affect the polarization of the image. The polarization preserving screens used in cinema applications tend to be high gain screens compared to the normal low gain screens used in most cinemas. This high-gain screen partially compensates for the low throughput of a 3D projector compared to a similar projector showing 2D material.

10.7.2 Stereoscopic 3D with Color Separation

Separation of the left and right image by color has a long history. The classic red/cyan anaglyphic comic book is perhaps the best-known example. While this technique is low cost to implement and can be used in projection applications, the image quality is poor and the image is normally not perceived in full color.

One modern technique does provide full color imaging when color separation is used for the left and right eye images. This technique, with the trade name Infitec, involves the use of slightly different primary colors for the left and right eyes (Jorke and Fritz 2006.) The wavelength separation of the primary color pairs is enough to allow separation by passive color filters, with a slightly different filter design for each eye.

For example, the green for the left eye could be centered on 535 nm and the green for the right eye could be centered on 523 nm. This 12 nm wavelength shift is sufficient to be separated by passive filters.

Each eye would have a triple-notch filter, with the filter on one eye passing one set of red, green and blue wavelengths and the filter on the other eye passing the other set of red, green and blue wavelengths.

These filters can be dichroic filters and dichroics were used in the initial tests of this system. The wavelengths of a dichroic filter change with tilt, as was discussed in Section 6.3.2.1. This causes two problems. First, if viewers rotate their heads the wavelengths of the filters shift and the left eye/right eye separation is degraded. The second problem is also caused by the angular sensitivity of dichroic filters. Dichroic filters of sufficient wavelength selectivity are difficult but not impossible to make on curved surfaces and historically they were made on flat surfaces. This gave the glasses a 'geeky' appearance felt to be unacceptable for cinema applications. The high reflectivity of the filters on a flat substrate also allowed images of bright objects in the theater such as exit signs to be seen by the viewer. Special techniques have been developed to deposit filters of sufficient wavelength resolution on spherical surfaces for the lenses, essentially eliminating the problems. Other filter technologies, such as the wavelength selective retarder technology described in Section 6.3.4 could also be used for the passive glasses. Since these filters require polarized light, they are not used unless the output light from the projector is already polarized.

The Dolby 3D cinema technology is based on this Infitec technology. Dolby installs a color wheel inside a standard digital cinema projector. Both halves of this color wheel appear nearly white, since they both transmit portions of the red, green and blue spectrum. These portions do not overlap, however. When the projector shows 2D material, the color wheel retracts from the optical path. For 3D material, the color wheel is inserted in the optical path and rotates in synchronous with the image content.

Like polarization separated systems and the eye sequential systems to be discussed in the next section, the throughput of a wavelength selective 3D system is typically very low, about 14 % of the throughput of the same projector showing 2D material. One advantage wavelength selection has over polarization selection is there is no need to install a special screen in the theater to show 3D. On the other hand, the low efficiency of the Infitec system may require use of a high-gain screen to provide sufficient brightness, requiring a screen change anyway.

Lasers have an obvious utility in a wavelength dependent 3D projector. Their narrow bandwidth will make it relatively easy to design filters to separate left and right eyes. They can also be switched on and off at the field rate, eliminating the light loss at the color wheel.

The simple version of this system will produce slightly different color gamuts for the two eyes. This problem can in most cases be ignored since the eye/brain combination has a remarkable ability to fuse the two images into a single full color 3D image. If necessary, normal color correction techniques can be used to ensure each eye's image covers the same color gamut.

10.7.3 Eye-sequential 3D Systems with Active Glasses

In an eye-sequential system, the images are presented at different times. This technique was used in a commercial cinema as early as 1924, using mechanical shutters built into the seats and synchronized with the projectors. In modern systems, the viewer wears active LCD glasses that alternately block one eye and then the other in synchronous with the image presentation. Normally, this is done with an IR synchronization signal broadcast from near the projection screen, so it can be received clearly by all sets of active glasses in the theater. Since the image is blocked by the active shutter glasses, there is no head tilt or audience space issue associated with this technology.

These active glasses are normally made with non-pixelated LCD cells. ECB, FLC and π-cells are three transmissive LC effects that are fast enough to be used in these types of glasses but other effects can be used too. Note that these shutter glasses have a polarizer before the LC cell and an analyzer after the cell. In unpolarized light, the cell will have at most a 50 % transmission since it can only use one polarization of light from the projector. Each eye is also blocked for at least 50 % of the time. This 50 % duty cycle plus the theoretical maximum 50 % transmission of the shutter glasses leads to an image luminance of a theoretical maximum of only 25 % of the 2D image luminance. Realistically, the luminance is more typically about 14 % of the 2D image luminance.

In order to avoid flicker and motion artifacts, each eye must see its corresponding image at a field rate of at least 48 Hz, with 60Hz or higher preferred. This means the projector must be capable of producing at least 96 distinct images per second, and preferably at 120 or more. The field rate of choice for 3D digital cinema is triple-flashing each frame of the film for each eye. This gives each eye a 72 Hz field rate, for a total of 144 displayed fields per second. DLP projectors can achieve this field rate easily. Some of the current generation of consumer DLP rear projection systems are sold as '3D ready'. Many normal consumer and professional DLP front projectors are also able to show 3D images (Woods and Rourke 2007).

References

Akimoto, O. and Hashimoto, S. (2000) High-resolution FLC Microdisplay. In M.H. Wu (ed.) *Proc. SPIE:* Projection Displays, **3954**, 104–10.
Apte, R.B., Sandejas, F.S.A., Banyai, W.C. and Bloom, D.M. (1993) Deformable grating light valves for high-resolution displays. *SID Intl. Symp. Digest of Technical Papers*, 807–8.
Apte, R.B., Sandejas, F.S.A., Banyai, W.C. and Bloom, D.M. (1994) Grating light valves for high resolution displays. *Solid State Sensors and Actuators Workshop*, Hilton Head Island, SC, June.
Bleha, W.P. (1995) Development of ILA projectors for large screen display. *Intl. Display Res. Conf. Record*, 91–3.
Bleha, W.P. (1997) Image Light Amplifier (ILA) technology for large screen projection. *J. SMPTE* **106**: 710–17.
Bleha, W. P. and Sterling, R. D. (1999) Digital cinema using the ILA projector. *Proceedings of the 6th International Display Workshops (IDW-99)*, 1009.
Bone, M. (2001) Method to improve contrast in an off-axis projection engine. *SID Intl. Symp. Digest of Technical Papers*, Paper 46.2, 1180–3.
Borel, T., Morvan, P., Sarayeddine, K., Rommeveaux, P., Becker, W. et al. (2006) 0.83″ Full HDTV liquid crystal on silicon component with in-pixel memory and 450Hz refresh rate. *SID Intl. Symp. Digest of Technical Papers*, Paper P-212, 1027–30.
Bowron, J., Baker, J. and Schmidt, T. (1996) New high-brightness compact LCD projector. In M.H. Wu (ed.), *Proc. SPIE: Projection Displays II*, **2650**: 217–24.
Bowron, J. and Schmidt, T. (1998) A new high resolution reflective light valve projector: Electrohome 1280. In M.H. Wu (ed.), *SPIE Proceedings: Projection Displays IV* **3296**: 105–15.
Brennesholtz, M.S. (2002) Color-sequential LCoS projector with a rotating drum. *SID Intl. Symp. Digest of Technical Papers*, Paper 51.4, 1346–9.
Brennesholtz, M.S. (2003) Optical and electronic addressing modes for single-panel display systems. *IDMC Proceedings* (SID Taipei chapter), Paper 24-3, 481–4.
Brennesholtz, M.S., McClain, S.C., Roth, S. and Malka, D. (2005) A single-panel LCOS engine with a rotating drum and a wide color gamut. *SID Intl. Symp. Digest of Technical Papers*, Paper 64.3, 1814–17.
Bridgwater, R.J., Luthra, A., Harding, J.R., Blackham, G.H. and MacPherson, I. (2006) Dynamic range enhancement of image display apparatus. US Patent 6,985,272, issued 10 January 2006, assigned to SEOS Ltd.
Brown Elliott, C.H. (2005) Color flat panel display sub-pixel arrangements and layouts. US Patent 6,950,115, issued 27 September 2005, Assigned to Clairvoyante, Inc.
Bruzzone, C.L., Ma, J., Aastuen, D.J.W. and Eckhardt, S.K. (2003) High-performance LCoS optical engine using Cartesian polarizer technology. *SID Intl. Symp. Digest of Technical Papers*, Paper 10.4, 126–9.
Chen, J., Robinson, M.G., Sharp, G.D. and Birge, J.R. (2006) Three-panel color management systems and methods. US Patent 7,002,752, issued 21 February 2006, assigned to ColorLink.
Chen, L.M. and Hasegawa, S. (1992) Visual resolution limits for colour matrix displays. *Displays Technology and Applications* **13**: 179–86.
Coates, D. (1993) Normal and reverse mode polymer dispersed liquid crystal devices. *Displays* **14**: 94–103.
DCI (2007) Digital Cinema Initiatives, LLC, *Stereoscopic Digital Cinema Addendum*, Version 1.0, issued 11 July 2007.
Doany, F.E., Singh, R.N., Rosenbluth, A.E. Chiu, G.L.-T. (1998) Projection display throughput: efficiency of optical transmission and light-source collection. *IBM J. of Research and Development* **42**(3/4): 387–400.
Fischer, F. (1940) Auf dem Wege zur Fernseh-Grossprojektion. *Schweiz. Arch. Angew. Wiss. Technik* **6**: 89–106.

Florence, J.M. and Yoder, L.A. (1996) Display system architectures for Digital Micromirror Device (DMD) based projectors. In M. Wu (ed.), *Proc. SPIE: Projection Displays II* **2560**: 193–208.

Gale, R.P. and Swanson, G.J. (1997) Efficient illumination of color AMLCD projection displays using binary optical phase plates. *SID Intl. Symp. Digest of Technical Papers*, 765–8.

GeoWall (2008) http://geowall.geo.lsa.umich.edu

Gleckman, P. (2001) Color *Projection System Incorporating Electro-Optic Light Modulator and Rotating Light-Reflective Element*, US Patent 6,266,105, issued 24 July 2001.

Glenn, W.E. (1958) New color projection system. *J. Opt. Soc. Am.* **48**: 841–3.

Glenn, W.E. (1970) Principles of simultaneous-color projection television using fluid deformation. *SMPTE Journal* **79**: 788–94.

Hamada, H., Nakanishi, H., Funada, F. and Awane, K. (1994) A new bright single-panel LC-projection system without a mosaic color filter. *Intl. Display Res. Conf. Record*, 422–3.

Harbers, G., Keuper, M. and Paolini, S. (2003) Performance of high power LED illuminators in color sequential projection displays. *Proceedings of the 10th International Display Workshops* (IDW), Fukuoka, Japan, (ITE & SID), Paper LAD3-4, 1585–8.

Hecht, E. (1990) *Optics*, 2nd edn. Reading, MA: Addison Wesley.

Huang, H.C., Leung, K.Y., Kwok, H.S., Zhang, B.L. and Chen, Y.C. (2007) Single-panel spatial-color and sequential-color LCOS projectors using LED lamps. *SID Intl. Symp. Digest of Technical Papers*, Paper P-188, 903–6.

Huibers, A., Richards, P.W., Grasser, R. (2006) New micromirror device for projection applications, *SID Int. Symp. Digest of Technical Papers*, Paper 50.5L, 1618–21.

Janssen, P.J. (1993) A novel single light valve high brightness HD color projector. *Intl. Display Res. Conf. Record*, 249–52.

Janssen, P.J., Shimizu, J.A., Dean, J. and Albu, R. (2002) Design aspects of a scrolling color LCoS display. *Displays* **23**: 99–108.

Johannes, H. (1989) *The History of the EIDOPHOR Large Screen Television Projector*. Zürich: Gretag, Regensdorf.

Jorke, H. and Fritz, M. (2006) Stereo projection using interference filters. In A.J. Woods *et al.* (eds), *SPIE Proceedings Stereoscopic Displays and Virtual Reality Systems XIII* **6055**, January.

Keuper, M.H., Harbers, G., Paolini, S. (2004) RGB LED illuminator for pocket-sized projectors. *SID Intl. Symp. Digest of Technical Papers*, Paper 26.1, 943–5.

Koma, N., Miyashita, T., Uchida, T. and Mitani, N. (2000) Color field sequential LCD using an OCB-TFT-LCD. *SID Intl. Symp. Digest of Technical Papers*, Paper P-28, 632–5.

Komanduri, R.K., Jones, W.M., Oh, C. and Escuti, M.J. (2007) Polarization-independent modulation for projection displays using small-period LC polarization gratings, *Journal of the SID* **15**(8): 589–94.

Ledebuhr, A.G. (1986) Full-color single-projection-lens liquid-crystal light valve projector. *SID Intl. Symp. Digest of Technical Papers*, 379–82.

Lipton, L. (2001) The stereoscopic cinema: from film to digital projection, *SMPTE Journal*, September, 586–93.

Loiseaux, B.A., Joubert, C., Delboulbé, A., Huignard, J.P. and Battarel, D. (1995) Compact spatio-chromatic single-LCD projection architecture. *Intl. Display Res. Conf.* (Asia Display '95) Proceedings, 87–9.

McKechnie, S., Goldenberg, J., Eskin, J., Shimizu, J., Bradley, R. *et al.* (1989) *Display System with Equal Path Lengths*. US Patent 4,864,390.

Meuret, Y. and De Visschere, P. (2003) Optical engines for high-performance liquid crystal on silicon projection systems, *Optical Engineering* **42**(12): 3551–6.

Park, J.M., Yoo, E.H., Kim, Y.W. and Brennesholtz, M.S. (2004) A 1080 line, single-panel LCOS rear projection system with a rotating drum. *SID Intl. Symp. Digest of Technical Papers*, Paper 12.4, 178–81.

Pentico, C., Newell, M. and Greenberg, M. (2003) Ultra high contrast color management system for projection displays. *SID Intl. Symp. Digest of Technical Papers*, Paper 10.5, 130–3.

Robinson, M.G., Korah, J., Sharp, G. and Birge, J. (2000) High contrast color splitting architecture using color polarization. *SID Intl. Symp. Digest of Technical Papers*, Paper 9.3, 92–5.

Robinson, M.G., Chen, J. and Sharp, G.D. (2006) Three-panel LCOS projection systems. *J. of the SID* **14**(3): 303–10.

Sampsell, J. (1993) An overview of the digital micromirror device (DMD) and its application to projection displays. *SID Intl. Symp. Digest of Technical Papers*, 1012–15.

Schadt, M. and Schmitt, K. (1997) Flat liquid crystal projectors with integrated cholesteric color filter/polarizers and photo-aligned optical retarders. *Intl. Display Res. Conf. Record,*. 219–22.

Shanley, J. (2007) LED/laser projection – a reality check. *Projection Summit 2007*, Session MS-2. Norwalk CT: Insight Media, 19 June 2007.

REFERENCES

Sharp, G.D. and Robinson, M.G. (2007) Enabling stereoscopic 3D technology, stereoscopic displays and virtual reality systems XIV. In A.J. Woods *et al.*, *Proc. of SPIE* **6490**.

Shikama, S., Kida, H., Daijogo, A., Maemura, Y. and Kondo, M. (1993) A compact LCD rear projector using a new bent-lens optical system. *SID Intl. Symp. Digest of Technical Papers*, 295–8.

Shimizu, J. (2001) Scrolling color LCOS for HDTV rear projection. *SID Intl. Symp. Digest of Technical Papers*, Paper 40.1, 1072–5.

Shimizu, J. and Janssen, P.J. (1996) *Compensation Plate for Tilted Plate Aberration.* US Patent 5,490,013.

Sperger, R., Edlinger, J., Rudigier, H., Schoch, H. and Weirer, P. (1993) High performance patterned all-dielectric interference color filters for display application. *Intl. Display Res. Conf. Proceedings*, 81–3.

Stanton, D., Shimizu, J. and Dean, J. (1996) Three-lamp single-device projector. *SID Intl. Symp. Digest of Technical Papers*, 839–42.

Takamatsu, T., Ogawa, S., Hamada, H., Funada, F., Ishii, M. *et al.* (1992) Single-panel LC projector with a planar microlens array. *Intl. Display Res. Conf. (Japan Display) Proceedings*, 875.

Teijdo, J. M., Ludley, F., Ripoll, O., Ueda, M., Oshima, Y. *et al.* (2006) Compact three-panel LED projector engine for portable applications. *SID Intl. Symp. Digest of Technical Papers*, Paper 73.2, 2011–14.

True, T. (1987) High-performance video projector using two oil-film light valves. *SID Intl. Symp. Digest of Technical Papers*, 68–71.

van Kessel, P.F. (2006) Options for enhanced colorimetry. *Proceedings of Projection Summit* 2006. Norwalk CT: Insight Media.

Wang, Y. (1997) Surface plasmon tunable color filter and display device. *SID Intl. Symp. Digest of Technical Papers*, 63–6.

Willemsen, O.H., Krijn, M.P.C.M. and Salters, B.A. (2005) A handheld mini-projector using LED light sources. *SID Intl. Symp. Digest of Technical Papers*, Paper 58.4, 1698–1701.

Williams, S.A. (1997) History of Eidophor projection in North America. In M.H. Wu (ed.), *SPIE Proceedings: Projection Displays III* **3013**: 7–13.

Woods, A.J. and Rourke, T. (2007) The compatibility of consumer DLP projectors with time-sequential stereoscopic 3D visualization. In A.J. Woods *et al.* (ed.), *Proc. of SPIE: Stereoscopic Displays and Virtual Reality Systems XIV* **6490**.

Yakovenko, S., Konovalov, V. and Brennesholtz, M. (2004) Lifetime of single-panel LCOS imagers. *SID Intl. Symp. Digest of Technical Papers*, Paper 6.2, 64–7.

11

Modeling Lumen Output

In this chapter, two models are developed to estimate the throughput of light-valve projectors prior to construction of a prototype. The simplified model employs a straightforward application of the effects of each of the components in the optical path. To account for the effects of color separation and color correction, a full colorimetric model is developed. The concept of étendue, used to estimate the collection efficiency of lamp/reflector combinations, is discussed in detail.

Lumen output of a projector is the property of the light engine and is not normally affected by factors external to the light engine. These external factors include video input format, format conversion techniques, or front vs rear projection. Neither do image size, projection screen technology, screen gain and screen contrast affect light engine throughput, although they obviously affect the perceived brightness of the image.

11.1 Simplified Model

In the projector, the total luminous flux from the lamp is only partially collected by the collection optics. As this collected light traverses the optical system, it is further attenuated by each of the optical elements in the path. A simplified model for this throughput is described by Equation (11.1).

$$\Phi_{out} = \Phi\eta_T = \Phi\eta_\Phi\eta_E\eta_{PC}\eta_I\eta_P\eta_{LV}\eta_A\eta_L\eta_{cc}\eta_S \qquad (11.1)$$

where:
- Φ_{out} is the output lumens from the projector,
- Φ is the total lumen output of the lamp,
- η_T is the total system efficiency,
- η_Φ is the efficiency of light collection by the reflector,
- η_E is the efficiency of the use of the collected lumens,

270   MODELING LUMEN OUTPUT

η_{PC} is the efficiency of polarization conversion,
η_I is the integrator efficiency,
η_P, η_A are the polarizer and analyzer efficiencies,
η_{LV} is the light-valve efficiency,
η_L is the projection lens efficiency,
η_{cc} is the efficiency of color correction and
η_S is the color separation efficiency.

The factor η_E refers to geometric limitations in the optical path and is discussed at length in this chapter.

These efficiency factors are illustrated in Figure 11.1, which shows a typical optical path in a projector. Optical elements, such as field lenses, are not included in this simplified picture. This simplified picture applies to a generic projector. In order to calculate the throughput of a conceptual or real projector, it is necessary to have at least a layout sketch that includes all of the optical elements.

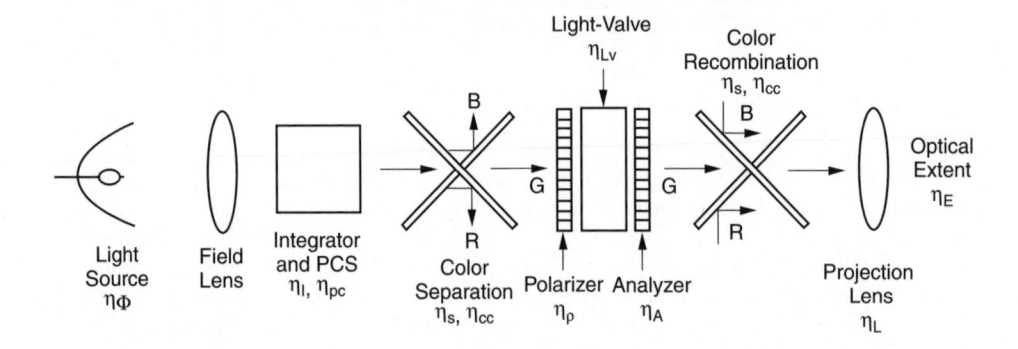

Figure 11.1 Schematic of one color channel in a lamp and microdisplay-based projector illustrating efficiency factors affecting light throughput. Color separation efficiency η_S and color correction efficiency η_{cc} are affected by both the separation and recombination dichroics as well channel gain. η_E is the efficiency associated with optical extent (étendue), discussed below.

The efficiency η_S describes the efficiency with which the light from the lamp/reflector combination is separated into the red, green and blue channels. The efficiency η_{CC} describes the efficiency with which the recombined white light output beam is color corrected to the desired white point. These two efficiencies can be combined into a single value, η_{color}, which describes the overall colorimetric efficiency of a projector. Derivations of numerical values for η_{CC} and η_S are discussed in Sections 11.5.2 and 11.5.3, respectively.

Each of these efficiency values in Equation (11.1) can be subdivided as necessary. For instance, η_{LV} the efficiency of the light valve is the product of the aperture ratio, coating transmission (e.g. ITO) and modulation efficiency.

Strictly speaking, Equation (11.1) only applies when every component in a system is neutral density, except for the color-related components included in η_{color}. In practical terms, however, most optical components are close enough to neutral density for this equation to give a satisfactory lumen throughput estimate, as long as η_{color} is known, and there have been no major changes to the projector design since the value for η_{color} was initially estimated using the techniques in Section 11.5.

Two complete numerical examples of the use of the simplified throughput model are given in Sections 12.1 and 12.2.

11.2 Light Collection and Étendue

The term η_E in Equation (11.1) is a geometrical light utilization term. It is related to the area of the light source (the arc size of a HID lamp or the area of an LED), the area of the limiting aperture in the optical system, and the f/# of the system. This geometric parameter is called 'étendue'. It can have profound effects on system throughput. Lasers typically have very low étendue, normally much smaller than the étendue of a HID lamp or a LED. This can make the design of a laser-based projector very different from the design of a projector for a HID lamp or LED. For laser illumination, normally the factor $\eta_E = 100\%$.

If one wishes to estimate the throughput of a projector prior to building a prototype, or prior even to detailed design of a prototype, one of the most difficult factors to estimate is the efficiency of utilization of the lamp output. For example, a deep parabola or ellipse can collect 85–90 % of the light emitted by an arc lamp. If the size of the resultant light beam is larger than the panel or has a lower f/# than the projection lens, some of this light will be lost at a limiting aperture.

These losses can become severe with small area light valves and/or higher system f/#s. With large light valves and low f/#'s this loss at the limiting apertures may be eliminated. However, the current trend in light-valve projector design is toward smaller light valves, with 0.9 inch (2.3 cm) diagonal light valves being about the largest microdisplays currently used in projectors designed for mass-market applications. 0.55 inch (1.4 cm) diagonal microdisplays are perhaps a typical size and smaller microdisplays are being developed for future systems. Still smaller microdisplays with diagonals as small as 0.21 inch (0.53 cm) and smaller have been proposed for use in pico-projectors.

One way to estimate the collection efficiency is to evaluate the proposed system with an optical ray tracing program that not only traces rays but keeps track of their intensities as well. This approach generally produces the most accurate throughput estimates, short of building and measuring a prototype system. It has the disadvantage, however, of requiring both a high degree of skill in the chosen optical modeling program and a detailed knowledge of the optical design of the proposed projector. An example of this approach is given in Schweyen *et al.* (1997).

For early concept stages, a simpler way to model throughput is desired. This model is partly dependent on the étendue concept. After a survey of multiple designs using this technique, a projector designer would then normally select one or a few promising designs that could be evaluated more fully using ray tracing, measurements with individual components and a prototype of the complete system.

In the remainder of this section, the étendue concept will be developed as a tool that will allow the system designer to estimate collection and light utilization efficiency of a system prior to construction of a prototype and without ray tracing a particular design.

11.2.1 Definition of Étendue

The optical engineer's version of the second law of thermodynamics is that the image luminance or energy density cannot exceed the source luminance or energy density. Étendue is a geometric property of the optics related to the beam divergence and the cross-sectional area of the beam. While étendue is strictly a geometric property, the second law of thermodynamics requires that the étendue of a ray bundle remains the same or increases as the ray bundle passes through an optical system. It can never decrease. If the étendue of a ray bundle decreased between the source and the image, the energy in the ray bundle would be squeezed into a smaller area and the luminance or energy density at the image would exceed that of the source. Étendue is also known as the 'optical extent' or the 'optical invariant'.

The formal definition of étendue is (Boyd 1983):

$$E = \iint \cos\theta \, dA \, d\Omega \quad (11.2)$$

272 MODELING LUMEN OUTPUT

where E is the étendue, and the double integral is over the area of interest and the solid angle of the light corresponding to that area. The parameters are illustrated in Figure 11.2. The angle θ is between the normal to the surface element dA and the centroid of the solid angle element dΩ. Note again that étendue is a geometric property: There is no reference in this equation to optical intensity. The units associated with étendue are mm^2 steradian, which again shows the geometric nature of étendue.

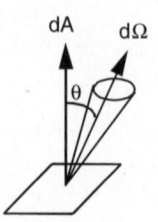

Figure 11.2 Relationship between θ, dA and dΩ.

When a beam is modified by a well-corrected optical element, étendue is preserved. For example, when a well-corrected lens focuses a collimated beam to a spot, the area of the beam is reduced but the divergence (convergence) angle of the beam increases, and the product of this area and the solid angle of the beam (the étendue) is conserved. In elements that are not well corrected or involve scattering or diffraction into multiple orders, étendue will increase.

Equation (11.2) can be evaluated at any surface in an optical system, and if the system contains only well-corrected elements, the same étendue value will be calculated. Due to the difficulty in evaluating the integral on an arbitrary surface, étendue is rarely calculated at any surface that does not have a high degree of symmetry. Étendue is most commonly calculated at object and image planes, on a surface perpendicular to the optical axis. It can also be calculated at the system pupil.

Étendue is an often misunderstood concept, partly due to the fact that it does not refer to the intensity of the light distribution of interest: it only refers to the distribution's geometric boundaries. These boundaries are often ill defined since at the spatial and angular edges of a light beam, there is normally a gradual fall-off of the light intensity, not a sharp line of demarcation. Another reason étendue can be a confusing concept is the difficulty of doing the integration in Equation (11.2) in the general case. Normally, it is necessary to integrate over two spatial and two angular dimensions, often in polar or spherical coordinates. For this reason, approximations are often used to calculate étendue, and they are sometimes used even under circumstances where they are not valid.

In order to calculate étendue, it is necessary to have knowledge of both area and angle. The area and angle are often determined by two separate components. For example, in a microdisplay or light-valve projector, to calculate étendue at the plane of the microdisplay, the area is determined by the area of the microdisplay and the angle is determined by the f/# of the projection lens or illumination system. If the microdisplay involves scattering or diffraction into multiple orders, the angular range after the microdisplay would be larger than the angular range before the microdisplay. Since the area remains the same, the étendue would increase at the plane of the microdisplay. This implies that two separate étendue values are needed to describe a system like this, an illumination path étendue and a larger image path étendue.

The étendue of some optical elements can be calculated without reference to any other optical element. For example, a polarizing beam splitter has a given area, depending on the size of the prism. Polarizing beam splitters normally also have a limited angular acceptance range, perhaps as low as ±5° for a MacNeille prism. Since both angular and area information are available from the single element, étendue can be calculated.

Somewhere in any projection system there is an optical element, or combination of optical elements, that has the lowest étendue value. Since étendue represents geometric boundaries of the light beam, the aperture where étendue is the lowest will pass the geometrically smallest optical beam. Any other aperture in the system, since it is geometrically larger, should, in principle, be able to pass the entire beam that passes through the limiting aperture. This limiting étendue determines the system collection efficiency, even if the limiting plane is remote from the lamp-reflector combination.

11.2.1.1 Étendue at a flat surface

One of the cases for which Equation (11.2) can be integrated in closed form is the case of a flat surface normal to the optical axis and the optical beam having a uniform divergence angle ($\theta_{1/2}$) over that surface, as is shown in Figure 11.3. This example would represent the calculation of the étendue for a microdisplay. In this case, the étendue is given by:

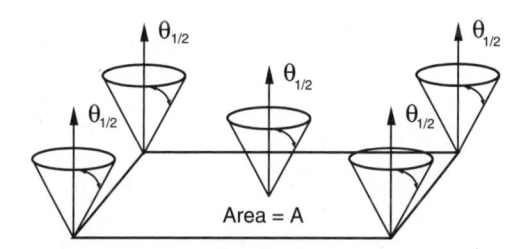

$$E = \pi A \sin^2(\theta_{1/2}) = \frac{\pi A}{4(f/\#)^2} \tag{11.3}$$

Figure 11.3 Flat surface with uniform divergence angle.

where A is the area of the surface, which can be a microdisplay or other optical aperture, and the f/# is measured at that surface. If, for example, one had a 30 mm by 40 mm flat panel device, capable of accepting a f/3.5 input beam, the étendue of the device is 76.9 mm² steradian.

In some projection systems, the pupil is rectangular instead of circular. Assuming the pupil is uniform over the flat surface and centered on the optical axis, Equation (11.2) can be integrated to give:

$$E = 4A \sin(\theta_{1/2}) \sin(\phi_{1/2}) \tag{11.4}$$

where $\theta_{1/2}$ and $\phi_{1/2}$ are the horizontal and vertical half angles, respectively.

For a light source where the emitted light is Lambertian, then $\theta_{1/2} = 90°$ and the étendue of the light source becomes:

$$E = \pi A \tag{11.5}$$

Typically, Equation (11.5) applies to the LEDs used for projection illumination sources. This equation applies whenever the light must be collected from a ±90° cone angle, even if the distribution within the cone is not Lambertian.

One of the two equations, (11.3) or (11.4), is the equation normally needed to calculate the limiting étendue of a projection system. When using these equations, care must be taken to ensure the assumptions made in deriving the equations are valid and the correct values for the angles and areas are used. Special care must be taken in projection systems with complex optical paths, such as angular color separation and scrolling color projectors. Care must also be taken for optical elements where étendue is not conserved, such as at scattering or diffractive elements, or low f/# elements with large aberrations such as reflectors and condensers in the illumination path.

Unlike scattering or diffraction into multiple orders, optical elements with large aberrations do not increase the 'true' étendue of an optical beam. They do increase the étendue if one of the approximations in Equation (11.3) or (11.4) is used to calculate the étendue. By inserting an element to correct the aberrations, it is possible to reduce the étendue calculated by Equation (11.3) or (11.4). In a complete étendue calculation using Equation (11.2) it would be found that the aberration inducing optical element did not increase the étendue and the aberration correcting element did not in fact decrease the étendue. As mentioned before, this calculation can be very difficult and is prone to error. Rather than go through this calculation, it would be more usual to do a complete ray trace of the system using a program such as ASAP to evaluate system throughput more directly.

11.2.2 Étendue Limited Systems

Generally, a projection system designer is mainly interested in the étendue of the light beam produced by the collection system, not the étendue of the arc itself. Reflectors or other high efficiency collection systems are rarely 'well corrected' in the optical sense. In order to calculate the light beam étendue it is necessary to know the arc size and shape, the geometry of the collection system and any scattering or reflections in the collection path.

With this knowledge, and the aid of a good ray-tracing program, the spatial and angular geometric distribution of light in any plane after the reflector can be calculated. Equation (11.2) can then be integrated numerically using this geometric distribution to give the étendue of the optical beam. The étendue of this beam is typically many times larger than the étendue of the arc alone, primarily due to aberrations introduced by the collection system.

The term 'étendue' can be applied to both the optical beam passing through the projector and the optical path of the projection system itself. When used to refer to an optical beam, a low étendue value is 'good' in that the entire optical beam can be coupled into a small area or high f/# microdisplay. When referring to a microdisplay or other element in the optical path, a high étendue value is 'good', because the optical element can then utilize the entire contents of a large optical beam.

A projection system is referred to as 'étendue-limited' when there is an optical element or combination of optical elements in the system whose étendue is lower than the étendue of the optical beam produced by the lamp/reflector combination. In this situation, the full number of lumens in the beam cannot be utilized by the projection system. If the étendue of a microdisplay, light valve or other limiting component in a projector is larger than the étendue of a light beam, the projection system would be capable of using the entire contents of that beam. In this case, the projection system is said to be 'not étendue-limited'.

Most practical étendue-limited projection systems are limited by the area of the microdisplay and/or the cone angle (f/#) of the projection lens. Other elements that can limit étendue in some systems are integrators, prisms, polarizing beam splitters or Schlieren stops. If a low-cost component such as a mirror or simple lens were the limiting element, it could be replaced relatively easily, increasing the projector throughput at relatively small cost. Alternatively, a smaller microdisplay or higher f/# projection lens could be used to match the étendue of the limiting component. This would reduce cost without affecting light throughput.

11.2.3 Lumen vs Étendue Function

As mentioned before, étendue describes the geometrical size of a light beam, not the amount of light contained within that beam. For example, if we have an f/3 light beam with a circular cross section 10 mm in diameter, the area of the beam is 78.5 mm^2, the half angle is 9.6° and the étendue of the beam is 6.9 mm^2 steradian. If this beam comes from a 5 W incandescent lamp, it may only contain 50 lumens but if it comes from a 1000 W xenon lamp it may contain 20,000 lumens.

If we now attenuate the beam, for instance by passing it through an aperture with an étendue of 5.6 mm^2 steradian (9 mm diameter and an acceptance angle corresponding to f/3), some of the light will be blocked by the aperture. While the étendue has been reduced by 19 %, typically the light throughput

LIGHT COLLECTION AND ÉTENDUE   275

will be reduced by less than 19%. This occurs because the edge of an optical beam normally is not as bright as the center of the beam. For example, we can assume that the beam only loses 10% of its light, and the beam from the xenon lamp now contains 18 000 lumens.

As one successively reduces the aperture size and remeasures the light, one will generate a set of data that represents one particular curve of lumens as a function of étendue for that particular lamp/reflector combination. An apparatus to perform these measurements is shown in Figure 11.4.

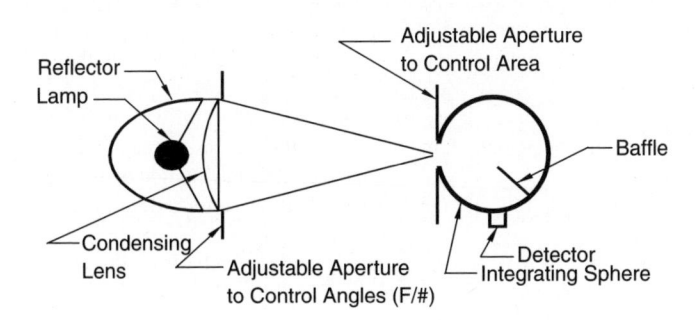

Figure 11.4 Apparatus for measuring lumens vs étendue function.

Several curves of the relative throughput as a function of étendue are shown in Figure 11.5. This figure relates to a short arc HID lamp in a parabolic reflector that collects over angles from 45° to 135°. This is a fictitious lamp, but it shows the same basic features seen for real lamps. 0° is defined as the apex of the parabola, 90° is perpendicular to the optical axis in Figure 11.4. Reducing the diameter of the aperture at the integrating sphere while leaving the aperture at the lamp fixed will give the curve for the 100% diameter (dashed curve) of the lamp aperture.

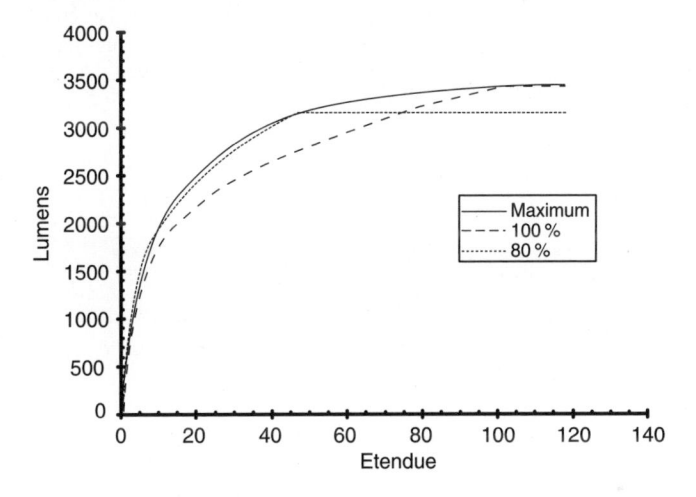

Figure 11.5 Variation of the percentage of usable lumens as a function of étendue for an example lamp. The efficiency of light utilization is the product of η_Φ and η_E.

Note that at étendues above about 100 mm² steradian, the amount of light collected in the 100% aperture case is a constant: essentially all the light from the lamp passes through the aperture and is measured by the sphere.

The dotted curve corresponds to reducing the diameter of the aperture at the lamp so light is only collected over an angular range of 45° to 125°. For this reflector, only 80% of the exit aperture is now being used. This reduced aperture diameter at the lamp increases the f/# of the beam entering the sphere.

276　MODELING LUMEN OUTPUT

With the aperture of the sphere opened to its maximum size, the system étendue is less than the system étendue when 100 % of the lamp aperture was used. As one would expect, this reduced diameter at the lamp means that less total light is collected.

However, as the aperture at the sphere is reduced in diameter, the shape of the lumens vs étendue curve becomes different than it was when the full lamp aperture was used. At étendues less than about 75 mm^2 steradian in this example, more light is collected in the sphere with the smaller lamp aperture than with the larger lamp aperture. This occurs because when the aperture at the lamp is reduced, reducing the solid angle, the aperture diameter at the sphere is increased in order to maintain the 75 mm^2 steradian étendue. This increased diameter at the sphere accepts more light than is eliminated by the reduced diameter at the lamp.

There are many combinations of reflector aperture and sphere aperture that will produce a given étendue. For one particular combination, a maximum amount of light will pass through the sphere aperture. This maximum light throughput for various étendues can be plotted and thought of as the definitive lumen vs étendue curve for a given lamp reflector combination. This curve for the example lamp is indicated as the solid line in Figure 11.5.

The lumens vs étendue function can be generated experimentally or by ray tracing with a program that keeps track of the intensity as well as the position and direction of a ray. If the projector design is far enough along, ray tracing can be done with the intended projector model. If the projector design is not yet done, or it is desired to compare two lamp/reflector combinations without reference to a specific projector, a 'virtual experiment' similar to the one in Figure 11.4 can be done by ray tracing to generate the lumens vs étendue data.

Many analytical models of the lumen throughput as a function of étendue can be developed. A simplified model used in the past (Nicolas 1992) assumes that, at optical path étendues higher than the arc étendue, all light can be utilized and there is no increase in light with increasing étendue. At optical path étendues lower than the arc étendue, the collection efficiency is considered to increase linearly with étendue.

A more accurate semi-empirical model was given by Brennesholtz (1996). The complete equation and the derivation of the equation is given in Appendix 3. This model is semi-empirical in the sense that it depends on both the geometric properties of the lamp and reflector such as arc length and reflector focal length and collection angles, and on the measured lumens vs étendue curve. This model was used to calculate the solid line in Figure 11.6 and both lines in Figure 11.7. This model is quite general and can be applied to all imaging collection systems.

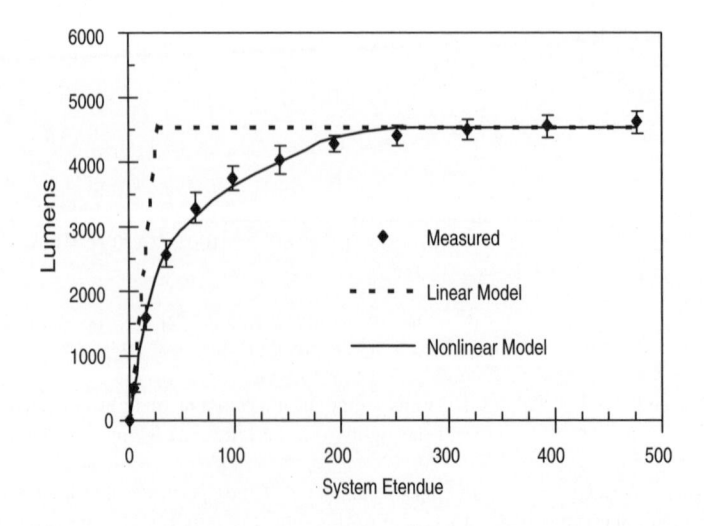

Figure 11.6　Two different functions fit to measured lumens vs étendue data (error bars are ±1 standard deviation).

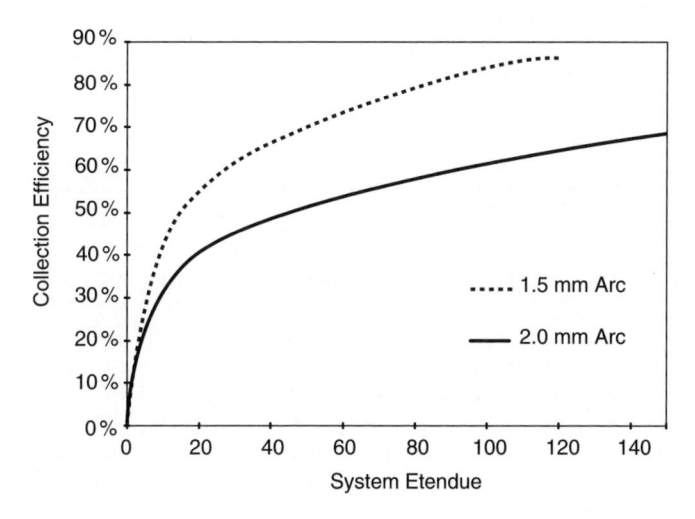

Figure 11.7 Lumens vs étendue curves for two similar lamps.

A model specific to the Philips UHP lamp was given by Moench (2000).

$$\Phi = 32P \arctan\left(\frac{E}{3.8d^2 + 0.9d + 0.8}\right) \quad (11.6)$$

In Equation (11.6),

- Φ is the total collected flux in lumens,
- P is the power of the lamp,
- E is the étendue of the microdisplay or other limiting element and
- d is the arc length of the lamp.

A third, extremely simple, analytical expression useful in calculating system throughputs of projectors illuminated by LEDs with Lambertian emission distributions is (Brennesholtz 2006):

$$C = C_{Max}\left(1 - e^{-kR}\right) \quad (11.7)$$

In Equation (11.7):

- C is the collection efficiency of the LED light into a microdisplay with a given étendue,
- C_{Max} is the maximum possible collection efficiency, dependent on the collection system coating and geometric properties,
- k is a constant determined by the optical system and
- R is the étendue ratio of the system, which is the ratio of the microdisplay étendue to the étendue of the LED source.

To calculate the étendue ratio of the system, the LED étendue can be calculated with Equation (11.5) and the microdisplay étendue can be calculated with Equation (11.3). C_{Max} values in the range of 70–80 % and k values in the range of 1.0–1.33 can be used to describe most high-performance collection systems for LEDs.

Figure 11.6 shows the measured and modeled lumens vs étendue curves for a 125 W short arc metal halide lamp in an elliptical reflector. The solid line was calculated using the parameter values in Table A3.1. The dashed line was calculated with a linear lumen vs étendue curve, where the assumption was made that the étendue of the light beam was the same as the étendue of the arc alone. As can be seen, the linear model provides a poor fit to the data for étendues between 20 and 200 mm^2-steradian by dramatically overestimating the collection efficiencies for étendue limited projectors.

Once the lumens vs étendue data are generated they can be utilized in a variety of forms. They may be left as tabular or graphical data or it can be fitted by an analytical expression, such as the lumen vs étendue relationship derived in Appendix 3. The advantage of an analytical expression is that many projector configurations can be evaluated quickly in a spreadsheet to determine which design is most likely to meet the marketing requirements. This projector model can take panel size, lamp type, f/#, etc. into account.

While the étendue of the arc alone cannot be used to determine the lumens vs étendue curve of a lamp/reflector combination, the arc size is still a very important parameter in determining collection efficiency. Figure 11.7 shows the collection efficiency curves for two similar lamps in identical parabolic reflectors. The lamps differ only in arc length: one is 1.5 mm long and the other is 2.0 mm long. The étendue of an arc 1.5 mm long with a collection range of 45–135° is approximately 10 mm^2 steradian.[1] For the 1.5 mm arc, all the collectable light, representing 85 % of the total lumen output, is collected into a beam with an étendue of about 110 mm^2 steradian. For the 2.0 mm arc, only about 60 % of the light is collected into a beam with this étendue. In order to collect the same 85 % of the lamp output it is necessary to go to an étendue of approximately 200 mm^2 steradian.

11.2.3.1 Étendue conserving transformations

Étendue conserving transformations are possible between various aperture sizes and shapes. For instance, a simple lens as shown in Figure 11.8 can convert a 10 mm diameter f/3 spot with an étendue of 6.9 mm^2 steradian into a 20 mm diameter f/6 spot, also with an étendue of 6.9 mm^2-steradian. While simple lenses have aberrations that theoretically increase the étendue, the aberrations are rarely severe enough to be noticeable at this level of accuracy. If the lens is made of high transmission glass and is properly AR coated, the lumens will also be conserved, and the lumens vs étendue function will be the same both before and after the lens.

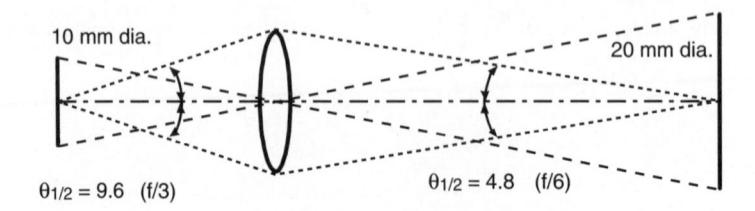

Figure 11.8 Conservation of étendue.

11.2.3.2 Shape conversion

Shape conversion from a circular light beam to a rectangular microdisplay can, in some cases, lead to a significant loss of light. Figure 11.9 illustrates why this loss occurs. In this figure, the 16:9 rectangle and the circle have the same area. Assuming the same f/#, they will then have the same étendue. However, it

---

[1] The equations needed to calculate the étendue of a cylindrical arc are given in Appendix 3.

can be seen that the areas of the circle that are outside the rectangle are closer to the center (the area of maximum brightness of the focused lamp) than the areas of the rectangle that are outside the circle.

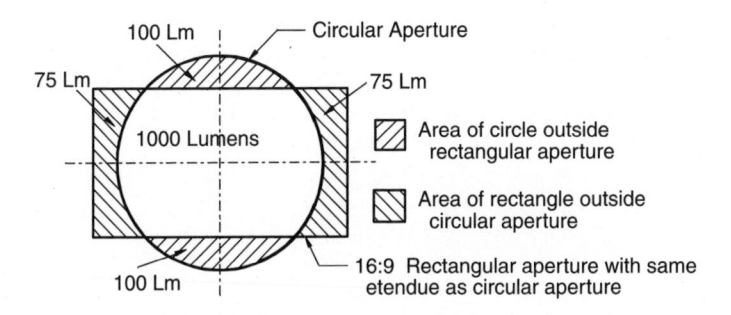

Figure 11.9 Circular vs rectangular apertures.

When a typical lamp is focused for maximum throughput, the contours of constant luminance are circles, with the highest luminance at or near the center. Therefore, areas of the circle that are outside the rectangle contain more light than the areas of the rectangle that are outside the circle. Thus, the rectangular aperture will collect less light than the circular aperture at the same étendue.

For example, let us assume we have a lamp/reflector combination that will illuminate the various portions of the field with the numbers shown in Figure 11.9. The main area, which is common to both the circle and the rectangle, contains 1000 lumens. The top and bottom portions of the circle each contain 100 lumens, for a total in the circular aperture of 1200 lumens. The two side portions of the rectangle, however, only contain 75 lumens each, for a total of 1150 lumens in the rectangular aperture. This represents a 4 % loss of lumens in the system.

Even if a non-conserving shape changing mechanism is chosen, such as imaging the round spot onto a rectangular aperture and ignoring the problem as in this example, the losses for 5:4 or 4:3 rectangles are rarely significant. Even for 16:9 apertures, the loss is usually less than 5 %. For projectors with aspect ratios of the illuminated area higher than 16:9 (1.78:1), shape conversion can be a significant loss of light that requires special attention from the designer. Electronic cinema has a native aspect ratio of 1.85:1, and sometimes it is used to show material with an aspect ratio as wide as 2.39:1. Scrolling color systems, such as the one described in Section 10.4.2.6, can have a stripe aspect ratio as high as 6.4:1. Depending on the system étendue ratio, this results in a throughput for the stripe of 60–80 % of the throughput for a circular spot with the same étendue.

Shape conversion from a circular aperture to a rectangular aperture while conserving étendue is possible, but requires more complex optical elements than a simple lens. Integrators can be designed specifically to conserve étendue and lumens while converting from the circle to a 4:3 or 16:9 rectangle (Shimizu and Janssen 1997). The design of integrators is discussed in detail in Section 6.4.

One way to avoid the aspect ratio conversion problem in estimating throughput is to use an aperture of the correct shape when measuring the lumen vs étendue function. For instance, if the target projector will use a 4:3 aspect ratio microdisplay, it would make sense to use a variable 4:3 aspect ratio aperture at the entrance to the integrating sphere in the apparatus in Figure 11.4. If this curve is then compared to the lumen vs étendue curve generated with a circular aperture, the designer will get an indication of exactly how much light is being lost by not including an étendue-conserving, aspect-ratio converting integrator.

11.2.3.3 Usable étendue

If one calculates the étendue of the panel and projection lens (or other limiting element) and uses that étendue as input to the lumens vs étendue function for the chosen lamp/reflector combination, one does not

280 MODELING LUMEN OUTPUT

always obtain the correct collection efficiency. The reason for this is that not all projector configurations utilize the full étendue of the limiting component.

The most common example of this is the use of a polarization conversion system (PCS) in an LCD projector, as sketched in Figure 11.10. In a polarization conversion system, the unpolarized light from the lamp is split into two orthogonally polarized beams, typically oriented horizontally and vertically. The polarization of one of the beams is rotated 90° by the polarization converter and both beams are used to illuminate the microdisplay. If étendue is ignored, this will double the amount of polarized light available at the LCD, assuming 100 % efficiency for all components. Polarization conversion systems are described in more detail in Section 6.5.3.

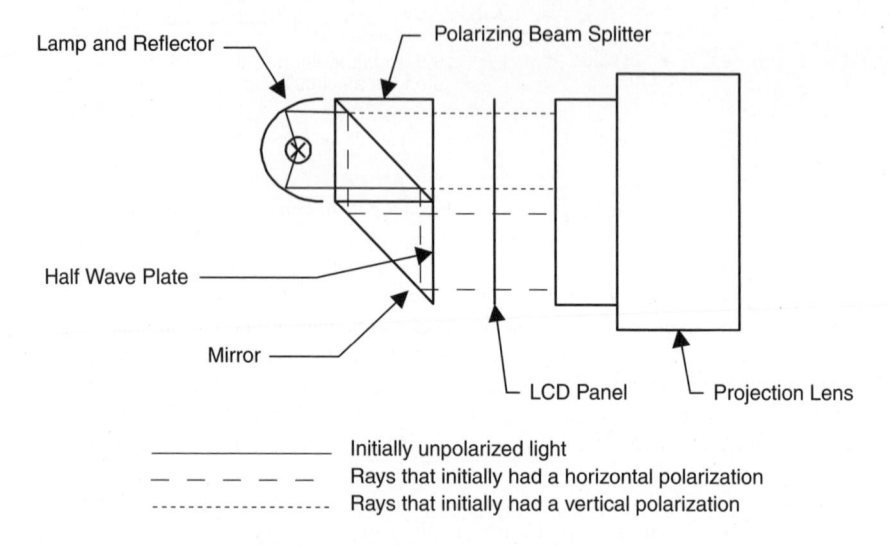

Figure 11.10 Basic polarization conversion system.

In practice, if the projection system is étendue limited, the amount of light gained will be less than this. The reason for this is each beam utilizes half of the étendue of the system. Figure 11.10 shows spatial recombination of the light, with each beam using half the area of the microdisplay and the full aperture of the lens. Alternatively, the beams could be recombined angularly, so that each beam illuminates the entire panel, and uses half of the pupil of the lens. In either case, only half of the étendue of the system is available to each beam. This then reduces the collection efficiency, so the gain in throughput is less than the expected factor of 2.

With an étendue limited projector and the linear portion of the linear lumens vs étendue model, shown in Figure 11.6 with a dotted line, there would be no gain in light at all: the gain in polarized light would be exactly offset by the loss in collection efficiency. In practice, there would be a net loss of light due to the additional losses at the extra optical surfaces. Note that the beams after the PCS cannot be recombined to reduce the étendue to the original value. While they were originally split by a PBS, the beams now have the same polarization and a PBS would either reflect or transmit both beams, rather than reflect one and transmit the other.

As another example, in the scrolling color system described in Section 10.4.2.6, any given color only illuminates 1/3 to 1/6 of the panel at a time. Therefore, the effective area and étendue is 1/3 to 1/6 that of the full panel. For example, take a 4:3 1.8" panel with a f/2 lens and color bars each covering 1/6 of the panel. The system has a full étendue of 197 mm^2 steradian. After this full system étendue is divided by 6, the usable étendue is 33 mm^2 steradian. This is the value that should be used to estimate collection efficiency.

These factors are multiplied to calculate usable étendue. If the panel in this example uses both scrolling color and polarization conversion, the usable étendue would be 1/12 the total étendue or 16 mm² steradian. This occurs because within each bar, half of the étendue is used by light that was originally vertically polarized, the other half by light that was originally horizontally polarized. Table 11.1 contains the ratio between total limiting étendue and usable étendue for various projector configurations.

Table 11.1 Étendue utilization factors.

Configuration	Correction Factor
Scrolling color	$\frac{1}{3} - \frac{1}{6}$
Polarization conversion	$\frac{1}{2}$
Pupil sharing among three colors (angular color separation)	$\frac{1}{3}$

11.2.3.4 Limitations of lumen vs étendue model

Any throughput estimate based on étendue suffers from a potentially serious flaw: the lumen vs étendue model predicts the *maximum* possible system throughput. This maximum can only be achieved when every component is designed to utilize the same geometrical portions of the light beam. If all the limiting apertures are the same shape, this is usually not a problem. If two apertures have the same étendue but have different shapes the situation becomes more difficult to analyze. In a projector, the two shapes normally are the circular aperture of the illumination system and the rectangular aperture of the microdisplay, as shown in Figure 11.9. While it is theoretically possible to optically map one aperture onto the other while conserving étendue, this is very difficult to achieve in a realistic projection system.

The simplest way to cope with this problem is to ensure that the étendues of all components except the limiting component are significantly larger than the étendue of the limiting component. This ensures that any light that will pass through the limiting component actually reaches that component and is neither prematurely blocked nor blocked after it passes through the limiting component.

This approach works well where there is only one high cost component such as the microdisplay whose low étendue must be accepted. With this approach, light will reach the microdisplay that cannot reach the projection screen. Light baffles should be employed to ensure that this light does not reduce the system contrast. If there are two high cost components whose size must be restricted, such as a microdisplay and a polarizing prism, the shapes and étendues of these components must be matched to each other to ensure that neither one blocks light that could pass through the other.

11.3 Integrators and Lumen Throughput

Use of an integrator inevitably leads to a loss of total light in a system compared to the same system with critical illumination and without an integrator. The system designer must be aware of the trade off between throughput and screen uniformity. Some of the factors causing this loss were covered in Section 6.4. This section will deal with the remainder of the factors in a qualitative way. Quantitative numerical examples will be given in Chapter 12.

There are several reasons for this loss of light. An integrator typically introduces 4 to 6 extra optical surfaces in the light path. At 0.75 % loss per surface, this represents a 3 % to 4.5 % loss. In a lenslet integrator, there are also losses associated with the spaces between lenslets where the shapes of the optical surfaces are not correct, as was described in Section 6.4.1.

282   MODELING LUMEN OUTPUT

Edge luminance values as low as 30 % of the center have been accepted in CRT-based video entertainment rear projection applications. If this is the only application for a given projector, the designer may decide to omit the integrator altogether. For a professional projector, the center to edge fall off should be much less because significant items such as icons and menus are often displayed near the edges of the display. If perfect uniformity is not possible in a data/graphics projector design, a good target for the edge luminance would be more than 85 % of the center luminance. Recognizing the loss of throughput associated with an integrator, the projector designer will need to consider whether an integrator is required at all, and if so, what type to use.

11.3.1   Overfill Losses

The light body produced by the integrator must be larger than the microdisplay. This requirement is independent of the type of integrator and is called overfill. Typically, the required overfill is 5–10 % by linear dimension, or 10–20 % by area.

The reason overfill is necessary is illustrated in Figure 11.11(a). If the light body is exactly the same size as the microdisplay, there is no margin of error for any system misalignment between the illumination path and the microdisplay. In this case, any misalignment would cause some pixels in the microdisplay not to be illuminated, so they would not appear in the final image. Figure 11.11(b) shows a second problem. Real integrators produce light bodies whose profiles are not perfectly square; their corners are somewhat rounded. If there is insufficient overfill, these areas of low illumination will be in the active microdisplay area, and the image on the screen will have dim edges.

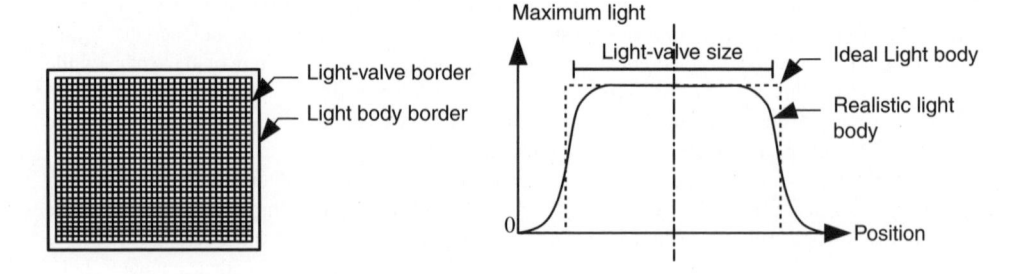

Figure 11.11   Light body and overfill with an integrator: (a) plan view; (b) light intensity profile.

In a system that is not étendue limited, overfill will reduce the light by the area fraction of the overfill. For example, in Figure 11.11(a), the light body is 10 % larger in each linear dimension than the microdisplay. Therefore the area of the light body is $1.1^2 = 121$ % of the area of the microdisplay. Assuming the light body is perfectly uniform, 1/121 % = 83 % of the lumens in the light body actually fall on the microdisplay and are useful to the system.

If the microdisplay/projection lens combination is the limiting component in an étendue-limited projector, The light lost due to overfill will be less than the area fraction. For example, assume the light body is 5 % larger than the microdisplay in each linear direction, or $1.05^2 = 110$ % larger in area. Assuming perfect uniformity in the light body, 91 % of the lumens in the light body will fall on the microdisplay itself. Collection efficiency, however, should be calculated on the basis of the area of the light body, not the microdisplay. Therefore, since the area is 10 % larger and the angles are unchanged, the usable étendue of the system is 10 % higher. If a linear lumen vs étendue curve is used, the increased collection efficiency exactly counteracts the fact that only a portion of the lumens in the light body are being used, and there is no lumen throughput reduction associated with overfill.

In a more realistic case, the lumens increase from collection efficiency is never enough fully to counteract the lumen decrease from overfill. However, the light loss is not as bad as would be expected from the simple area fraction.

11.3.2 Integrator Étendue and Collection Efficiency

If a lenslet array integrator is used, the étendue of the integrator should match the étendue of the panel and projection lens or other étendue limiting component. Lenslet integrators have a defined étendue because they define both angle and area. The overall shape of the first array determines area by intercepting some or all of the original beam from the lamp. The second integrator plate determines angle, since it serves as the system pupil.

In a well-designed lenslet integrator, the collection efficiency at the design étendue should remain the same as in a similar system with no integrator, except for losses due to the extra optical surfaces. However, the lenslet integrator will alter the lumens vs étendue curve for the system. This modified lumens vs étendue curve is shown in Figure 11.12. A perfect lenslet integrator was assumed for this curve, one that has no losses at its surfaces, no gaps between lenslets and produces a uniform light body. This integrator was designed to work at an étendue of 20 mm^2 steradian.

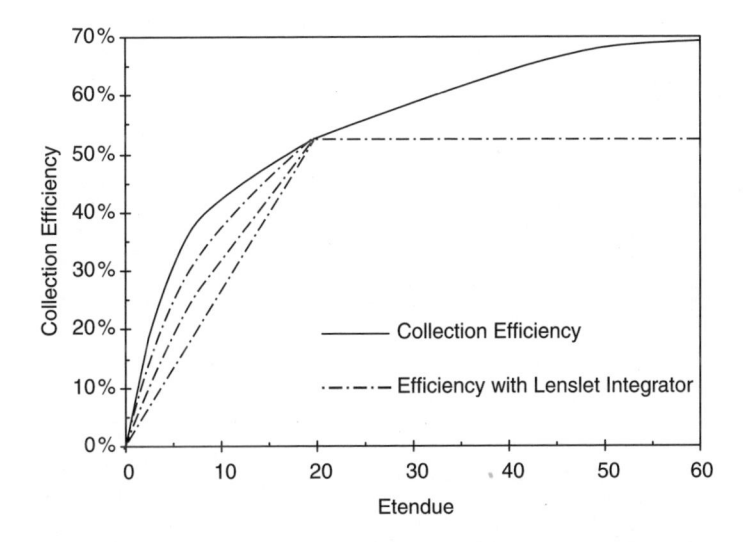

Figure 11.12 Lumens vs étendue with and without a lenslet integrator.

At the design étendue, the system has the same collection efficiency with or without the integrator. If the panel has a higher étendue than the design étendue of the integrator, the collection efficiency does not increase because the integrator produces a light body with a fixed étendue that does not change when the illuminated device changes. To expand the light body to fill a larger device, the cone angle will decrease and the étendue of the light body will remain constant, as will the total amount of light in the light body.

When a lenslet integrator has a lower étendue than the design value, the details of how the integrator is designed and apertured to produce the lower étendue light body become important. In the worst case, with a completely uniform light body, and the aperturing done at the light body, the lumens vs étendue curve becomes linear. This case is identical to the overfill problem, where only a fraction of the light body is used. This would occur, for example, if a smaller microdisplay is simply put in an optical system intended for a larger device. If the aperture is inserted before the first integrator plate, one of the intermediate curves may be achieved. This will require changing focal length of the auxiliary lens shown in Figure 6.28 in order to get the correct size light body at the microdisplay. It is unlikely that the full collection efficiency possible without the integrator will be maintained. A detailed ray trace of the new integrator/microdisplay combination would be necessary to determine if the performance would be satisfactory. If the performance is not satisfactory, there is no alternative to redesigning and retooling a new integrator for the new étendue.

Rod type integrators do not suffer from this problem because a rod integrator does not have a defined étendue. The cross section of the rod defines the area, but there is no part of a rod integrator that defines angles, up to the maximum acceptance angle of the rod. If the input cone angle from the lamp/reflector combination has a smaller cone angle than the maximum acceptance angle of the rod, the cone angle is preserved through the rod. At the output end of the rod, the cone angle and area are the same as at the input end so the étendue is preserved through the rod. The overfill issues described above apply equally to rod and lenslet integrators.

Cone prism (axicon) integrators deliberately change the angles of the light passing through them. Therefore any analysis of a cone prism based on étendue must be done very carefully. Analysis of a cone prism with a ray-tracing program is much safer. One thing to note about a cone prism is that in some circumstances they can actually improve the lumens vs étendue curve of a lamp/reflector combination. For example, in Figure 6.37, there is a dark hole in the center of the angular light distribution of the cone of light from the lamp/reflector combination. This central hole is a common feature of light distributions from reflectors with lamps on the reflector axis. This hole has a finite cone angle, and therefore a finite étendue. A properly designed cone prism can ensure that none of the panel/projection lens étendue is wasted on this dark étendue region because these angle/area combinations are not imaged into the system pupil.

11.4 Microdisplay and Light-valve Properties

The properties of microdisplays or light valves, as they affect lumen throughput, are discussed in this section. Other optical components, including dichroic filters, film polarizers, reflective polarizers, and integrators, were discussed in Chapters 6 and 7.

Two main microdisplay properties affect system throughput: transmission (or reflection) and modulation efficiency. The best source of information on either of these properties is from the microdisplay manufacturer. If samples of the target microdisplay are available, the manufacturer's data should be confirmed by measurement. When comparing two competitive microdisplays, make sure the two different manufacturers are including all the same components. For example, for a TN-LCD microdisplay, a quoted transmission may or may not include the transmission of the polarizer and analyzer.

11.4.1 Panel Transmission

The transmission or reflection of a microdisplay assembly is the product of the transmissions of several different efficiency factors, given in Equation (11.1) by η_P, η_{LV}, and η_A. η_{LV} can be described of the product of several efficiency terms. For example, the total panel efficiency of a transmissive active matrix TN-LCD has contributions from the following factors, including polarizer/analyzer efficiencies:

1. Polarizer transmission,

2. Reflection off various optical surfaces,

3. Substrate transmission,

4. ITO transmission (2 ITO-glass and 2 ITO-LC/alignment layer surfaces, plus absorption in the ITO),

5. Active matrix aperture ratio, and

6. Analyzer transmission.

In counting optical surfaces for item 2, it is necessary to count all surfaces. For example, if a polarizer with a laminated AR layer is bonded to an LCD, the surfaces must include the glass/glue, glue/polarizer, polarizer/glue, glue/AR layer and AR/air interfaces.

In a reflective microdisplay, there is only one layer of ITO that the light must pass through. However the light passes through this layer twice, so any estimate of microdisplay reflectivity must also include the effects of two layers of ITO. In a reflective microdisplay, there is also a mirror, often aluminum, that must be taken into account.

ITO reflection plays a special role in LCoS cells with a compensation film, such as the ones described in Section 4.2.5. In this case, the light reflected by the ITO passes through the compensation film, but not through the LC cell that is being compensated. The ITO reflection does not contribute to the dark-field luminance in an uncompensated cell, since reflection from the ITO does not affect the polarization of the light. It does contribute to the dark field in a compensated cell since the compensation film changes the polarization and some of the light changes to the polarization state for the bright field. In compensated cells, index matched ITO (IMITO) must be used to obtain good contrast.

Polarizer transmission, surface reflections, substrate transmission and ITO transmission do not vary too much from one LV manufacturer to another, and are independent of the size of the LV. Data on these items are available from the manufacturers or in the scientific literature. Data on certain key coatings are given in Chapter 6.

Aperture ratio is the ratio of the transmissive or reflective area of the pixel to the total area of the pixel. The total includes the opaque areas. It can vary considerably from one microdisplay design to the next. It depends very strongly on the active matrix technology and pixel architecture, as well as the resolution and size of the microdisplay. Aperture ratios range from less than 25 % for a small high-resolution transmissive microdisplay to more than 80 % for very large transmissive microdisplays or light valves. Reflective microdisplays or light valves typically have aperture ratios greater than 92 %. Obviously, it is impossible accurately to estimate the throughput of a projector without a good estimate of the aperture ratio.

In estimating the lumen output of a projector, wherever possible the measured aperture ratio of actual panels should be used. If this ratio is not available, it is possible to generate a good estimate using Equation (11.8), which refers to the pixel configuration in Figure 11.13. In this figure, the constant K_2^2

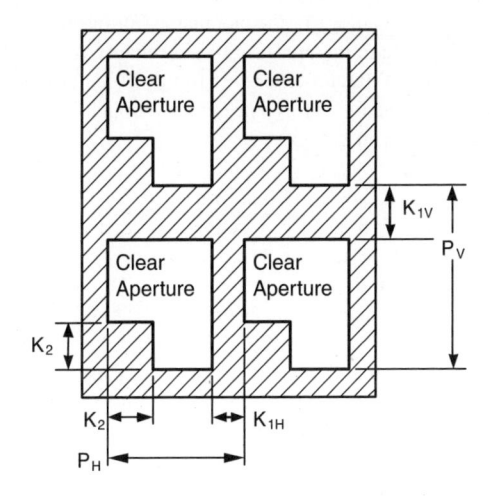

Figure 11.13 Schematic of active matrix pixel layout.

accounts for both the TFT and the capacitor areas that block the clear apertures. K_1 and K_2 are constants only for a specific pixel design with a specific manufacturing process.

$$R = \frac{(P_h - K_{1h})(P_v - K_{1v}) - K_2^2}{P_h P_v} \tag{11.8}$$

where:

- R is the aperture ratio,
- P_h is the horizontal pitch in mm,
- P_v is the vertical pitch in mm,
- K_{1h} is the width of vertical lines,
- K_{1v} is the width of horizontal lines and
- K_2^2 is the capacitor and/or transistor area.

The constants K_{1h} and K_{1v} can be derived in a variety of ways. A dimensioned layout of the pixel is not normally published by a panel manufacturer. However, they will sometimes publish a microphotograph of a single pixel in a technical journal or conference proceedings. If this is done, one may be able to scale the photo and calculate the parameters for that particular panel. Alternatively, microscopy can be employed directly to measure these dimensions. On the assumption that these constants represent the limits of the manufacturer's technology, they could then be used to estimate what aperture ratio the manufacturer could achieve with that manufacturing technology at different horizontal and vertical pitches.

While increasing the pixel pitch decreases the effects of the interpixel region and capacitor and increases the aperture ratio, the improvement in aperture may be less than would be predicted from this simple model. Larger pixels require larger capacitors for satisfactory operation, increasing the value of K_2 and reducing the aperture ratio. However, part of the capacitor may be hidden behind the gate lines (rows), which would reduce the value of K_2. It is only the part of the capacitor that is not under the row line that limits the aperture ratio. In addition to larger capacitors, it may be necessary to increase the row or column line width to improve the conductances of these lines in order to obtain the desired temporal properties in the larger microdisplay.

Equation (11.8) can also be used to estimate the aperture ratio of reflective pixelated panels. Since reflective panels can hide the capacitor, transistor and other electronics behind the mirror, the K_2 value is normally zero, or at least very small. The interpixel gap, K_1, is determined by the process used to make the reflective panel and the technology of the modulator. For instance, K_{1h} and K_{1v} values of 1μm are appropriate for the DMD™. Due to the presence of the center post, a K_2 value of 1μm, or less, is appropriate for the DMD. For the DMD, however, due to the difficulty in changing the pixel pitch, it is not normally necessary to use Equation (11.8) since the pixel pitch can only be changed after a long development period at Texas Instruments.

Each specific active matrix technology has pixel pitches P_h and P_v, below which the aperture ratio goes to zero. For example, taking $K_{1H} = K_{1V} = 0.009$ mm and $K_2 = 0.006$ mm, values that should be achievable with most technologies for transmissive active-matrix panels, the aperture ratio goes to zero at 0.015 mm pitch. (Some x-Si or p-Si active-matrix technologies can produce smaller K_{1H}, K_{1V} and K_2 values than these.) The aperture ratio as a function of pitch (assuming square pixels) is plotted in Figure 11.14 for this specific set of structural dimensions.

The aperture ratio for a reflective system with $K_{1H} = K_{1V} = 0.001$ mm, $K_2 = 0$ mm and square pixels is also plotted in Figure 11.14. Note that the aperture ratio with these parameters is very much higher than it is for transmissive devices with the same pitch. As a result, smaller reflective microdisplays can be used to produce the same lumen throughput as larger transmissive devices with the same pixel count. It should be noted that reflective microdisplays usually require more complex and expensive optical systems, which were described in Chapter 10.

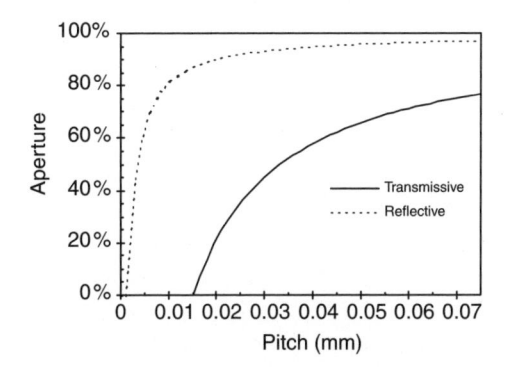

Figure 11.14 Active matrix aperture ratio.

11.4.2 Modulation Efficiency

The modulation efficiency of a microdisplay is the percentage of the light that passes through or is reflected off the microdisplay and is modulated with the desired image. For example, in an ECB-LCD microdisplay, the polarization after the LC cell is elliptical, not linear. Typically about 70–90 % of this light passes through the analyzer (output polarizer) to make the bright field. The remainder of the light is blocked by the analyzer. Therefore, the cell would have a modulation efficiency of 70–90 %.

Typically, the modulation efficiency for transmissive TN-LCD microdisplays can be taken to be 100 %. Light valves based on other technologies can have lower efficiency. The modulation efficiency for the Talaria oil-film light-valve projector was at best about 60 %, perhaps the lowest value of any commercially viable light valve. The modulation efficiency of the microdisplay or light-valve targeted for a projector must be known in order to make an accurate estimate of the projector throughput.

11.4.3 Duty Cycle

Color sequential or bit sequential devices normally have a temporal duty cycle that is less than 100 %. For example, there may be a blanking period while the dividing line between two segments of a color wheel passes through the optical beam in order to avoid illuminating the microdisplay with a mixture of the two colors. This temporal duty cycle may be included in the device modulation efficiency. If appropriate, duty cycle may be included as a separate term in Equation (11.1).

The efficiency based on the duty cycle in color-sequential systems varies depending on whether the light source can be pulsed or not. In a system with a white lamp that cannot be pulsed, duty cycle may be 33 % or less and 2/3 of the light generated by the lamp must be discarded. If a light source can be efficiently pulsed and can generate separate red, green and blue light, this color-sequential light loss is a total light output issue but not necessarily an efficiency issue. For example, in a laser or LED-based color sequential system, only the red source would be on when the projector is showing the red field. The green and blue are turned off, saving the power they would otherwise use.

11.5 Full Colorimetric Model of the Projector

This section assumes that the reader has a basic understanding of the science of colorimetry, particularly as it applies to additive color systems. If this is an unfamiliar topic, it is recommended that *Appendix 2, Colorimetry* or one of the standard texts on color (Wyszecki and Stiles 1982; Ohta and Robertson 2005) be studied prior to continuing with this chapter.

288 MODELING LUMEN OUTPUT

There are spectral effects in color projectors which are not included in the simplified throughput model, described earlier in this chapter. An example of this is the three spectra shown in Figure 11.15.

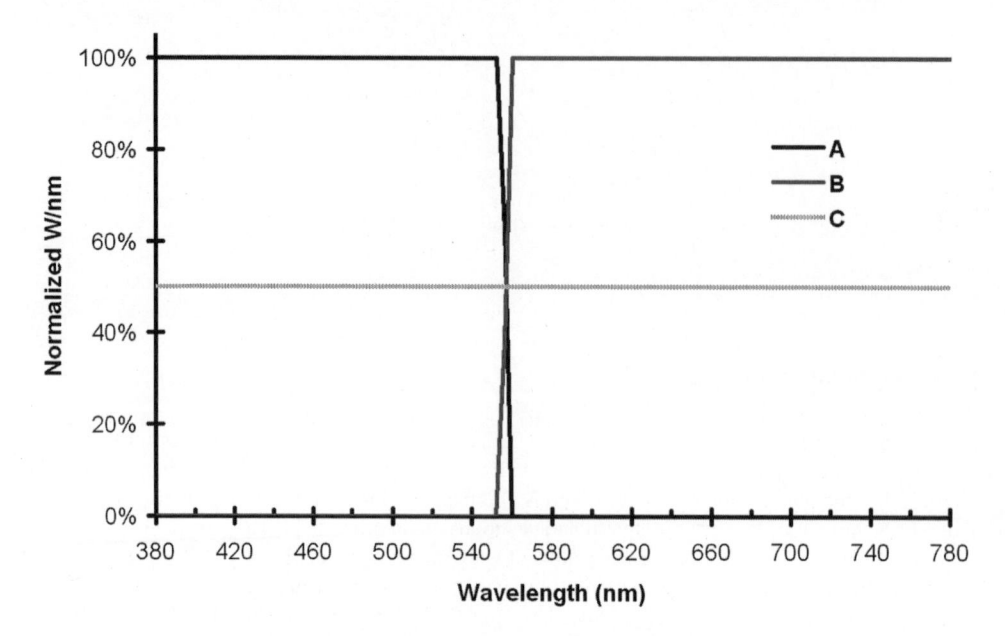

Figure 11.15 Transmission spectra of three sample filters.

Each of these filters would have $\eta_i = 50\ \%$ photometric transmission when calculated with Equation (6.2) and the spectrum of a Philips UHP lamp, yet clearly these filters are different. Filter A would be a cyan filter, with 100 % transmission in the blue and short wavelength green regions and 0 % transmission at longer wavelengths. Filter B is a yellow filter, with 0 % transmission at short wavelengths and 100 % transmission for long-wavelength green, yellow and red wavelengths. Filter C is neutral density filter with 50 % transmission at all wavelengths. According to the simplified model, as exemplified by Equation (11.1), if these filters were used in pairs, each pair would have 50 % × 50 % = 25 % transmission. Clearly this is not correct. If A and B were used together, the total photometric transmission would be 0 %. If A were used with another A, or B were used with another B, the total transmission would be 50 %. If C were used with any of the three types of filter, A, B or C, the total photometric transmission would be the expected 25 %, but the color would be different in each cases. These spectral effects require the use of the full colorimetric model of the projector to analyze and understand the performance of a system correctly.

While the color filters in a projector are typically in three bands (red, green and blue) compared to the two bands (cyan and yellow) shown in Figure 11.15, the color filters used in projectors tend to have similar properties to filters A and B; 100 % transmission in one wavelength band and 0 % transmission in other wavelength bands. A full colorimetric model is developed here to permit accurate modeling of the projector colorimetry, throughput and white point.

Most projection systems use additive color to produce the full color gamut on the screen. The discussion in this section is oriented around these additive color projectors, but the underlying colorimetric principles could easily be applied to monochrome or subtractive color projectors. The simplified model from Section 11.1 does not apply to subtractive color systems or systems where the color primaries can be varied dynamically. A full colorimetric model must be used to estimate throughput of these types of systems.

FULL COLORIMETRIC MODEL OF THE PROJECTOR

Spectral variations are only included in the simplified model (Equation (11.1)) as the parameters η_s and η_{cc}. The values of these two color efficiency terms cannot be derived within this simple model and they are assumed to be known for the calculations in Section 11.1. A more complete model is necessary to describe these spectral effects and derive η_s and η_{cc}. In the following sections, this full colorimetric model is developed.

A complete numerical example of the use of the full colorimetric model is given in Section 12.3.

11.5.1 White Light Throughput prior to Color Correction

The first step in estimating the lumen throughput of a projector is to gather all the necessary data. These data will include:

1. A complete sketch of the planned projector, including every optical component,
2. Spectral distribution of the lamp output,
3. Lumens vs étendue function for the lamp/reflector combination and
4. Spectral transmittance or reflectance of every component in the projector.

In a normal color projector, there are three separate paths, one each for red, green, and blue. In multiple primary color projectors, there can be six or more separate paths in the projector. Typically, some components are common to all paths and others are specific to a particular path. For example, the lamp/reflector combination is normally common to all paths. Some components can serve different roles in different paths. As an example, a dichroic filter can be transmissive in one path and reflective in another. In some cases, the paths can differ by only a single element. In a color wheel projector, for example, the paths are identical except for the insertion of three or more different color filters by the rotating color wheel.

The second step is to calculate the tristimulus values of the throughput for each of the three colors separately, using the following equations:

$$X_i = \int_{380}^{780} E(\lambda) \left[\prod_{j=1}^{n_i} \eta_{i,j}(\lambda) \overline{X}(\lambda) \right] d\lambda$$

$$Y_i = \int_{380}^{780} E(\lambda) \left[\prod_{j=1}^{n_i} \eta_{i,j}(\lambda) \overline{Y}(\lambda) \right] d\lambda \qquad (11.9)$$

$$Z_i = \int_{380}^{780} E(\lambda) \left[\prod_{j=1}^{n_i} \eta_{i,j}(\lambda) \overline{Z}(\lambda) \right] d\lambda$$

In these equations,

i	corresponds to red, green or blue, plus additional colors in a multiple color projector with more than three colour paths,
$E(\lambda)$	is the spectral energy density of the lamp,
n_i	is the number of optical elements in path I,
$\eta_{ij}(\lambda)$	is the efficiency of the jth component in the ith channel where i would be red, green or blue in a three channel projector,
$\overline{X}, \overline{Y}$ and $\overline{Z}$	are the CIE 1931 2° functions, and
λ	is the wavelength.

290 MODELING LUMEN OUTPUT

For a normal, three primary color projector, $i = (1,2,3)$. For multiple color systems, the range of i covers all colors in the system, including if appropriate, a 'white' color corresponding to a clear color wheel segment. For the purposes of this chapter, however, discussion will be limited to three-color systems.

In essence, Equation (11.9) calls for the calculation of the throughput of the projector using the simplified model as shown in Equation (11.1) at each wavelength for each of the separate color paths in the projector and then calculating the colorimetric and photometric properties. The simplified model reverses this and calculates the photometric properties first for each element, ignoring the colorimetric properties, and then combines elements.

The X_i, Y_i and Z_i values calculated in Equation (11.9) are called the tristimulus values. They are always written in capital letters to distinguish them from the CIE color coordinates x, y and z.

For many of the components, $\eta_{ij}(\lambda)$ in these equations will be common to all three color channels. Examples of items that are normally common are the reflector, integrator and projection lens. The $\eta_{ij}(\lambda)$ for the dichroic or other color filters will of course be different for the different color paths. Microdisplay efficiency can also be different for different colors, even when a single microdisplay is used. For example, this occurs in an LCoS microdisplay in a single-panel system if the microdisplay is driven with different bias voltages for each color.

Since $\overline{X}$, $\overline{Y}$ and $\overline{Z}$ are tabulated functions, as is often the case with $\eta_{ij}(\lambda)$, these integrals are normally evaluated numerically as summations. The output of Equation (11.9) for a three-color projector is a set of nine tristimulus values that, taken together, fully describe the colorimetry and lumen throughput of the projector. The actual lumen throughput of an individual red, green or blue channel is:

$$\Phi_{R,G,B} = 683 Y_{R,G,B} \tag{11.10}$$

The 683 in Equation (11.10) comes from the fact that there are 683 lumens/watt at the wavelength for which the CIE $\overline{Y} = 1$. The lumen throughput of the projector, prior to color correction is just:

$$\Phi_{Uncorrected} = \Phi_R + \Phi_G + \Phi_B \tag{11.11}$$

Throughout the remainder of this section, the subscript 'u' will be used for values before color correction, and the subscript 'c' will be used for values after color correction. The CIE 1931 x-y color coordinates of the uncorrected white point produced by this projector are derived from the following matrix equation:

$$\begin{bmatrix} X_u \\ Y_u \\ Z_u \end{bmatrix} = \begin{bmatrix} X_R & X_G & X_B \\ Y_R & Y_G & Y_B \\ Z_R & Z_G & Z_B \end{bmatrix} \begin{bmatrix} 1 \\ 1 \\ 1 \end{bmatrix} \tag{11.12}$$

The 1 in the column matrix on the right indicates that each color channel is driven to its maximum (100%) light throughput.

The uncorrected CIE x and y terms are then given by Equation (11.13).

$$x_u = \frac{X_u}{X_u + Y_u + Z_u}$$
$$y_u = \frac{Y_u}{X_u + Y_u + Z_u} \tag{11.13}$$

If the lamp is close to the desired white point and the three channels are reasonably in balance with each other in terms of efficiency, these x-y color coordinates may be close enough to the desired white point to make additional color correction unnecessary.

11.5.2 Color Correction to the desired White Point

Normally, there is a target white point for a projector, which frequently is based on marketing or competitive factors. It can, for example, be color D65, as specified for use in most standard- and high-definition television applications. Other applications such as digital cinema and graphic arts have different target white points. This D65 color roughly corresponds to the color of sunlight, which in turn is close to the color of a black body at 6550 °K. The CIE x-y color coordinates of D65 are $x = 0.313$, $y = 0.329$. Consumer television display manufacturers, however, tend to set white at a higher color temperature, such as 8200 K, with CIE $x = 0.293$, $y = 0.303$. Even higher temperature white points are sometimes used, depending on marketing requirements and other factors. 'Whiter whites' are perceived at higher color temperatures by most viewers, but other displayed colors are distorted.

This distortion occurs because no analog or digital television transmission standard broadcasts absolute colors in terms of the CIE x-y values. Instead, what is broadcast is the color difference between the defined white point and the color intended by the content creator. If the white point of the display does not match the white point expected by the content creator then all colors are displayed incorrectly by the display unless the display includes color correction circuits to adjust the colors. The issue of television broadcast standards is beyond the scope of this book and the reader is directed to standard texts on the subject such as (Poynton 2003).

For the remainder of this section, it is assumed a corrected white color point has been chosen, with color coordinates x_c and y_c. Some sections of the discussion apply to three panel projector architectures. The modifications to the equations for single-panel systems will be discussed in the next section.

Depending on the spectra of the lamp, the dichroic filters and the other optical components in the system, as well as the chosen white point, it is normally necessary to reduce the throughput in one or two channels to produce the required white. Leaving the third channel at the maximum throughput will maximize lumen throughput of the system for this white point. The question is: how much luminance is lost by discarding light to produce the required white color point?

Figure 11.17 illustrates an example in which the uncorrected white point, x_u, y_u, has too little red relative to the blue and green. If x_u and y_u are achieved when each of the three channels (red, green and blue) are driven to the maximum light throughput, light in one or two channels will have to be discarded in order to achieve the target white. Since all three channels are by definition at their maximum throughput, it is not possible to increase the amount of red without a redesign of the projector. In this example, it is necessary to reduce the amount of blue and green in order to produce the required white color.

Let us assume for the time being that we will be able to achieve the same luminance after color correction as before. This will give a color and luminance after color correction of x_c, y_c and Y_u as given by Equation (11.11). When the x, y and Y values for a color are known, i.e., color and luminance, the CIE colorimetric equations can be inverted to give the tristimulus values:

$$X_{c'} = \frac{x_c Y_u}{y_c}$$

$$Y_{c'} = Y_u \qquad (11.14)$$

$$Z_{c'} = \frac{(1 - x_c - y_c) Y_u}{y_c}$$

Note that these are not yet the true X_c, Y_c and Z_c values since they incorporate the uncorrected value Y_u.

Since it will be necessary to adjust the gain of one or more channels to get the desired white point, Equation (11.12) for the color corrected tristimulus values can be rewritten:

$$\begin{bmatrix} X_c \\ Y_c \\ Z_c \end{bmatrix} = \begin{bmatrix} X_{R'} & X_{G'} & X_{B'} \\ Y_{R'} & Y_{G'} & Y_{B'} \\ Z_{R'} & Z_{G'} & Z_{B'} \end{bmatrix} \begin{bmatrix} k_R \\ k_G \\ k_B \end{bmatrix} \qquad (11.15)$$

292 MODELING LUMEN OUTPUT

The values k_R, k_G and k_B in the right column matrix are the gains to be set into the red, green and blue channels, respectively. These gains can be computed by inverting the matrix of tristimulus values and applying Equation (11.14):

$$\begin{bmatrix} k_R \\ k_G \\ k_B \end{bmatrix} = \begin{bmatrix} X_{R'} & X_{G'} & X_{B'} \\ Y_{R'} & Y_{G'} & Y_{B'} \\ Z'_R & Z'_G & Z'_B \end{bmatrix}^{-1} \begin{bmatrix} X_c \\ Y_c \\ Z_c \end{bmatrix}$$

$$= \begin{bmatrix} X_{R'} & X_{G'} & X_{B'} \\ Y_{R'} & Y_{G'} & Y_{B'} \\ Z_{R'} & Z_{G'} & Z_{B'} \end{bmatrix}^{-1} \begin{bmatrix} \dfrac{x_c Y_u}{y_c} \\ Y_u \\ \dfrac{(1 - x_c - y_c) Y_u}{y_c} \end{bmatrix} \quad (11.16)$$

There is one remaining problem: at least one, and possibly two of the constants k_R, k_G and k_B computed by this method, is larger than 1.0. In the example shown in Figure 11.16, the k_R would be larger than 1.0, and the k_G and k_B values would be smaller than 1.0. This can be solved by normalizing to the largest value:

$$k'_{R,G,B} = \frac{k_{R,G,B}}{\max(k_R, k_G, k_B)} \quad (11.17)$$

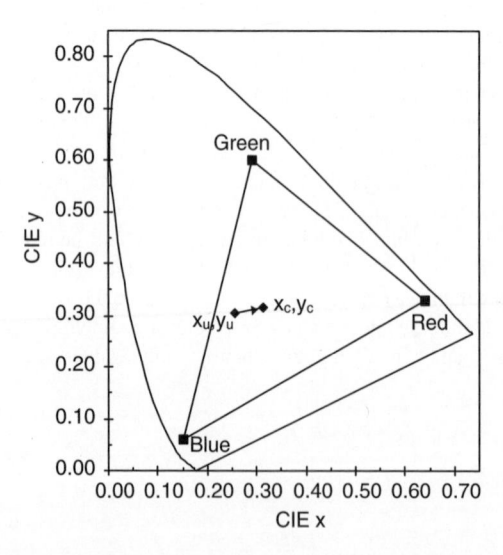

Figure 11.16 CIE diagram showing corrected and uncorrected white points.

One (or more) of these k' values will be equal to 1.0, the other(s) will be less than 1.0. Any channel with a k' value less than 1 must have its throughput reduced in order to pass less light into the white field. This can be done electronically, by reducing the amplitude of the video signal for that channel, or it can be done optically with neutral density filters, reduced apertures or some other method. The dichroic filters may also be redesigned to pass less light and produce a more saturated primary color. If this last option is chosen, Equations (11.9) through (11.14) must be applied iteratively since the change in the dichroics will change the X_i, Y_i and Z_i values in Equation (11.9).

Regardless of how the gain is reduced, the lumen throughput for the color corrected projector will be less than the throughput for the uncorrected projector. The throughput for the color corrected projector can be calculated by:

$$\Phi_C = k'_R \Phi_R + k'_G \Phi_G + k'_B \Phi_B \tag{11.18}$$

Color correction efficiency can then be defined simply by:

$$\eta_{CC} = \frac{\Phi_C}{\Phi_U} \tag{11.19}$$

where Φ_U is defined in Equation (11.11). While this color correction efficiency is strictly accurate only for the single projector for which it was derived, it can actually be used with a whole family of similar projectors, as long as the basic colorimetry is the same. For instance, if one wishes to understand the effect of f/# or panel size on the lumen throughput of a projector when the lamp, dichroic filters and architecture remain the same, η_{CC} can be used in conjunction with the simplified lumen throughput model, given in Equation (11.1).

The value of η_{CC} is sensitive to the relative efficiency of the red, green and blue channels of the projector. Small variations in these relative efficiencies can strongly affect η_{CC}. The changes need not be in a dedicated red, green or blue element. For example, if the design of the reflector coating is modified to reduce the amount of blue and UV in the light beam to reduce the aging of a LCD panel, this can reduce the throughput of the blue channel without affecting the red or green. This reduced blue actually represents very few lumens, since human vision is not very sensitive to blue in terms of brightness. However, if the blue is reduced, the color balance is changed and the amount of red and green will need to be reduced to return to the correct white point. This reduction in green and to a lesser degree red is what will most affect the total lumen output of the projector.

11.5.3 Single-panel Color Sequential Projectors

The assumption behind Equation (11.17) that the value of k_i cannot be greater than 1.0 is not correct for most color sequential projector designs. By adjusting the time allocated to each color, it is possible to increase the amount of the color(s) in short supply and decrease the amount of the color(s) that is (are) too bright. In the example shown in Figure 11.16 where red is in short supply leading to a $k_r > 1$ and a k_g and $k_b < 1$, the time allocated to red can be increased and the times allocated to blue and green decreased. In a color wheel system, this is done by adjusting the color wheel angles. In a color wheel with two cycles of red, green and blue filters around the periphery of the wheel, the equal-angle situation would allocate 60° to each color filter and produce the same time for each color. If Equation (11.17) gave the values

$$k_R = 1.15$$
$$k_G = 0.95 \text{ and}$$
$$k_B = 0.90.$$

then color wheel angles of 69°, 57° and 54° respectively would give the desired color balance to the target white point. These values of k would then be used in Equation (11.18) to calculate the system color corrected white luminance. In this case where green is decreased and red increased, there would be a net loss in lumen throughput, but the loss would not be as large as it would be if the gains calculated by Equation (11.17) were used. If green were the color in short supply, as is normally the case in projectors with LED illumination, the lumens of the color corrected projector would actually be higher when this color correction technique was used than when Equation (11.17) was used.

There is a limit to how small a color wheel angle can be used, imposed by the limitations on data loading time, gray scale requirements, etc. For example, in the DLP device, the reduced time allocated to green would reduce the number of bit planes that could be displayed.

294 MODELING LUMEN OUTPUT

An alternative approach is to keep the color wheel angles constant but increase the lamp power by 15 % during the red segment and reduce it by 5 % and 10 % respectively during the green and blue segments. This technique requires a special ballast and was discussed in Section 8.2.3.1.

For color sequential projectors where three separate light sources are used, i.e. lasers or LEDs, either of these techniques can be applied. The duty cycles of the red, green and blue sources can be adjusted, or the drive power to the sources can be adjusted, or some combination of the two can be used. In LED projectors, one can also use a larger area of LED die for the color in short supply, normally green. This larger area for green would have a different étendue and therefore a different collection efficiency than red or blue. This would need to be taken into account when estimating throughput.

11.5.4 Colorimetric and Throughput Issues with Projectors with more than Three Primary Colors

In a three-color projector, determining the amount of each primary needed to produce the target white is deterministic in that there is only one possible solution. This is shown in Equation (11.16) where the 3×3 matrix of tristimulus values is inverted. In a multiple color projector with four colors the matrix is 4×3: there are four colors, each with three tristimulus values, and four values of k_i are needed. This incomplete set of linear equations has an infinite number of solutions, one of which must be chosen by the projector designer.

Perhaps the most common solution is to choose the design where the four (or more) values of k_i are closest to each other in value. This means all the channels are driven at nearly the same power, a situation that normally optimizes the design in terms of color quality, efficiency and cost.

Another solution is to choose the colors and the k_i values so the display can reproduce the spectrum of the target color, not just the CIE x-y values (Ben-Chorin 2007). Normally six or more colors are needed to achieve a good spectral representation of natural colors.

11.5.5 Color Separation Efficiency

The color separation efficiency is the other parameter in the simplified model (Equation (11.1)) that needs to be calculated using the full colorimetric model. This number represents the efficiency with which the white light from the lamp is broken into red, green and blue light for modulation by the microdisplay(s) and then is recombined into a full color image for the viewer. Ideally, this efficiency should be about 100 %. Realistic three-panel projectors with dichroic filters for the splitter and combiner can achieve 90–95 %. Color wheel single-panel projectors typically are about 33 %. There are even a few architectures where color separation efficiencies (as defined in this section) greater than 100 % can be achieved (Sonehara 1992).

This color separation efficiency value has two uses. First, it serves as a guide to the projector designer to this important property of the system. Secondly, it can be used in the simplified model to do a survey of similar projector designs in order to focus attention on the best option to meet the marketing need. Color separation efficiency also includes the effect of color recombination, and can be computed for each of the three channels with the following equation:

$$\eta_{S_i} = \frac{\int_{380}^{780} E'(\lambda)\eta_{i,S}(\lambda)\,\eta_{i,R}(\lambda)\,\bar{Y}(\lambda)\,d\lambda}{\int_{380}^{780} E'(\lambda)\,\bar{Y}(\lambda)\,d\lambda}$$

where E' is defined by:

$$E'(\lambda) = E(\lambda) \prod_{j=1}^{n} \eta_j(\lambda)$$

(11.20)

In these equations,

- i   denotes red, green and blue,
- $E(\lambda)$ is the spectral energy density of the lamp/collector,
- η_j is the efficiency of the jth non-neutral density element,
- $E'(\lambda)$ is the spectral energy density of the lamp/collector as modified by all the spectrally varying components except the dichroic filters
- $\eta_{i,S}(\lambda)$ is the dichroic efficiency as seen by each color channel during separation,
- $\eta_{i,R}(\lambda)$ is the dichroic efficiency in each channel during recombination and
- $\bar{Y}(\lambda)$ is the CIE luminance matching function.

The values η_R, η_G and η_B are simply the fraction of the total lumens that pass through the red, green and blue portions of the projector, respectively. For systems such as the three panel projectors described in Sections 10.2 and 10.3, where the three channels operate in parallel, the color separation efficiency is:

$$\eta_S = \eta_{S_R} + \eta_{S_G} + \eta_{S_B}. \qquad (11.21)$$

Ideally, in a projector intended for NTSC video applications,

$$\eta_{S_R} \approx 30\%, \eta_{S_G} \approx 59\%, \text{ and } \eta_{S_B} \approx 11\%.$$

When these conditions are all satisfied, color correction problems for the projector are minimized. In a three-panel projector, the value of η_S can be close to 100 %, depending on the lamp and the chosen primary colors. In systems where the three color channels do not operate in parallel, such as most of the single-panel systems described in Section 10.4, the color separation efficiency is:

$$\eta_S = \frac{\eta_{S_R} + \eta_{S_G} + \eta_{S_B}}{3}. \qquad (11.22)$$

In a typical color wheel time sequential system, the color separation efficiency is at most about 33 %. This low color separation efficiency for time sequential projectors, as compared with three panel projectors, is one of the reasons why three panel projectors dominate the high-performance projector market.

Equation (11.22) also describes the color separation in color matrix systems, as described in Section 10.4.1. Here the efficiency can be significantly less than 33 %, because the absorption filters commonly used in matrix projectors are significantly less efficient than the dichroic filters normally used in field sequential projectors. While dichroic filters can be used in these color microfilter systems (Fuhrmann et al. 1997), it is not commonly done, primarily due to the cost and yield of the process.

The color separation efficiency of angular color separation (ACS) projectors, such as the one described in Section 10.4.1.2, is described by Equation (11.21). Since an ACS projector uses dichroic filters, it can have a color separation efficiency near to 100 %, and potentially high lumen throughput. However, this type of projector can suffer from very low collection efficiency due to a low étendue, unless the panel is very large. These systems use pupil sharing in the projection lens and thus require very low f/# projection lenses, another disadvantage. ACS systems are also prone to color cross-talk in the filters, reducing the color purity of each color.

Scrolling color projectors, discussed in Section 10.4.2.6, are also described by Equation (11.21) and have almost 100 % color separation efficiency. These systems have a much lower usable étendue than a three-panel projector with the same size panels. To get the same throughput in a scrolling color projector as in a three-panel projector, typically about the same panel area is needed. In other words, the scrolling color projector needs one large panel but the three-panel projector needs three small panels. Since they generally do not use pupil sharing, the lens situation is not as bad as it is for ACS projectors.

Like the color correction efficiency η_{CC}, color separation efficiency η_S, is strictly accurate only for the projector for which it was derived. However, a value for η_s can be used in estimating the throughput of a whole series of projectors with similar colorimetry. Color separation efficiency is, however, less subject to minor variations in component properties than is color correction efficiency.

11.6 Problems with Lumen Throughput Calculations

Calculations of the lumen throughput of a projector often overestimate the amount of light at the projection screen. This overestimate can be by a factor ranging from 10 % to more than 100 %, depending on how carefully the estimate is made. Major factors affecting the accuracy of the calculation are:

- Incorrect calculation of η_{CC} or η_S. The value of η_{CC} is particularly prone to problems, as discussed above.

- No simple equation or analytical model handles the issue of étendue correctly. The proper way to handle étendue is with a ray trace of the complete optical system, but this is not feasible early in the concept or design stage of a projector. When the system is étendue limited, optical elements may increase beam étendue without the designer being aware of it. For instance, there may be more spherical or chromatic aberration in a relay lens than assumed when estimating throughput. This extra étendue in the optical beam will reduce total system throughput.

- An étendue estimate that assumes a uniform cone angle will over-estimate the system étendue and collection efficiency if there is vignetting in the projection lens or any relay lenses in the system.

- Transmission and efficiency of optical components is normally measured on-axis. Off-axis values can be less than the on-axis values due to either longer optical paths or increased reflection due to higher angles of incidence.

- Estimated values of efficiencies, when measured values are not available, are often too optimistic.

- Efficiency errors are cumulative. For example, if there is a 1 % error in each of 5 elements with efficiencies of 0.50, the error will be $1 - (0.49^5)/(0.50^5) \approx 10\%$ error.

- In systems with polarizing beam splitters, i.e., reflective LCD systems, the η_{CC} and η_S values can change quite rapidly with f/#. The full colorimetric model should be used at every f/# of interest.

11.7 Lumen Output Variation in Production

Equation (11.1) gives the expected value of lumen output when a single projector is built out of components, all of which have been measured. This is, perhaps, a good model of what happens in a research lab, but not a good model of what happens in a projector factory. In the factory, a large number of projectors will be assembled out of interchangeable components. Each individual component will vary from the average value μ_i for that component type by some small amount. Typically, this variation can be described with a standard deviation value σ_i. The relative standard deviation for each component type could be defined as:

$$s_{i,rel} = \frac{\sigma_i}{\mu_i} \qquad (11.23)$$

where the subscript i refers to the ith component in the optical path. A Gaussian distribution within the production lot of the specific characteristic is assumed.

To a first approximation, the relative standard deviation of the projector lumen output is given by:

$$s_{lumens} = \sqrt{s_{lamp}^2 + \sum_i s_{i,rel}^2} \qquad (11.24)$$

where:
- s_{lumens} is the relative standard deviation of the lumens at the screen for all projectors,
- s_{lamp} is the relative standard deviation of the lamp output, and
- $s_{i,rel}$ is the relative standard deviation of the efficiency for the ith component.

Note that s_{lumens} is cumulative: the more components and the larger the variation in each component, the larger the variation in the total projector lumen output. For example, if the lamp, reflector and dichroics each has a relative standard deviation of 10 %, and there are 10 other components each with a relative standard deviation of 2 %, the lumens standard deviation is 18 % according to Equation (11.24). If the brightest projector in a production run is 2 standard deviations above average and the dimmest projector is 2 standard deviations below the average, then the luminous ratio between the brightest and the dimmest is:

$$\frac{(1+2s)}{(1-2s)} = \frac{[1+(2)(0.18)]}{[1-(2)(0.18)]} = 2.1 \qquad (11.4)$$

where s is the relative standard deviation calculated from Equation (11.24) (18 % in this example). In other words, the brightest projector in the production run will be more than twice as bright as the dimmest projector!

The simple combination of standard deviations, as shown by Equation (11.24), is only an approximation that does not apply exactly when:

- The relative standard deviations are large ($s_{i,rel}$ > about 10 %),
- The total relative standard deviation is large (s_{lumens} > about 15 %),
- The distributions are not Gaussian, or
- Equation (11.1) does not apply because the spectral transmission of the components is not neutral density.

Since at least one of these qualifications would normally apply to the components in a projector, more sophisticated statistics will give a more accurate picture. A MonteCarlo approach is a good way to estimate luminance and color variations (Pettitt et al. 1996). In this method, one assumes that there is a statistical distribution of the spectral transmission and other optical properties for each individual component in the projector. These distributions do not necessarily have to be normal distributions. A computer program then selects a complete set of projector parts at random, selecting each part from its own statistical distribution.

The efficiency of the projector is then calculated at each wavelength from the performance of the individual parts. The primary colors, white color and lumen throughput is then calculated from the spectral data. The program then iterates, selects another set of parts, and recalculates the performance. After several thousand sample projectors (computer iterations), the average, range and distribution of projector performance values becomes quite well known. One advantage of this MonteCarlo approach is it provides an excellent model of what happens on a projector assembly line.

This is not just a theoretical exercise, because for any given projector design, a 2:1 range in projector lumen output when all individual projector components are within their specifications is a serious quality control and marketing issue. There are several solutions to this problem, none of which is fully satisfactory:

- Reduce the standard deviation of each component,
- Sort the finished projectors into different lumen output grades or
- Sort the components into different grades.

Reducing the standard deviations of the components, especially of the components with high relative standard deviations is the 'correct' solution. However, this can be expensive, and is not always physically possible. For example, in the making of dichroic filters, achieving a cutoff tolerance of much better than 1 % of the cutoff wavelength may not be practical.[2] Thus, these filters may not always be accurate enough. If too tight a specification is required, the only way the dichroic maker can meet the specification is by sorting, at very high cost.

If the finished projectors are sorted into different lumen outputs, then the brightest can be sold for a premium. However, this can create a difficult inventory control issue. By mixing and matching components, rather than selecting components at random from all available ones, all projectors could be built with a lumen output near the average. This would severely complicate production, to say the least. Both of these 'solutions' lead to problems later, when projectors are serviced. If the replacement components do not match the original components, the luminance of the projector could change dramatically.

While there is no good solution to this problem, it can be reduced by careful design and quality control. When a component is being considered for inclusion in a projector, it should be examined for standard deviation and statistical distribution as well as average and high and low limit values. If a component high and low limit specifications are enforced by sorting, that component will have a much higher relative standard deviation than if the limits were met because the component was made by an inherently accurate production process that does not need sorting to meet the component specification.

This sorting vs accurate manufacturing leads into the question of statistical process control, a topic much too important and detailed to be covered briefly in this book. This reader is referred to one of the many standard texts on the subject of quality design such as (Baker 1990) or (Oakland 2007).

References

Baker, T.B. (1990) *Engineering Quality by Design: Interpreting the Taguchi Approach*. New York: Marcel Dekker, Inc.
Ben-Chorin, M. and Eliav, D. (2007) Multi-primary design of spectrally accurate displays. *J. of the SID* **15(9)**: 667–77.
Boyd, R.W. (1983) *Radiometry and the Detection of Optical Radiation*. New York: John Wiley & Sons, Inc., 89–93.
Brennesholtz, M.S. (1996) Light collection efficiency for light valve projection systems. In M. Wu (ed.), *SPIE Proceedings: Projection Displays II* **2650**: 71–9.
Brennesholtz, M.S., Nicholson, W. and Chinnock, C. (2006) *LED Projection Systems: A Study of the Use of LEDs as an Illumination Source for Projection Systems*. Norwalk, CT: Insight Media, February.
Fuhrmann, J., Sautter, B., Kallfass, T. and Lueder, E. (1997) Enhancement of the light efficiency of LC projection systems by the use of dichroic color filters. *SID Intl. Symp. Digest of Technical Papers*, 761–4.
Moench, H., Derra, G., Fischer, E. and Riederer, X. (2000) UHP lamps for projection systems. Paper *IDW 2000*, SID, Paper LAD2-2.
Nicolas, C.N., Loiseaux, B., Huignard, J.P. and DuPont, A. (1992) Efficient optical configuration for polarized white light illumination of 16/9 LCDs in projection display. *Japan Display Proceedings*, 121–4.
Oakland, J.S. (2007) *Statistical Process Control*, 6th edn. Oxford: Elsevier.
Ohta, N. and Robertson, A.R. (2005) *Colorimetry: Fundamentals and Applications*, Chichester: John Wiley & Sons, Ltd.
Pettitt, G.S., DeLong, A. and Harriman, A. (1996) Colorimetric performance analysis for a sequential color DLP^{TM} projection system. *SID Intl. Symp. Digest of Technical Papers*, Paper P-28, pp. 510–13.
Poynton, C. (2003) *Digital Video and HDTV: Algorithms and Interfaces*. The Morgan Kaufmann Series in Computer Graphics. San Francisco: Elsevier.
Schweyen, J., Garcia, K. and Gleckman, P. (1997) Geometrical optical modeling considerations for LCD projector display systems.M. Wu (ed.), *SPIE Proc: Projection Displays III* **3013**: 126–40.

---

[2] For example, a red filter with a cutoff of 600 nm would be difficult to manufacture in volume with a tolerance of less than ±6 nm.

Shimizu, J. and Janssen, P. (1997) *Integrating Lens Array and Image Forming Method for Improved Optical Efficiency.* US Patent 5,662,401.

Sonehara, T. (1992) Dichroic Optical Elements for Use in a Projection Type Display Apparatus, US Patent 5,098,193, issued 24 March 1992.

Wyszecki, G. and Stiles, W.S. (1982) Color Science: Concepts and Methods, Quantitative Data and Formulas, 2nd edn. New York: John Wiley & Sons, Inc.

12

Projector Lumen Throughput

In this chapter, a number of issues related to the lumen output of video projectors will be examined. The first two sections will be examples of estimating projector lumen output using the simplified model from Chapter 11. Section 12.3 gives an example of a complete colorimetric analysis of a projector.

This chapter will only deal with estimating lumen throughputs for projectors. The measurements of lumen throughputs are discussed in the next chapter.

12.1 Throughput of a Simple Transmissive Projector

The light engine of the basic, simplest projector consists of only a light source, a transmissive LCD panel incorporating color microfilters, and a projection lens. This projector is illustrated in Figure 12.1. The light valve, of the type described in Section 10.4.1.1, has three sub-pixels per pixel, with a red, green or blue microfilter covering each sub-pixel. Use of an integrator is optional in this type of projector. Projectors with this configuration have been marketed as low performance, low cost systems. These systems are often based on panels originally developed for hand-held direct-view systems and for laptop computers. Since these panel sizes vary from large to very large, the projector is rarely étendue limited, even with relatively high f/# projection lenses. Due to the large panel size, these systems normally have a field lens at the panel. The field lens may be a two element lens, with one element before the panel and the other after, allowing the light to be telecentric at the panel itself. One or both of these elements may be a Fresnel lens.

The throughput will be predicted using the simple model described in Section 11.1. The first step in calculating projector throughput is to calculate the system étendue, usually with Equation (11.3). For this example, a panel diagonal of 3.6" (91.4 mm), 4:3 aspect ratio, and a f/3.8 projection lens is assumed. The system étendue is then 218 mm^2 steradian. Since there is no polarization conversion, or any of the other factors from Table 11.1 that affect usable étendue, the total étendue and usable étendue are the

---

Projection Displays, Second Edition   M. Brennesholtz, E. Stupp
© 2008 John Wiley & Sons, Ltd

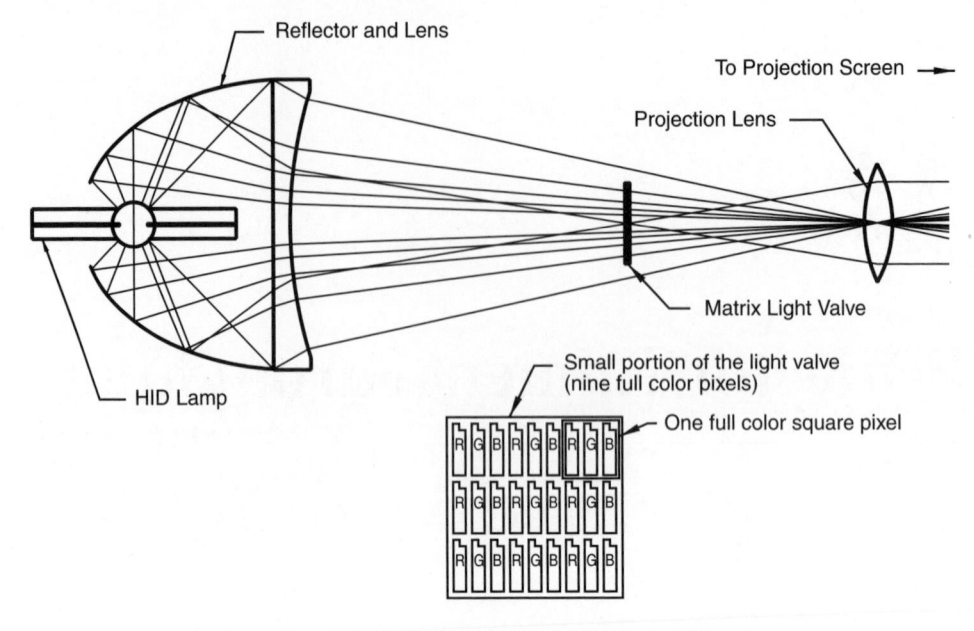

Figure 12.1 Simple projector based on color microfilter light valve.

same. At this étendue, the 125 W metal-halide lamp described in Table A3.1 is not étendue limited and produces 4423 lumens. This represents only 57.1 % of the light emitted by the arc, due to the limitations of the reflector geometry and the reflector coating. Table 12.1 is a summary of all the efficiencies and other parameters used in this projector example.

Table 12.1 Throughput of a simple projector.

Panel geometric properties	
Diagonal (mm)	91.4
Sub-pixels horizontally	1920
Pixels vertically	480
Sub-pixel pitch H (mm)	0.038
Pitch V (mm)	0.114
Width (mm)	73.2
Height (mm)	54.9
Panel area (mm^2)	4013
Half angle at panel (degrees)	7.6
Projection lens f/#	3.80
Total étendue (mm^2 steradian)	218.3
Lamp properties	
Watts	125
Lumens/watt	62
Total lumens	7750
Collection efficiency (%)	57.1
Collected lumens	4423
Projector efficiency (%)	
Color separation	28.2
Polarizer	44.0

ITO 2 layers	81.0
LCD aperture	76.1
Analyzer	87.2
Modulation efficiency	100
Color correction	100
Projection lens (zoom)	80.0
Overall projector efficiency (%)	5.3
Calculated lumens	**236**
Lumens/watt	1.89

The projector efficiency is the product of a number of factors (Equation 11.1). A colorimetric analysis done in accordance with Section 11.4 using typical absorptive color filters and the spectrum of the metal halide lamp gives a color separation efficiency of 28.2 %. This value is quite low compared with the η_{cc} in a three-panel projector. The low color separation efficiency in this type of projector is the result of two main factors:

1. About two-thirds of the incident light on each sub-pixel has the wrong color. For example, red and green light falling on the blue pixel is lost.

2. The blue and green absorptive filters normally used in this type of projector have peak transmissions of about 65 %.

This second factor can be improved by accepting less saturated colors. This in fact would be considered the default choice for this sort of low-cost projector. Color filter research, mostly aimed at direct-view LCD systems (Sugiura 2001) has also shown that higher transmission filters are possible and, with the right light source, they can produce good colorimetry. Sugiura reports peak transmissions for blue and green about 85 % with a red filter peak transmission of about 95 %. Since there is no polarization conversion, the polarizer at the LCD absorbs essentially all of the light of the wrong polarization, plus part of the light of the right polarization, for a transmission of 44 %. The indium-tin oxide (ITO) transmission is assumed to be 81 %, total for both layers. The aperture ratio of 76 % is calculated with Equation (11.5) using the constants $K_{1H} = K_{1V} = 6.1\,\mu m$ and $K_2 = 12.1\,\mu m$. The pixel pitches are calculated from the pixel counts, panel diagonal and aspect ratio. Note that the individual sub-pixels have 3:1 aspect ratios to give a square full color pixel. The polarization analyzer after the LCD transmits 87 % of the light of the correct polarization. The panel is a TN panel and 100 % modulation efficiency is assumed.

For this particular projector, the absorptive filters were designed to match the spectrum of the metal halide lamp well enough that no color correction was necessary (100 % color correction efficiency). Finally, the zoom lens on the projector was assumed to have a transmission of 80 %. Multiplying all these efficiency factors together gives a total projector efficiency of 5.3 %. Combining this with the 4423 lumens collected from the lamp gives a total throughput of 236 lumens. This gives an efficacy for the projector of 1.89 lumens/watt, a respectable value for this type of projector but much lower than the value possible with more sophisticated projector architectures which can produce up to 10 lumens/watt.

This lumens/watt value only includes the watts consumed by the lamp. Obviously there will be other power consumed by the finished projector, including dissipation in the circuitry and fan, and power used by the lamp ballast that never makes it to the lamp itself.

Since there is no integrator in the system in Table 12.1, Kohler type illumination of the panel would be used to ensure there was adequate uniformity (see Section 6.4). The estimate of 236 lumens for throughput is based on the complete utilization of all light from the system, i.e. the projector was not étendue limited. Kohler type illumination achieves its uniformity by overfilling the panel. Therefore, there would be some light beyond the edge of the panel so the figure of 236 lumens is an overestimate. This is an example of the statement in Section 11.2.3.4 that the lumens vs étendue formulation can only estimate the maximum

304 PROJECTOR LUMEN THROUGHPUT

possible throughput for the case where étendue is conserved at each optical element. In this case, étendue was deliberately not conserved in order to avoid the need for an integrator. With critical illumination it would be possible to achieve the 236 lumen forecast but the image at the screen would be very strongly peaked at the center.

12.2 Throughput in a Three-panel Projector

The second example projector was chosen in order to highlight the difference between total system étendue and usable étendue. The projector is a three-panel transmissive TN-LCD equal path projector, as shown in Figure 12.2. An integrator is used to improve the uniformity of the panel illumination. The throughput analysis is done for this system both with a polarization conversion system and without this PCS. The data and results for both calculations are in Table 12.2.

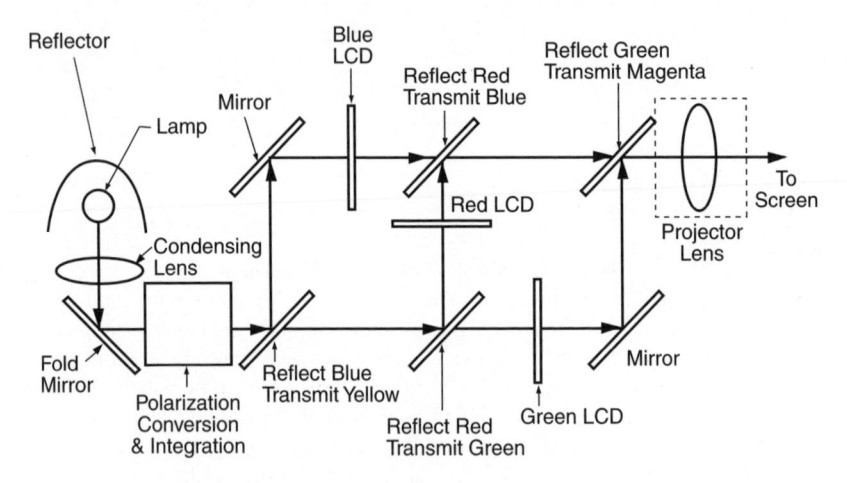

Figure 12.2 Three panel equal path projector.

Table 12.2 Throughput of a three-panel projector.

Panel geometric properties	No PCS	With PCS
Diagonal (inches)	0.75	0.75
Diagonal (mm)	19.1	19.1
Pixels horizontally	1024	1024
Pixels vertically	768	768
Horizontal pitch (mm)	0.015	0.015
Vertical pitch (mm)	0.015	0.015
Width (mm)	15.2	15.2
Height (mm)	11.4	11.4
Area (mm^2)	174.2	174.2
Overfilled area (mm^2)	193.0	193.0
Half angle (degrees)	12.0	12.0
f/#	2.40	2.40
Total étendue (mm^2 steradian)	26.2	26.2
Usable étendue (mm^2 steradian)	26.2	13.1

Lamp properties		
Watts	100	100
Arc length (mm)	1.0	1.0
Total lumens	6000	6000
Collection efficiency	80.2 %	65.1 %
Collected lumens	4813	3909
Projector efficiencies		
Cold mirror	97.0 %	97.0 %
PCS		92.0 %
Integrator	85.0 %	85.0 %
Field and relay lenses (3)	94.1 %	94.1 %
Polarizer	44.0 %	92.0 %
ITO (2 layers)	81.0 %	81.0 %
LCD aperture ratio	54.6 %	54.6 %
Analyzer	88.0 %	88.0 %
Projection lens (zoom)	80.0 %	80.0 %
Color separation efficiency	84.0 %	84.0 %
Overfill	90.3 %	90.3 %
Color correction efficiency	92.2 %	92.2 %
Projector efficiency	7.4 %	14.3 %
Total projector lumens		
Total lumens	358	559
Lumens/watt	3.58	5.59

The polarization recovery system works by separating the vertically polarized light from the horizontally polarized light into two beams. The polarization of one beam is rotated 90° by a half wave plate or some other method. The beams are then recombined to illuminate the panels. Polarization recovery systems are described in more detail in Section 6.5.3. The recombined beam has an étendue of twice what the original beam had. Alternatively, one could say half of the étendue of the system is being used for the vertically polarized light and the other half for the initially horizontally polarized light.

The XGA panels used in this example are assumed to have a 0.75" diagonal, a 4:3 aspect ratio and a 54.6 % aperture ratio. Aperture ratio was calculated with a $K_{1H} = K_{1V} = 3.0$ and $K_2 = 4.5\,\mu$. The area of each panel would then be 174.2 mm². With 5 % horizontal and vertical overfill, the overfilled area would be 193 mm². An f/2.4 projection lens is assumed, giving a total étendue of 26.2 mm² steradian, according to Equation (11.3).

Because of the polarization recovery system, only half of this étendue is used by the unmodified incoming beam in the version of the projector with a PCS. The other half is used by the beam with the rotated polarization. This gives a usable étendue of 13.1 mm² steradian for each beam. At this étendue, the UHP-type lamp with a 1.0 mm arc will produce usable 3909 lumens. Taking into account the efficiency of all the other components necessary in a three-panel projector gives a projector efficiency of 14.3 %. Combining the lumens with the projector efficiency gives a total lumen output of 358 lumens without the PCS and 559 lumens with the PCS.

Note that this 0.75" diagonal panel is considered large by today's standards. As such, the system is only modestly limited by étendue. This allows the PCS to give a 56 % increase in throughput. If a smaller panel were used, the gain from the PCS would typically be in the 20–50 % range.

If a projector designer is working with systems with a relatively low étendue, the size of the arc is a critical parameter in the throughput of the projector. If this projector were to use an otherwise identical lamp with a 1.3 mm arc instead of the 1.0 mm arc used in Table 12.3, the throughput would only be 399 lumens with the PCS instead of 559, a decrease to 71 % of the original output. This decrease shows the importance of short arc lamps in étendue limited projection systems.

12.3 Throughput Estimate using the Full Colorimetric Model

A single-panel color wheel projector based on the DMD™ from Texas Instruments will be used as an example of the use of the full colorimetric model as described in Section 11.5. The process used in applying the model will be stressed. Note that the projector used as the example in this section does not represent any of the real DMD projectors currently in production.

Rather than take the values of η_{cc} and η_s as specified quantities, they will be calculated from the spectral data of the various components in the projector. The operation of the DMD is explained in Section 5.1. A single-panel, color-wheel-based architecture, as shown in Figure 12.3, will be used in this example.

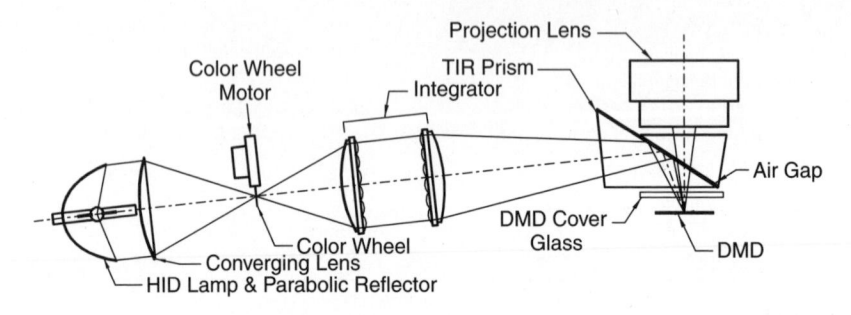

Figure 12.3 Color sequential DMD projector.

The following spectral elements will be accounted for in doing the color model of this projector:

1. The spectrum of the UHP lamp, shown in Figure 8.4,

2. A parabolic reflector with a cold mirror coating having characteristics as shown in Figure 6.23,

3. Color wheel with red, green and blue transmissive dichroic filters as shown in Figure 6.5 and a high efficiency AR coating on the other side,

4. A lenslet integrator with all four surfaces coated with high efficiency AR coating at normal incidence as shown in Figure 6.12 and

5. A TIR prism with all surfaces coated with a high efficiency AR coating. The prism-air gap-prism interface is treated as two high efficiency AR coatings with a 45° incidence angle, as shown in Figures 6.12a and 6.20.

6. The DMD will be treated as an aluminum mirror and two surfaces of high efficiency AR coating on the cover glass.

7. The projection lens will be treated as six high efficiency AR coatings on normal glass and two high efficiency AR coatings on high index glass, as shown in Figure 6.12.

8. A 800×600 DMD with 17 μm pixels, 10 % area overfill and a f/3 lens has an étendue of 13.4 mm^2-sr. For the UHP, the collected lumens at this étendue will be approximately 2800. Of course, with the DMD, no PCS is needed.

The spectral calculations are done over the range of 400–700 nm. All integrals are done in a spreadsheet program with the following numerical integration:

$$\int_{400}^{700} f(\lambda)d\lambda = \sum_{i=0}^{75} f(400 + i^*\Delta\lambda)\Delta\lambda$$

where $\Delta\lambda = 4\,\text{nm}$ (12.1)

The power spectrum of the lamp $\Phi(\lambda)$ is specified in units of watts/nm. The usable lumens from the lamp are obtained by convolving $\Phi(\lambda)$ with the CIE 1931 $\overline{Y}$ curve, as indicated in Equation (12.2).

$$\Phi_V = 683 \int_{380}^{780} \Phi(\lambda)\overline{Y}(\lambda)d\lambda \qquad (12.2)$$

It is assumed that the spectrum $\Phi(\lambda)$ is measured in an apparatus similar to the one shown in Figure 8.26 and with an aperture that has the same étendue as the DMD/projection lens combination.

Figure 12.4 shows the spectra of the spectrally varying components in the projector. The solid black line is the product of all the AR coatings in the system, 26 surfaces in all. In a reflective system, some surfaces must be passed through twice, once in each direction. These surfaces must of course be included twice in the system transmission. The black dashed line is the reflectivity of typical uncoated aluminum. This reflectance was not measured on a DMD. The red, green and blue curves are the spectral transmissions of the dichroic filters used in the color wheel.

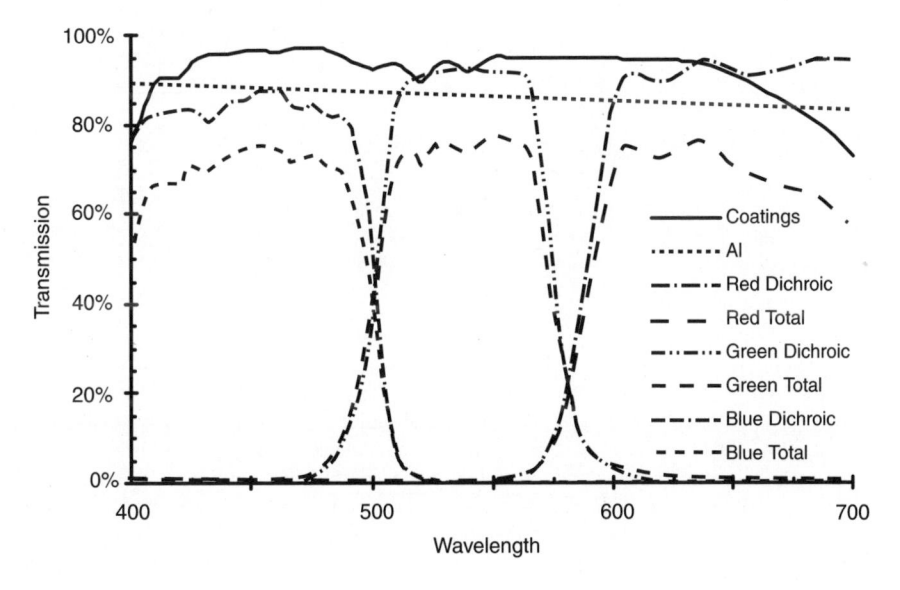

Figure 12.4 Transmission spectra of spectrally varying components.

There are a number of factors that do not vary with wavelength that must be included in order to calculate the total system output. The DMD efficiency is taken to be 69.5 % (Hornbeck 1997). This includes 89 % aperture ratio, 15 % diffraction loss (85 % efficiency) and 92 % temporal efficiency. Next, the integrator has about a 93 % aperture ratio. In order to minimize mechanical alignment problems, the integrator will be designed so the light body will overfill the DMD, for an overfill efficiency of 90 %. Finally, in a color sequential system there will also be a 33 % temporal duty cycle for each color. These wavelength independent factors multiply together to give an efficiency of 19.3 %.

308 PROJECTOR LUMEN THROUGHPUT

In Figure 12.5, the solid black line is the spectrum of the UHP lamp (Figure 8.4) being considered in this projector. The finely dashed line is the lamp spectrum, as corrected by all the spectrally selective components except for the dichroic filters (i.e. the AR coatings, the cold mirror and the aluminum in the DMD). The red, green and blue curves in this figure are the spectra produced by the dichroic filters and the corrected lamp. These will be the spectra actually observed at the screen.

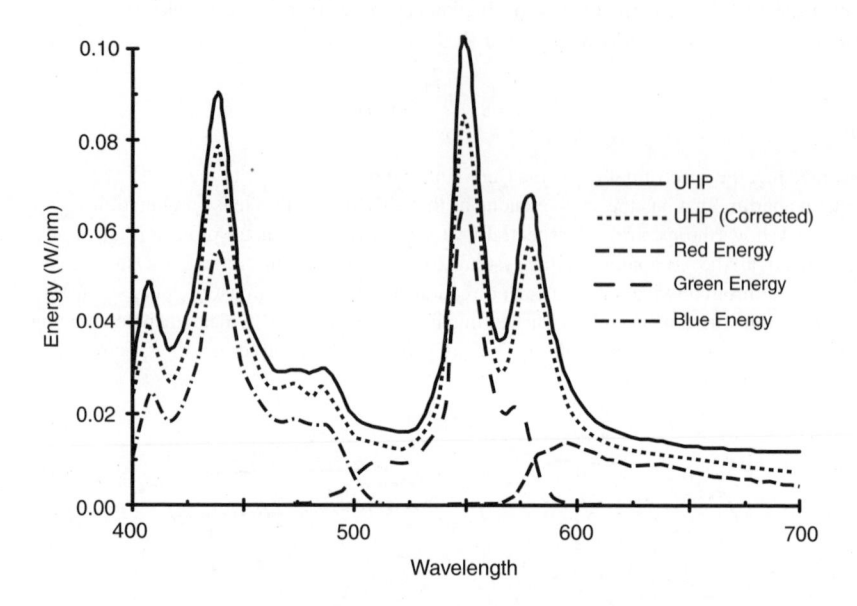

Figure 12.5 Channel energies in color-wheel DMD projector.

It is now possible to evaluate the color separation efficiency η_s using Equation (11.20). The denominator of this equation is just the dashed curve from Figure 12.6 convolved with the CIE 1931 $\overline{Y}$. The red, green and blue numerators of this equation are the red, green and blue curves convolved with the $\overline{Y}$ curve.

The integrations give color separation efficiencies of $\eta_{s,red} = 12.7\%$, $\eta_{2,green} = 50.4\%$ and $\eta_{s,blue} = 4.3\%$. Applying Equation (11.22) to get overall color separation efficiency for the projector gives $\eta_s = 12.7\% + 50.4\% + 4.3\%/3 = 22.4\%$ Since a complete colorimetric model is being developed for this projector, this efficiency factor is not actually needed at the present time. This color separation efficiency factor will be useful in the future if one ever wished to analyze a similar projector using the simplified model (Equation (11.1)). 'Similar' in this case means one with the same lamp and dichroic filters and a similar architecture but perhaps with a different pixel count or size, a different integrator or a different projection lens.

It is useful, however, to compare the color separation efficiency to the nominal 33% for a color sequential projector. One reason for the low value of this factor is the relatively low peak transmission of the dichroic filters used. A second reason can be seen by comparing the green and red curves with the corrected lamp spectrum in Figure 12.5. It can be seen that much of the energy in the yellow emission line of the lamp is lost in the wavelength gap between the red and green filters. It is desirable to eliminate this yellow emission; including it in the red channel produces an orange-red color and including it in the green channel produces a yellow-green. Since the lamp emission in the range of 575–590 nm represents about 25% of the lumen output of the lamp, losing even a portion of this light has a significant effect on total lumen output. Multiple color systems with a yellow channel can recover much of this lost light.

Comparing the Y_i portion of Equation (11.9) to the numerator of Equation (11.20) shows the two are identical. Therefore, it is straightforward to modify the spreadsheet that calculated Equation (11.20) to calculate X_i and Z_i as well as Y_i. This integration gives the values necessary for Equation (11.12):

$$\begin{bmatrix} X_R & X_G & X_B \\ Y_R & Y_G & Y_B \\ Z_R & Z_G & Z_B \end{bmatrix} = \begin{bmatrix} 0.134 & 0.162 & 0.104 \\ 0.083 & 0.328 & 0.028 \\ 0.001 & 0.014 & 0.562 \end{bmatrix} \qquad (12.3)$$

Table 12.3 Calculated color coordinates and luminance.

Color	x	y	Lumens	CCT	MPCD
Uncorrected white point					
Red	0.617	0.379	56		
Green	0.321	0.651	224		
Blue	0.150	0.040	19		
White	0.283	0.310	299	8843K	+22
Corrected white points					
White (D 65)	0.313	0.329	222	6550K	+7
White (8200 K)	0.293	0.303	241	8200K	0

These values are used to calculate the color coordinates and luminance for the uncorrected white channel. To do this, the coordinates and luminance of each channel are individually determined as shown in the upper part of Table 12.3. The uncorrected white point is then calculated by adding together the red, green and blue tristimulus values from Equation (12.3) as shown in Table A2.1 and calculating the white color and luminance. The uncorrected white point has a correlated color temperature (CCT) of 8843 K and is 22 minimum perceptible color difference units (MPCD) above the black body line. While this CCT would probably be acceptable for video applications, this high a MPCD value indicates the entire image will have a greenish cast which would probably not be accepted. If this uncorrected white point is acceptable, it is possible to stop here and not do the next step of estimating the color corrected white. It will be assumed, however, that it is desired to have a D65 white, as specified for television broadcasts, with CIE $x = 0.313$ and $y = 0.329$. In this case, Equation (11.16) becomes

$$\begin{bmatrix} 0.134 & 0.162 & 0.104 \\ 0.083 & 0.328 & 0.028 \\ 0.001 & 0.014 & 0.562 \end{bmatrix}^{-1} \begin{bmatrix} 0.417 \\ 0.438 \\ 0.477 \end{bmatrix} = \begin{bmatrix} k_R \\ k_G \\ k_B \end{bmatrix} = \begin{bmatrix} 1.35 \\ 0.93 \\ 0.82 \end{bmatrix} \qquad (12.4)$$

As can be seen, the k_R value is greater than 1.00, and k_G and k_B are less than 1.00. Since the gain on a light-valve projector cannot normally be increased (maximum throughput corresponds to gain $= 1$), it is necessary to apply Equation (11.17) to get the final channel gains:

$$\begin{bmatrix} k'_R \\ k'_G \\ k'_B \end{bmatrix} = \begin{bmatrix} 1.00 \\ 0.69 \\ 0.61 \end{bmatrix} \qquad (12.5)$$

Using Equations (11.18) and (11.19) will give color corrected white throughput of 222 lumens, corresponding to a color correction efficiency of 74%.

If instead, a corrected white color temperature of 8200 °K ($x = .293$, $y = 0.303$) is desired, the uncorrected color and lumens as given in Table 12.3 still apply. It is only necessary to substitute the CIE x-y values for 8200 °K for the D65 values in the spreadsheet and recalculate. The new color correction efficiency is 81%, for a lumen throughput of 241.

These calculations have assumed the white point would be corrected by reducing the light throughput in the green and blue channels to get color balance. If the color wheel angle change technique as described in Section 10.4.2.2 is used instead, the color wheel angles for red, green and blue for a two cycle (RGBRGB) color wheel would be 78° for red, 54° for green and 48° for blue. This would result in a total of 290 D65 white lumens for a color correction efficiency of 97 %. This is a significant improvement over the equal-angle color wheel case.

As mentioned in Section 11.5.3, a 54° color wheel angle may not allocate sufficient time to green to achieve a satisfactory gray scale. If this occurs, then it will be necessary to allocate more time to green at the expense of red and/or blue. If this is done, then it will be necessary to discard some of the lumen gain and the color correction efficiency will be between the 97 % achievable by pure angle adjustment and the 74 % achieved by the equal angle case and use of pure throughput reduction techniques.

References

Hornbeck, L.J. (1997) Digital Light Processing™ for high-brightness, high-resolution applications. In M.H. Wu (ed.), *SPIE Proceedings: Projection Displays III* **3013**: 27–40.
Sugiura, T. (2001) EBU color filters for LCDs. *SID Intl. Symp. Digest of Technical Papers*, 146–9.

13

Characteristics and Characterization

Projection displays are designed to achieve specific performance levels for a range of applications. When all is said and done, how well the display fulfills the intended requirements determines how successful the design is. The characteristics that are most critical in describing the display performance and the methods for measuring them are discussed here. Among the parameters used to specify a display's performance are its luminance, resolution, colorimetry contrast, and short-term temporal characteristics.

Included in this discussion are some of the characteristics of the human visual system. In some situations, the ability of the viewer to see the information is the limiting factor, rather than the display itself. This discussion is not intended to be exhaustive but rather to elucidate a few of the more important characteristics of the eye–brain system which may affect the perceived qualities of the display. There is a significant literature available on the characteristics of the human visual system (Cornsweet 1970; Bass 1995).

Characterization of a projected image is mostly done using static images. Much of the following discussion is based on measurements of this type of image. It is assumed in these measurements that the temporal characteristics of the display do not generate significant artifacts in the image. This would be achieved when the temporal characteristics of the display and the human visual system, together or separately, do not produce unwanted distortions of the perceived signal information. Temporal artifacts and other sources of image degradation are discussed in the next chapter.

The Video Electronics Standards Association (VESA) Flat Panel Display Measurement Standard (FPDM) (VESA 01) provides recommended measurement techniques for virtually every display parameter discussed in this chapter and artifacts discussed in Chapter 14. This standard covers not just direct-view flat panel displays but also direct-view CRT systems, projection CRT systems and projection systems based on microdisplays or light valves. It is recommended that anyone interested in display measurements obtains a copy of this standard which can be ordered from http://www.vesa.org/Store/buystandards.htm. Another guide specifically oriented toward projector users is Jeffreys (2005).

Projection Displays, Second Edition M. Brennesholtz, E. Stupp
© 2008 John Wiley & Sons, Ltd

13.1 Characteristics of the Human Visual System

Among the important visual characteristics affecting the display are the eye resolution, temporal response, luminance response, and color sensitivity. The descriptions of luminance and chrominance are part of the subjects of photometry and colorimetry, respectively. Descriptions of the quantifications of these subjects are given in Appendices 1 and 2, respectively.

It is necessary to understand the limitations of the visual system in order to design a projection system that matches the perceptual limitations of the intended audience. Designing a projector with higher image resolution than can be perceived by the audience does not add value, but it typically does add cost. Figure 13.1 shows the nominal contrast sensitivity of the human visual system as a function of spatial frequency (Boff and Lincoln 1988; Robson 1966). Contrast sensitivity is the reciprocal of the contrast threshold for detecting the presence of a given spatial frequency. The higher the contrast sensitivity the more sensitive the eye is to that frequency. Note that the amplitude and the spatial frequency for peak sensitivity are dependent on several parameters, including brightness and the temporal frequency.

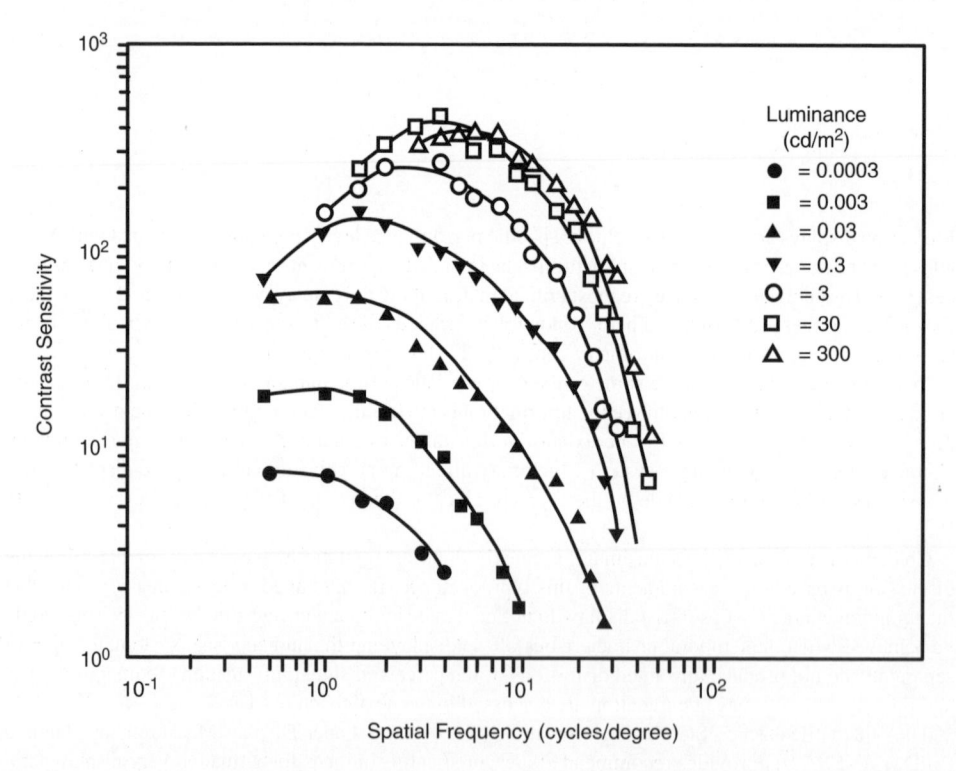

Figure 13.1 Contrast sensitivity of human visual system as a function of spatial frequency for various luminance levels (reproduced with the permission of the Optical Society of America (Robson 1966)).

The contrast sensitivity in Figure 13.1 can be translated to a limiting angular resolution of a feature of about 0.3–0.4 milliradian. In addition to the temporal and spatial frequencies, the visibility of a feature is dependent on the luminance and modulation depth at the spatial frequency component being viewed. Based on this, the smallest feature that can be resolved is 0.9–1 mm at the typical television viewing distance of 3 m.

An NTSC signal can contain signal frequencies up to 440 television lines/picture width (TVL/PW). If this minimum feature is taken to be equal to one TVL, the NTSC signal can be just resolved on a display of 20–26 inch diagonal by a viewer 3 m from the screen. A 16:9 display capable of showing the 1280 × 720 resolution elements of an HD image will require a diagonal of 52–69 inch for these pixels to be resolved. A 1920 × 1080 HD display requires a screen with a diagonal of 70–90 inches for the pixels to be resolvable at a 3 meter viewing distance. All of these estimates are dependent on the visual acuity of the viewer.

The contrast sensitivity of the eye follows the Weber-Fechner law (Schreiber 1991). A simplified version of this law is given by Equation (13.1).

$$\frac{\Delta L}{L} = const. \qquad (13.1)$$

where ΔL is the change in luminance which can be detected at luminance L.

This equation indicates that the ratio of the minimum detectable gray scale step to the luminance is approximately constant over a relatively broad luminance range. It also implies that the eye is more sensitive to luminance variations in the darker regions of an image. This has significant implications for the perception of artifacts, including those resulting from too few gray scale levels employed in the signal processing. These issues related to visual artifacts will be discussed further in the next chapter.

13.2 Spatial Characteristics of the Image

Resolution is the primary spatial characteristic of a display and it determines how much information the display can show. Among the spatial artifacts a designer must be concerned with are Moiré, aliasing and the screen door effects discussed in the next chapter.

Two common methods are used to describe display resolution. Pixel count and its cousins such as display addressability describe the number of horizontal rows and vertical columns of individual picture elements, or pixels. Modulation transfer function and its cousins, such as television lines of resolution, relate to how much modulation the display is capable of as a function of spatial frequency.

13.2.1 Pixel Count

Pixel count is typically quoted as a horizontal x vertical pair of numbers. For instance, VGA is 640 × 480 pixels, for a total of 307,200 pixels. Individual pixels display only intensity information: They have (or should have) no spatial features. All spatial detail is shown by collections of pixels. In full color displays in which dedicated elements are used for R, G, and B, these elements are called sub-pixels and the resolution is often quoted with a third number, such as 640 × 480 × 3. This indicates that there are three sub-pixels (red, green and blue) for each full color pixel. The performance characteristics of a projection display using pixelated microdisplays is discussed below.

Display addressability indicates the number of separate addresses where distinct pixel information values can be directed. Display addressability is especially appropriate for describing video cards, where there is an actual memory location for each addressed pixel. The achievable display resolution may be less than the addressability due to design and/or component limitations.

For example, using a poor projection lens with a high-resolution pixelated microdisplay having one-to-one mapping with the memory locations on the video card will provide far less high frequency information than would be expected from the addressability. Under these conditions, the full resolution on the video card would not be displayable on the projector. It is common for manufacturers of CRT systems to specify addressability of their displays rather than the highest frequency information that can be seen.

13.2.2 Modulation Transfer Function

The resolution of a display is the measure of the viewer's ability to resolve detail in the image. The most common measure for resolution is the modulation transfer function (Kingslake 1983), or MTF. This function describes the relative modulation as a function of spatial frequency. From an empirical perspective, modulation at a given spatial frequency v_S is given by:

$$MOD(v_S) = \frac{L_B(v_S) - L_D(v_S)}{L_B(v_S) + L_D(v_S)} \quad (13.2)$$

L_B and L_D are the high and low system luminance, respectively, produced in response to an input sine-wave signal at spatial frequency v_S. High and low modulations are illustrated in Figure 13.2 for a system displaying a spatial frequency v_S. It is assumed in this figure that the same input signal is used for both systems and that this signal amplitude will produce 100 % modulation in a very high contrast, high resolution system. For the high contrast system, the percentage modulation is shown in Figure 13.2(a) to be 100 %. However, the lower contrast system has much reduced modulation, which results in reduced visual perception of contrast and image sharpness.

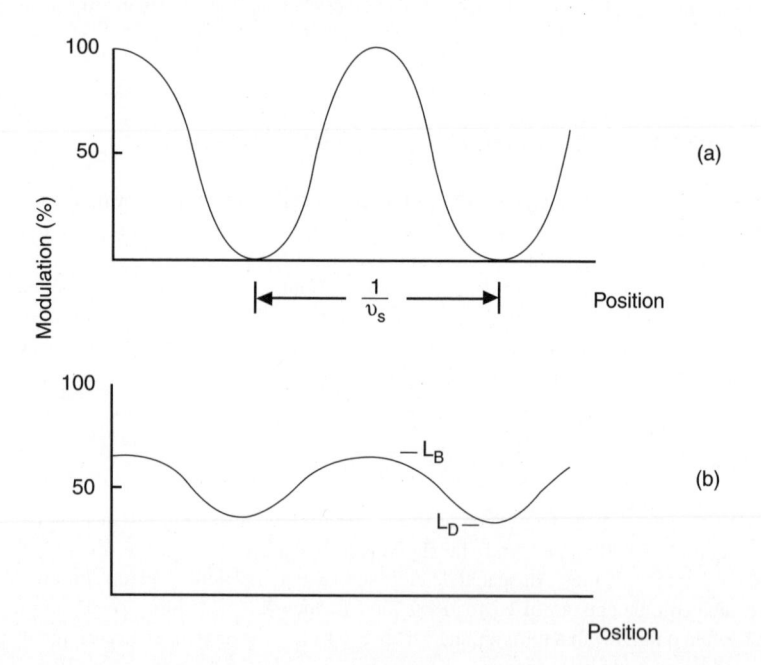

Figure 13.2 Output modulation of system with sinusoidal response to sine wave input at frequency v_S: (a) high-contrast system; (b) low-contrast system.

Strictly speaking, the MTF is only defined for linear systems where the spatial properties are independent of position. Linearity and spatial independence imply that a sinusoidal input will produce a sinusoidal output. When these conditions are met, this permits multiplication of the MTFs of successive subsystems to achieve the overall response of the system. As a practical matter, there are almost always some spatially dependent elements in projection optical systems. Some of the display components that are spatially dependent include digital video input, pixelated microdisplays, the CRT raster in the vertical direction, and a structured projection screen. It is important to identify the conditions that apply to the analysis. The mathematical theories for the frequency responses of linear optics and electronics are very similar. Consequently, the MTF concept is also applied to describing the frequency responses of analog circuits.

Mathematically, the MTF is the modulus of the ratio of the Fourier transforms of the normalized output and input signals and, as such, it is insensitive to the phase of the signal. For an optical system with a signal describing a point source, the output energy distribution in the image is called the point-spread function. The Fourier transform of the point-spread function is the spatial frequency response of the optical system and is called the optical transfer function. The phase-independent magnitude of the optical transfer function is the MTF.

If $f_I(x)$ is the input signal producing an output $f_O(x)$ and $F_I(\nu)$ and $F_O(\nu)$ are the respective Fourier transforms of these signals, then the MTF as a function of spatial frequency ν_S is given by

$$M(\nu_S) = \left| \frac{F_O(\nu_S)}{F_I(\nu_S)} \right| \qquad (13.3)$$

where the Fourier transform of f(x) is

$$F(\nu_S) = \int_{-\infty}^{+\infty} f(x) e^{-i 2\pi \nu_S x} dx \qquad (13.4)$$

The MTF is dimensionless and is applied to systems in which the outputs and inputs are measured in the same units, e.g. volts. For electro-optic devices where the input is typically measured in electrical units and the output is measured in optical units a pragmatic system MTF is used; viz., the ratio of MOD(ν_S) to MOD(0) is reported as the measure of the modulation. For lower contrast systems or systems in which the maximum response is not at zero frequency, use of this ratio could give an artificially high value for MTF at frequency ν_S.

13.2.2.1 MTF of raster-scanned displays

The MTF of a raster-scanned CRT is defined only in the direction of scan, typically the horizontal direction. In the vertical direction, the display is sampled. The effects of this are discussed further in the next section. For a two-dimensional spatial frequency characteristic, the MTF is the Fourier transform of the point spread function. For a one-dimensional spatial frequency, the MTF is the Fourier transform of the line spread function. The line-spread function of a CRT frequently has a Gaussian distribution. Using this distribution, the CRT MTF (Barten 1995) can be given by Equation (13.5). In this formulation, the MTF is given in terms of the electron beam size and image size.

$$MTF(\nu_S, \Delta x) = \exp \frac{\pi^2 \Delta x^2 \nu_S^2}{4 w^2 \ln(\frac{i}{i_0})} \qquad (13.5)$$

where
- i/i_0 is the ratio of the current density i at which the diameter is measured to the peak current density i_0,
- ν_S is the spatial frequency in line pairs/scan width,
- Δx is the diameter of the spot in mm at the current density i, and
- w is the scan width of the raster in mm.

Figure 13.3 shows the MTF of a CRT as an example. There are two things to note about this figure. First, it only applies to the horizontal resolution of the CRT, parallel to the scan lines. In the vertical direction, the system is a sampled system and normal MTF theory does not apply. Secondly, this figure and Equation (13.5) only take into account the MTF due to the electron beam spot size and assume the spot size is a Gaussian. Non-Gaussian spot shapes are often encountered in CRT and other raster-based displays, including scanned laser displays. Additionally, scattering of the electrons as they enter the phosphor layer leads to an increase in the apparent beam size.

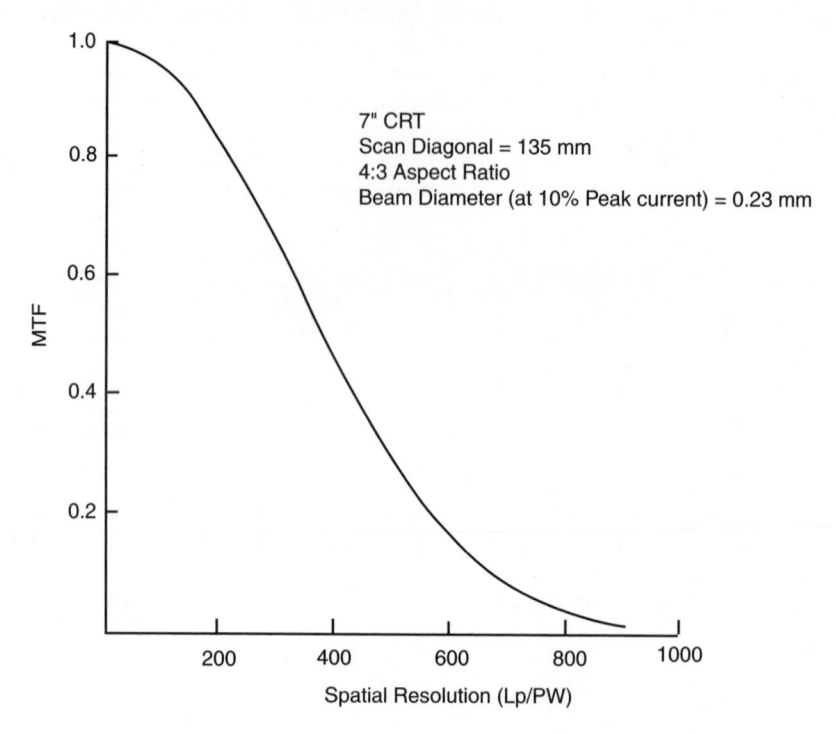

Figure 13.3 The MTF of the electron-beam in a CRT with a Gaussian electron spot size of 0.23 mA.

13.2.2.2 MTF of sampled systems

In a pixelated display, the signal level is essentially constant within one pixel, unlike the continuous change in the scan direction with a CRT. The pixel signal is generated from the input signal by a sampling process. Sampled systems are not spatially invariant and strictly speaking the MTF is not defined for these systems. A change in the phase (spatial shift) of the signal relative to the sampling can give markedly different results. Figure 13.4 illustrates the phase sensitivity of the output at the Nyquist frequency for a pixelated display. A $\pi/2$ (90°) phase shift of the signal with respect to the pixel structure results in 50 % gray signal on all pixels (each pixel receives white for half time and black for half time). This situation thus results in no modulation, while a 0 radian (0°) phase shift produces 100 % output.

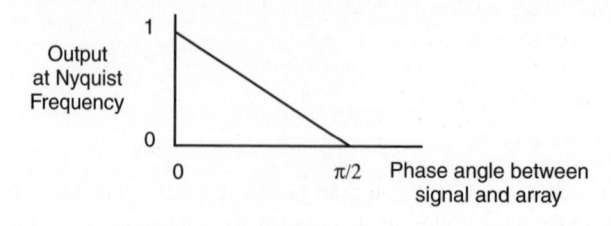

Figure 13.4 Output at the Nyquist frequency as a function of the phase angle between the signal and the active-matrix sampling system. A square-top sampling profile is assumed.

Various attempts have been made to define an MTF for the spatially non-independent sampled displays (Balram and Olson 1996a,b; Ohishi 1971). While this approach is not mathematically correct (Sluyterman 1993), it can provide useful insights into the resolution of sampled systems. The approach is to generate a function for sampled systems which provides a good approximation to a Gaussian MTF over the useful frequency range of the device. The sampling process is equivalent to a convolution of the samples with a block-shaped line-spread function. The Fourier transform of this function is the 'MTF' of this display, given by the absolute value of the sinc function (Equation (13.6)).

$$M(v_S) = \left| \frac{\sin(\pi k v_S)}{\pi k v_S} \right| \qquad (13.6)$$

In this equation, k is the width of the display elements, which is the reciprocal of the spatial sampling frequency.

The MTF (Equation (13.6)) is plotted in Figure 13.5 as a function spatial frequency in units of the sample frequency 1/k. Also plotted in Figure 13.5 is the Gaussian MTF (Equation (13.5)) with the standard deviation of the block function ($\sigma = k/(12)^{1/2}$) (Barten 1991). The agreement between the MTFs for a Gaussian and Equation (13.6) is quite good up to the aliasing limit. Sampled systems can only restore the input signal up to the Nyquist frequency, equal to one half the sampling frequency.

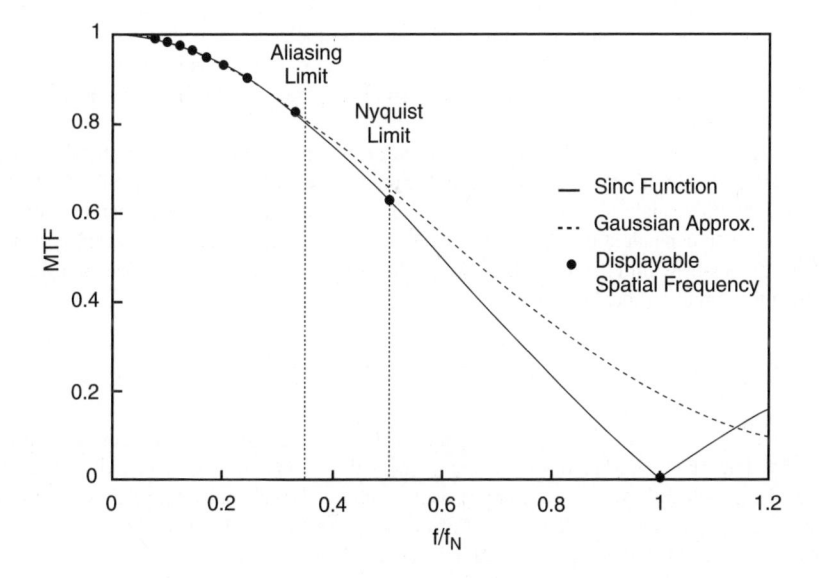

Figure 13.5 MTF of a display with a sampling structure and the Gaussian approximation. The dots indicate the frequencies that can be generated with a pixelated display. The horizontal axis is the frequency, relative to the Nyquist frequency. (Reproduced from Barten 1991; with permission of SID.)

These systems can only reproduce specific frequencies, given by 1/k, 1/2k, 1/3k, 1/4k, ... The display of the Nyquist limit (1/2k) requires two pixels. The next lower frequency that can be displayed is 1/3k, which uses three pixels. Signal frequencies between these displayable frequencies cannot be displayed on the sampled system. At very high spatial frequencies, this can significantly affect the displayed image by introducing phase-dependent artifacts into the image. At intermediate and low spatial frequencies, the density of the displayable frequencies is high enough that there is little difference between the 'MTF' or the perceived image quality of the continuous and the sampled image.

Performance of displays at medium and low spatial frequencies are so important that special tests have been prescribed to measure them, most notably ANSI and sequential contrast. Sequential contrast (actually, 1−(1/contrast)) can be thought of as the MTF of a display system at a spatial frequency of 0 lines per picture height and 1−(1/(ANSI contrast)) can be thought of as the average horizontal and vertical MTF of a display at 2 cycles per picture height and width.

Interference between the signal frequencies and the sampling structure can give rise to aliasing effects. These artifacts are discussed in the next chapter. To avoid these, the signal frequency should be limited to about 0.7 times the Nyquist frequency. The factor 0.7 is sometimes called the Kell factor (Hsu 1986). The Kell factor in general is used to relate the resolution achievable by an unsampled system (i.e. parallel to the CRT raster lines) to the resolution achievable by a sampled system (i.e. perpendicular to the CRT raster lines). While the Kell factor was a useful approximation in CRTs, it had little basis in the theory of resolution or the human visual system (HVS) and is little used now that virtually all display systems are sampled in both the horizontal and vertical directions.

Nevertheless, the sinc function representation of the MTF of pixelated displays provides a good approximation for usable spatial frequencies. The argument of the sinc function is determined by the number of pixels in each direction. An appropriate description of the resolving ability of such displays, therefore, is the pixel count. Thus, a VGA display has 640 × 480 pixels and can be used to display a continuously varying image such as a TV image containing spatial frequencies below the Nyquist frequency.

In the horizontal direction, the Nyquist frequency for the VGA device is 320 cycles/picture width (c/pw, with two TV lines equal to one cycle).[1] Accounting for the Kell factor, the non-aliasing frequency range is up to about 225 c/pw. An NTSC signal contains video information at frequencies up to about 220 c/pw and thus a pixelated display with VGA resolution is reasonably suited for displaying NTSC signals.

From Figure 13.5, the lowest MTF of the pixelated display for frequencies up to the alias frequency is calculated to be 0.81. This MTF is typically higher than that obtained with a projection CRT and, with all other things being equal, the image on the pixelated VGA display will appear sharper. Note that the resolution limit is independent of the dimensions of the image. This is valid to the approximation that the physical processes within the microdisplay or light valve, such as electrical leakage between pixels, do not affect the signal sampled by the pixel.

13.2.2.3 MTFs of other elements

Until now, the discussion has been limited to the MTF of the image-generating component(s) of the projection system. The total MTF of the system includes the effects of the signal bandwidth, lenses and other optical components, convergence errors, veiling glare, screens, and degradation due to ambient illumination. The system MTF is the product of the MTFs of the subsystems, within the limits of the assumptions made in MTF theory, i.e. spatial independence and linearity.

The signal bandwidth is the electrical bandwidth of the video amplifiers in an analog, unsampled display system. It is often straightforward to design amplifiers with the requisite bandwidth to pass the electrical signals giving rise to the image with minimum degradation. Excessive bandwidth should be avoided to minimize system noise and cost. Frequently, the bandwidth is set to correspond to the aliasing limit.

Display systems with digital electronics are nearly universal today, even when the video signal is analog. Usually, the first step in the display is to sample and digitize the video signal. For example, when a digital video signal is generated by a computer it is sometimes converted to an analog signal to transmit

---

[1] This is a common measure of video resolution and it equals the resolution in cycles/unit length multiplied by the width of the picture in the same length units.

to the monitor via a VGA cable. The monitor then re-digitizes the image for processing and display. Modern interfaces such as DVI avoid this digital-to-analog-to-digital process altogether by transmitting the signal in digital form. While this does not necessarily simplify the evaluation of the resolution of the display, the added analog conversion processes at least are not present to confuse the issue or degrade the image.

The intensity profile of a single spot in the projected image consists of the desired image produced by the lens plus undesired degrading effects due to scattered light and ambient illumination. This is illustrated in a simplified manner in Figure 13.6. The profile can be considered the point-spread function P(x,y) of the optical system. It can be separated into three components: the spot profile S(x, y), the scattered light profile H(x, y), and a very low frequency component A, often due to ambient light. The MTF corresponding to this point spread function is

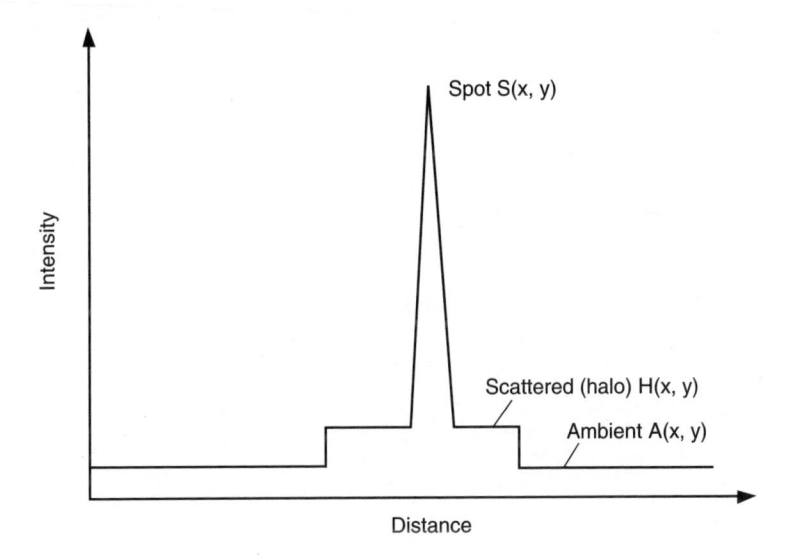

Figure 13.6 Intensity profile of 'spot' on the TV screen including contributions from the beam, scattered light, and the very low frequency background due to reflected ambient light (McKechnie *et al.* 1985).

$$M(\mu, \nu) = \frac{1}{T} \left| \int\int_{-\infty}^{+\infty} [S(x, y) + H(x, y) + A(x, y)] e^{-i2\pi(\mu x + \nu y)} dx dy \right| \quad (13.7)$$

where T is a normalizing constant representing the total energy contained in the point-spread function P(x,y). If E_S, E_H and E_A are the energies contained the spot, the scattered light, and the ambient reflections, respectively, we have

$$T = \int\int_{-\infty}^{+\infty} P(x, y) dx dy = E_s + E_H + E_A \quad (13.8)$$

The contributions of the three terms to the optical system MTF are illustrated in Figure 13.7. The effects of the ambient illumination are discussed below and are dependent on the screen. The effect of the ambient is to remove energy from the higher frequencies and add it to the DC term of the point-spread function of the optical system. Scattering is represented simplistically as a low-frequency term.

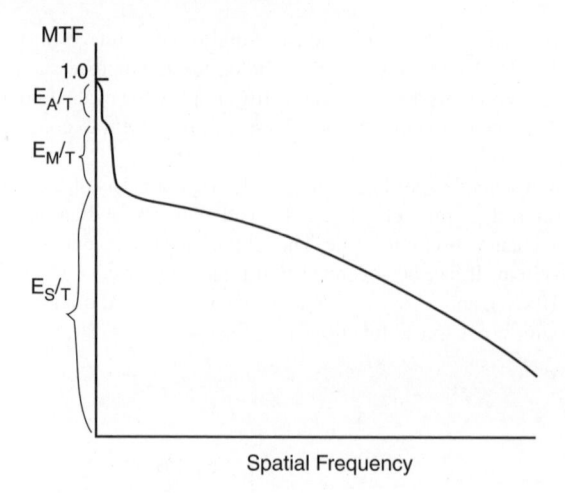

Figure 13.7 MTF contributions from point spread, scattering and ambient to the system MTF (McKechnie *et al.* 1985).

Optical technologies permit the design and fabrication of projection lenses with high MTF for the resolution ranges of projection systems,. The design tradeoffs are based principally on economics and the desired system performance. Among the parameters that the lens designer has include materials (glass vs plastic, with high index materials normally more expensive), number of lens elements (increasing the number of elements can be effective in reducing aberrations, distortions and improving off-axis resolution performance), and lens coatings (multiple layer coatings on each lens surface can minimize unwanted reflections, resulting in reduced veiling glare).

13.2.2.4 Convergence

When a projection system depends on three separate microdisplays or CRTs to produce red, green and blue, it is necessary to overlap the red, green and blue images on a pixel-by-pixel basis. This process is called convergence. The color sub-images are additive and the human visual system fuses them into a full color image. This convergence can take place at the projection screen, as in CRT and other three lens projectors, or it can take place internal to the projector, prior to the single projection lens, as is common in current three-panel LCD or DMD projectors.

Projector architectures which are based on a single non-subpixelated microdisplay do not normally suffer from convergence problems. However, chromatic aberrations in the projection lens and other imaging components can look to the casual viewer like a convergence problem. At the edge of the image, chromatic aberrations can cause red, green and blue pixels to not overlap at the screen. One sign that it is a chromatic aberration problem rather than a convergence problem is to look at another corner. If it a chromatic aberration problem in the projection lens, the pixels will not overlap there either, but the errors will be in the opposite direction. At the center of the screen in a single-panel projector of this type, there should be no apparent misconvergence.

For a subpixelated microdisplay, there is one-third of a pixel misconvergence of the three colors inherent in the microdisplay design. The visibility of this misconvergence is dependent of the viewing distance. The lower resolution of the eye in the red region and the much lower resolution in the blue region contribute to making this misconvergence normally not visible.

For a pixelated microdisplay projector with multiple microdisplays, precise six-axis adjustment mechanisms are required for at least two of the three light valves. In the case of the CRT projectors, highly accurate and stable raster circuits and deflection components are needed. Also, the projection lenses must be well matched both with regard to distortion and chromatic qualities. Obviously, no matter what architecture is used, the system should be mechanically and thermally stable. In either case, a reasonable convergence target is about half a pixel at the center of the image. Typically, convergence to a few pixels accuracy is accepted at the corners of most display systems.

To illustrate the difficulty of internal convergence in a pixelated three-microdisplay projector, consider the following example: if the panels have a pixel pitch of 30 μm, the relative position of the three LCD cells would need to be maintained to within 15 μm in both the horizontal and vertical dimensions to achieve the half pixel registration. Assume these cells are mounted on a steel baseplate with an expansion coefficient of 11.7×10^{-6} cm/(cm C). If the center-to-center distance of the cells is 5 cm, a temperature change of 25.5 C would lead to a relative motion between cell centers of 15 μm!

Stress in the mechanical mounting system can result in thermally induced bending which will add to the relative shifts of the sub-images. To these misregistrations must be added the effects of chromatic errors in the optical system. Solutions based on temperature control, low expansion materials, and novel mechanical structures for mounting the panels have been employed to minimize this problem. Architectures in which the panels are mounted as close as possible to each other, i.e. using a recombination cube in non-equal path systems, also minimize this problem.

One-half pixel misconvergence in a device with a pixel count of 1024 × 768 represents a linear displacement of less than approximately 0.05 %. The lenses in a three-lens system can have focal length variations of as much as ±5 %, resulting in different projected image sizes for the three colors. Distortion over the image area in the three lenses must also match each other within this same 0.05 %. This is one of the reasons that pixelated displays typically have a single-lens architecture. Raster control in CRT, light-amplifier, and Eidophor/Talaria-type projectors can be used to compensate for these problems as well as for the keystoning that results from having two of the three light generating subsystems off-axis.

Misconvergence of the three colors leads to a degradation of the MTF. Without convergence errors, or if only one color is present, the MTF [M(μ,ν)] for a CRT is given by Equation (13.5) and by Equation (13.6) for a pixelated display. The total MTF with misconvergence is M_T, given by:

$$M_T(\mu,\nu) = M(\mu,\nu) \cdot M_C(\mu,\nu) \qquad (13.9)$$

$M_C(\mu,\nu)$, the MTF degradation factor due to misconvergence, is given by Equation (13.10) (McKechnie et al. 1985).

$$MTF_C(\mu,\nu) = \frac{1}{R+G+B} \left\{ \begin{array}{l} G^2 + R^2 + B^2 + 2R \cdot B \cdot \cos 2\pi [\mu(x_R - x_B) + \nu(y_R - y_B)] \\ +2 \cdot G \cdot R \cdot \cos 2\pi(\mu x_R + \nu y_R) + 2 \cdot G \cdot B \cdot \cos 2\pi(\mu x_B + \nu y_B) \end{array} \right\}^{\frac{1}{2}}$$

(13.10)

In these equations,

μ,ν   are the frequency components of the Fourier transform, in units of 1/x or 1/y (1/mm),

R,G,B   are the relative luminosities of the red, green and blue bands, respectively. For white light, G:R:B ≈ 6:3:1; for green light, G:R:B = 1:0:0, and

x_R, y_R, x_b, y_b are the horizontal and vertical misconvergence values in mm for red and blue from the green band, measured from the center of the green position to the center of the red or blue positions..

If only one color is present, MTF_C in Equation (13.10) is equal to unity. With only one color, misconvergence is impossible and no correction is needed.

Automatic systems have been designed to converge CRT projection systems (Kanazawa and Mitsuhashi 1989). These systems can be used in the factory for initial setup, or be designed into the projector to correct errors that arise after the system leaves the factory. One such system was described in Section 9.1.3.

322 CHARACTERISTICS AND CHARACTERIZATION

13.2.2.5 MTF of projection screens

Several of the screen types used with projection systems based on a variety of optical designs were described in Section 7.2. The screens can be either transmissive or reflective. Both types can have internal diffusing elements and/or micro-optical structures to control the angles of the exiting light. Additionally, the screens can incorporate elements, on the surface or internally, which decrease the reflection of ambient light.

The double lenticular rear projection screen (Bradley *et al.* 1985) used for consumer projection TVs has embossed cylindrical lenses on the front and back surfaces of the screen to control the horizontal viewing angles. Optical scattering off multiple small embedded particles, called diffusion, is used to control the vertical angles. Other front and rear projection screen types with structures are described in Section 7.2. Regardless of the details of the structure, if the structure is oriented in the vertical direction, an effect of the structure is to sample the image in the horizontal direction. Thus, the MTF of the screen in the horizontal direction has a $\sin(x)/x$ character. Of course, if the screen has a 2D structure rather than just vertical structures, the vertical resolution would also have a $\sin(y)/y$ form as well. In order to avoid aliasing effects and resolution reduction caused by the screen structure, the pitch of the sampling structure must be sufficiently fine so that the Nyquist frequency of the screen is well above the Nyquist frequency for the pixelation in the image. Typically this means that the pixel at the projection screen must be larger than 3 or 4 of the sampling structures.

Diffusion is added to the screen to increase the vertical viewing angle. It also increases the horizontal angles but these effects are relatively small compared with the optical effects of the lenticular array. Diffusion consists of imbedding particles throughout the volume of the screen or embossing the surface of the screen. The particles are index of refraction mismatched to that of the host material of the screen. This or the embossing leads to scattering.

For screens thicker than about the minimum feature size in the image, i.e. one pixel, adding diffusion to a screen will broaden the point spread function. This factor is discussed in Section 7.3.1. and shown in Figure 7.34. As a result of the lateral scattering of the light the amplitude of the MTF is reduced. This occurs when the diffusion is either surface diffusion on both surfaces or volume diffusion through the body of the screens. For these screens, the amount of diffusion added must be limited to avoid excess degradation of the image. On the other hand, these screens are also the best for reducing speckle in the image, so these two factors must be balanced against each other. For thin screens, or thick screens with diffusion embossed on only one surface, far less lateral scattering takes place and the screen does not degrade the MTF of the image.

13.2.2.6 MTF of electronic images vs film

Cinema film has long been held up as the 'gold standard' of image quality, but modern electronic projectors are already equalling or exceeding film in image quality, especially in terms of resolution and MTF. For cost and image quality reasons, cinema film is currently being supplanted by electronic projection in theaters. Contrast of electronic projection is comparable to the contrast of film, although the best (and most expensive) electronic projectors can far exceed the contrast achieved by film. Color is an example where the typical film has a wider color gamut than the typical electronic projector, even one intended for cinema projection. However, at the high end of the colorimetry scale with lasers, multiple primary colors or other color gamut expanding techniques, electronic projectors can have larger color gamuts than even the best of film systems.

With the traditional film process the film went through four generations of optical reproduction before the viewer saw it in the theater (Baroncini *et al.* 2004). These generations were called the camera negative, the inter-positive, the inter-negative and the release print. Measured MTF of these four generations are shown in Figure 13.8.

SPATIAL CHARACTERISTICS OF THE IMAGE   323

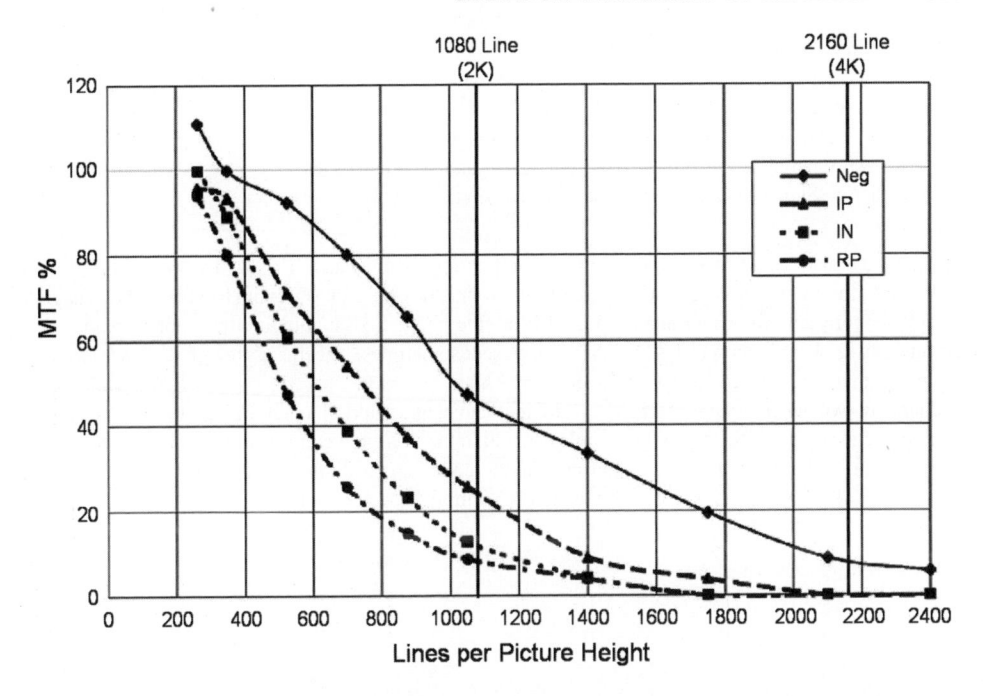

Figure 13.8 Horizontal resolution of four generations of cinema film, camera negative, interpositive, internegative and release print (reproduced by permission of © 2004 SMPTE).

This figure has also been marked by the pixel count of 2 K and 4 K digital cinema projectors. Note that at the limiting resolution of a 2 K digital cinema projector, the consumer would only see less than 10 % MTF from a release print in a local multiplex, and may see less. With the next generation of 4 K digital cinema projectors, film will no longer be even an acceptable acquisition medium, since the camera negative has less than 10 % MTF at the 4 K resolution limit.

13.2.3 Image Quality Metrics

There have been many attempts through the years to produce a single quantitative metric that correlates with human perception of 'image quality'. These metrics have become increasingly sophisticated and have increasing correlation with perceived image quality. Unfortunately, they have become increasingly difficult to measure and compute. In practical terms, the current generation of image quality metrics cannot be used by a display engineer without help.

This section alone will not go into the details of how these metrics are computed and even the citations given would probably not be enough to allow calculation of many of these metrics. The discussion here is meant to guide the projection engineer in the correct direction when evaluating one of these metrics for usefulness in designing and evaluating a projection system. At this point, the projection engineer should consult with a human factors specialist.

Some image quality metrics a display engineer might encounter include:

MTF-A (Modulation Transfer Function-Area) (Snyder 1988)
ICS (Integrated Contrast Sensitivity) (van Meeteren 1973)
SQF (Subjective Quality Factor) (Granger and Cupery 1972)
SQRI (Square Root Integral) (Barten 1990)

GC (Grain Contrast) (Goldenberg *et al.* 1997) (see Section 7.3)
VFGC (Visually Filtered Grain Contrast) (Watson 2006)
MSE (mean square error) (Pratt 1991)
VMSE (Visually weighted mean square error) (Pratt 1991)
S-CIELAB (spatial extension to CIELAB) (Zhang and Wandell 1996)
SSO (Spatial Standard Observer) (Fiske *et al.* 2007)

The first four of these metrics are evaluated and compared in (Farrell 1999). The correlation found between the metric and subjective image quality was found to be best for SQF, with ICS and MTF-A having acceptable correlation for continued use. SQRI had the poorest correlation and was not recommended for future use. Many other metrics have been proposed; sometimes, it seems as though there are as many metrics as there are human factors engineers.

There are two general classes of metrics: single-channel and multi-channel (Farrell 1999). The single-channel metric assumes there is only a single channel in the eye/brain combination that processes spatial information. The most common channel used is the contrast sensitivity function (CSF) (Mannos and Sakrison 1974), as shown in Figure 13.9, which is a simplified version of Figure 13.1.

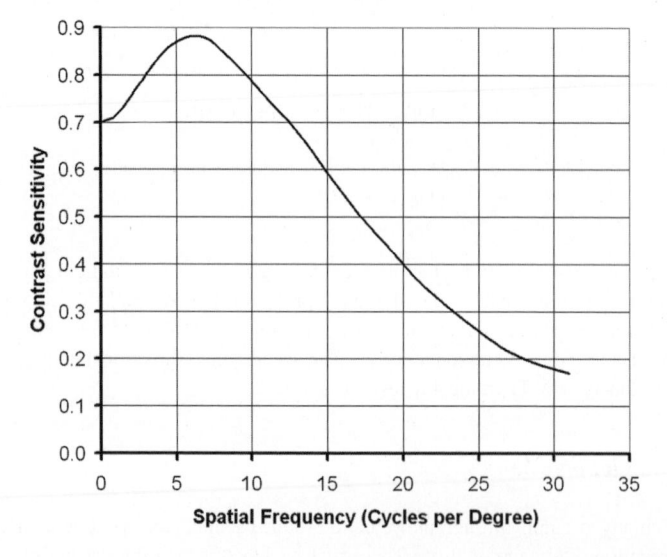

Figure 13.9 Contrast sensitivity function of the human visual system as used in some single channel image quality metrics.

No single channel filter can successfully predict the visibility of all types of image content. For example, the CSF is tuned to how the human visual system (HVS) detects sine wave objects. Therefore, the properties of any object that can be described satisfactory by its Fourier transform (i.e. described as a collection of sine waves) can be evaluated by a single-channel model using the CSF. Other systems, such as film grain, that are difficult to describe as a collection of sine wave objects, are more difficult to evaluate using a model based on the CSF. This, in part, explains the proliferation of image quality metrics. Each metric is tuned to accurately evaluate a single type of image defect, and the numerical value of the metric correlates well with the perceived intensity of that defect. For example, MTF-A correlates well with the perceived 'sharpness' of an image and VMSE correlates well with the visual impact of image defects introduced by digital signal processing. GC, also known as speckle contrast, and its much improved version, VFGC, are specifically designed to evaluate speckle at projection screens.

The HVS has multiple filters in it, not just the CSF. Multichannel metrics combine multiple filters in a systematic way in an attempt to emulate the processing of the HVS. A well-constructed multichannel metric will correlate with perceived image quality for a much wider variety of images and image defects than any possible single channel model.

The spatial standard observer (SSO) is a multichannel metric and is the only one that will be described in more detail here. The SSO model was developed in Watson and Ahumada (2005) by testing the correlation between a wide variety of model metrics and perception studies of a collection of 43 standard images called the ModelFest stimuli. The models tested used a variety of analytical models for the CSF function, aperturing, pooling and channels. The final model involved all of these factors and became the SSO. The SSO has been used for a variety of applications, including evaluation of noise such as speckle introduced by projection screens (Fiske *et al.* 2007), detection of Mura and evaluation of motion blur in LCDs (Watson 2006).

Since the SSO is based on the ModelFest stimuli and the HVS data produced by those stimuli, it is possible the SSO will be supplanted by a still newer, better and more complex image quality metric. The creators of the 43 images in the ModelFest stimuli image set admit it does not cover all possible image types. In particular, it does a poor job of evaluating diagonal resolution in the HVS. It also only has a single image related to random noise, plus a single image that represents a real image.

To properly evaluate the SSO model, or for that matter any of the other image quality metrics, it is necessary to input a large amount of image data into the computer model. In the past, scanning photometers have been used to acquire this data. Now, it is possible to acquire this data much more conveniently with a calibrated video camera (Jiang *et al.* 2006). With proper calibration, a camera of this sort can acquire spatial color data as well as photometric data, for use in color-based models such as S-CIELAB or to evaluate the red, green and blue color fields separately (Dallas *et al.* 2007).

13.3 Luminance, Contrast and Color

The projector qualities of luminance, contrast and color are among the more important descriptors of the system performance. Luminance and its units are among the subjects belonging to the subject of photometry. A brief review of photometry is given in Appendix 1. The luminance of a projector is a measure of its light output. Contrast is a specification of the dynamic range of the projected image, i.e. the difference between the highest and lowest illumination levels the projector can deliver. Color is important in describing how well the system reproduces the hues and saturation of the original image. Colorimetry summarized in Appendix 2.

Colorimetry is the science and art of specifying, measuring and evaluating colors in terms of how they are perceived by the human visual system (HVS). Photometry is closely related to colorimetry except that it only deals with the perception of brightness by the HVS.

The scientific and engineering equivalents of colorimetry and photometry are spectroscopy and radiometry, respectively. These sciences deal only with the properties of the light itself, not the properties of the HVS. The typical units used in spectroscopy and radiometry involve energy and power units while colorimetry and photometry involve lumen-based units. The lumen was designed to correlate with 'brightness' as perceived by a human. The correlation is not perfect, however, and sometimes a brighter image as perceived by a person will contain fewer lumens than a second image perceived as dimmer.

13.3.1 Luminance and Brightness

Brightness is the perceived attribute according to which more or less light appears to be coming from a given area. Luminance is a physical quantity measurable with appropriate instrumentation. The luminance scale, designed to match the human eye response, is based on the CIE 1931 eye response curves (CIE 1932). The correlation between perceived brightness and measured luminance is not perfect for several reasons. First, the CIE 1931 eye-response curves have known errors in them. Attempts to correct these errors (CIE 1964) have not gained widespread acceptance in the display community and the CIE 1931 curves dominate all display colorimetry and photometry. The second reason brightness and luminance do not always correlate with each other is that the human visual system is very complex and cannot be described by three simple spectral curves such as the CIE 1931 curves. The final reason why there is not a 1:1 correlation between colorimetry and photometry on one hand and perceived color and brightness on the other hand is variation from person to person. The CIE 1931 and CIE 1964 curves are average or typical curves and do not necessarily apply to all individuals.

The relation between brightness and luminance can be non-linear. Normally a display perceived as twice as bright contains more than two times the luminance. Regardless of the problems with photometry and colorimetry, it is appropriate to specify light output of projectors using lumens, luminance and other photometric units because they are readily measured properties with a long history of use in the display industry (Walsh 1965; Boyd 1983).

13.3.1.1 Measurement of luminance

The total optical flux from the projector is measured using an integrating sphere, as illustrated in Figure 13.10. The sphere homogenizes the input light to provide a uniform illumination of its internal measurement detector. Hence, no information is obtained from this measurement on the distribution of light within the image produced by the engine. However, projected images most often have non-uniform distributions, with the center usually having higher luminance that the corners of the display. While a center-to-corner luminance ratio greater than 3:1 had been accepted in the past from CRT projectors for consumer projection, it is now more common for the corner to be at least 80 % of the center luminance.

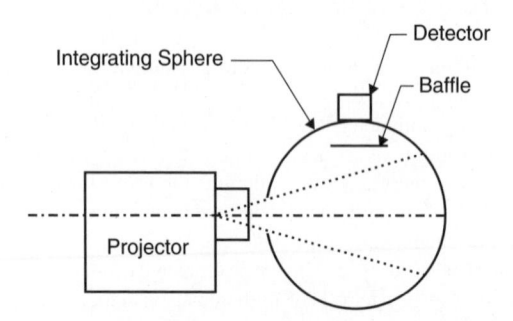

Figure 13.10 Integrating sphere method of measuring total lumen output.

Unfortunately, the integrating sphere method has a number of practical difficulties. One very important consideration is that most people who desire to measure the output of a video projector are not equipped with an integrating sphere with a large enough entrance port to capture all the light from the projection lens. On rear projection systems, larger front projection systems and fixed installation projectors, it is often difficult or impossible to position the sphere in front of the projection lens.

Another problem with this method is difficulty in using it with three-lens projection systems. When used with three-lens systems, it is necessary to measure the output of each lens separately, and add the values together. In systems such as the Talaria oil-film projector, where there is a bright border or halo around the image that does not carry any of the image information, this method cannot be used at all. If it were used, the halo would be included with the modulated light and give a high but incorrect lumen output reading.

Luminance measured with an integrating sphere or some other method gives the total light output of a projector and ignores the screen properties. The image brightness at the screen produced by the projector is properly measured in a luminance unit, which depends on the screen properties, including total reflectivity, size and gain. There is confusion, and sometimes errors, in the common descriptions and measurements of the luminance of the images produced by projection displays. One of the display industry accepted methods for measurement of projection systems is to use luminance meters which were calibrated with a reflecting Lambertian surface (see Appendix 1). The meters provide readings in candela/m^2 (cd/m^2, formerly called 'nits', a term that is no longer used), the SI unit of luminance. This unit is defined for both Lambertian and non-Lambertian surfaces. The foot-Lambert, a related but not identical unit, is formally defined only for Lambertian surfaces[2] but is often measured on non-Lambertian surfaces as well. The foot-Lambert unit is now obsolete, and it is recommended all measurements of this type be done in cd/m^2. The luminance of the screens used in projection systems often do not have Lambertian distributions, as discussed in Section 7.2.

The relationship between lumens and luminance in cd/m^2 is:

$$\Phi = \pi H W L \tag{13.11}$$

where Φ is the flux in lumens, H and W are the height and width of the projection screen and L is the luminance measured reflected from a gain $= 1.0$ surface.

13.3.1.2 ANSI lumens

There are some in the display industry who have taken advantage of the higher center luminance of the image to create 'better' performance specifications. It was not uncommon to have only the center luminance quoted without specifying the uniformity of the image. In an attempt to provide consistency in the specifications of image luminance, an ANSI standard (ANSI 1992) for the measurement of projector output was approved. The current version of this specification is actually an IEC specification (IEC 2002) but the measurement is still commonly called 'ANSI lumens'. It consists of measurements at a nine-points, as illustrated in Figure 13.11. The ANSI luminance is the average of these measurements. The flux Φ in Equation (2.1) is then measured in units called ANSI lumens. Note that in the phrase 'ANSI lumens', the ANSI part refers to the test method used. The lumens part refers to the same lumen as defined by the CIE.

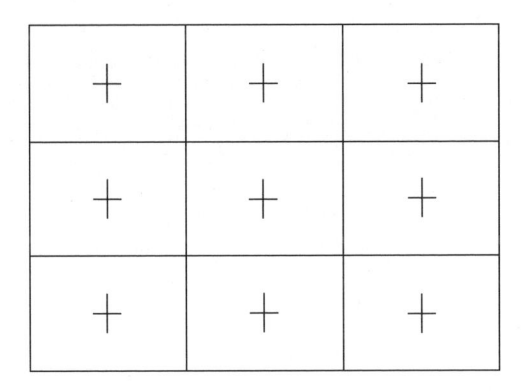

Figure 13.11 Points for measurement of luminance, according to ANSI IT7.215-1992.

---

[2] Foot-Lamberts are commonly converted to cd/m^2 with the scale factor 3.426 cd/m2 = 1 foot-Lambert.

The ANSI test method for video projectors contains not only the test method for luminance and the set up conditions, but also the test methods for contrast and resolution. A copy of this test method is a required part of any video projection system engineer's library. One of the key requirements for testing projectors according to the ANSI method is that the same setup conditions be used for measuring lumen output and resolution. This is not normally an issue with pixelated projectors such as LCD or DMD systems. With CRT projectors, on the other hand, this can be very important.

Prior to the establishment of the ANSI test method, it was common for CRT projector manufacturers to measure lumen output with very high beam currents, in order to get the highest possible lumen output. Because of electron spot blooming, however, the resolution at these very high beam currents would be poor. They would then lower the beam current to the value of current for best spot focus and measure resolution with this lower current. The difference in luminance levels between the maximum luminance and the luminance for maximum resolution could be as much as a factor of 10 or more.

This could be very misleading to an end-user who bought a projector based on the specified luminance and the specified resolution. The ANSI test method does not prohibit the evaluation of a projector under both high and low drive conditions. It does require, however, that the resolution as well as luminance be reported for the high drive condition, and the luminance as well as resolution be reported for the low drive condition.

In order to evaluate the ANSI lumens, a solid white video image is used. The screen luminance is then measured in the center of each rectangle shown in Figure 13.11. A convenient way to do this for front projection systems is with a plaque of white, gain 1.0 reflecting material and a luminance meter. The plaque is placed at the center of each rectangle and the reflected light is measured. Measuring directly off the projection screen is rarely satisfactory, since most screens have a gain >1.0. This makes accurate measurements difficult since the exact gain of the screen is rarely known. Even if it were known, the gain depends on the angles between the normal to the screen, the projector and the luminance meter.

After the centers of the nine rectangles have been measured, the ANSI lumen value is calculated by the equation:

$$\Phi_{ANSI} = \pi HW \frac{\sum_{1}^{9} L_i}{9} \tag{13.12}$$

where:

- H is the height of the screen,
- W is the width of the screen and
- L_i is the measured luminance at the center of each of the 9 rectangles.

The equation is just the area of the screen times the average luminance in the 9 rectangles and is an extension of Equation (13.11). One of the advantages of the ANSI method is it typically correlates fairly well with lumens measured in an integrating sphere. It also provides some quantitative data on the light distribution in the image, although it does not measure close enough to the edge of the screen for many purposes.

If each rectangle is measured twice, once with the gain 1.0 plaque and a second time directly off the projection screen, gain can be calculated with the equation:

$$G_i(\theta_P, \phi_P, \theta_M, \phi_M) = \frac{L_{i,screen}(\theta_P, \phi_P, \theta_M, \phi_M)}{L_{i,plaque}} \tag{13.13}$$

where:

- θ_P, ϕ_P are the angles between the projector center line and the normal to the screen,
- θ_M, ϕ_M are the angles between the meter center line and the normal to the screen,
- $L_{i,screen}$ is the luminance measured off the screen and
- $L_{i,plaque}$ is the luminance measured on the plaque.

13.3.1.3 Center weighted lumens

Occasionally one encounters a lumen specification called 'Center Weighted Lumens' or 'Center Weighted Peak Lumens'. Measurements based on these concepts can seriously overestimate the lumen output of a CRT projector. The luminance measured by these methods have limited value to a projector designer or a projector buyer. The methods are occasionally used by projector marketing departments to provide optimistic values of the lumen output of a projector. The ANSI test method has done much to eliminate the use of center-weighted lumen measurements.

Center-weighted lumens involves measuring the luminance only at the center of the projection screen, rather than at the 9 point grid specified by ANSI. This single value is then multiplied by the area of the screen to give a lumen output value. If the white field was perfectly uniform with no luminance fall off from center to edge, this would in fact give an accurate lumen output value. However, there is almost always a fall-off from center to edge. For example, if the corner luminance is 30 % of the center luminance and the luminance varies as $1-R^2$, where R is the distance to the center of the screen, a center weighted lumen value would be approximately 141 % of the sphere lumens.

The ANSI lumen measurement itself is slightly center weighted. With the same center to edge gradient and 30 % corner luminance value, the ANSI measurement would give approximately 104 % of the value measured with an integrating sphere.

The difference between center weighted lumens and ANSI lumens is greatest for CRT projectors. As CRT projectors became less common, the use of a center-weighted measurement has fallen out of favor as a marketing tool.

13.3.1.4 Luminance patch tests

The luminance patch test is not a test method prescribed by any standards organization such as ANSI, SMPTE or the IEC. However, it correlates well with human perception when displaying entertainment-type material on a CRT projector. It is used throughout the consumer electronics industry, and it elucidates an important feature of a CRT projector. This test is not normally performed on a microdisplay or light-valve projection system.

The projection CRT has the potential to produce a small area luminance higher by a factor of eight, or more, compared with the luminance of a full white field. This ability is limited by the maximum allowed thermal dissipation in the tube faceplate, the average current that can be generated by the high-voltage power supply and the loss of resolution at higher beam currents. CRT projector manufacturers design their products to allow the luminance to be driven to very high levels over an area which is a small fraction of the total. This 'punch' or 'sparkle' effect is measured with a patch test, as illustrated in Figure 13.12. Since there is no standardization in the patch test method, data from two different CRT manufacturers usually can not be compared directly against each other.

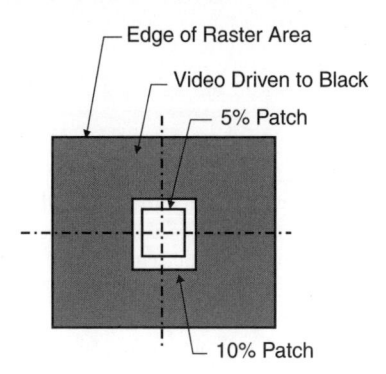

Figure 13.12 Patch test pattern.

For the patch test, the measurement units are *Center Weighted Peak Lumens*. These are similar to center weighted lumens in that the luminance at the center of the screen is multiplied by the area of the screen to produce a lumen value. In this case, however, instead of displaying 'white' on the entire screen, only a patch is illuminated, as shown in Figure 13.12. Since the CRT projector cannot produce these peak lumens on the entire screen, multiplying by the area of the screen is inappropriate. The correct thing to do would be to multiply the patch luminance times the area of the patch, to estimate the lumens in the patch.

Figure 9.8 shows this for a consumer CRT rear projector system in which the luminance is increased about four times at small excited areas, with the output normalized to the luminance of the full white screen. For large patches, driving the maximum current into the tube would exceed the tube average power limit, so the electronics automatically reduces the video drive so as to not exceed the limit. This function is called current limiting, and is implemented in various ways in all CRT projectors. For patch sizes up to about 20 % of the total area, the power limit is not reached and the current limiter allows higher drive levels.

For CRT data display systems, this effect makes little difference. Frequently, data applications use reverse video, with large white areas driven by the video signal, and black characters representing undriven areas. In this high duty cycle application, there is little difference between the peak and the average brightness. Entertainment applications (television, movies, etc.) are very different. Here the average luminance is about 20 % of the peak luminance. In a microdisplay projector, peak light is generated at all times and this extra 80 %, on average, of the light is absorbed and wasted. In a CRT projector, this extra energy is used to drive additional current into the highlight areas. This gives the sparkle to highlights in a video image that a microdisplay or light-valve projector does not have.

There is no difference between peak and average lumens in a light-valve projector. Hence, center weighted peak lumens is never cited for these systems. Because of the extra local luminance provided by the ability of a CRT projector to produce small areas of high output, it is necessary for microdisplay projectors to produce higher total lumens than a CRT projector to produce the same perceived image brightness when showing video material. This is why a typical microdisplay-based RPTV needs to produce 300–450 ANSI lumens to equal the image produced by a CRT projector with only 150 ANSI lumens.

13.3.2 Contrast

The contrast ratio of a display is the attribute which describes luminance gamut from the maximum L_V(white) that can be displayed to the minimum L_V(black). There are several definitions of contrast in use. The simplest of these is called the full-on/full-off or sequential contrast ratio (Infante 1993), and is defined as

$$C = \frac{L_V(white)}{L_V(black)} \quad (13.14)$$

As the name implies, the contrast ratio is determined by measuring the luminance in the center of the screen when the entire display is driven with a signal to produce the 100 % white level and dividing this by the value measured when the entire display is driven by the signal to produce the black level (0 % white). For example, if the peak luminance is 600 cd/m^2 and 0.40 cd/m2 remain when the display is black, the contrast ratio is 600/0.4 = 1500:1. When only one contrast ratio number is given, it usually is the contrast measured in the direction normal to the screen.

The definition given in Equation (13.14) does not include any effects of ambient illumination, i.e., the measurement is made in complete darkness. Since displays are rarely used with these conditions, an expanded definition of full-on/full-off contrast ratio is made for the situation when ambient illumination is incident on the screen of the display.

$$C_A = \frac{L_V(W) + RL_A}{L_V(B) + RL_A} \qquad (13.15)$$

L_A is the ambient illumination reflecting from the screen, measured reflecting off a 100 % reflectivity white plaque. R is the reflectivity of the screen and $L_V(W)$ and $L_V(B)$ have the same meaning they did for Equation (13.14). If the screen has gain, the value of R will depend on both the angle of the observer and the angle of the incident light. Therefore, for this type of screen, observers in different locations will see different contrast, and identical screens in different rooms with different illumination patterns will show different contrast.

When L_A is zero in Equation (13.15), i.e. there is no ambient light and the room is dark, this equation gives the same contrast as Equation (13.14). To continue this example, if there is a luminance of 20 cd/m² due to ambient light reflecting off a gain 1 projection screen with a reflectivity of 95 %, the display contrast would be $(600 + 95\%*20)/(0.4 + 95\%*20) \approx 32 : 1$ contrast. Human factors experiments have shown most task-oriented applications only require about 3:1 contrast, although for pleasing video, pictures require at least 10:1 contrast. If a contrast enhancing screen with 65 % reflectivity and the same 600 cd/m² image were used in ambient illumination that provided 100 cd/m² at the screen, the image would have 10:1 contrast. With 450 cd/m² ambient, this image from a 1500:1 projector would produce a perceived contrast of 3:1!

Because of the variations produced by the measuring conditions, standard ambient lighting conditions have been specified by ANSI(ANSI 1992). The effect of ambient illumination on the contrast is very much a function of the screen design. Some screens are more effective in absorbing ambient illumination than others and, with the same projector, will give a higher system contrast. Screen designs to enhance contrast for both front and rear projection were discussed in Chapter 7.

When a contrast ratio is specified by a microdisplay manufacturer for their product, it is normally a full on/full off value. Since the illumination path f/# has a significant effect on the contrast ratio of most microdisplays and light valves, it is necessary to know the f/# of the illumination path in the manufacturer's test setup. If the manufacturer's f/# is higher that the planned f/# of the projector, a lower device contrast ratio can be expected from the projector than is specified by the manufacturer.

The second contrast method is called the 'checkerboard' method, and is prescribed by ANSI (ANSI 1992; IEC 2002) for the measurement of finished systems. In this method, a test pattern generator is programmed to generate a 4 × 4 checkerboard pattern of black and white rectangles that covers the entire image area, as illustrated in Figure 13.13. The luminance at the center of each rectangle is measured. The eight white values are averaged together, as are the eight black values. The contrast ratio is then calculated by using the average white value and the average black value in Equation (13.14) or (13.15), as appropriate. Typically this contrast value is much lower than the value given by the sequential contrast method. The 1500:1 projector used as an example might have 150:1 ANSI contrast. The difference is normally due to scattering in the projection optics. Scattering can include light scattered internally in optical elements and light scattered by mechanical components such as aperture stops. Some of this scattered light goes directly to the wrong position on the screen while some of it goes backwards to the microdisplay to be reflected into the image area.

A modified version of this test is sometimes used. For some displays, there may be measurement issues related to asymmetric positional dependencies of the black and white rectangles. For these circumstances, each of the rectangle measurements are made as indicated in Figure 13.13. The video signal is then changed so that the white and black rectangles are interchanged in the image. All 16 rectangles are remeasured. The 16 white rectangles from these two sets of measurements are averaged as are the 16 black rectangles. The ratio of these two averages is the contrast.

332 CHARACTERISTICS AND CHARACTERIZATION

Figure 13.13 ANSI checkerboard pattern for contrast measurement.

CRT projectors can have checkerboard contrast ratios of less than 40:1 to about 60:1. These low values are due in part to the very low f/# projection lenses used, which accept scattered light over wide angles. It is also due to the fact that CRT phosphors are typically white diffuse scattering materials, so any light scattered backwards to the phosphor layer will contribute to the poor ANSI contrast. On the other hand, CRT projectors can provide very high full on/full of ratios, since no light is generated by the CRT in the black state.

If desired, ambient light can be taken into account in the checkerboard contrast method as well as in the full on/full off method. Note that when measuring contrast, the ambient illumination includes the image light reflected by a front projection screen or transmitted through a rear projection screen. This light will scatter off the walls of the room and some of it will make its way back to the screen to reduce contrast. This ambient light is, of course, content dependent which makes the ANSI method with 50 % image luminance particularly vulnerable to error. Therefore, in making contrast measurements, it is important to ensure the measurement facility and hardware does not degrade the measurements. The measurements of high contrast systems can easily be degraded by reflections from the room and screen (Florence *et al.* 1997), or even light colored clothing being worn by the tester. A baffled, absorptive environment is recommended to minimize this interference with the measurement.

13.3.3 Colorimetry

13.3.3.1 The specification of color

The 1931 CIE chromaticity diagram (CIE 32) for the standard colorimetric observer is given in Figure 13.14, and in Appendix 2. The horseshoe shaped curve is the gamut of pure spectral colors that could be produced by single wavelength illumination such as light from a laser. The wavelengths are shown along the horseshoe at their appropriate positions. The straight line across the bottom is the mixture of the shortest violet wavelength that is visible to the human eye, about 384 nm, with the longest red wavelength that is visible, about 764 nm. The amount of each in the mixture is inversely related to the distance in this diagram from the position of the pure wavelengths. The common names of colors, as used by display engineers, are given in their approximate locations. A more complete diagram of color names is given in Plate I. If you compare the names given in Figure 13.14 to the names given in Plate I, you will see there is a mismatch between the terms used by display engineers and the terms used by colorimetry experts. The black-body line and the three white points will be discussed in Section 13.3.3.3

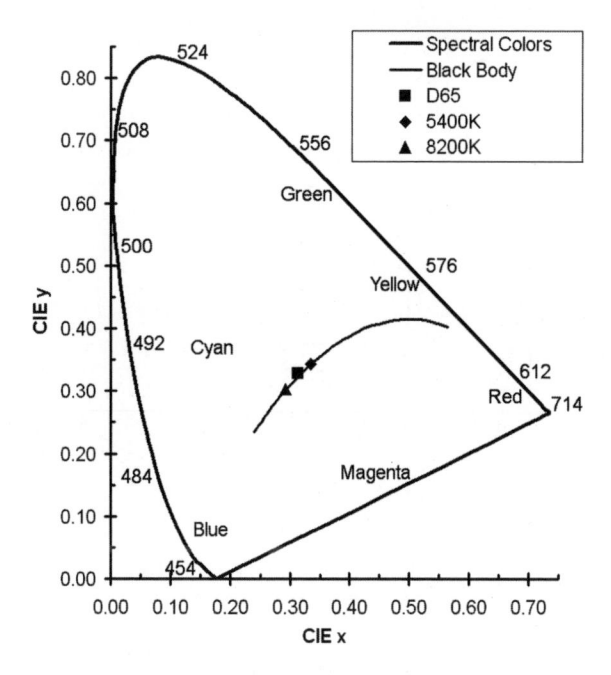

Figure 13.14 1931 CIE chromaticity diagram. The Planckian locus is discussed in the text.

Moving from the white point toward the horseshoe boarder represents increasing color saturation. A changing angle between the line which includes the white point and the color point and the horizontal axis represents a changing hue. An outline of the concepts used in deriving this representation is given in Appendix 2. Color science is covered well in a number of reference works (MacAdam 1981; Wyszecki and Stiles 1982).

Within this maximum definable color space, a display system typically has a number of defined primary colors. Most commonly there are three primary colors in the red, green and blue regions, but displays with more than three colors are becoming more common. By mixing the primary colors in the appropriate ratio, any color inside the triangle or polygon defined by the primary colors can be made. This system restricts a display to showing only a portion of the full color gamut. For example, HDTV displays can only show about 32 % of the colors in u'-v' space that the human eye is capable of perceiving. The measurement of color gamut size will be discussed in more detail in Section 13.3.3.5.

13.3.3.2 The gamut of real colors

The limitation on reproducible colors is not as severe as it might superficially seem. The colors of a wide variety of real reflective objects, including natural objects, paint chips and fabrics, were measured by Pointer (1980). The colors of all these objects were within the real color gamut, shown in Figure 13.15. A similar experiment was reported in Vrhel *et al.* (1994). Vrhel and his colleagues measured about 354 real objects, including the complete Munsell color cascade and samples of all Dupont paints available. This experiment gave about the same color gamut as Pointer.

Two versions of the Pointer gamut are shown in Figure 13.15. With real, reflective objects, any object that has a very saturated color also has a very low reflectivity and therefore appears very dim to the human eye. The human visual system, when looking at dim objects, sees the colors desaturated, compared to the actual location of the color on the x-y diagram. This brightness effect is documented by a relative

334 CHARACTERISTICS AND CHARACTERIZATION

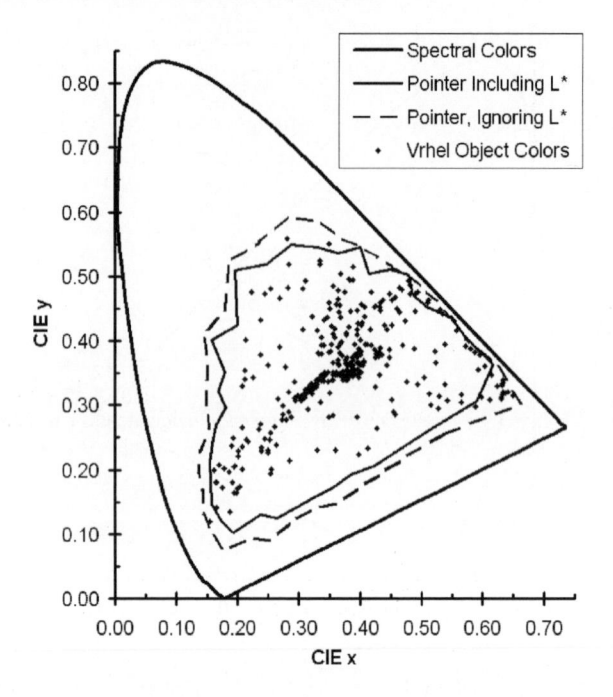

Figure 13.15 The gamut of real surface colors.

luminance factor L*, which is discussed in Appendix 2. When the L* factor is ignored and only the color is plotted, the colors measured by Pointer all fell within the dashed line. When the L* factor is taken into account and the colors are corrected to the x-y location where they would be perceived, the colors all fall within the solid line. Since this solid line surrounds all colors the HVS would perceive in the real world, this could be taken as the absolute minimum color gamut a display can use and still present a realistic view of the world.

Note that Pointer and Vrhel measured real reflective objects. Transmissive objects, such as the red filter in a traffic light or for that matter the dichroic filter in a projector, can lie outside either gamut. Emissive objects such as CRT phosphors, LEDs and lasers can lie well outside these gamuts, right out to the edge of the CIE diagram.

13.3.3.3 White point of displays

Most applications have a specification for the white point of the display. Techniques for correcting the white point of a projector to this target were discussed in Section 11.5.2. This section, however, will concentrate on what the white point should be for different applications and how that white point is specified. Video, for example, normally has a specified target white point of D65, close to the color of sunlight. Other white points are commonly used ranging from about 3000 K up to about 11 000 K. White points are always close to or slightly above the black-body line. This line represents the color emitted when a black-body is heated to a given temperature. For example, incandescent lamps such as the ones described in Section 8.2.4 normally have black-body emission spectra in the 2700 K to 3100 K region. While the black body curve spans a very large region of the color diagram, the human visual system perceives most colors along the curve as 'white'. Black-body radiators emit a broad spectrum of wavelengths, with the peak wavelength of the emission varying as 1/T. As the temperature increases, the color of the black-body emission changes from a deep red at a temperature of about 700 K to white in the 2900 K to 9300 K region, becoming more blue-shifted at higher temperatures. This line is sometimes called the Planckian locus, although it is more commonly referred to as the black-body curve.

The black-body curve and a few selected white points are shown on the CIE diagram in Figure 13.14. 5400 K, for example, is the white point that is typically used in film-based cinema while D65 is used in most video applications. Digital cinema does not have a defined white point but presumably digital movies are edited to look as much like filmed movies as possible. While D65 is defined for video, many consumer electronics manufacturers actually set up their display products to produce a higher color temperature, such as 8200 K. Color temperatures that are higher than the D65 specified for television are used because the bluish-white makes white objects look 'whiter-than-white'.

When a white point is near the black body line but not exactly on it, a value called the correlated color temperature (CCT) is used to describe the white point (Canty and Kirkpatrick 1961). This corresponds to the closest point on the black body curve to the white point in the now-obsolete CIE u-v color coordinate system. The calculation of CCT is the only current use of the u-v system, to the best of the authors' knowledge. Numerous methods have been given to calculate CCT from the CIE x-y values, including, for example, Krystek (1985). The CCT is only a 1-dimensional unit in a two-dimensional color space. The distance from the black-body line is often reported in minimum perceptible color difference (MPCD) units. A discussion on the computation of CCT and MPCD units is given in Appendix 200 of (VESA 2001).

13.3.3.4 Color gamuts of displays

In video systems, the color gamut of the signal is specified in the standards for the transmission system. Figure 13.16 shows the maximum range specified by the original NTSC system, SMPTE 'C' (SMPTE

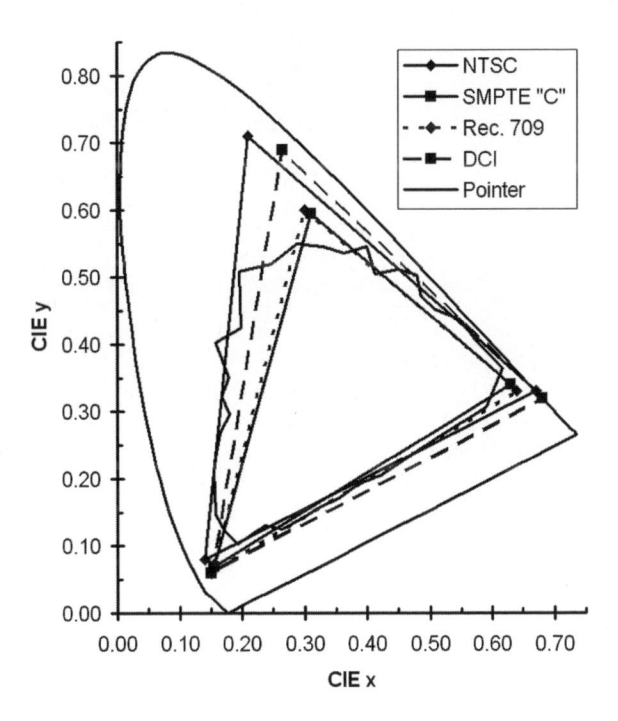

Figure 13.16 Color gamuts of principal electronic image formats compared to real colors.

336 CHARACTERISTICS AND CHARACTERIZATION

Table 13.1 Primary colors in video systems.

	Red		Green		Blue	
	x	y	x	y	x	y
NTSC (1953)	0.670	0.330	0.210	0.710	0.140	0.080
SMPTE 'C'	0.630	0.340	0.310	0.595	0.155	0.070
EBU	0.640	0.330	0.290	0.600	0.150	0.060
Rec. 709 (HDTV)	0.640	0.330	0.300	0.600	0.150	0.060
Digital Cinema (DCI)	0.680	0.320	0.265	0.690	0.150	0.060

1994), HDTV (Rec. 709) (ITU 2002) and digital cinema (DCI 2007) together with the Pointer gamut of colors that includes the L^* effect. For clarity, the EBU standard, used for standard definition TV in Europe, is omitted from this figure (EBU 1975; Pritchard and Gibson 1980). Table 13.1 gives the coordinates for the four principal transmission standards. Since the dichroic filters and lamp in light-valve projectors determine the gamut, these systems have the potential for a wider color gamut than CRT projectors. In CRTs, the achievable coordinates are limited by the phosphors.

The NTSC primaries allowed for good coverage of the gamut of real colors, but were never really seen in consumer electronics display devices (DeMarsh 1993). Historically, this was due primarily to the low efficiencies and luminance of CRT phosphors that match the specified NTSC color points, particularly for green emission. In practice, American broadcasters today transmit standard definition signals with primaries given by the SMPTE 'C' standard. EBU primaries are similar to those of SMPTE. A range is actually specified for each EBU color but the table only shows the nominal point. In practice, the Rec. 709, SMPTE 'C' and EBU gamuts are close enough to each other that only a video purist would worry about the color distortions produced by showing HD material on a SD display or vice versa.

The transmissions of video signals are color balanced to produce a white corresponding to sunlight, which approximates the color coordinates of a black body at 6550 K. For NTSC, this balance produces about 30 % of the white luminance from the red primary, 59 % of the white from the green and 11 % from the blue. For EBU primaries, the nominal luminance percentages are 21 % red, 71 % green and 7 % blue.

The color gamut in CRT projectors is determined by the spectral characteristics of the phosphors in the three tubes modified by any color filtering used to improve the purity of the emissions. The color gamut in microdisplay projectors is determined by the convolution of the spectral characteristics of the lamp and the transmission or reflection spectra of the optical filters used first to separate the light from the lamp into red, green and blue, and then to recombine these after modulation by the light valves. A numerical example of this is given in Chapter 12. This process imposes two requirements on the light: (1) the individual red, green and blue colors must match the target primary colors for the system and (2) when the red, green and blue primaries are recombined into white, they must have the correct relative intensities to produce the desired white point. If the desired white point is not achieved, color correction is needed in the system. Typically this reduces the total system throughput. Techniques to achieve this color correction are given in Chapter 11.

Two colors close to each other in the x-y or u'-v' color space may be perceived as the same color or as different colors, depending on a variety of factors. These factors include the brightness of the colors, the distance between them and the angle between them relative to the horizontal axis. Numerous experiments have been performed to determine the sensitivity of the HVS to small color differences. Figure 13.17 shows just-noticeable-difference (JND) ellipses as determined by MacAdam (MacAdam 1981). These MacAdam ellipses are probably the most commonly used color difference ellipses. If two colors are

LUMINANCE, CONTRAST AND COLOR

within the same ellipse, they are probably indistinguishable by an average viewer. The ellipses shown in Figure 13.17 are actually shown 10 times actual size for clarity.

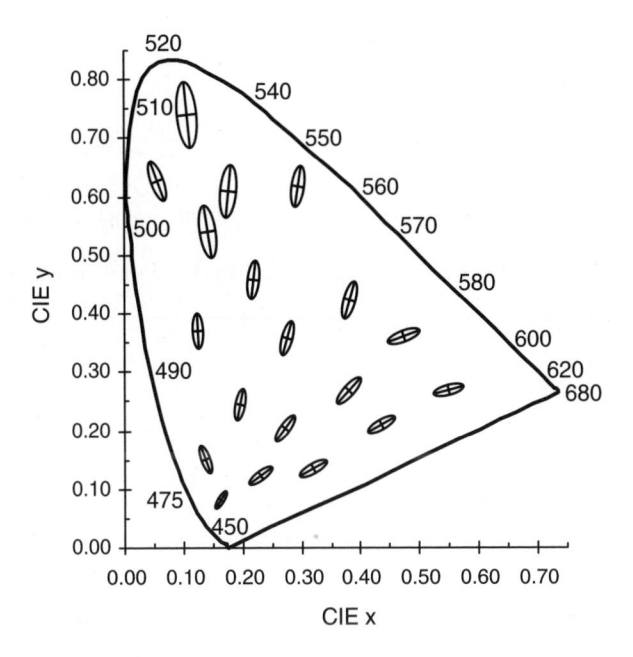

Figure 13.17 MacAdam JND ellipses plotted on the 1931 CIE diagram. The ellipses are enlarged 10 × in this figure. (From MacAdam 1981; with permission of Springer-Verlag.)

13.3.3.5 Expanded color gamuts and displays with more than three primary colors

There is no formal definition of what an 'expanded color gamut' projector is. For this book, we will define an expanded color gamut display as one that has a color gamut larger than the ITU-R Recommendation BT-709 color gamut defined for HDTV, commonly called either Rec. 709 or BT-709 (ITU 2002). Note that by this definition, both the DCI and the original NTSC color gamuts are 'expanded color gamuts'.

There are two main techniques to get expanded color gamuts. First, a projector with three primary colors can have an expanded color gamut if each primary color is more saturated than the primary specified by Rec. 709. In particular, this will occur when laser or LED illumination is used. Laser wavelength requirements to match Rec. 709 were discussed in Section 8.3.1. When these wavelength requirements are met, the laser projector does not just meet the requirements for HDTV, it far exceeds them. Typically, LED illumination sources with dominant wavelengths that meet the requirements for lasers in Table 8.3 will also have expanded color gamuts.

The second technique to produce expanded color gamuts is through the use of more than three primary colors. These types of projectors are called multiple primary color (MPC) projectors and were described in Chapter 10.

The main measurement and characterization issue associated with expanded color gamut projectors is defining the size of the color gamut. What does a projector manufacturer mean when he says his projector has a color gamut 120 % of NTSC? When a manufacturer claims his projector has an expanded color gamut that is 88 % of NTSC, many people are rightfully confused.

The only standard the authors know of that define color gamut size is the same one that defines ANSI luminance and ANSI contrast (IEC 2002). This standard requires color gamut be defined in terms of square u'-v' units as a percentage of the total area of the CIE diagram in u'-v' units, which is 0.1952[3]. This color gamut measurement technique has three major problems:

1. It is contained in an 'Informative' appendix and is therefore not binding.
2. It can produce seriously misleading results.
3. To the best of the author's knowledge, no manufacturer follows this definition.

Many manufacturers do follow a closely related procedure of defining the color gamut size in terms of some other video color gamut. The most commonly used gamut is the original 1953 NTSC gamut, which has an area of 0.0744, or 37 % of the total visible gamut.

Instead of u'-v' units, some manufacturers specify color gamut size in x-y units. In x-y units, the full CIE curve has an area of 0.3352 and the 1953 NTSC gamut has an area of 0.1582 or 47 % of the total visible gamut. Specifying the gamut size in x-y units is bad science and bad human factors and should never be done. The reason for this can be seen in the MacAdam ellipses in Figure 13.17. The green ellipses are much larger than the ellipses in the red region, which in turn are much larger than the ellipses in the blue region. The green ellipse in x-y units closest to the green primary for HDTV is 5.5 times larger than the ellipse for the red primary and 62 times larger than the ellipse corresponding to the blue primary (Brennesholtz 2007a). This means that measuring gamut sizes in x-y units is essentially a measure of the gamut size in the green region with little effect from the red region and essentially no effect from the blue region. In u'-v' units, this effect is diminished. The red ellipse is now the largest but it is only 1.4 times larger than the green ellipse and 3.0 times larger than the blue ellipse. Therefore in a u'-v' color gamut size measurement, the red and green are given roughly equal weight and the blue is not ignored completely. While this is still not a good measure of gamut size, it is better than measuring it in x-y units.

Another problem occurs when the projector primaries do not coincide with the video signal primaries, as illustrated in Figure 13.18. In this figure, the measured color gamut of a color-wheel-based DLP projector with a UHP-type lamp is compared to the color gamut defined by SMPTE 'C'. This RPTV set was intended primarily for the display of standard definition video in the US, which is broadcast assuming the receiver will have SMPTE 'C' colorimetry. The DLP color gamut is 87 % of NTSC in u'-v' units, while the SMPTE 'C' gamut is 77 % and the Rec. 709 HDTV gamut is also 87 %. Therefore, by this simple measure, this gamut is equal to the HDTV gamut and exceeds the requirement for standard definition. If you examine Figure 13.18 closely, you will see this is not quite true. Due to the lack of red in UHP-type lamps, the designers of this projector decided to desaturate the red in order to increase the total system lumen output of the system. This means that the system cannot display all the red colors contained in a standard definition signal, let alone all the colors in HDTV. One of the complaints about the SMPTE 'C' colors, and to a lesser extent about the HDTV color gamut, is the red is too orange to produce good system colorimetry. This projector, with an even more orangy-red than SMPTE 'C' will perform poorly when displaying video material containing red objects. Yet, its color gamut area is significantly larger than SMPTE 'C' and equal to HDTV.

Improved color gamut measurements have been proposed (Brennesholtz 2007b; Ben Chorin et al. 2007). The Brennesholtz method compares the color gamut to the gamut that is needed to produce accurate colorimetry. In this example, the green yellow and blue colors outside the SMPTE 'C' color gamut are not counted in the area since they are not needed to produce accurate colors on the display. The small area in red needed by SMPTE 'C' but not produced by the projector reduces the total area. Using this rating, the projector would produce 94 % of the colors in u'-v' space that are required for good SMPTE 'C' colorimetry. While the single number does not indicate exactly what colors cannot be reproduced by the projector, at least it indicates there is a deficiency in the projector.

---

[3] One of the authors (Brennesholtz) calculated the area of the full CIE color gamut in u'-v' units as 0.2007, or a 3 % difference from the value used by ANSI/IEC

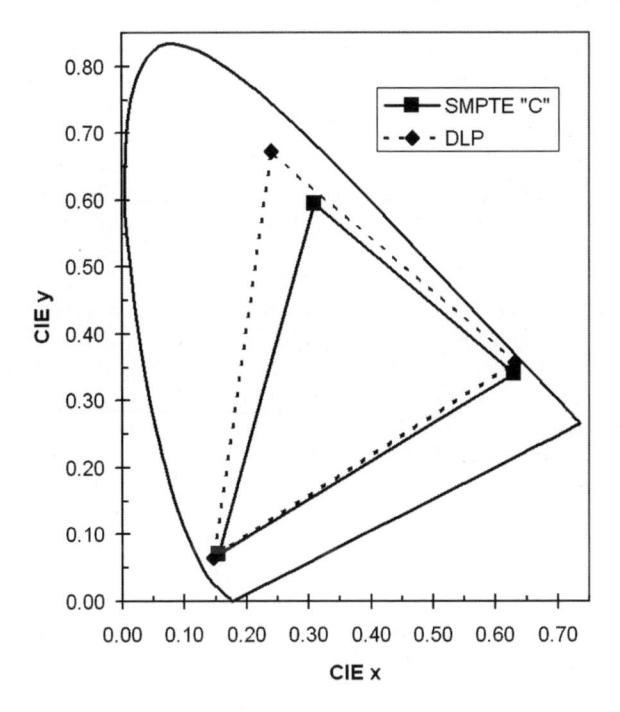

Figure 13.18 Comparison of the color gamut produced by a DLP projector with a UHP lamp to the SMPTE 'C' color gamut.

This method could be improved further by using a color space other than u'-v' that does not have the 3:1 error built into it that is inherent in the u'-v' space (Wyszecki and Stiles 1982, Figure 4(5.4.1)). Such spaces have been developed, but to the best of the authors' knowledge, they have not been applied to the measurement of color gamut size.

Ben Chorin takes a slightly different approach and bypasses the errors inherent in color spaces. Instead of comparing the color gamut to a defined color gamut, he compares the color gamut to the colors demanded by a 'typical' video signal. In a sense, this is defining a new color space, one corresponding to video signals, not the human visual system. While this approach may prove fruitful in the future, considerable development is required before it can be implemented in practice. In particular, some standards body such as the SMPTE or the ITU would need to define the statistical properties of the colors contained in a typical video signal.

There is one problem in the existing color gamut measurement method from the IEC and the proposed methods by Brennesholtz and Ben Chorin: they only evaluate the two-dimensional area of a color gamut. Color gamut should in fact be measured as a three-dimensional volume, including not only the color gamut in the color plane but the luminance achievable for each color. Unfortunately, the mathematics of this is somewhat complex and will be left to specialists in color measurement and specification.

This problem occurs in both MPC and white-segment systems, and to a lesser extent in expanded color gamut systems where the red, green and blue colors are more saturated than the specified primaries. In a three primary system where the primary colors exactly match the specified primaries and there are no serious non-linearities, the measured white lumen output of the projector is equal to the sum of the lumen outputs of the red, green and blue primary colors. In projectors with either a white segment or multiple primary colors, the white lumen output can be significantly higher than the sum of the red, green and blue lumen outputs. In practice, this means you can display both a very bright white and fully saturated colors, but the saturated colors are not as bright as one might expect from the total system lumen output.

This, in fact, is not a problem if one only wants to display natural images in reflected light, since saturated natural objects typically are also very dim objects. If one wishes to display transmissive or emissive objects such as traffic lights or lasers, the saturated colors will appear dimmer than the content creator intended. This is also a problem with computer generated images such as PowerPoint presentations, where the presentation will often contain fully saturated color shown at maximum luminance. These types of images, if they are created to look right on a desktop monitor, will look different on a projector with a white segment or multiple primary colors. Different, in the view of a content creator, is always undesirable.

A simple test has been proposed to evaluate this discrepancy has been proposed as a modified ANSI luminance measurement test method (Lang 2007a, 2007b). Lang, working in cooperation with the 3LCD consortium, has proposed the lumen output of a projector be measured twice, once with the projector driven with a white signal and then with individual red, green and blue signals. The luminances measured with the red, green and blue primary signals are then added together and ratioed to the measured white lumen output. A three primary color projector where the primary colors match the specified primary colors will give a value of about 100%. On these projectors, all content, regardless of the source, will look the same on a calibrated desktop monitor and on the projector. MPC or white segment projectors will have a ratio of less than 100%. The lower the ratio, the more differences there are likely to be between the appearance of the content on a desktop monitor and on the projector.

13.3.4 Measurement of the Luminance and Color of Projection Systems

The Electronics Industries Association (EIA) provides a standard for measuring the color of CRT systems (EIA 88). This standard discusses the advantages and disadvantages of spectroradiometers and colorimeters, and gives the procedures for calibrating a colorimeter against a spectroradiometer. While the standard was designed for direct-view CRT systems, it can be applied to projection displays as well.

Instruments to measure color and luminance are discussed in more detail in Appendix 2. These instruments either can be designed to measure the light incident upon them from any direction (cosine corrected) or can have a narrow field of view, typically 1°–2°. Cosine corrected instruments are normally intended to detect incident light, i.e. the instrument would be placed at the screen facing the projector. This arrangement is shown in Figure 13.19(a). With this type of instrument, one must be careful that there is no other source of light in the room except for the projector. Alternatively, a shield can be provided to restrict the field of view of the sensor so it sees only the projector, not a full 2π steradians. Color measured in this manner depends only on the projector, not the properties of the projection screen. Narrow angle instruments are typically used to measure the light coming from the projection screen, as shown in Figures 13.19(b) (rear projection) and 13.19(c) or 13.19(d) (front projection). In all of these cases, the spectral properties of the screen are included in the measurement of the projector. For rear projection systems, this is especially appropriate, since the projector and the screen commonly come as a mated pair.

For front projection systems, a projector can be used with many different screens and it is normally desired to have the color of the projector itself, without including the properties of the screen. This can be done in two ways. First, if the gain of the screen as a function of wavelength $G(\lambda)$ is known, the spectral dependence of the radiance $L(\lambda)$, measured by the method shown in Figure 13.19(c) with a spectroradiometer, can be corrected with the equation:

$$L'(\lambda) = \frac{L(\lambda)}{G(\lambda)} \qquad (13.16)$$

If the screen has gain, as most front projection screens do, $G(\lambda)$ can be greater than 1, and is dependent on the angle θ. The reflectance of the screen is included in $G(\lambda)$. In a screen with no gain (G = 1), $G(\lambda)$ is just the reflectance of the screen as a function of wavelength. After the luminance is measured and

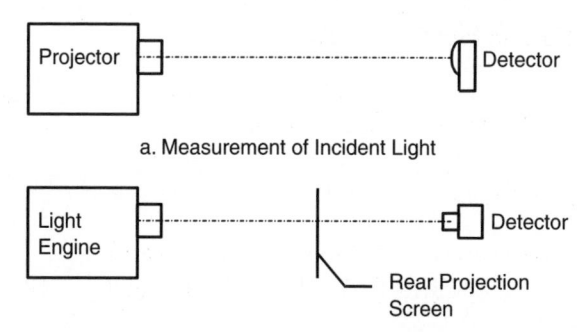

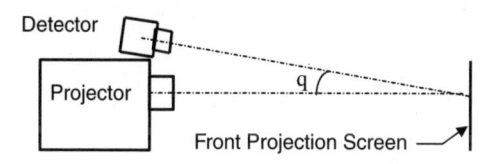

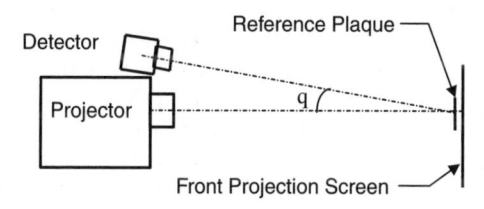

LUMINANCE, CONTRAST AND COLOR

Figure 13.19 Projector colorimetry measurement setup: (a) measurement of incident light; (b) measurement of a rear projection system; (c) measurement of a front projection system; (d) measurement of front projection system with reference plaque.

corrected (if necessary) with Equation (13.16), the normal colorimetric equations can then applied to calculate tri-stimulus values and color coordinates.

The second approach to measuring front projection systems is to use a small target placed in front of the main projection screen. If a target is used where $G(\lambda) = 1$ for all wavelengths, no correction to $L(\lambda)$ is required. This approach is shown in Figure 13.19(d). If the target is also fully diffusing, with a Lambertian angular dependence, the angle θ will have no effect on the measured value. One suitable reflectance target material is Spectralon, manufactured by Labsphere.

Given an accurate instrument, color is much easier to measure accurately than luminance. For example, a small error in measuring the distance to the screen and screen size may have a significant effect on the estimate of total lumen output if output is measured using the technique shown in Figure 13.19(d). In this case, the $1/R^2$ effect of the difference between the plaque and screen distance should be taken into account. For color, however, this effect will be normalized away when the x-y color coordinates are calculated.

13.3.5 Display Gamma and Gray Scale

In television transmissions, the amplitudes of transmitted signals V_t have a power-law-compressed relationship to the video signal V_s according to the relationship

$$V_t \propto V_s^{\gamma_s} \tag{13.17}$$

where γ_s is 1/2.2. The compression was originally based on the characteristics of CRT displays. CRT electron guns typically have a current/drive voltage relationship described by a power law, viz.,

$$I = kV^{\gamma_d} \tag{13.18}$$

where
- I is the current from the electron gun
- k is a constant
- V is the drive voltage (proportional to the transmitted signal, V_t.) applied to the gun, and
- γ_d is the gamma of the electron gun.

The transmission standard presumes γ_d to be 2.2 to achieve a tube current proportional to V_s. Actual direct-view color CRTs have γ_d in the range from 2.31 to as high as 3.1. The gamma of the gun is nearly a constant but it is known to vary from gun-to-gun and to change as the tube ages. Lumen output of a CRT is (nearly) directly proportional to the current from the gun, so the final relationship between the video drive and lumen output has this same γ relationship. (Some phosphors, particularly blue emitters, exhibit saturation at high drive levels. This effect will distort the relationship between video drive and the lumen output.) Combining γ_s of the source (= 1/2.2) with a display γ_d of 2.8 gives an overall system γ of about $2.8/2.2 = 1.27$, instead of the 'ideal' value of 1 for linear imaging. Tests show consumers prefer an overall γ in the 1.2–1.4 range, rather than the linear 1.0 (Barten 1996; Roufs *et al.* 1994; Mitsubayashi *et al.* 1996; Shimodaira *et al.* 1995).

Video encoding does not in fact use a pure gamma law as shown in Equation (13.17) or (13.18). Standards bodies have added a short linear tail at low signal levels. See the appropriate video standard such as (ITU 2002) for HDTV for the exact encoding equation for a given video format.

The transmission or reflectivity of a microdisplay typically shows a different drive/luminance relationship than a CRT. Since all video standards are encoded to match CRT characteristics, it is necessary to do non-linear processing on the signal in order to get the desired luminance out from the input video level. This non-linear processing is typically called 'gamma correction'. 8-bit gamma-encoded video signals are mostly free from an artifact called contouring, where a change of one bit in the video input results in a visible step in the screen luminance (Brennesholtz 1998). With a microdisplay technology with drive/luminance relationship far from the $\gamma = 2.2$ value of a CRT, 8 bits of drive is not sufficient to avoid visible contouring. Linear devices ($\gamma = 1.0$) such as the DLP and other devices with bit-plane based grayscale are particularly prone to this problem, especially at low luminance levels. In these cases, 11 bits drive per color or more is required to avoid visible contouring. In cases where 11 bits drive per color is not possible, such as single-panel DLP systems, image processing which introduces dithering, can reduce the contouring to acceptable levels. This image processing can introduce visible noise in the image for dark scenes and reduce the resolution in moving images.

Eight bits of gamma-corrected video is sufficient to avoid most, but not all contouring problems. Experiments on digital cinema projectors (Cowan *et al.* 2004) showed that with an encoding $\gamma = 2.6$, some individuals could see contouring in some special images unless 12 bits of encoding were used. Ten bits of encoding allowed a few individuals to see contouring in special images designed to maximize the visibility of contouring. Current digital cinema encoding uses 12 bits per color and $\gamma = 2.6$ (DCI 2007).

13.4 Image Content-dependent Adaptive Processes

Adaptive processing is image processing intended to provide the best settings for the projection display in order to match the displayed output to the video signal. This image processing is different than the more common processing intended to modify the video signal in order to match the optical properties of a projection display. The reason adaptive processes are discussed in the chapter on characteristics and characterization is that these processes are intended to change the normal video signal/displayed image process. This often leads to a debate about the characterization of a display with adaptive processes. For example, is the contrast achieved with a dynamic aperture 'real' contrast, or are the numbers 'phony' and only of value to the marketing department?

Many properties of a projected image can be improved by adaptive processes that modify the properties of the system in real- time depending on the content of the image. Several of these processes have already been discussed in terms of CRT projection:

1. Current limiting of the tube on bright scenes to avoid overdriving the power supplies,
2. Dynamic blue defocus to increase the output of the blue CRT and
3. Scan velocity modulation to enhance edges in the image.

Similar processes can enhance the images generated by microdisplay-based projectors.

One of the basic enhancements used by many projection systems showing video is a dynamic aperture to enhance the perceived contrast of the system (Iisaka et al. 2003; Toyooka et al. 2004). In this type of system, the software examines the video signal to look for low-brightness scenes. In these scenes, the diameter of the aperture is reduced to reduce the total light output of the projector. To maintain the intended brightness at the screen, the video is scaled to compensate for the reduced system brightness. Take for example, a video scene where the brightest pixel has a digital value of 136 out of 255. In a system with a $\gamma = 2.2$, this pixel should be displayed at 25 % of the maximum luminance. If the aperture diameter is reduced to make the lumen output of the system 25 % of the maximum output and the pixel value is scaled to 255, the pixel will be displayed at the correct luminance. Of course all the other pixels must also be scaled so they show at their intended luminance.

The advantage of this system is that by reducing the aperture diameter and the light at the microdisplay, the dark level is reduced, in the example, by a factor of 4. This reduced dark level is perceived as increased contrast. The actual contrast improvement may be even by a factor of more than 4. This may occur because the contrast of most microdisplays is dependent on the system f/# and reducing the aperture diameter will increase the f/#, increasing the fundamental contrast of the microdisplay.

In systems with individually controllable red, green and blue light sources (i.e. systems with laser and LED illumination) the light source can be dimmed on a per color basis. For example, if the brightest red video value is 255 while the brightest blue and green video values in a scene are 136, then the red source can be left at full brightness while the blue and green sources dimmed to 25 % of their maximum value. This will not only increase the apparent contrast, but it will save power to the light sources, an important factor in a battery-powered projector. Dimming the light source does not affect the f/#, so the additional contrast improvement associated with the increased f/# is not available.

For direct-view LCD displays, dimming can be done spatially as well as temporally (Seetzen et al. 2005; Shirai et al. 2006). This involves making a low-resolution image with the backlight and adapting the bit level of each pixel according to the local backlight luminance. While this has not been used to the authors' knowledge in a projector to date since it does not appear to be practical when a single lamp is used, the introduction of new technology light sources may allow 1D or 2D dimming in projectors in the future. The four-panel LCoS projector described in Section 10.3.2.1 is similar to this system carried to its logical conclusion, with each individual pixel dimmed as appropriate to enhance contrast.

There is no standard technique for measuring the contrast of a projection system with dynamic contrast of this sort. For example, if the ANSI contrast test method as shown in Figure 13.13 is used, this system will show no improved contrast. This occurs because the bright squares of the ANSI checkerboard are driven to maximum white (i.e. bit value 255 in an 8 bit system). Since the video calls for maximum white, it is not possible to reduce the aperture diameter and the black level of the black squares in the checkerboard is not reduced and the measured contrast is not changed. On the other hand, if the full on/full off contrast measurement method is used, and the aperture is capable of closing completely, the screen will be completely dark and 0 luminance will be measured in the dark state. When Equation (13.14) is applied, the system appears to have infinite contrast.

The use of image-content dependent adaptive processing is expected to increase in the future so the image measurement and characterization issues raised are not likely to go away.

References

ANSI (1992) ANSI IT7.215-1992. *Data Projection Equipment and Large Screen Displays – Test Methods and Performance Characteristics.*

ANSI (1993) ANSI/SMPTE 196M-1993. *Standard for Motion-Picture Film – Indoor Theater and Review Room Projection – Screen Luminance and Viewing Conditions.*

Balram, N. and Olson, W. (1996a) Multi-valued modulation transfer function for flat panel displays. *SID Intl. Symp. Digest of Technical Papers*, 429–32.

Balram, N. and Olson, W. (1996b) Vertical resolution of monochrome CRT displays. *SID Intl. Symp. Digest of Technical Papers*, 211–14.

Baroncini, V., Mahler, H. and Sintas, M. (2004) The image resolution of 35mm cinema film in theatrical presentation. *SMPTE Motion Imaging*, 113(2&3): 60–6.

Barten, P.G.J. (1990) Evaluation of subjective image quality with the square-root integral method. *Journal of the Optical Society of America* 7(10): 2024–31.

Barten, P.G.J. (1991) Resolution of liquid crystal displays. *SID Intl. Symp. Digest of Technical Papers*, 772–5.

Barten, P.G.J. (1995) MTF, CSF, and SQRI for image quality analysis. *SPIE International Symposium on Intelligent Systems and Advanced Manufacturing*, Short Course SC06, Philadelphia, 1995.

Barten, P.G.J. (1996) Effect of gamma on subjective image quality. *SID Intl. Symp. Digest of Technical Papers*, 421–4.

Bass, M., van Stryland, E.W., Williams, D.R. and Wolfe, W.L. (eds) (1995) *Handbook of Optics*, 2nd edn. New York: McGraw-Hill.

Ben Chorin, M., Eliav, D., Roth, S., Pagi, A. and Ben-David, I. (2007) New metric for display color gamut evaluation, *SID Intl. Symp. Digest of Technical Papers*, Paper 21.3, 1053–6.

Boff, K.R. and Lincoln, J. E. (eds) (1988). *Engineering Data Compendium, Human Perception and Performance*, Vol.1. Harry G. Armstrong Aerospace Medical Research Laboratory, Wright-Patterson Air Force Base, Ohio.

Boyd, R.W. (1983), *Radiometry and the Detection of Optical Radiation*. New York: John Wiley & Sons, Inc.

Bradley, R., Goldenberg, J.F. and McKechnie, T.S. (1985) Ultra-wide viewing angle rear projection television screen. *IEEE Trans. Consumer Elect.* **CE31**: 185–93.

Brennesholtz, M.S. (1998) Luminance contouring at the display in digital television. In M. Wu (ed.), *SPIE Proceedings: Projection Displays IV* **3296**: 62–70.

Brennesholtz, M.S. (2007a) Expanded color gamut displays – Part 1: Colorimetry for video signals and display systems. *Information Display* **22**(9): 24–8.

Brennesholtz, M.S. (2007b) Expanded color gamut displays – Part 2: Wide color gamut displays. *Information Display* **22**(10): 14–17.

Canty, B.R. and Kirkpatrick, G.P. (1961) Color temperature diagram. *Journal of the Optical Society of America* **51**(10): 1130–2.

CIE (1932) *CIE Proceedings 1931*. Cambridge: Cambridge University Press,.

CIE (1964) *10 Degree Photopic Photometric Observer*. Originally published in 1964 by the Commission Internationale de l'Eclairage (CIE). Current version is CIE 165: 2005.

Cornsweet, T.N. (1970) *Visual Perception*. New York: Academic Press.

Cowan, M., Kennel, G., Maier, T. and Walker, B. (2004) Contrast sensitivity experiment to determine the bit depth for digital cinema. *SMPTE Journal* **113**: 281–92.

REFERENCES

Dallas, W J., Roehrig, H. and Krupinski, E.A. (2007) Image quality analysis of a color LCD as well as a monochrome LCD using a Foveon color CMOS camera. In F.P. Doty, H.B. Barber and H. Roehrig (eds), *Proc. of SPIE: Penetrating Radiation Systems and Applications VIII* **6707**.

DCI (2007) Digital Cinema Initiatives, LLC, *Digital Cinema System Specification, Version 1.1*, 12 April 2007.

DeMarsh, L. (1993) TV displays phosphors/primaries – some history. *SMPTE Journal*, 1095–9.

EBU (1975) European Broadcasting Union Technical Centre, Brussels, *EBU standard for chromaticity tolerances for studio monitors*, August 1975; re-edited in 1981. Tech.3213-E.

EIA (1988) Electronic Industries Association. *Measurement of the Color of CRT Screens*. TEP105-11-A.

Farrell, J.E. (1999) Image quality evaluation. Chapter 15 in L.W. MacDonald and M.R. Luo (eds), *Colour Imaging: Vision and Technology*. Chichester: John Wiley & Sons, Ltd, 285–313.

Fiske, T.G., Silverstein, L.D., Hodgson, S. and Watson, A.B. (2007) Visual quality of high-contrast projection screens. Part I: Visibility of screen-based artifacts and noise. *Journal of the SID* **15**(6): 409–19.

Florence, J.M., Miller, R. and Bartlett, T.A. (1997) Contrast ratio in DMD-based projection systems. *SID Intl. Symp. Digest of Technical Papers*, 920–2.

Goldenberg, J., Huang, Q. and Shimizu, J.A. (1997) Rear projection screens. In M. Wu (ed.), *Proc. SPIE: Projection Displays III* **3013**.

Granger, E.M. and Cupery, K.N. (1972) An optical merit function (SQF), which correlates with subjective image judgements, *Photographic Science and Engineering* **16**(3): 221–30.

Haan, G. de, Loning, A. and Loore, B. de (1996) IC for motion-compensated 100 Hz TV with natural-motion moviemode. *IEEE Trans. Consumer Elect.* **CE42**: 165–74.

Hsu, S. (1986) The Kell factor: past and present. *Journal of the SMPTE* **89**: 206–14.

IEC (2002) *Electronic projection – Measurement and documentation of key performance criteria – Part 1: Fixed resolution projectors*, IEC standard 61947-1, dated 2002-8.

Iisaka, H., Toyooka, T., Yoshida, S. and Nagata, M. (2003) Novel projection system based on an adaptive dynamic range control concept. *IDW '03, Proceedings of the 10th International Display Workshops, 4 Dec. 2003 Fukuoka, Japan*, ITE & SID, Paper LAD1-4, 1553–6.

Infante, C. (1993) CRT display measurements and quality. *SID Intl. Symp., Seminar F-6 Lecture Notes*, Seattle.

ITU (2002) *Recommendation ITU-R BT.709-5 Basic Parameter Values for the HDTV Standard for the Studio and for International Programme Exchange*. Geneva: International Telecommunications Union, April 2002.

Jeffreys, G., Christensen, J. et al. (2005) Specifying and Assessing Projected Image Quality, Infocomm White Paper, June 2005. Download from: http://www.infocomm.org/cps/rde/xbcr/infocomm/Projected_Image_Quality.pdf

Jiang, Q., Cheng, J., Li, J., Lin, Z. and Ran, Q. (2006) An improved CRT projection tube resolution measurement system. In X. Hou, J. Yuan, J.C. Wyant, H. Wang and S. Han (eds), *2nd International Symp. on Adv. Optical Manufac. and Testing Tech.: Optical Test and Measurement Tech. and Equi. Proc. of SPIE* **6150**.

Jorna, G.C. (1993) MTF measurements, image-quality metrics and subjective image quality for soft copy color images, *SID Intl. Symp. Digest of Technical Papers*, Paper 26.4: 404–7.

Kanazawa, M. and Mitsuhashi, T. (1989) Automatic convergence correction systems for projection displays. *Japan Display (IDRC) Proceedings*, 264–7.

Kingslake, R. (1983) *Optical System Design*. New York: Academic Press.

Krystek, M. (1985) An algorithm to calculate correlated colour temperature. *Color Research and Application* **10**(1): 38–40.

Lang, K. (2007a) Analysis of the proposed IEC 'Color Illuminance' digital projector specification metric, using CIELAB gamut volume. *15th Color Imaging Conference in Albuquerque*, NM. (Can be downloaded from www.lumita.com)

Lang, K. (2007b) Proposal to modify International Standard IEC 61947-1 'Electronic projection – Measurement and documentation of key performance criteria – Part 1: Fixed resolution projectors', 30 December 2007. (Can be downloaded from www.lumita.com)

MacAdam, O.L. (1981) *Color Measurement*, 2nd edn. Berlin: Springer-Verlag.

McKechnie, T.S., Grant, M. and Stanton, D. (1985) MTF Characterization of Projection TV Systems. Philips Laboratories – Briarcliff Research Internal Report #776.

Mannos, J.L. and Sakrison, D.J. (1974) The effects of a visual fidelity criterion on the encoding of images. *IEEE Trans. Information Theory* **20**: 525–36.

Mitsubayashi, T., Shimodaira, Y., Washio, W., Ikeda, H., Muraoka, T. et al. (1996) Gamma range of reproduced pictures having quality above acceptable limits. *16th Intl. Display Res. Conf. ('96 Eurodisplay)Proceedings*, 257–60.

Ohishi, I. (1971) Spatial frequency characteristics of display with mosaic-array-elements. *Seminar on Solid-State Image Transducers*. Inst. of TV Engineers (Japan), Document #47.

Pointer, M.P. (1980) The gamut of real surface colors. *Color Research and Development* **5**; 145–55.
Pratt, W.K. (1991) *Digital Image Processing*, 2nd edn. New York: John Wiley & Sons, Inc.
Pritchard, D.H. and Gibson, J.J. (1980) Worldwide color television standards – similarities and differences. *Journal SMPTE* **89**: 111–20.
Robson, J.G. (1966) Spatial and temporal contrast-sensitivity functions of the visual system. *J. Opt. Soc. Amer.* **56**: 1141–2.
Roufs, J.A., Koselka, V.J. and van Tongeren, A.A. (1994) Global brightness contrast and the effect on perceptual image quality. *Proc. SPIE* **2179**: 80–9.
Schreiber, O.W.F. (1991) *Fundamentals of Electronic Imaging Systems. Some Aspects of Image Processing*, 2nd edn. Berlin: Springer-Verlag.
Seetzen, H., Ward, G. and Whitehead, L. (2005) High dynamic range imaging pipeline. *ADEAC '05*. Proceedings published by the SID, Paper 6.3, 59–62.
Shimodaira, Y., Muraoka, T., Mizushina, S., Washio, H., Yamane, Y. *et al.* (1995) Acceptable limits of gamma for a TFT-liquid crystal display on subjective evaluation of picture quality. *IEEE Trans. Consumer Elect.* **CE41**: 550–4.
Shirai, T., Shimizukawa, S., Shiga, T., Mikoshiba, S. and Käläntär, K. (2006) RGB-LED backlights for LCD-TVs with 0D, 1D, and 2D adaptive dimming. *SID Intl. Symp. Digest of Technical Papers*, Paper 44.4, 1520–3.
Sluyterman, S. (1993) Resolution aspects of the shadow-mask pitch of TV applications. *SID Intl. Symposium Digest*, 340–3.
SMPTE (1994) SMPTE C color monitor colorimetry. *SMPTE RP*, 145.
Snyder, H.L. (1988) Image quality. In M. Helander (ed.), *Handbook of Human-Computer Interaction*, pp 437–74, New York: Elsevier.
Toyooka, T., Yoshida, S. and Iisaka, H. (2004) Illumination control system for adaptive dynamic range control. *SID Int. Symp. Digest of Technical Papers*, Paper 12.3, 174–7.
van Meeteren, A. (1973) *Visual aspects of image intensification*, Ph.D. dissertation, University of Utrecht, Utrecht, The Netherlands.
VESA (2001) Flat Panel Display Measurements Standard Version 2.0, FPDM2 VESA-2001-6, Video Electronics Standards Association (VESA), Milpitas, CA.
Vrhel, M.J., Gershon, R. and Iwan, L. S. (1994) Measurement and analysis of object reflectance spectra. *Color Research and Application* **19(1)**: 4–9.
Walsh J.W.T. (1965) *Photometry*. New York: Dover Publications.
Watson, A.B. (2006) The spatial standard observer: A human vision model for display inspection. *SID Intl. Symp. Digest of Technical Papers*, Paper 31.1, 1312–15.
Watson, A.B. and Ahumada, Jr., A.J. (2005) A standard model for foveal detection of spatial contrast. *Journal of Vision* **5**(9): **6**, 717–40.
Wyszecki, G. and Stiles, W.S. (1982) *Color Science: Concepts and Methods, Quantitative Data and Formulae*. 2nd edn. New York: John Wiley & Sons, Inc.
Zhang, X.M. and Wandell, B.A. (1996) A spatial extension to CIELAB for digital color image reproduction. *SID Intl. Symp. Digest of Technical Papers*, 731–4.

14
Image Artifacts

Artifacts in projection displays are the (undesired) qualities of an image which were not part of the original signal used to create that image. These errors can be a result of the system design, the result of the physical characteristics of the display, interactions between the signal and the display, and/or interactions between the human visual system and the display.

As an example, many readers are familiar with the cross-color artifact in NTSC images. This system artifact is frequently visible as multi-hued colors across a pattern of a repetitive high frequency black-and-white signal, e.g. the signal produced by a finely striped garment in the image. This specific error in the image results from the architecture of the NTSC transmission standard. Other artifacts result from the display itself. Some are the result of the sampling processes of pixelated or raster-scanned displays, others can result from temporal characteristics and/or system architectures. The artifacts produced by sampling can be inherent in a specific system implementation.

In the following sections, some of the more common display artifacts are discussed. We do not discuss here the image errors resulting from misconvergence of the color images from multiple image sources. These errors are discussed in Section 13.2.2.4.

14.1 Spatial Artifacts

Spatial artifacts are visible in the plane of the image. Normally they are visible in static images, unless the system has special image processing to eliminate them staticly. They can be due to the properties of a single component (e.g. screen door effect) or they can be due to the interaction of multiple components (e.g. Moiré). This section will not discuss spatial artifacts that are due to the video signal such as NTSC cross color or blockiness in MPEG compressed video.

14.1.1 Moiré: Alias and Beat Frequencies

Moiré patterns result from two generic sources, viz., alias frequencies and beat frequencies. Alias frequencies are generated when the sampling frequency f_S is less than twice the frequency f_0 of the pattern being sampled, i.e. when the Nyquist criterion is not satisfied (Watkinson 1994; Hentschel 1997, 2001). With aliasing, frequencies are generated in the output which were not part of the input signal. This results in the generation of low frequency modulation of the output.

Note the term 'sampling' does not just refer to sampling of an analog signal by digital electronics. Spatial sampling in projection systems occurs at pixelated microdisplays and projection screens with lenticular structures. Temporal sampling can be due to color field rates or bit-plane drive schemes.

When the Nyquist criterion is met, i.e. the sampling frequency is more than twice the highest signal frequency, the spurious frequencies are not generated. However, it is possible to get lower frequency modulation of the signal by interaction of the signal and sampling frequencies. These Moiré beat frequencies are most apparent when the signal and sample frequencies become close to each other. Harmonics of these two frequencies can also be involved, due to the sampling process. The low frequency visibilities of beating and aliasing are the same. This is discussed further below.

Mathematically, the beat frequencies arise from the convolution of the two frequencies, as indicated in Equation (14.1). The frequencies can be spatial or temporal.

$$cos(nk_1 x)cos(mk_2 x) = \frac{1}{2}[cos((nk_1 + mk_2)x) + cos((nk_1 - mk_2)x)] \qquad (14.1)$$

where
- n, m = 1,2,3,... define the fundamental and harmonics of the signal and sampling frequencies, and.
- k_1, k_2 are the fundamental spatial or temporal frequencies of the signal and sampling processes.

The sampling process generates higher order terms, even in the absence of these frequencies in the input. This is discussed below. The sum frequency, $nk_1 + mk_2$, is most often a high spatial (or temporal) frequency and is normally not visible. The difference frequency, $nk_1 - mk_2$, can have a significant range, from very low frequencies to higher frequencies. In particular, it can produce a visible modulation of the input signal when this difference frequency is sufficiently low to be within the spatial (or temporal) range detectable by the human eye (Boff and Lincoln 1988).

One of the consequences of sampling is the replication of the spectrum of input frequencies around integer multiples of the sampling frequency. Figure 14.1 illustrates the one-dimensional input and output frequencies in a system in which there is idealized square wave sampling of incoming sine wave signals,. The output data $s(n)$ are discrete time samples of the input signals $s(t)$, separated by a spatial (or temporal) distance of $1/f_S$, where f_S is the sampling frequency. The spectrum of the input signal is repeated by the sampling frequency and all integer multiples of it, in both the positive and negative frequency directions.

Each of the signals in the repeat spectra has the potential to interact with other frequencies to produce Moiré. The integers n and m in Equation (14.1) need not have the same value. Beats can be generated by interactions between terms with $n = m$ as well as different order terms. The fundamental range ($n = m = 1$) is sometimes called the baseband range.

When the signal frequency is less than one-half the sampling frequency, aliasing does not occur. The original signal frequency appears in the output, albeit with low frequency spatial or temporal modulation at the difference frequency between the signal and sampling frequencies. Figure 14.2 illustrates the case of the signal frequency equal to 0.47 of the sampling frequency. The difference between these frequencies (0.53) is above the Nyquist frequency. The sampled signal contains the input frequency and no alias signal, but it does have the beat frequency modulation. Sharp cutoff filtering can, in principle, reduce this modulation with only amplitude effects on the signal, dependent on the cut-off characteristics of the

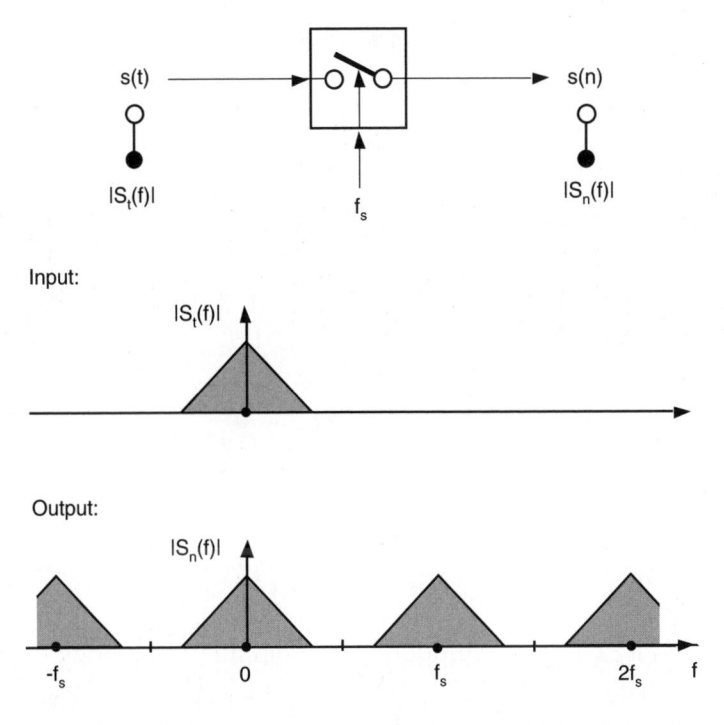

Figure 14.1 Sampling of a continuous (analog) input signal s(t) (Hentschel 1997).

filter. Alternatively, keeping f_0 less than ≈ 0.67 times the Nyquist frequency will essentially eliminate the visibility of beat modulation.

Figure 14.3 illustrates a case in which there is aliasing. The signal frequency f_0 here is 0.53 times the sampling frequency f_S. The sampled signal shows two effects. First, there is a low frequency modulation, and second, the signal period has changed. The repetition spectrum gives a frequency line at $0.47 f_S$, which is lower than the incoming signal frequency. The alias frequency ($0.47 f_S$) cannot be removed without removing signal information. Due to the fact that the alias and signal frequencies are close together, the frequency difference $f_b = |f_0 - (f_S - f_0)| = |f_S - 2f_0|$ can be seen as a beat frequency, causing a low frequency modulation of the (aliased) sampled signal. The beat frequency is minimized by filtering out the higher physical frequency. In this case, this is the signal frequency and only the alias frequency would remain. Similar filtering for the case where the Nyquist criterion is satisfied (Figure 14.2) leaves the information at the signal frequency intact.

14.1.1.1 Origins of Moiré

The origins of the sampling processes which give rise to Moiré in projectors can be electronic, optical, mechanical or combinations of these. For example, optical and electrical frequencies in the system can give rise to aliasing. Figure 14.4 illustrates the severe artifacts in the image that would be generated by a sine wave electrical input frequency f_0 interacting with the optical sampling at temporal frequency f_S of the panel, when $f_0 = 0.9 f_S$. Moiré that results from purely optical interferences would not be affected by electrical filtering. Moiré can also result if an image comes from a digital source that does not have the same pixel count as the display and is not correctly interpolated to fit on the display.

Similarly, the rear projection lenticular screen is a sampling device which could produce artifacts if the signal frequencies projected onto the screen become too close to the screen sampling frequency

350 IMAGE ARTIFACTS

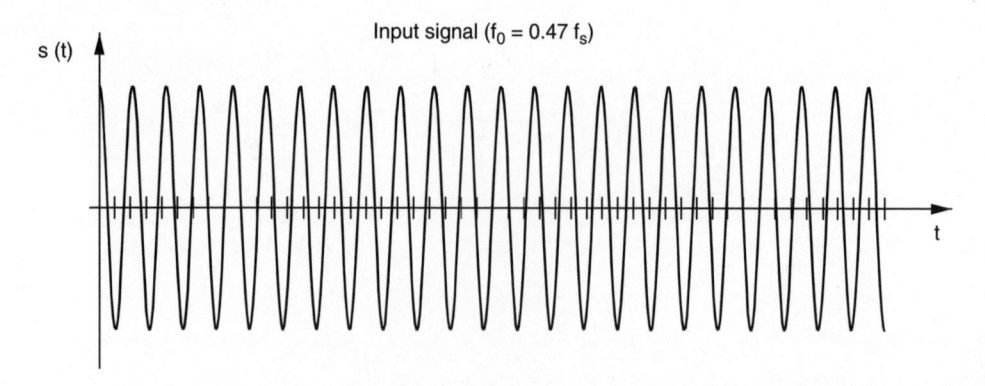

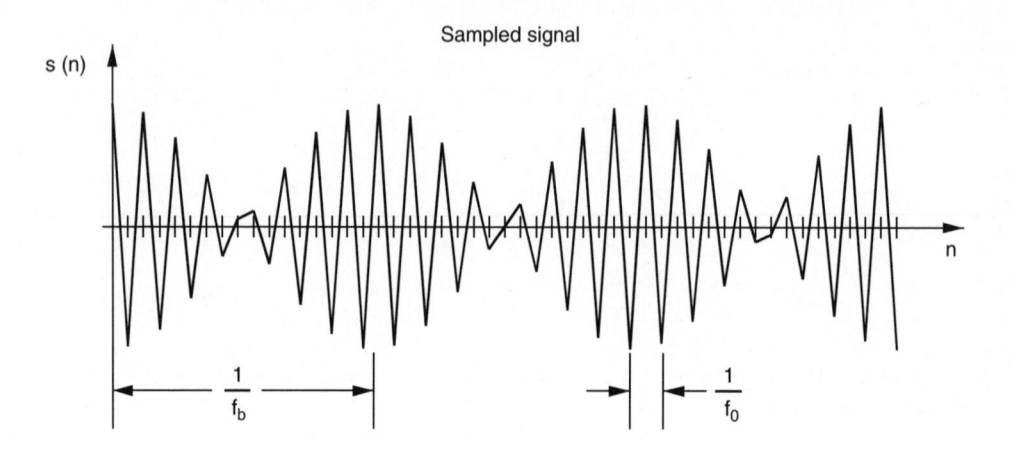

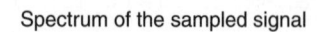

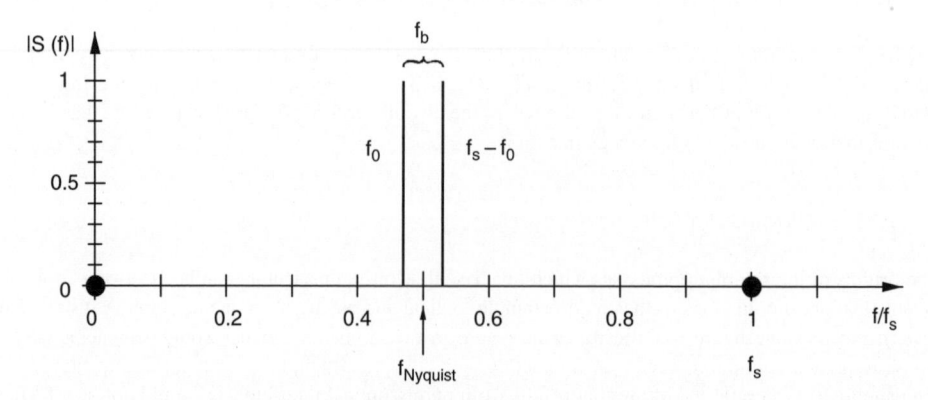

Figure 14.2 Origin of beat frequencies in a sampled system satisfying the Nyquist criterion (Hentschel 1997).

SPATIAL ARTIFACTS 351

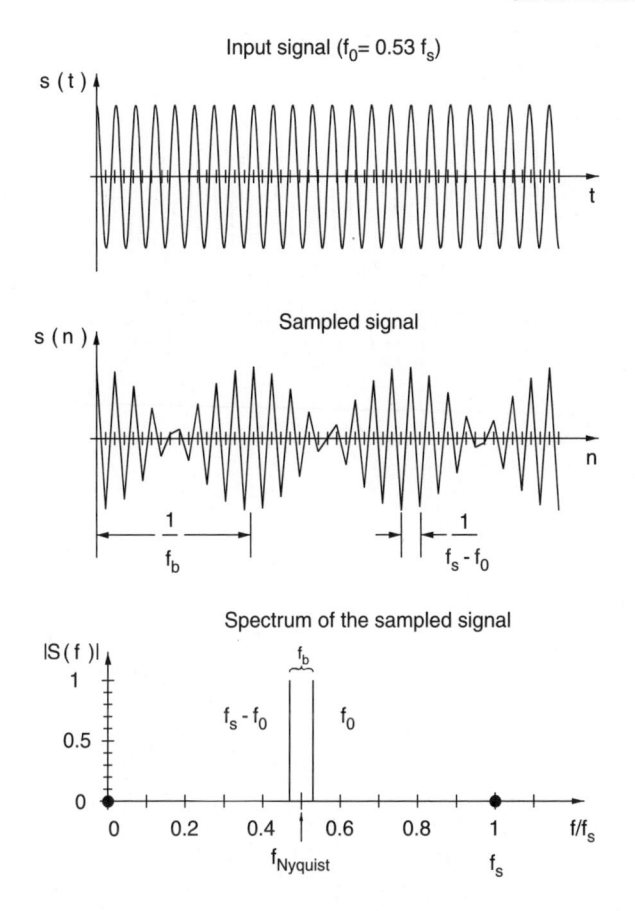

Figure 14.3 Origin of alias frequencies for case of $f_0 = 0.53\ f_s$ (Hentschel 1997).

($f_s = 1$/pitch). In a direct-view (non-projection) system, the signal in a raster scanned CRT incorporating a shadow mask is vertically sampled by the beam and two-dimensionally sampled by the shadow mask, a process which may result in visible Moiré in the image.

A rear projection screen, described in Section 7.2.3, typically has two different sampling elements: The lenticular high gain screen provides horizontal sampling while the Fresnel lens that is part of the screen provides radial sampling. Both of these samplings are optical in character. Since the spatial frequency is a vector quantity in the 2-D case, the Fresnel lens frequency vector and the screen frequency vector rotate with respect to each other, depending on azimuthal angle. At some position on the screen, the vector difference between the two is a spatial frequency that can become visible. The Moiré can be at any angle: horizontal, vertical or diagonal. It can also be at any location on the screen. One of the arts in screen design is the minimization of the visibility of this artifact.

With microdisplays, the higher the aperture ratio of the device, the less serious is the Moiré. For example, the light emanating from a fully-on one-dimensional pixelated display has the spatial characteristic shown in Figure 14.5. Its intensity distribution can be described by the one dimensional cosine series given by Equation (14.2) (Spiegel 1974).

$$L_x = L_0 \left\{ \frac{W}{P} + \frac{2}{\pi} \sum_{n=1}^{\infty} (-1)^n \sin\left(\frac{n\pi W}{P}\right) \cos\left(\frac{2n\pi x}{P}\right) \right\} \qquad (14.2)$$

352 IMAGE ARTIFACTS

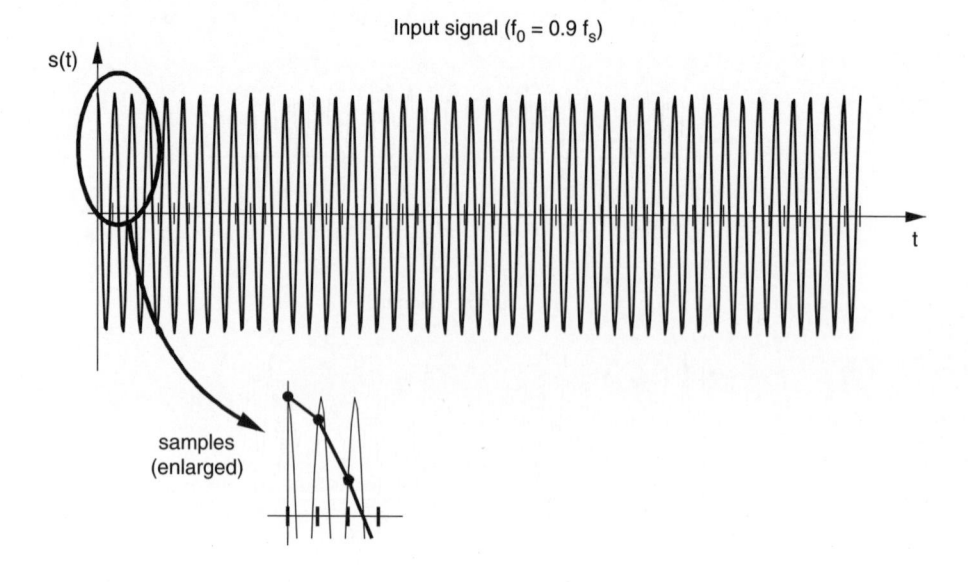

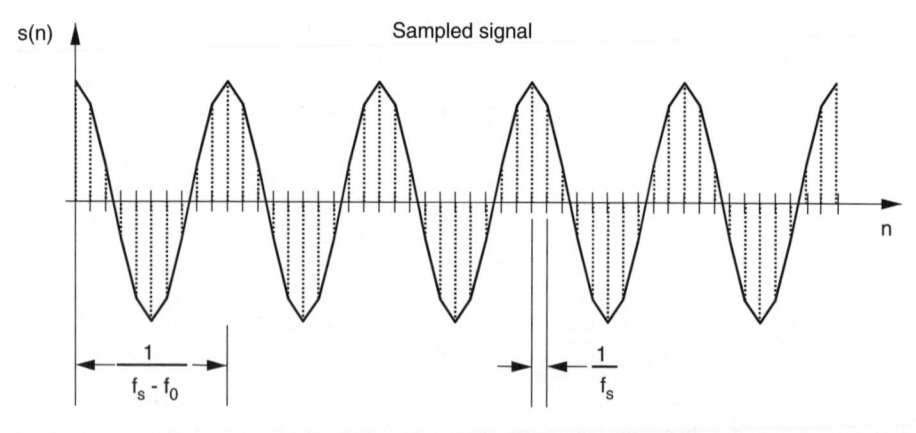

Figure 14.4 Aliasing produced by the sampling of a sine wave electric frequency f_0 by a pixelated display with a sampling frequency f_S, when $f_0 = 0.9 f_S$. Low pass filtering allows for connection of the individual samples (Hentschel 1997).

where
- L_x is the luminance of the microdisplay as a function of position along a horizontal line when displaying a white field,
- L_0 is the maximum luminance of the microdisplay, obtained with a 100 % aperture ratio (W = P),
- x is the horizontal position,
- W is the width of a pixel,
- P is the pixel pitch, and
- W/P is the 1 dimensional aperture ratio.

When W = P (100 % aperture ratio), the $\sin(n\pi W/P)$ term in each harmonic of Equation (14.2) goes to zero, and the DC term goes to 1. If these non-DC frequencies were the interfering sources in the projection system (Equation (14.1)), there would be no Moiré. As the aperture ratio is reduced

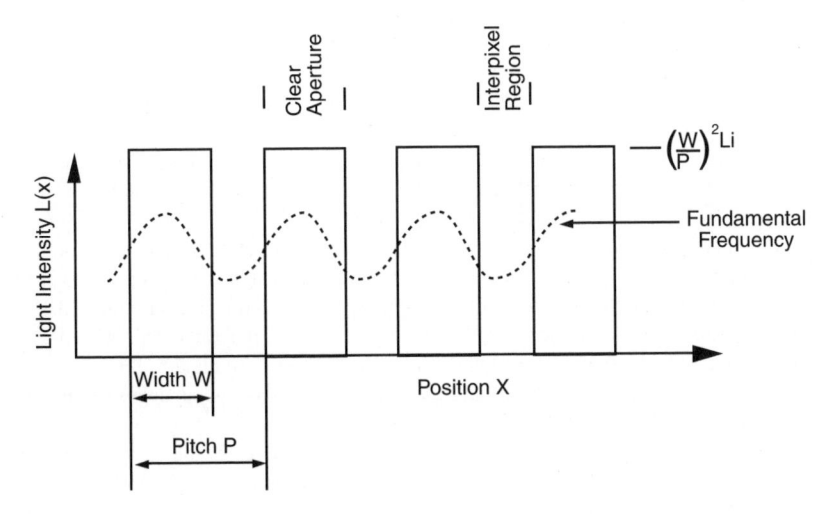

Figure 14.5 Spatial distribution of the light output from a fully-on 1-D pixelated display. For high spatial frequencies, the luminance pattern is low-pass filtered by the visual system, leaving only the fundamental frequency.

from 100 %, these higher frequency terms take on non-zero amplitudes and may now contribute to Moiré.

The observer can only perceive Moiré when its amplitude becomes large enough. The sensitivity of the eye to this modulation is a function of temporal frequency and spatial frequency, discussed below. Other factors, such as average luminance, also affect the sensitivity. It is necessary for the modulation to exceed a threshold and have a spatial frequency within the range visible to the eye to be visible.

The following methods would, therefore, minimize Moiré resulting from optical sampling:

1. Maximize aperture ratio of the sampling component in order to minimize the energy content in the alias and/or beat frequencies to reduce their visibility, and
2. Ensure all difference frequencies are high enough in spatial frequency so that the sensitivities for these frequencies by the human visual system are very low.

One way to ensure this latter condition in pixelated displays is for all components after the microdisplay to have as high a sampling frequency as possible (fine pitch). As an example, experience has shown that Moiré originating between these sources will not be significant if the pitch of the screen is < 0.4 times the apparent pitch of the microdisplay after it is projected onto a rear projection screen. Unfortunately, with high resolution devices on smaller screens, as might be encountered in a projection workstation monitor, it may be very difficult to meet this condition due to technological limitations in manufacturing fine pitch screens. If this becomes an intractable problem, it is possible the only solution would be to use a non-sampling screen technology.

14.1.2 Screen Door Effect or Pixelation

The screen door effect is the name given to the effect produced by the visibility of the interpixel regions in the image, i.e., the image appears to be viewed through a screen, or mesh, door. The spectrum of the screen luminance with this effect is described by the two dimensional version of Equation (14.2), taking into account the spatial modulation in both the x and y directions. However, no second spatial sampling element is involved in the screen door effect, as is the case with Moiré.

When a white picture is projected and the microdisplay has a 100 % aperture ratio, the viewer will perceive a uniform white field at the screen. If the aperture ratio is less than 100 %, the total average throughput decreases as $(W/P)^2$, assuming a square pixel with equal x and y pitches. In addition, the spectrum of the image on the screen will now contain all the frequencies corresponding to multiples of the pixel pitch.

For a specific microdisplay design, the visibility of the screen door effect is dependent on the microdisplay pitch, the magnification of the projected image, the viewing distance, the amplitudes of the cosine terms in Equation (14.2), and the sensitivity of the eye to see these frequencies (Figures 13.1 and 13.9). Smaller aperture ratios result in higher modulation amplitudes and make the screen door effect more visible. If the angular extent of a pitch cycle in the image is less than about 0.7 milliradian, as measured at a given viewing distance, it will probably not be visible to the viewer and no degradation of the image will be perceived.

When the angular extent of the image of the pitch subtends an angle resolvable by the viewer, this effect can noticeably degrade the visual quality of a display without necessarily degrading its nominal image resolution. The maximization of aperture ratio is normally done anyway to maximize light throughput. One way to get a high aperture ratio is to use a reflective microdisplay, since in this approach the structures that would be between pixels in transmissive devices can be hidden beneath the reflective mirror with minimal surface area requirements.

14.1.2.1 Depixelation

Several schemes have been proposed for decreasing the visibility of the screen door effect by depixelation of the image (Jutamulia *et al.* 1995). These methods include low pass optical filtering to remove the higher order spectra, methods to remove the phase coherence among the higher Fourier orders, and optical expansion of the aperture until it overlaps the aperture of the adjacent pixel.

Microlens arrays (one lens per pixel) have been proposed to enlarge the image of the clear aperture of each pixel so that the images generated by adjacent pixels are touching (Dolgoff 1992), which reduces the visibility of the interpixel regions. However, these latter techniques are often costly to implement since they require accurately registered optical microstructures. As pixel pitch is reduced, implementation of this method becomes difficult.

Low pass optical filtering is the simplest depixelation technique to implement. It can be done by slightly defocusing the image on the screen by moving the image slightly forward of the plane of the screen. This works on both Moiré and the screen door effect because the amplitude of the higher harmonics decreases faster than the fundamental (pixel spatial frequency) as the lens is defocused. Therefore, the amplitude and visibility of the Moiré and the screen door effect are reduced while affecting the image resolution from the projector only slightly. This method is rarely used, however, because viewers have shown a preference for best focus, even if this maximizes the screen door effect.

One of the more satisfactory depixelization methods involves spatial phase scrambling in the Fourier transform plane of the projection lens (Kelly 1992; Yang *et al.* 1996). This removes the pixel images with very little effect on the MTF at the pixel spatial frequency and essentially no effect on the projector throughput. While this technique would remove the pixelation, it is rarely used. The direction of product development is toward systems with higher pixel counts. With the advent of 1080 line microdisplays, screen door effect is an issue only with the largest of projected images, and then only when viewers are seated close to those images. As the resolution of the display is increased, the visibility of the screen door effect in an image of a given area is reduced due to the increase in the sampling frequency. The need for depixelation is thus reduced and the cost of implementing it becomes more difficult to justify.

In summary, among the ways to reduce the screen door effect are:

- Increase the aperture ratio of the microdisplay or other sampling component, or
- Increase the number of pixels in the microdisplay(s), or
- Reduce the system magnification so the individual pixels are no longer resolvable by the human visual system, or
- Depixelation of the image.

14.2 Temporal Artifacts

There are several general classes of temporal artifacts that have been observed in projection systems: flicker, motion artifacts, smear, color breakup on stationary images and color breakup on moving images. Flicker is the artifact that results from the ability of the eye to resolve, partially or completely, each address of a pixel. Motion artifacts include judder, which results from displaying a single image for too long a time period on the display. As a consequence, the display presents successive images of objects in motion which are too far apart to permit the eye to synthesize continuous motion and the motion appears discontinuous. Smear occurs when the response of the microdisplay is too slow and the image from one frame affects the displayed appearance of the next frame.

Color breakup on stationary images appears primarily in color-field sequential displays with single-panel architectures. This breakup is frequently explained by saccadic motion of the eye, which cause the color components of an image to be separated on the retina. Breakup on moving images are also seen with these architectures as well as in CRT projectors, where the three phosphors can have significantly different decay times. These artifacts are discussed in the following sections.

The perception of these artifacts depends on the temporal response of the eye. This is, in turn, a function of the screen luminance and spatial frequency, among other parameters. Figure 14.6 illustrates the temporal contrast sensitivity as a function of temporal frequency, for various spatial frequencies (Robson 1966). The eye response is reduced at lower luminance (not shown in this figure) and higher spatial frequencies.

14.2.1 Flicker

Flicker is a frequently observed artifact in display systems in which the screen illuminance is non-constant between information refreshes. For example, in CRT-based systems, the information is rewritten at the frame or field rate. When the phosphor is excited by the electron beam, a peak light output is achieved. The light output then decays with a characteristic determined by the phosphor until the electron beam once more excites the phosphor. The light output is thus modulated at the frame or field rate. Figure 14.7 illustrates the output from a CRT pixel as well as from a pixel in an active-matrix LCD light valve.

The temporal response of human visual system makes the observer sensitive primarily to the fundamental and mean of the Fourier spectrum (Pearson 1991). A parameter which quantifies this is the ripple ratio, which is the ratio of the amplitude of the fundamental term to the DC term in the temporal Fourier series of the luminance. The various terms in the ripple ratio are illustrated in Figure 14.7. This ratio can be high in CRT systems due to the large variation in output during a frame or field time.

Much of the ripple in LC cells is due to electrical leakage, either in the LC itself or the switching device in the active-matrix pixel. With low leakage, the light intensity passed by a well-made LC cell will be nearly constant during the frame. In microdisplay projection systems, the frame rate is usually high enough so that significant leakage does not occur between rewrites and the ripple ratio, and hence the perception of flicker, is low.

356 **IMAGE ARTIFACTS**

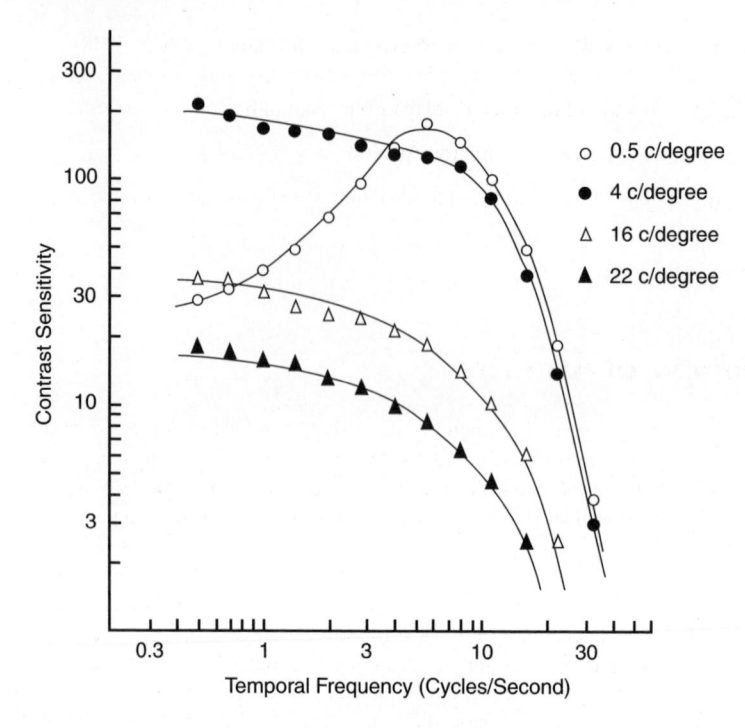

Figure 14.6 Temporal contrast sensitivity (reciprocal of threshold contrast) functions for different spatial frequencies (Robson 1966; reproduced with permission of Optical Society of America).

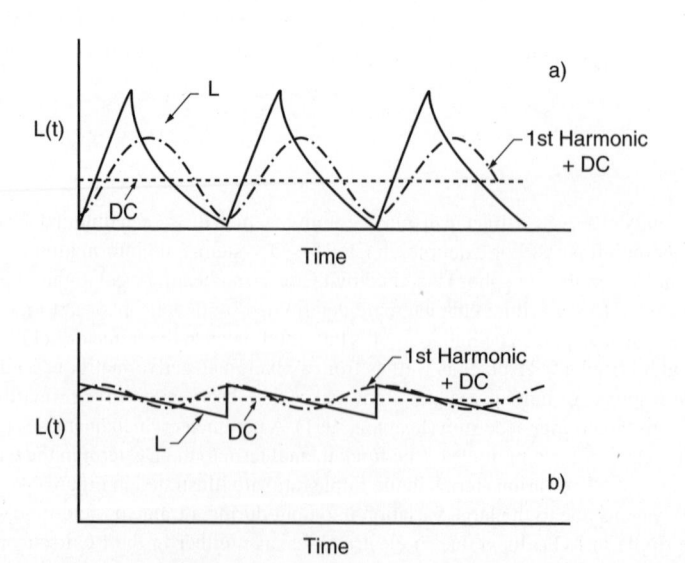

Figure 14.7 Illustration of the DC and fundamental frequency amplitudes of (a) an impulse-type luminous output as would be obtained with a CRT; and (b) a low electrical-leakage LC microdisplay projector (not to scale).

The flicker artifact is well known and there are many parameters in addition to the Fourier spectrum which affect its perceptibility (Boff and Lincoln 1988). Among the other variables, which are frequently not independent, are spatial frequency and target size, background intensity and type of surround, light adaptation, foveal vs peripheral viewing, and age and sex of the viewer. The primary factor that results in perceived flicker that the projector engineer must be concerned with is the dependence on screen luminance. In general, the higher the screen luminance the higher the percentage of the population that will perceive flicker at a given temporal frequency.

Figure 14.8 shows the contrast sensitivity for flicker as a function of the writing frequency, with the luminance level as a parameter (Boff and Lincoln 1988; Kelly 1961). In this plot, the higher the sensitivity the more easily flicker is observed. The contrast sensitivity decreases and the maximum sensitivity threshold shifts to lower frequencies as the average luminance is reduced.

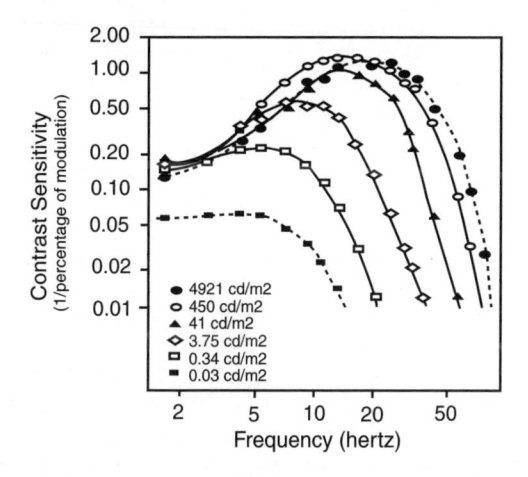

Figure 14.8 Sensitivity to flicker as a function of average luminance level and temporal modulation rate (Kelly 1961; reproduced with permission of the Optical Society of America).

Many of those involved in displays have rules of thumb about acceptable screen luminance to avoid flicker. The cinema industry limits the screen brightness of their film-based projection systems to about 60 cd/m^2. These projectors repeat each frame twice during the normal frame time of 1/24 second, producing a displayed frame rate of 48 per second. Operation at higher luminance at this frequency would produce unacceptable flicker. Computer display manufacturers know that to eliminate flicker in a CRT monitor for essentially all display users at all normal luminance levels, a refresh rate of up to 85 Hz may be needed.

Much of the data on the occurrence of flicker in displays is based on CRT systems. There is limited published data on flicker perception with LC microdisplay projection systems. In principle, these should show much reduced flicker since the ripple ratio of the projected image is much reduced compared with CRT projectors, as indicated in Figure 14.8. In addition, most projectors operate at 120 Hz or higher frame rates. The 120 Hz operation of a LC-based projector is necessary for the LC inversion, discussed in Section 4.2.6, to operate correctly. Single-panel DLP projectors operate at still higher field rates to avoid color breakup, to be discussed in Section 14.2.3.2. These high field rates mean flicker is unlikely to be perceived in DLP systems.

For a CRT projector the ripple ratio can be reduced by using a longer persistence phosphor. However, too long a persistence phosphor cannot be used because residual information from the previous frame would still be visible when the present frame is written. This would introduce smear in the displayed image. Experience has shown that to avoid smear in a CRT display, the phosphor light output must decay to less than 40 % of its peak value in a field time. In addition, the decay should be described by a single (exponential) time constant. It should not exhibit a change in time constant during the decay, which can be manifested as a very slow decay tail.

14.2.2 Image Smear, Judder and Motion Blur

Visible artifacts due to motion in the image are collectively called motion blur (Shimodaira 2003). Motion blur is the visible degradation of a moving image due to the temporal properties of a display. The most common manifestation of motion blur is reduced visible resolution of a moving object compared to the same object when stationary. Two particular examples of motion blur problems have long been present in video: smear and judder. This section will only discuss motion blur generated by the display. Other sources of motion blur such as the video or film camera or the video compression codec will not be discussed.

There are two factors to the temporal response of a display (Klompenhouwer 2006). First is the temporal response of the electro-optical modulator. When the temporal response of the modulator is too slow, the visible artifact is commonly called smear. Smear was a special issue in CRT projectors where the green phosphor could have a long decay time and a moving image would have a faint green tail following it.

This response time factor is normally a non-issue in MEMS devices such as the DLP or a MEMS linear array in a laser projector. Response speed of MEMS devices is typically measured in microseconds, but smear is an effect that has a time scale in the 5–30 mS or longer range.

Any LC effect fast enough for single-panel operation is also fast enough for motion blur not to be a result of the LC response time. Smear has been reported in a VAN LCoS microdisplay with nominal LC response time as short as 6 mS (Cuypers *et al.* 2003). In this cell, disclinations between the dark area and the light area at a moving edge extended the response time beyond what would be normal for this cell. The evaluation of motion blur in LC systems is also confused by the fact that the response time of the liquid crystal depends on the two drive levels involved. Typically, for a normally white display, the response time from white to black is the shortest and the transition time from black to white is longer. For a normally black display, the response time from black to white is shorter and from white to black is longer. When response time of a LC is cited, the value cited is often the total time from black to white and back to black again. Transitions between intermediate levels take longer than these full-scale transitions. In any event, it is commonly believed that LC response speeds in the 4 mS range and below will not contribute to image smear.

The second and currently the most common cause of motion blur is the sample-and-hold nature of modern microdisplay-based projection systems. This type of motion blur occurs because the eye typically tracks the average position of a moving object in an image. During the frame time with a microdisplay, the image is stationary on the screen, but since the eye is moving, the image is moving on the retina. This motion on the retina smears the image out on the retina and degrades the perceived resolution.

Judder is the term that specifically relates to the motion artifact derived from image sample-and-hold where very low field rates such as 24 or 30 Hz are upconverted to 60 Hz or higher by field or frame repeat (de Haan 1999). At these very low frequencies, the images are not smeared by the human visual system (HVS). Instead, the HVS at least partially perceives each individual image and the motion is perceived as jerky, rather than smooth. Judder has been described as the temporal equivalent of the 'jaggies' in an image.

One particularly severe form of judder occurs when 24Hz film material is converted to 60Hz video with a technique called 3:2 pull down. In this case, one frame of the film is shown for three video field times (50 mS) while the next frame of the film is shown for two video fields (33 mS). The total time to show these two frames of the movie is correct, 83 mS, but one of the frames is held in the image for an unusually long time, 50 mS. When the film is shown in its native format, the total time used to show one frame is 42 mS.

Impulse drive, where any individual pixel is only illuminated for a brief instant, does not normally show judder. This is the situation for an interlaced CRT projector showing video material in its native format that was derived from an interlaced camera. Since the pixel is only illuminated for an instant, the image is sharp on the retina even if the eye is tracking the motion and no smear occurs. Although the eye tracks the moving object while the picture is written to other pixels, the original pixel area is dark.

CRT projectors can suffer from smear due to the long persistence of the phosphor used, especially in the green tube, even when showing native interlace material. As discussed above, they also suffer from judder when displaying material where the source was not synchronized with the display. This occurs, for example, when cinema material is shown with 3:2 pull-down. A moving object is shown first at three separate instants in the same location, and then for two separate instants in a second location.

Note that a laser projector with either linear arrays and a single scanning mirror or a two-axis scanned laser system is an impulse drive type of display, like a CRT projector. Therefore these types of displays are expected to have low image smear, comparable to CRT projectors with very fast phosphors. Material with 3:2 pull down will show judder even with laser projectors, unless motion-compensated upconversion is used.

Moving picture response time (MPRT) (Igarashi *et al.* 2003; Klompenhouwer 2006) has become the normal metric of motion blur including image smear and judder. This involves the measurement of the duration of the impulse of the light at any individual pixel. Additional methods used include Temporal Impulse Response and the Temporal Point-Spread Function (Temporal PSF) (Pan *et al.* 2006). These methods are closely related to MPRT and typically give the same results as the MPRT.

MPRT is not necessarily the ultimate answer to evaluation of motion blur. Reportedly, it correlates well with black-white transitions and moving edges but does not correlate well with intermediate transitions (Sluyterman *et al.* 2007). It is also difficult to measure on real displays, especially displays with scanning backlight systems. MPRT appears to be especially important as a theoretical tool that can be used to predict whether a new drive scheme can reduce motion blur or not. Promising drive schemes ultimately need to be built and tested with human observers.

Motion blur reduction currently is a topic of intense research. For example, at SID '07, there were five sessions dedicated to various aspects of motion blur reduction in LCDs. Session 11 looked at backlight design, Sessions 12 and 17 looked at fast LCD effects, Session 26 looked at the measurement of motion blur and Session 61 looked at high frame rate display drive systems. There were additional papers in other sessions, including the poster session, related to motion blur reduction. While most of the current research into motion blur reduction is oriented toward direct-view LCD television systems, much of this research also applies directly to projection systems.

There are two main techniques to reduce image motion blur caused by the sample-and-hold nature of a microdisplay-based projector, black data insertion (BDI) and high frame rate driving (HFR). A third technique, scanning backlights, currently has only been applied to direct-view LCD although it could, in principle, be applied to the illumination path of a projector as well (Schmidt 1995).

Black data insertion involves inserting black fields between video fields. This reduces the hold time of an individual field and therefore reduces the motion blur. This is essentially doing what CRT and scanned laser projectors do naturally, write the picture only briefly and have the screen dark in-between images. Film projectors, for example, do this to a minor extent. Since showing a film at 24 Hz would cause severe flicker, typically each frame of the film is shown twice, with a 50 % duty cycle. In other words, a frame of the film is shown for 10.4 mS, black is shown for 10.4 mS, the frame is shown again for another 10.4 mS and then black is shown again for the final 10.4 mS, for a total of 41.7 mS. This cycle is then repeated for the next frame of the film. The hold time in this sample-and-hold display is not, however, 41.7 mS. The hold time in terms of MPRT starts when the frame is first shown and ends when the frame finally disappears for the second time 31.3 mS later. In video terms, this would be black frame insertion of 25 % (10.4 mS out of 41.7 mS) and would significantly reduce visible judder.

The problem with black data insertion with lamp-based projection systems is the reduced duty cycle. In a lamp-based system where the lamp is on continuously, a 25 % black frame insertion period represents a 25 % loss of light. In the film case, it represents a 50 % light loss, since an additional black frame is inserted in the middle of the displayed field. This additional frame insertion is needed for flicker reduction and does not help with motion blur reduction.

A better solution from a projector point of view is high frame rate driving with motion compensated upconversion. Some projectors or video sources such as DVD players already have this built into them in the form of a 'Movie Mode'. A projector without a movie mode shows incoming film material with the 3:2 pull-down of the video source. In the movie mode, however, the 3:2 pull-down is automatically detected by the video signal processing circuits, the 3:2 pull down is removed and the resulting 24 Hz video is upconverted to 60 Hz progressive. For this to be useful in reducing motion blur, this up-conversion must be done with motion compensation.

Judder may still visible in both 50 Hz and 60 Hz displays, even with motion compensated up-conversion from the incoming material to 50 Hz or 60 Hz progressive. Most visible motion blur due to judder is eliminated if motion compensated up-conversion to 100 Hz or 120 Hz is used. If still higher frame rates are needed, for example to eliminate the color breakup discussed in the next section, this additional upconversion can normally be done by frame repeat and without motion compensation.

14.2.3 Artifacts in Color-field Sequential Systems

Projectors that use a single microdisplay or other image generation device to generate a full color image with color sequential technology (see Section 10.4.2) suffer from an artifact called color breakup. This artifact is also called the color flash effect, color rainbow effect and color fringing. This has long been recognized as a problem (Arend *et al.* 1994) and for a time it was believed the solution would involve image processing (Baron *et al.* 1996). Unfortunately, there are two root causes of color breakup and image processing only affects the lesser of them, color breakup associated with motion in the image. The major source of color breakup is motion of the eye and occurs even with still images. This motion of the eye, called saccadic motion, is very rapid and occurs at unpredictable times. Therefore, this problem cannot be solved with image processing.

14.2.3.1 Color breakup due to motion in the image

The first color-field sequential artifact is color fringing on moving edges. When there is a bright object moving in a field sequential display that has one RGB cycle per field, color fringes may be visible on the leading and trailing edges of the object. These fringes represent a color version of the judder problem as discussed above. The 'taking' camera, whether film or electronic, takes all three colors simultaneously. The field sequential display displays them at slightly different times. As the eye tracks the moving object, one color slightly leads the eye, another is centered at the point of the eye's gaze and the third slightly lags the eye. Therefore, the eye sees the red, green and blue images on slightly different portions of the retina, and for the moving object, the images may not be converged.

Observing this artifact is difficult and it is not normally noticed unless special video source material with bright moving objects is used in the demonstration. The artifact can also be seen on CRT projection systems, where it is due to the differences in persistence between the green phosphor and the red and blue phosphors. The visibility of this effect is suppressed in complex and/or colorful images.

If suppression is desired, a technique can be employed in which successive color fields are presented in forward and then reverse order, e.g., BGRRGB. As the eye tracks the moving object, it sees each color in both a leading and lagging position, relative to the nominal eye track position. The eye averages the leading and lagging fields together to give the same location to all three colors, minimizing color fringing at the edges. Motion dependent image processing can also reduce or eliminate the visibility of this artifact (Kurita and Kondo 2001).

In practice, this artifact is normally ignored by projector designers. The frame rates required to reduce color breakup due to saccadic motion of the eye to acceptable levels are high enough to render motion dependent color breakup nearly invisible.

14.2.3.2 Color breakup due to saccadic motion of the eye

In sequential systems showing a stationary image, the stationary eye would be presented with red, green, and blue images in time sequence at the same position on the retina. If the frame rate is sufficiently high, i.e. high enough to supress the perception of flicker, these three images would be fused by the eye/brain system into a single full color image with no color fringes. However, the eye is not stationary but rather moves in jumps called saccades. This movement can result in the time-separated colors being spatially separated on the retina, resulting in visible red, green and/or blue fringes at the edge of bright objects.

This simple explanation would indicate that color break-up is a severe problem in all color sequential projectors. There is, however, a mechanism called saccadic suppression in the eye/brain combination to suppress vision during saccades (Yohso and Ukai 2006). When these mechanisms are fully functional, nothing is perceived during a saccade, including the expected separation of red, green and blue. The mechanism of saccadic suppression is not perfect and varies by individual, allowing most people to see some color breakup and a few individuals to see severe color breakup. Overall, color breakup is not as severe a problem as might otherwise be expected. There are a few individuals, however, where color breakup is a serious problem and prevents them from watching most color sequential displays.

The visualization of these artifacts is enhanced by a moving object, such as a moving hand passing through the space between the viewer's eye and the screen. Other ways to see them is by deliberately darting the eye to view another part of the screen. A startling event, such as the sudden ringing of the telephone, can also cause the artifact to become visible.

A number of factors related to color breakup were documented by Philips to guide potential customers in the design of single-panel LCoS systems (Brennesholtz 2002):

1. Extremely high color field rates are needed to totally eliminate color breakup. (>1500 color fields per second) (Post *et al.* 1998; Wada *et al.* 1999).

2. Color breakup is more severe for some images than for other images. Color breakup is more severe for the high-contrast, stationary images typical of computer monitors than it is for lower contrast moving images typical of video. White on black is the worst case.

3. A 180 Hz frame rate (540 Hz color field rate) is high enough so that many people do not notice color breakup at all in video images and few seriously object to it. Most people who do not initially perceive color breakup can be trained to see it. Satisfactory images can also be generated at 150 Hz (450 color field rate). Philips did not recommend the use of 120 Hz (360 Hz field rate) or less for color sequential LCoS systems.

4. At 180 Hz frame rate (540 Hz color field rate) color breakup is due entirely to motion of the eye, not motion of the object in the scene. Therefore, image processing to compensate for object motion will do little to reduce perceived color breakup.

5. Color breakup is more severe at high screen luminance and low ambient brightness (Mori *et al.* 1999).

6. Color breakup occurs whether at any given instant there is one color on the screen (most color wheel systems, as described in Section 10.4.2.2), two colors on different portions of the screen (most drum systems, as described in Section 10.4.2.5) or three colors on different portions of the screen (Philips scrolling color as described in Section 10.4.2.6). No system has a documented advantage over the others.

7. Multiple primary color projectors at the same color field rate as three primary color projectors have less visible color breakup (Eliav *et al.* 2005).

It is considered impossible to eliminate color breakup completely in a color sequential projector that uses a color wheel because of the limitations on the rotation rates of the color wheel and the losses when the spoke between colors is in the panel region. LED and laser based illumination systems, on the other hand, can easily be pulsed at 1500 Hz or even faster if necessary. Initial LED-based rear projection systems have had little or no visible color breakup. Reportedly these LED systems have used color field rates up to 5000 Hz. This field rate is possible in the DLP and similar MEMS microdisplays by treating each bit-plane or small group of bit-planes as a separate color field (Masumoto et al. 2003). By interleaving the red, green and blue bit planes and synchronizing them with the LED illumination, very high effective field rates can be achieved. TN-LCoS is not fast enough to give many more than 540 color fields per second. While FLC-LCoS can achieve speeds comparable to the DLP, most of the available devices of this type use pulse width modulation to achieve gray scale instead of bit plane addressing. If color breakup proves to be a problem in single-panel LCoS projectors, perhaps bit-plane addressed LCoS will appear in the future to reduce this effect.

Quantitative evaluation of color breakup is possible (Järvenpää 2004; Lee et al. 2006) but the techniques are not well defined yet. In general, these measurement techniques involve imaging the system with a moving camera or mirror to simulate a saccade. Since the camera does not have the saccadic suppression mechanism, this measurement will give only the worst case and will not represent the typical viewer experience. To the best of the authors' knowledge, no standards body has issued an accepted test method for color breakup.

References

Arend, L., Lubin, J., Gille, J. and Larimer, J (1994) Color breakup in sequentially scanned LCDs. *SID Intl. Symp. Digest of Technical Papers*, 201–4.

Baron, P., Monnier, P., Nagy, A.L., Post, D.L., Christianson, L. *et al.* (1996) Can motion compensation eliminate color breakup of moving objects in field-sequential color displays? *SID Intl. Symp. Digest of Technical Papers*, 843–6.

Boff, K. and Lincoln, J. (eds) (1988) *Engineering Data Compendium. Human Perception and Performance*, Harry Armstrong Laboratory, Wright -Patterson Air Force Base, Ohio. Vol. I, Section 1.5 provides a summary and reference list for temporal effects in human vision.

Brennesholtz, M.S. (2002) *Color breakup in color sequential projectors*. Philips LCoS Application Note AN011, December.

Cuypers, D., Van Doorselaer, G., Van Den Steen, J., De Smet, H. and Van Calster, A. (2003) WXGA LCOS projection panel with vertically aligned nematic LC. *Proc. of the 10th Intl. Display Workshops (IDW)*, Fukuoka, Japan, Paper LAD1-1, 1541–4.

de Haan, G. (1999) Video Format Conversion, *SID Intl. Symp. Digest of Technical Papers*, Paper 7-2, 52–5.

Dolgoff, E. (1992) Optical depixelation for electronic image projection. E. Schlam and M. Slusarczuk (eds), *Proc. SPIE: High Resolution Displays and Projection Systems* **1664**: 160–71.

Eliav, D., Langendijk, E.H.A., Swinkels, S. and Baruchi, I. (2005) Suppression of color breakup in color-sequential multi-primary projection displays. *SID Intl. Symp. Digest of Technical Papers*, Paper 47.3, 1510–13.

Hentschel, C. (1997) *Moiré in monitors*. Philips Research-USA Technical Report TR-97-014.

Hentschel, C. (2001) Video moiré cancellation filter for high-resolution CRTs. *IEEE Transactions on Consumer Electronics* **47**(1): 16–24.

Igarashi, Y., Yamamoto, T., Tanaka, Y., Someya, J., Nakakura, Y. *et al.* (2003) Proposal of the perceptive parameter motion picture response time (MPRT). *SID Intl. Symp. Digest of Technical Papers*, Paper 31.2, 1039–41.

Järvenpää, T. (2004) Measuring color breakup of stationary images in field-sequential-color displays, *SID Intl. Symp. Digest of Technical Papers*, Paper 7.2, 82–5.

Jutamulia, S., Toyoda, S. and Ichihashi, Y. (1995) Removal of pixel structure in liquid crystal projection display. *Proc. SPIE* **2407**: 168–75.

Kelly, D.H. (1961) Visual response to time-dependent stimuli. I. Amplitude sensitivity measurements. *J. Opt. Soc. Am.* **51**: 422–9.

Kelly, S. (1992) Removing the pixel structure from projected images using a transparent spatial filter. *Proc. SPIE* **1664**: 153–9.

Klompenhouwer, M.A. (2006) Comparison of LCD motion blur reduction methods using temporal impulse response and MPRT. *SID Intl. Symp. Digest of Technical Papers*, Paper 54.1, 1700–3.

REFERENCES

Kurita, T. and Kondo, T. (2001) Effect of motion compensation on color breakup reduction and consideration of its use for cinema displays. *Proc. SID Asia Display-IDE*, Paper VHF-1, 1629–32.

Lee, J., Jun, T., Lee, J., Han, J., Souk, J.H. (2006) Noble measurement method for color breakup artifact in FPDs. *IMID/IDMC Digest*, Society for Information Display, Paper 5-3, 92–7.

Masumoto, Y., Wada, M. and Fushimi, Y. (2003) Simplified projection TV optics using DMD. , *Proc. of the 10th Intl. Display Workshops (IDW '03)*, Fukuoka, Japan, Paper LAD3-1, 1573–6.

Mori, M., Hatada, T., Ishikawa, K., Saishouji, T., Wada, O. et al. (1999) Mechanism of color breakup in field-sequential-color projectors, *SID Int. Symp. Digest of Technical Papers*, 350–3.

Pan, H., Feng, X.-F. and Daly, S. (2006) LCD motion blur analysis and modeling based on temporal PSF. *SID Intl. Symp. Digest of Technical Papers*, Paper 54.2, 1704–7.

Pearson, R.A. (1991) Predicting VDT flicker. *Information Display*, # 7 & 8, 22–6.

Post, D.L., Nagy, A.L., Monnier, P. and Calhoun, C.S. (1998) Predicting color breakup on field-sequential displays: Part 2. *SID Intl. Symp. Digest of Technical Papers*, 1037–40.

Robson, J.G. (1966) Spatial and temporal contrast-sensitivity functions of the visual system. *J. Opt. Soc. Amer.* **56**: 1141–2.

Schmidt, J.H. (1995) *Transmissive Polygonal Optical Scanning of Illumination for Light Valve Video Projector*, US Patent 5,428,467, issued 27 June 1995.

Shimodaira, Y. (2003) Fundamental phenomena underlying artifacts induced by image motion and the solutions for decreasing the artifacts on FPDs. *SID Intl. Symp. Digest of Technical Papers*, Paper 31.1, 1034–7.

Sluyterman, A.A.S. and van der Poel, W.A.J.A. (2007) Motion-fidelity improvement at a frame rate of 120 Hz via the use of a scanning backlight. *SID Intl. Symp. Digest of Technical Papers*, Paper 11.1, 127–30.

Spiegel, M.R. (1974) *Theory and Problems of Fourier Analysis, with Applications to Boundary Value Problems*. Schaum's Outline Series, McGraw-Hill.

Wada, O., Nakamura, J., Ishikawa, K. and Hatada, T. (1999) Analysis of color breakup in field-sequential color projection system for large area displays. *Proc. of the Sixth Intl. Display Workshops (IDW '99)*, Japan, 993–6.

Watkinson, J. (1994) *The Art of Digital Video*, 2nd edn. Oxford: Focal Press.

Yang, X., Li, N. and Jutamulia, S. (1996) Liquid crystal image depixelation by spatial phase scrambling. Ming Wu (ed.), *Proc. SPIE: Projection Displays II*, **2650**: 149–59.

Yohso, A. and Ukai, K. (2006) How color break-up occurs in the human visual system: mechanism of color break-up phenomenon, *SID Intl. Symp. Digest of Technical Papers*, **XXXVII**, Paper 25.2, 1225–8.

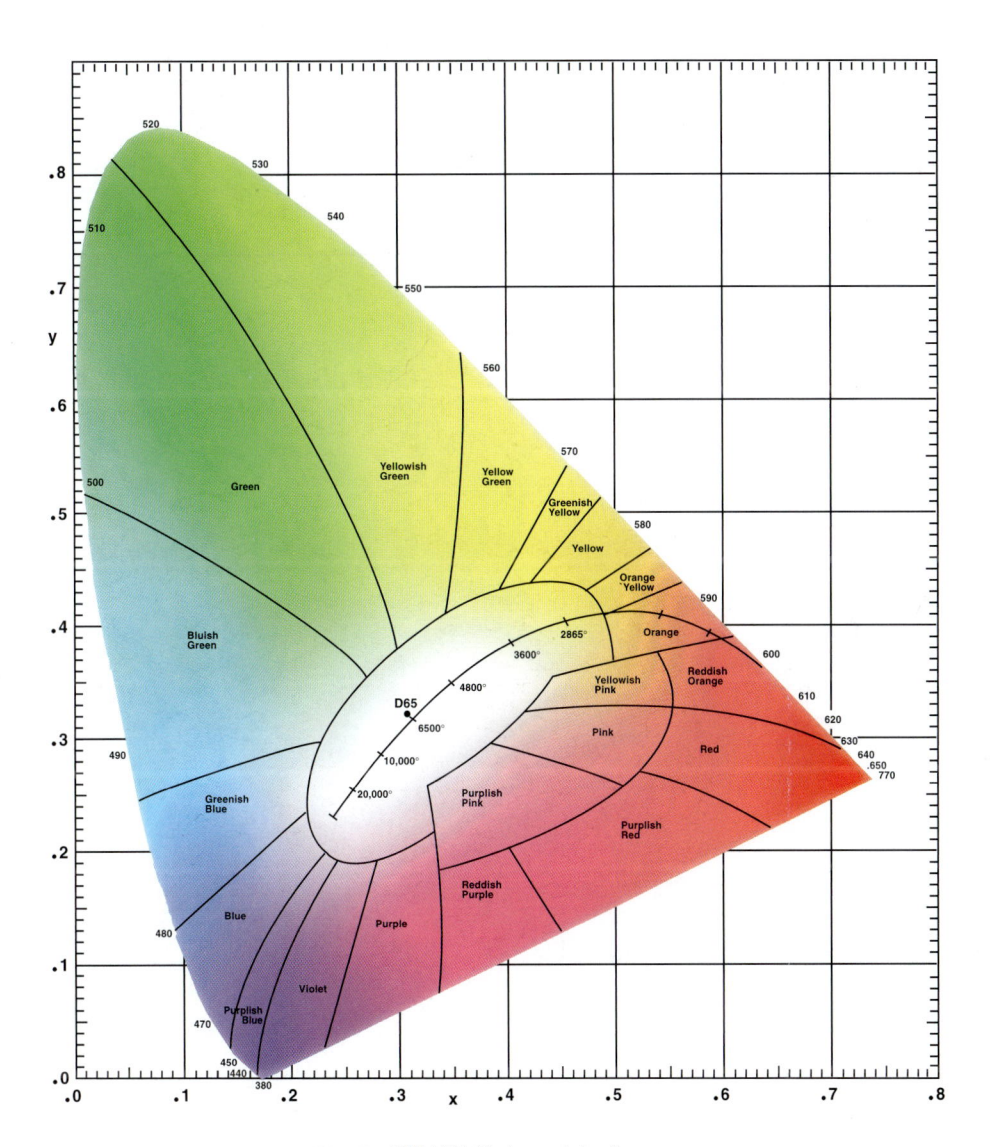

Plate I. CIE 1931 2° chromaticity diagram.

Appendix 1 Radiometry and Photometry

Radiometry is the subject concerned with the measurement of the energy content in optical radiation fields and how this energy flows through optical systems. Photometry is the subject of the measurement of the energy content in optical radiation fields that can induce a visual response and the quantification of that response. In this appendix, the measures and units of the various quantities used in radiometry and photometry will be described.

The sensitivity of the human visual system has a strong dependence on the wavelength of radiant energy. Figure A1.1 shows the spectral response of the visual system when the eye is light adapted (Kingslake 1965). This adaptation, called photopic vision, has a peak response at 555 nm. When the eye is dark adapted, the spectral response of the eye is changed and the adaptation is called scotopic vision. Photopic response occurs for luminance levels $\geq 3\,\text{cd}/\text{m}^2$ and scotopic response occurs for luminance levels $\leq 3 \times 10^{-5}\,\text{cd}/\text{m}^2$. (These units are discussed below). Mesopic vision occurs if the visual system is adapted to a luminance between these two levels.

Table A1.1 summarizes many of the quantities and units used in radiometry and photometry, and their equivalencies, using a notation similar to Boyd (Boyd 1983). Definitions and discussions of these terms follow. Each of the quantities defined in this table can be a function of wavelength. While both photometric and radiometric quantities can have spectral dependencies, the wavelength dependence is usually only given for the radiometric terms. The spectrally dependent terms would have an additional unit of $(\text{nm})^{-1}$, signifying the measurement over a limited wavelength. For example, Q, with units of joules, would go to $Q(\lambda)$, with units of joules/nm.

Energy and flux: The basic measure of radiometric energy Q is the joule (J) and that of radiometric flux, or power (Φ) is the watt (W). The equivalent luminous measures are the talbot (Q_v, no abbreviation) for luminous energy and the lumen (Φ_v, abbreviated lm) for luminous flux. The lumen is the luminous flux corresponding to 1/683 W of radiant flux, or power, at the wavelength of peak photopic vision,

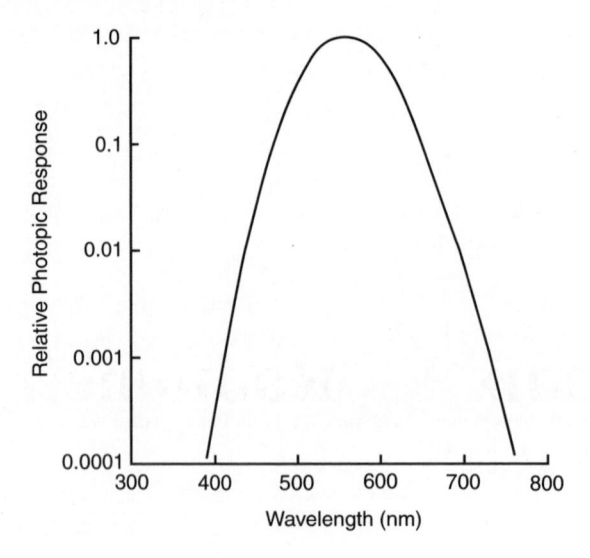

Figure A1.1 Relative response of human visual system as a function of wavelength.

Table A1.1 SI Radiometric and photometric units and quantities and their equivalencies.

Radiometric units/quantities				photometric units/quantities			
Quantity	Symbol	Definition	Units	Quantity	Symbol	Definition	Units
Radiant energy	Q	$\int \Phi dt$	J	Luminous energy	Q_V	$\int \Phi_v dt$	talbot = lm s
Radiant energy density	U	$\dfrac{dQ}{dV}$	$\dfrac{J}{m^3}$	Luminous density	U_V	$\dfrac{dQ_v}{dV}$	$\dfrac{lm\ s}{m^3}$
Radiant flux	Φ	$\dfrac{dQ}{dt}$	W	Luminous flux	Φ_V	$\dfrac{dQ_v}{dt}$	lm
Irradiance	E	$\dfrac{d\Phi}{dA}$	$\dfrac{W}{m^2}$	Illuminance	E_V	$\dfrac{d\Phi_v}{dA}$	$\dfrac{lm}{m^2} = lx$
Radiant exitance	M	$\dfrac{d\Phi}{dA}$	$\dfrac{W}{m^2}$	Luminous exitance	M_V	$\dfrac{d\Phi_v}{dA}$	$\dfrac{lm}{m^2} = lx$
Radiant intensity	I	$\dfrac{d\Phi}{d\Omega}$	$\dfrac{W}{sr}$	Luminous intensity	I_V	$\dfrac{d\Phi_v}{d\Omega}$	$\dfrac{lm}{sr} = cd$
Radiance	L	$\dfrac{d^2\Phi}{dA_{proj}d\Omega}$	$\dfrac{W}{m^2\ sr}$	Luminance	L_V	$\dfrac{d^2\Phi_v}{dA_{proj}d\Omega}$	$\dfrac{cd}{m^2} = nt$

555 nm. Thus, 1 W$_{555nm}$ = 683 lm. At wavelengths above and below 555 nm additional radiometric watts are required to elicit the same photometric response as does 1 lm at 555 nm. Thus, at a wavelength of about 475 nm, where the relative response is about 0.1 that of the peak response, 1 W$_{475nm}$ ≈ 68.3 lm. For a continuous spectrum, the luminous flux Φ_V is related to the spectrally varying radiant flux $\Phi(\lambda)$ by

$$\Phi_V = 683 \int \overline{Y}(\lambda)\Phi(\lambda)d\lambda \tag{A1.1}$$

where $\Phi(\lambda)$ is the spectral radiant flux contributing to Φ_v and $\overline{Y}(\lambda)$ is the relative spectral luminous efficiency function for photopic vision, shown in Figure A1.1.

Irradiance and illuminance: The irradiance E is the flux per unit area received from a source by or passing through a surface element and it is measured in units of W/m^2. It is alternately called the radiant incidance and it is expressed by

$$E = \frac{d\Phi}{dA} \quad (A1.2)$$

The photometric equivalent E_V, or illuminance, is measured in lm/m^2 or lux (lx). In English units, the unit is the foot-candle (1 fc $\equiv$ 1 lm/ft^2).

Radiant and luminous exitance: The exitances M and M_V are the radiant and luminous fluxes per unit area leaving the surface of a source of radiation. The mathematical relation between M and Φ is similar to that between E and Φ (Equation A1.2). The surface can be an active emitter or it can be passive, i.e. a reflector and/or transmitter of incident radiation. For a passive, non-emitting surface the exitance is less than the irradiance or illuminance by a factor given by the absorption of the radiation. The quantities M and M_V have no angular dependencies. The unit for luminous exitance is the same as for illuminance, i.e. lx or foot candle.

Radiant and luminous intensity: The angular dependencies of radiation from a surface are contained in the radiant and luminous intensities, I and I_V, and the radiance and luminance, which is discussed in the next section. The intensities are the measures of the flux emitted by an entire source per unit solid angle and are given by

$$I = \frac{d\Phi}{d\Omega} \quad (A1.3)$$

A similar equation defines the relationship between I_V and Φ_V. The intensities I and I_V describe the amount of radiant and luminous flux, respectively, leaving a surface (active or passive) into a solid angle $d\Omega$. The unit for luminous intensity is the candela, or lm/sr. I and I_V are particularly useful descriptors when the dimensions of the emitting area are small compared with distance between this area and the detector of the radiation. Under these conditions, this radiation source approximates a point source. For a point source, the irradiance E of an elemental area dA illuminated at an angle θ with respect to the normal by an emitting surface with intensity I is

$$E = \frac{I \cos \theta}{r^2} \quad (A1.4)$$

i.e. the irradiance varies as $\cos(\theta)$ and $1/r^2$.

Radiance and luminance: These quantities are the measures of the fluxes, radiant and luminous, respectively, per unit solid angle per unit area leaving an active or passive source. Figure A1.2 illustrates the

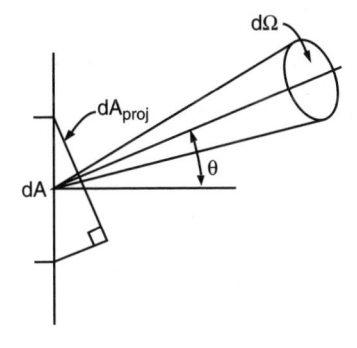

Figure A1.2 Geometry of source of radiation as a function of viewing angle.

geometry of the source when the observation is made at an angle θ with respect to the normal to the source surface.

The effective elemental area, also called the projected area dA_{proj} of the source when viewed at an angle θ is

$$dA_{PROJ} = dA \cos\theta \qquad (A1.5)$$

The radiance is the flux from a source at an angle θ per unit projected area (Equation (A1.5)) per unit solid angle.

$$L = \frac{d^2\Phi}{dA_{PROJ}d\Omega} = \frac{d^2\Phi}{dAd\Omega\cos\theta} \qquad (A1.6)$$

The SI unit for luminance (L_V) is the candela per square meter, abbreviated cd/m^2 (1 cd/m^2 = 1 lm/m^2/sr). The name 'nits' has historically been used for this unit. While it has been officially discarded by the CIE, the term nits is sometimes encountered by the working projection engineer. In the Imperial system, there is no corresponding luminance unit and luminance is measured in cd/ft^2. This unit is often (inappropriately) called the foot-Lambert (see below) and is the only Imperial unit likely to be encountered by a projection engineer. While there is not an exact 1:1 equivalence in defining the units, the conversion factor of 1 foot-Lambert = 3.426 cd/m^2 is commonly used.

As indicated in Section 13.3.1, radiance or luminance is sometimes improperly called brightness. Brightness properly refers to the amount of light as perceived by the human visual system (HVS). Luminance is the equivalent unit measured by an instrument. When the luminance scale was created in 1924, the intention was to have a 1:1 correspondence between brightness and luminance. Due to the complexity of the HVS, the correlation is not perfect and the differences must occasionally be taken into account by the display engineer.

Lambertian sources: A Lambertian surface is one which is equally luminous in all directions, i.e. its radiance or luminance is independent of viewing angle. Many incoherent sources of radiation, including scattering surfaces, approximate Lambertian surfaces. For a Lambertian surface, Equation (A1.6) for the Lambertian radiance or luminance becomes

$$L_l = \frac{d^2\Phi_l r^2}{dA_l dA_r} \qquad (A1.7)$$

where Φ_l is the flux from a Lambertian surface of elemental area dA_l incident on a receiver of elemental area dA_r at a distance r. The exitance M_l for a Lambertian surface is related to the luminance L_l by

$$M_l = \pi L_l \qquad (A1.8)$$

Special photometric units are employed for Lambertian luminance: the foot-Lambert (fL), the apostilb (asb), and the Lambert (L).

$$fL \equiv \frac{1}{\pi} cd/ft^2$$

$$asb \equiv \frac{1}{\pi} cd/m^2$$

$$L \equiv \frac{1}{\pi} cd/cm^2$$

In general these units should not be used, in part because most surfaces are not Lambertian. As mentioned above, foot-Lambert is still occasionally used and converted to cd/m^2 with a conversion factor of 3.426 even when non-Lambertian surfaces are used.

References

Kingslake, R. (1965) Applied Optics and Optical Engineering, Vol.1. Light: Its Generation and Modification. New York: Academic Press.

Boyd, R.W. (1983) Radiometry and the Detection of Optical Radiation. New York: John Wiley & Sons, Inc.

Appendix 2 Colorimetry

A2.1 Introduction

Projection systems can be broken down into three categories, according to their ability to reproduce color:

1. Monochrome,
2. Partial color and
3. Full color.

Monochrome systems (i.e. black and white) produce only a single color, normally white. The intensity of this color is spatially modulated to produce the image, but the color does not change. Monochrome systems normally have gray scale, i.e. they can reproduce intermediate levels of brightness between total black and maximum white. The number of levels of gray scale available determines if the system can reproduce accurately black and white photographs. The use of 8 bits of gamma-corrected video with 256 levels is normally considered the minimum acceptable for full gray scale reproduction. Monochrome systems can be made with very high spatial resolution and are used for applications like medical imaging or document retrieval.

Partial color systems can produce a small number of colors, either in discrete steps or as part of a limited range. The complete color range cannot be produced, and photo-realistic images cannot be produced. An example of this type of system is one that could produce red, yellow and white, but not green or blue. Color in a system like this most often used to highlight critical information to bring it to the attention of an operator, as in an air traffic control situation.

Most modern projection systems are full color systems: they produce a wide range of colors, and all colors can be produced with a variety of brightnesses. Photorealistic images can be produced. Television is an example of a full color system. A minimum of 8 bits per color for a total of 24 bits, about

---

Projection Displays, Second Edition   M. Brennesholtz, E. Stupp
© 2008 John Wiley & Sons, Ltd

372 COLORIMETRY

16 million possible combinations, are required for full color operation. However, 16 bits, when partitioned among the three colors in a computer graphic application, produce satisfactory results for many applications.

This appendix will cover the basic principals of color, from the point of view of how color principals are actually used by video projection engineers. For a more general discussion of color, including non-CIE colorimetry systems, the reader is referred to one of the many excellent references on color (Wyszecki and Stiles 1982).

A2.2   CIE 1931 2° Color Matching Functions

In order to understand color in a video system, it is first necessary to be able to quantify it. The normal method for quantifying color in video work was established by the Commission International de l'Eclairage (CIE) in 1931. This established three tabular functions, based on psychophysical experiments, called $\overline{X}$, $\overline{Y}$ and $\overline{Z}$ (pronounced x-bar, y-bar and z-bar). The CIE standard has been updated over the years to include narrow (2°) and wide (10°) fields of view and separate functions for low light levels (scotopic) and high light levels (photopic).

The CIE also updated the functions in 1964 to correct a known problem in the blue region of the $\overline{Y}$ function. This problem causes the CIE 1931 functions to understate the response of the human eye to blue light. The video industry, however, in virtually all cases, continues to use the original 1931 2° photopic functions. These are the functions used throughout this appendix and book. They are shown graphically in Figure A2.1 and are tabulated at 5 nm intervals in Section A2.8.

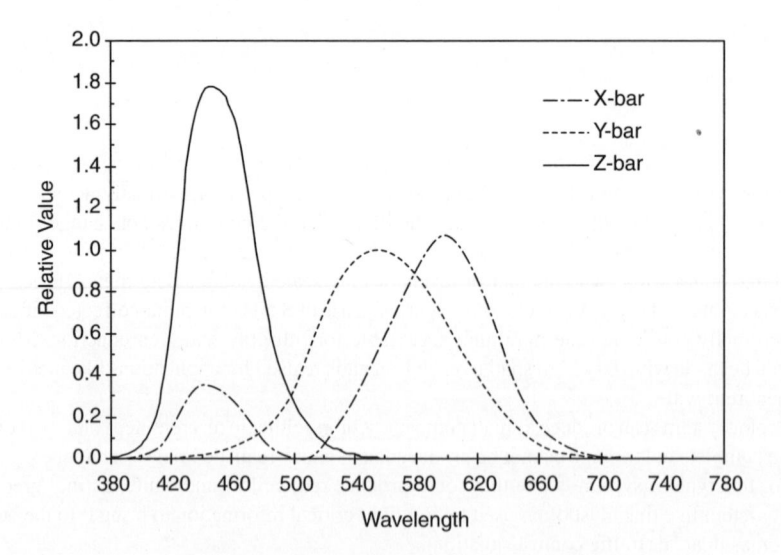

Figure A2.1   CIE 1931 2° color-matching functions.

While the CIE functions are tabulated over the wavelength range of 360–830 nm, the human eye has extremely low sensitivity to light outside the range from 380–730 nm. Even if color calculations are done over the still more restricted range of 400–700 nm, this will rarely introduce significant errors in color calculations related to video projection systems.

The CIE $\overline{Y}$ curve is also the luminance matching curve. Luminance is discussed in Appendix 1. When the 1931 CIE color system was derived, it was specifically designed to match the pre-existing $\overline{Y}$ curve from 1924 (Fairman *et al.* 1997).

A2.3 Calculation of Color

The CIE color matching functions are used in the following equations:

$$X = \int_{360}^{830} \overline{X}(\lambda) \Phi(\lambda)\, d\lambda$$

$$Y = \int_{360}^{830} \overline{Y}(\lambda) \Phi(\lambda)\, d\lambda \quad \text{(A2.1)}$$

$$Z = \int_{360}^{830} \overline{Z}(\lambda) \Phi(\lambda)\, d\lambda$$

where $\Phi(\lambda)$ is the spectral energy distribution (SED) of the light whose color is being calculated. $\Phi(\lambda)$ is in units of power per unit wavelength (watts/nm) and may have an associated geometric unit (per unit area and/or per unit steradian). The calculated X, Y and Z values are called tri-stimulus values, and the unit is the lumen, with the same associated geometric units. For example, the equations will transform watts/nm-mm^2 into lumens/mm^2.

Since the $\overline{X}$, $\overline{Y}$ and $\overline{Z}$ functions are tabulated functions, the tristimulus values, X, Y and Z, are most often calculated with a digital computer in numerical form using the equations:

$$X = \sum_{360}^{830} \overline{X}(\lambda) \Phi(\lambda)\, \Delta\lambda$$

$$Y = \sum_{360}^{830} \overline{Y}(\lambda) \Phi(\lambda)\, \Delta\lambda \quad \text{(A2.2)}$$

$$Z = \sum_{360}^{830} \overline{Z}(\lambda) \Phi(\lambda)\, \Delta\lambda$$

where $\Delta\lambda$ is the wavelength step in the tabulated $\overline{X}$, $\overline{Y}$, $\overline{Z}$ and $\Phi(\lambda)$ functions.

CIE color coordinates x, y and z are calculated from the tristimulus values with the equations:

$$x = \frac{X}{X+Y+Z}$$

$$y = \frac{Y}{X+Y+Z} \quad \text{(A2.3)}$$

$$z = \frac{Z}{X+Y+Z}$$

As can be seen from the definitions of x, y and z in Equation (A2.3),

$$x + y + z = 1$$

374 COLORIMETRY

For this reason, z is rarely given, normally only x and y values are quoted. While the x − y scale is used throughout this book, that is primarily the personal preference of the authors. The u′ − v′ scale is also often encountered in video work, and is derived from the X, Y and Z tristimulus values with the following equations:

$$u' = \frac{4X}{X + 15Y + 3Z}$$
$$v' = \frac{9Y}{X + 15Y + 3Z}.$$
(A2.4)

Alternatively, u′ and v′ can be derived from the x and y values with these equations:

$$u' = \frac{4x}{-2x + 12y + 3}$$
$$v' = \frac{9y}{-2x + 12y + 3}.$$
(A2.5)

If one wishes to derive x–y values from given u′–v′ values, the inverse equations are used:

$$x = \frac{9u'}{6u' - 16v' + 12}$$
$$y = \frac{4v'}{6u' - 16v' + 12}.$$
(A2.6)

Since the u′–v′ scale is derived from the x–y scale, it is also ultimately dependent on the CIE 1931 2° matching functions. Therefore, any fundamental problems associated with the CIE matching functions would apply to the u′–v′ scale as well as the x–y scale. The use of the u′–v′ scale will be discussed further in Section A2.6.

Conventionally, color coordinates such as x–y or u′–v′ are written with small letters. These color coordinates are independent of the brightness of the source and depend only on its color. The tri-stimulus values X, Y, Z are always written in capital letters and the value of each tri-stimulus value is directly proportional to the output of the source.

The intensity dependent X, Y, Z tristimulus values are additive when beams of different colors of light are mixed together. For example, if the three beams of light listed in Table A2.1 are mixed together, their X, Y and Z tri-stimulus values will be added together to give the X, Y and Z values for the light in the resultant beam. Equation (A2.3) can then be applied to these total X, Y and Z values to give the x–y color coordinates of the resultant beam: x = 0.242, y = 0.273. Note that it is the tristimulus values that can be added together, not the x–y color coordinates. The color names are given in the Kelly chart, to be discussed in Section A2.5.

Table A2.1 Example of additive colors.

	X	Y	Z	x	y
Color 1	4	2	1	0.571	0.286
Color 2	2	6	1	0.222	0.667
Color 3	+2	+1	+14	0.118	0.059
Result	8	9	16	0.242	0.273

A2.4 Color Temperature

The human eye and brain perceive fairly wide varieties of colors as 'white'. These white colors all lie on or near the blackbody curve. Each point on this curve represents the color of light emitted by a blackbody at a given temperature. The spectrum of black body emission is given by the Plank equation and is a well-known function of temperature. A particularly clear explanation of the Plank equation and color temperature is given by Rea (1993). The formula is also given in most other standard works on television, color, physics or optics. The temperature is normally cited in Kelvin, not Celsius or Fahrenheit.

For example, an ordinary incandescent light bulb for household use has a color temperature of about 2800 K. This is on the yellow end of the 'white' region. Sunlight has a color temperature of about 6550 K and is right in the middle of the 'white' region. The nominal color of sunlight is assigned the designation D65 by the CIE.[1] Video projection systems are often set up for higher color temperatures than this, with values up to 10,000 K, or higher, sometimes used. 10,000 K has a distinctly bluish cast to the eye, especially if it can be seen in comparison to an object with a lower color temperature.

The black-body line is a single line through the middle of the white region. When a whitish color is not exactly on the blackbody line, one can talk about its 'correlated color temperature' (CCT) (Krystek 1985). This is the color temperature of the nearest point on the black-body curve. 'Nearest' varies, depending on the system of color coordinates used. Commonly, CCT means the nearest point in the u-v color coordinate system (Robertson 1968). The u-v system is related to the u'-v' system and is not used anywhere else nowadays except in the calculation of correlated color temperature.

A2.5 The Chromaticity Diagram

The x-y color coordinates are often displayed on a two-dimensional chart called a chromaticity diagram. The basic diagram is shown in Plate I. The horseshoe-shaped portion of the curve is the gamut of pure spectral colors, running from the shortest wavelength the human eye can see, about 380 nm, to the longest wavelength the eye can see, about 770 nm. The straight portion across the bottom is where light of these two wavelengths are blended together. The border plus the interior of this curve represents all colors that are visible to the human eye. Combinations of x and y that are outside this curve are not physically possible as long as $\Phi(\lambda) \geq 0$ for all wavelengths, and do not occur when Equations (A2.1) and (A2.3) are applied to real spectra.

Names are assigned to the various regions of the chromaticity diagram, which is also called the Kelly Chart (Kelly 1943; Keller 1983). These names plus the black body curve are shown in Plate I.

A2.6 Color-luminance Difference Formulas

Color-luminance difference formulas are used to indicate whether two different colors or brightnesses are perceptually different from each other. They also give an indication of how different they are. There are many color-luminance difference formulas available in the literature, but only two will be discussed here: $L^*u^*v^*$ and $L^*a^*b^*$. Both formulas are specified by the CIE. More details on these and other formulas can be found in Wyszecki and Stiles (1982).

---

[1] The actual color of sunlight varies with season, time of day, location on earth, air clarity, etc.

The L*u*v* formula is derived from the u'–v' color coordinates plus luminance. The formulas for L*u*v* are:

$$L^* = 116\left(\frac{Y}{Y_n}\right)^{1/3} - 16 \quad \text{for} \quad \frac{Y}{Y_n} > 0.008856$$

$$L^* = 903.3\frac{Y}{Y_n} \quad \text{for} \quad \frac{Y}{Y_n} \leq 0.008856 \quad (A2.7)$$

$$u^* = 13L^*(u' - u'_n)$$

$$v^* = 13L^*(v' - v'_n)$$

where Y is the CIE tristimulus value corresponding to luminance, as defined in Equation (A2.1) and u'–v' are the color coordinates as defined in Equation (A2.5). The subscript n stands for the properties of the standard, reference white light. The unsubscripted values of Y, u' and v' are for the color under test.

The L*a*b* formulations are defined by the equations:

$$L^* = 116\left(\frac{Y}{Y_n}\right)^{1/3} - 16 \quad \text{for} \quad \frac{Y}{Y_n} > 0.008856$$

$$L^* = 903.3\frac{Y}{Y_n} \quad \text{for} \quad \frac{Y}{Y_n} \leq 0.008856 \quad (A2.8)$$

$$a^* = 500\left[f\left(\frac{X}{X_n}\right) - f\left(\frac{Y}{Y_n}\right)\right]$$

$$b^* = 200\left[f\left(\frac{Y}{Y_n}\right) - f\left(\frac{Z}{Z_n}\right)\right]$$

where the function $f(x)$ is defined by:

$$f(x) = x^{\frac{1}{3}} \quad \text{for} \quad x > 0.008856$$

$$f(x) = 7.787x + 0.1379 \quad \text{for} \quad x \leq 0.008856.$$

X, Y and Z are the tristimulus values as defined in Equation (A2.1) and the subscripted and unsubscripted values have the same meaning as they do for L*u*v*. As can be seen, the L* function is the same in either the L*u*v* or L*a*b* formulation. If one calculates L*u*v* and L*a*b* for the reference white itself, you will always get L*= 100 with u*= v*= a*= b*= 0.

If one wishes to calculate the difference between two colors, one would first calculate the L*u*v* or L*a*b* values for each test color compared to some reference white and then calculate:

$$\Delta E = \sqrt{(\Delta L^*)^2 + (\Delta u^*)^2 + (\Delta v^*)^2} \quad \text{or}$$

$$\Delta E = \sqrt{(\Delta L^*)^2 + (\Delta a^*)^2 + (\Delta b^*)^2}. \quad (A2.9)$$

The units of ΔE are called by various authors, 'just noticeable differences (JNDs)'; 'minimum perceptible color differences (MPCDs)', or 'color-luminance differences (CLDs)'. The intention of this formulation is that a ΔE of 1.0 would be just barely perceptible to some people. For reliable differentiation by most people, a ΔE of 3 is necessary.

A user should be aware that the ΔE scale, whether based on L*u*v*, L*a*b* or some other color scale, is not perfect. The perceptibility of $\Delta E = 1.0$ depends on a number of factors, including color, luminance, angular size and texture of the stimulus and adaptation of the viewer. Perceptibility of $\Delta E = 1.0$ also strongly depends on the surrounding area, including its color, luminance, etc.

While more complex color difference formulations than $L^*u^*v^*$ and $L^*a^*b^*$ are available that reduce some of the errors in these two systems, one reaches a point of diminishing returns. There are three main reasons for this:

- Even the most complex formulations contain some residual problems,
- The more complex formulations require measuring more data than is normally taken by projector engineers and
- The statistical variability in human beings is large enough to make the extra accuracy of the more complex formulations unusable in many cases.

A2.7 Measurement of Color

The Electronics Industries Association (EIA) provided a standard for measuring the color of CRT systems (EIA 1988). While the standard was designed primarily for direct view CRT systems, it can be applied to projection displays as well. This standard discusses equipment issues, calibration issues and issues related to system set up conditions necessary to get accurate color and luminance measurements. Keller (1997) and VESA (2001) give a complete account of display measurement techniques and instrumentation, including those required for color measurement.

There are two basic instruments used to measure the color of display systems: a spectroradiometer and a colorimeter. Each type of instrument has its own applications and uses. Anyone who does research or development work involving color will normally need a spectroradiometer. Colorimeters are often used in production environments for testing and quality control. Colorimeters are also often very portable.

The spectroradiometer is a system designed to measure energy as a function of the wavelength of light. The output of a spectroradiometer is spectral energy density (SED) curve of the light source being measured. The units of the output are watts/nanometer, with additional geometrical units (e.g. per cm^2 or per steradian) that depend on the input optics. Once the SED curve of a light source has been measured, Equations (A2.1) or (A2.2) can be applied to calculate the tristimulus values and Equation (A2.3) can be applied to derive the x–y color coordinates. Normal evaluation of a projector would involve separate measurements of the red, green, and blue primary colors plus the white point of the system being measured.

The spectroradiometer has four major subsystems: (1) input optics, (2) monochromator, (3) detector/radiometer, and (4) data recording/reduction. The elements of a spectroradiometer are shown schematically in Figure A2.2.

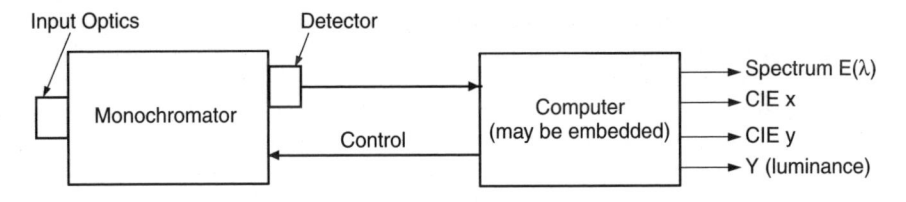

Figure A2.2 Schematic representation of a spectroradiometer.

The input optics are designed to get the desired light from the outside world to the input slit of the monochromator. A wide variety of optics are required for the various tasks facing a projector designer, including, among others, a collimating lens, an integrating sphere, a fiber optic probe and a cosine detector. Note that for each set of input optics, a different calibration is required for the system.

The monochromator must be capable of operating from at least 400 nm to 700 nm, with a range of 380–720 preferred. Some systems extend the range up to 1100 nm (1.1 μm) or beyond and are useful for projector thermal management studies as well as colorimetry. The spectral bandwidth of a monochromator intended for colorimetry work is typically several nanometers wide. For example, a monochromator with a 4 nm bandwidth measuring light at 480 nm is actually measuring light in the bandwidth from 478 nm to 482 nm. This bandwidth is much larger than the bandwidth of a monochromator intended for spectroscopy, where a bandwidth of 0.01 nm is not uncommon.

Modern, compact spectroradiometers use silicon photodiodes for detection of the radiation. If the monochromator selects wavelength by mechanical scanning, a single detector is used. If all the optics in the monochromator are stationary, a CCD array is used as a detector. Spectroradiometers for very low light level work still use photomultiplier tubes because of their greater sensitivity.

The spectroradiometer must be calibrated in absolute units against a known (standard) light source. The data/reduction section of the system applies the calibration data to the raw data from the detector and A/D converter to generate $\Phi(\lambda)$, the spectral energy density. It then calculates the tristimulus values X, Y, and Z from the SED and, from these, the color coordinates x, y. If $u'-v'$ values are desired, Equation (A2.5) is also applied.

A properly calibrated spectroradiometer under ideal conditions can be expected to read the x, y coordinates to an accuracy of ± 0.003 and should be repeatable within ± 0.0005. Luminance measurements should be accurate to 5 % and repeatable to 2 % or better. Since NIST does not certify absolute luminance standards any more accurately than 4.1 % (Burns 1973), it is impossible for any instrument to have an absolute luminance accuracy better than 4.1 %. The PR-650 from Photo-Research is an example of one of the many commercially available instruments capable of doing accurate colorimetry.

A colorimeter is an instrument with three or four separate photodetectors. Custom filters tailor the spectral sensitivity of each photodetector to match the $\overline{X}$, $\overline{Y}$ or $\overline{Z}$ curve. In a colorimeter, the integrals in Equation (A2.1) are done in analog form by the filter/photocell combination.

In all colorimeters, two of the photocells are tailored to match the $\overline{Y}$ and $\overline{Z}$ functions. However, examination of Figure A2.1 shows that the $\overline{X}$ curve is bi-modal, with one peak in the blue region and a second peak in the red region with the minimum between the two peaks at 504 nm. These two regions are typically called $\overline{X}_A$ and $\overline{X}_B$, where the A and B stand for amber and blue, respectively. It is very difficult to tailor a single filter/photocell combination that duplicates the complete $\overline{X}$ function. Therefore, the two regions of the $\overline{X}$ function come from two different photocells.

In all colorimeters, one of the photocells has its response tailored to the $\overline{X}_A$ function. In a 4 photocell colorimeter, there is a separate photocell tailored to the $\overline{X}_B$ function. This arrangement is shown in Figure A2.3(a). In the more common three photocell colorimeter the $\overline{Z}$ photocell is used to produce the $\overline{X}_B$ value. This three filter arrangement is shown in Figure A2.3(b). An example of a commercially available three-filter colorimeter used by video projection workers is the Minolta CL-100.

The input optics of a colorimeter may have as many different forms as the input optics of a spectroradiometer. Since the three or four photocells are physically separated in space, the optics must have some sort of integration to ensure that luminance non-uniformities in the source do not lead to color errors. If, for example, due to a luminance non-uniformity in the source, the $\overline{X}_A$ photocell is seeing more light than the other photocells, the colorimeter will respond as if there is more red light ($\overline{X}_A$) than actually is present. Therefore the measured color value will have a higher x value than it should and will be reported as being redder than it should be.

The outputs of the photocells are directly proportional to the X, Y and Z tristimulus values, if the match between the photocell spectral sensitivity and the CIE functions is good. Unfortunately, the match between the photocells and the CIE functions is rarely good enough for a colorimeter to be considered an accurate reference instrument. A colorimeter should always be calibrated against a spectroradiometer for each separate light source of interest. For instance, separate calibrations are required for CRTs from different manufacturers, because they may use different phosphors with different spectral energy distributions.

MEASUREMENT OF COLOR 379

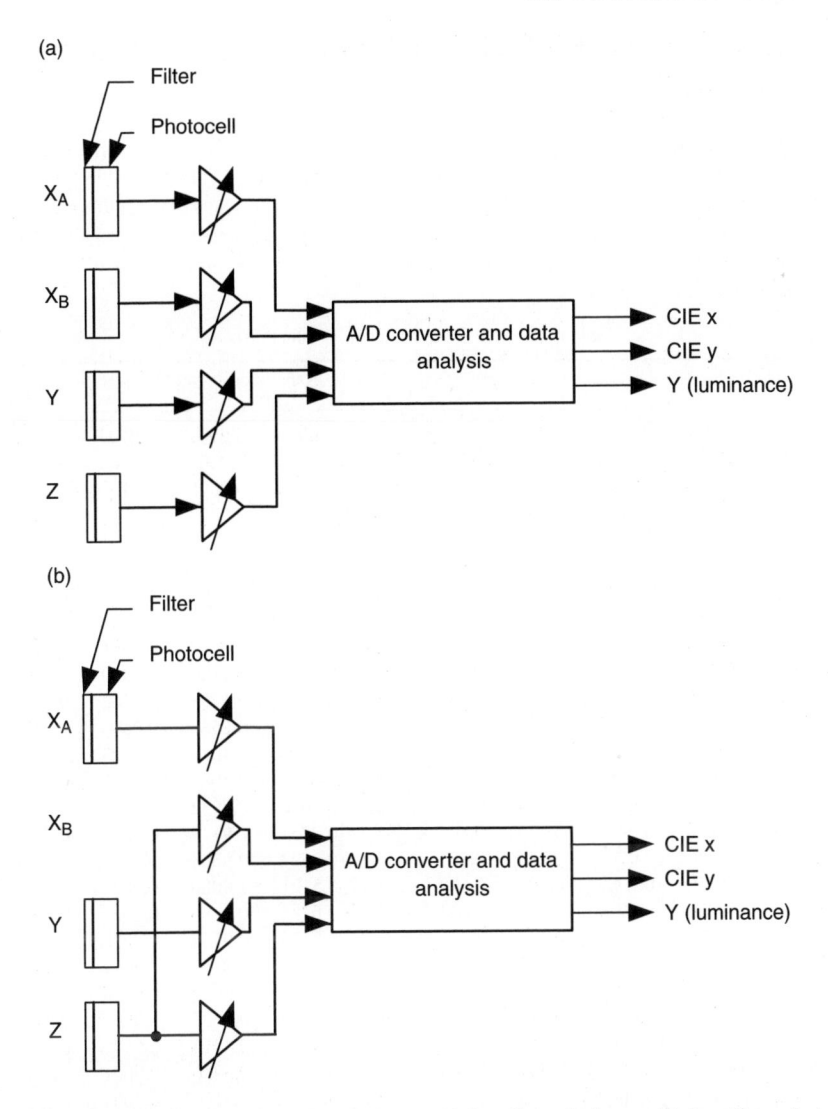

Figure A2.3 Schematic representation of a colorimeter: (a) four filter colorimeter; (b) three filter colorimeter.

A well-calibrated colorimeter should be accurate to ±0.006 in x and y and should be repeatable to ±0.001. When used to measure luminance, the colorimeter can be accurate to ±5 % and repeatable to ±2 %.

Four-filter colorimeters have fallen out of favor for practical applications. If additional accuracy is needed beyond what a three-filter colorimeter can provide, the user would normally move to a spectroradiometer, not to a four-filter colorimeter. The speed advantage a four-filter colorimeter had over scanned monochromator-based spectroradiometer is lost when a modern spectroradiometer with a CCD array and no moving parts is used.

The main value of a colorimeter is for survey work where absolute accuracy is not required. For example, a colorimeter could be used to map out the color and luminance distribution of a projector at the screen to see if the non-uniformity is acceptable. Colorimeters are also useful when all light sources to be measured are nominally identical, e.g. on a projector production line.

A2.8 Tabulated CIE 1931 2° Photopic Color Matching Functions

Table A2.2 gives the color matching functions with 4 digits of precision and 5 nm intervals. It is the abridged table but it is normally sufficiently accurate for projector work. The full table to 7 significant figures at 1 nm intervals is available from Wyszecki and Stiles (1982). The $x(\lambda)$ and $y(\lambda)$ values are the CIE x–y color coordinates for monochromatic light of the specified wavelength.

This table to higher precision, and many other color related tables, are available online at http://cvision.ucsd.edu/index.htm.

Table A2.2   CIE 2° color matching functions.

WL	$\bar{X}$	$\bar{Y}$	$\bar{Z}$	$x(\lambda)$	$y(\lambda)$	WL	$\bar{X}$	$\bar{Y}$	$\bar{Z}$	$x(\lambda)$	$y(\lambda)$
380	0.0014	0.0000	0.0065	0.174	0.005	385	0.0022	0.0001	0.0105	0.174	0.005
390	0.0042	0.0001	0.0201	0.174	0.005	395	0.0076	0.0002	0.0362	0.174	0.005
400	0.0143	0.0004	0.0679	0.173	0.005	405	0.0232	0.0006	0.1102	0.173	0.005
410	0.0435	0.0012	0.2074	0.173	0.005	415	0.0776	0.0022	0.3713	0.172	0.005
420	0.1344	0.0040	0.6456	0.171	0.005	425	0.2148	0.0073	1.0391	0.170	0.006
430	0.2839	0.0116	1.3856	0.169	0.007	435	0.3285	0.0168	1.6230	0.167	0.009
440	0.3483	0.0230	1.7471	0.164	0.011	445	0.3481	0.0298	1.7826	0.161	0.014
450	0.3362	0.0380	1.7721	0.157	0.018	455	0.3187	0.0480	1.7441	0.151	0.023
460	0.2908	0.0600	1.6692	0.144	0.030	465	0.2511	0.0739	1.5281	0.136	0.040
470	0.1954	0.0910	1.2876	0.124	0.058	475	0.1421	0.1126	1.0419	0.110	0.087
480	0.0956	0.1390	0.8130	0.091	0.133	485	0.0580	0.1693	0.6162	0.069	0.201
490	0.0320	0.2080	0.4652	0.045	0.295	495	0.0147	0.2586	0.3533	0.023	0.413
500	0.0049	0.3230	0.2720	0.008	0.538	505	0.0024	0.4073	0.2123	0.004	0.655
510	0.0093	0.5030	0.1582	0.014	0.750	515	0.0291	0.6082	0.1117	0.039	0.812
520	0.0633	0.7100	0.0782	0.074	0.834	525	0.1096	0.7932	0.0573	0.114	0.826
530	0.1655	0.8620	0.0422	0.155	0.806	535	0.2257	0.9149	0.0298	0.193	0.782
540	0.2904	0.9540	0.0203	0.230	0.754	545	0.3597	0.9803	0.0134	0.266	0.724
550	0.4334	0.9950	0.0087	0.302	0.692	555	0.5121	1.0000	0.0057	0.337	0.659
560	0.5945	0.9950	0.0039	0.373	0.624	565	0.6784	0.9786	0.0027	0.409	0.590
570	0.7621	0.9520	0.0021	0.444	0.555	575	0.8425	0.9154	0.0018	0.479	0.520
580	0.9163	0.8700	0.0017	0.512	0.487	585	0.9786	0.8163	0.0014	0.545	0.454
590	1.0263	0.7570	0.0011	0.575	0.424	595	1.0567	0.6949	0.0010	0.603	0.396
600	1.0622	0.6310	0.0008	0.627	0.372	605	1.0456	0.5668	0.0006	0.648	0.351
610	1.0026	0.5030	0.0003	0.666	0.334	615	0.9384	0.4412	0.0002	0.680	0.320
620	0.8544	0.3810	0.0002	0.691	0.308	625	0.7514	0.3210	0.0001	0.701	0.299
630	0.6424	0.2650	0.0000	0.708	0.292	635	0.5419	0.2170	0.0000	0.714	0.286
640	0.4479	0.1750	0.0000	0.719	0.281	645	0.3608	0.1382	0.0000	0.723	0.277
650	0.2835	0.1070	0.0000	0.726	0.274	655	0.2187	0.0816	0.0000	0.728	0.272
660	0.1649	0.0610	0.0000	0.730	0.270	665	0.1212	0.0446	0.0000	0.731	0.269
670	0.0874	0.0320	0.0000	0.732	0.268	675	0.0636	0.0232	0.0000	0.733	0.267
680	0.0468	0.0170	0.0000	0.733	0.267	685	0.0329	0.0119	0.0000	0.734	0.266
690	0.0227	0.0082	0.0000	0.734	0.266	695	0.0158	0.0057	0.0000	0.735	0.265
700	0.0114	0.0041	0.0000	0.735	0.265	705	0.0081	0.0029	0.0000	0.735	0.265
710	0.0058	0.0021	0.0000	0.735	0.265	715	0.0041	0.0015	0.0000	0.735	0.265
720	0.0029	0.0010	0.0000	0.735	0.265	725	0.0020	0.0007	0.0000	0.735	0.265
730	0.0014	0.0005	0.0000	0.735	0.265	735	0.0010	0.0004	0.0000	0.735	0.265
740	0.0007	0.0002	0.0000	0.735	0.265	745	0.0005	0.0002	0.0000	0.735	0.265
750	0.0003	0.0001	0.0000	0.735	0.265	755	0.0002	0.0001	0.0000	0.735	0.265
760	0.0002	0.0001	0.0000	0.735	0.265	765	0.0001	0.0000	0.0000	0.735	0.265
770	0.0001	0.0000	0.0000	0.735	0.265	775	0.0001	0.0000	0.0000	0.735	0.265
780	0.0000	0.0000	0.0000	0.735	0.265	785	0.0000	0.0000	0.0000	0.735	0.265
790	0.0000	0.0000	0.0000	0.735	0.265						

References

Burns, V.I. and McSparron, D.A. (1973) *Optical Radiation Measurements: Photometric Calibration Procedures*, National Bureau of Standards Technical Note 594–3, Washington D.C.

Electronic Industries Association (1988) *TEP105-11-A, Measurement of the Color of CRT Screens*, December.

Fairman, H.S., Brill, M.H., Hemmendinger, H. (1997) How the CIE 1931 color-matching functions were derived from Wright-Guild data. *Color Research and Application* **22(1)**: 11–23.

Keller, P. (1983) CIE-USC chromaticity diagram with color boundaries. *Proc SID* **24(4)**: 317–21.

Keller, P. (1997) *Electronic Display Measurement: Concepts, Techniques, and Instrumentation*. New York: John Wiley & Sons, Inc.

Kelly, K.L. (1943) Color designations for lights. *J. Research of the Natl. Bureau of Standards* **31**: 271–8.

Krystek, M. (1985) An algorithm to calculate correlated color temperature. *Color Research and Applications* **10**: 38–40.

Post, D.L., Lippert, T.M. and Snyder, H.L. (1983) Quantifying color contrast. *SPIE Proc.* **386**: 12–19.

Rea, M.S. (ed.) (1993) *Lighting Handbook*, 8th edn. Illuminating Engineering Society of North America.

Robertson, A.R. (1968) Computation of correlated color temperature and distribution temperature. *Journal of the Optical Society of America* **58**(11): 1528–35.

VESA (2001) *Flat Panel Display Measurements Standard, Version 2.0, FPDM2* VESA-2001-6, Video Electronics Standards Association (VESA), Milpitas, CA.

Wyszecki G. and Stiles, W.S. (1982) *Color Science: Concepts and Methods, Quantitative Data and Formulae*, 2nd edn. New York: John Wiley & Sons, Inc.

Appendix 3 Lumen vs Etendue Parametric Model

A3.1 Introduction

This appendix will derive an analytical model for the lumens vs étendue function for lamp/reflector combinations intended for use in projection systems. This model is useful in calculating the lumen throughput of projectors, as described in Chapter 11, before construction of a prototype. The discussion in this appendix follows (Brennesholtz 1996).

The parameters in this model are intended to have physical meaning, such as arc length and diameter, and reflector size and shape. The values of these parameters would be derived semi-empirically in actual practice, by fitting the lumens vs étendue data with values close to the actual properties of the system. While the lumen vs étendue data would normally be measured, it may also be derived from ray tracing the lamp reflector combination in a 'virtual experiment'.

An alternative way to derive an analytical lumens vs étendue curve from the measured data is to use a curve-fitting program, such as TableCurve, to find an arbitrary function that fits the data. The fit found in this case is in fact often better than the fit using the function in this appendix.

The disadvantage of using an arbitrary function is that the fitting parameters do not have any obvious physical meaning. This means that a new set of measurements or a new ray trace and a new fit must be done for every lamp and reflector combination to be considered. This can be tedious, or impossible, if one is interested in calculating the throughput with a lamp that the manufacturer promises will be available next year. With a semi-empirical fit and with each parameter having a physical meaning, the parameters can be changed with a reasonable hope of realistic results.

For example, sometimes one will have a lamp where the arc length is 1.5 mm. One can then measure this lamp and construct a semi-empirical model for this particular lamp/reflector combination. If the manufacturer promises that next year the arc length will be reduced to 1.2 mm but all other lamp and reflector properties will remain the same, it would be very desirable to be able to estimate the effect of this arc length change on projector throughput. With a semi-empirical model it is relatively easy to change the parameter related to arc length and leave all the other parameters the same. With a fit to an arbitrary function, such as a ratio of polynomials, this is not possible.

A3.2 Definition of Étendue

Étendue is a parameter of an optical beam related to the beam divergence and the cross-sectional area of the beam. Étendue is also known as 'optical extent' or the 'optical invariant'. When a beam is modified by a well-corrected optical element, étendue is preserved. For example, when a well-corrected lens focuses a collimated beam to a spot, the area of the beam is reduced but the divergence (convergence) angle of the beam increases, and étendue is preserved. In optical elements that are not well corrected for optical aberrations,[1] involve scattering or have diffraction into multiple orders, étendue will increase. Étendue is related to the second law of thermodynamics, where entropy can remain constant or increase, but never decrease. The same is true of the étendue of a light beam: it can remain constant or increase, but never decrease.

The formal definition of étendue is (Boyd 1983):

$$E = \iint \cos\theta \, dA \, d\Omega, \tag{A3.1}$$

where E is the étendue, and the double integral is over the area of interest and the solid angle of the light corresponding to that area, as shown in Figure A3.1. The angle θ is between the normal to the surface element dA and the centroid of the solid angle element $d\Omega$. This angle takes into account the projected area of the area element dA, rather than the actual area. Note that étendue is a geometric property: there is no reference in this equation to optical intensity. The units associated with étendue are mm^2-steradian, which again shows the geometric nature of étendue.

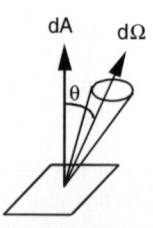

Figure A3.1 Relationship between θ, dA and $d\Omega$.

Étendue is a somewhat misunderstood concept, partly due to the fact that étendue does not refer to the light intensity within the distribution of interest: it only refers to the distribution's geometric boundaries. Another reason étendue can be a confusing concept is the difficulty of doing the integration in Equation (A3.1) in the general case. Normally, it is necessary to integrate over two spatial and two angular dimensions, often in polar or spherical coordinates. For this reason, approximations such as Equations (A3.2) and (A3.3) are often used to calculate étendue, even under circumstances where these approximations may not be valid.

---

[1] In some cases involving poorly corrected aberrations in optical elements, étendue can *appear* to decrease. See Section 11.2.1.1 for a discussion of this issue.

A3.3 Étendue at a Flat Surface

Equation (A3.1) can be integrated at a flat surface where the optical beam has a uniform divergence angle ($\theta_{1/2}$) over that surface. In this case, the étendue is:

$$E = \pi A \sin^2(\theta_{1/2}) = \frac{\pi A}{4(f/\#)^2} \tag{A3.2}$$

where A is the area of the surface, which can be a light valve or other optical aperture. If, for example, you have a 30 mm by 40 mm flat panel device, capable of accepting a f/3.5 input beam, the étendue of the device is 76.9 mm²-steradian.

A3.4 Étendue of a Cylindrical Surface

The arc of many arc lamps can be modeled as a cylinder of a length Z_o and a diameter D_0, with all the light being emitted from the curved surface. Most reflectors can be modeled as azimuthally symmetric, with an inner collecting angle θ_1 and an outer collecting angle of θ_2. This geometry is shown in Figure A3.2. In this case, the integral in Equation (A3.1) can be evaluated in closed form, giving the result:

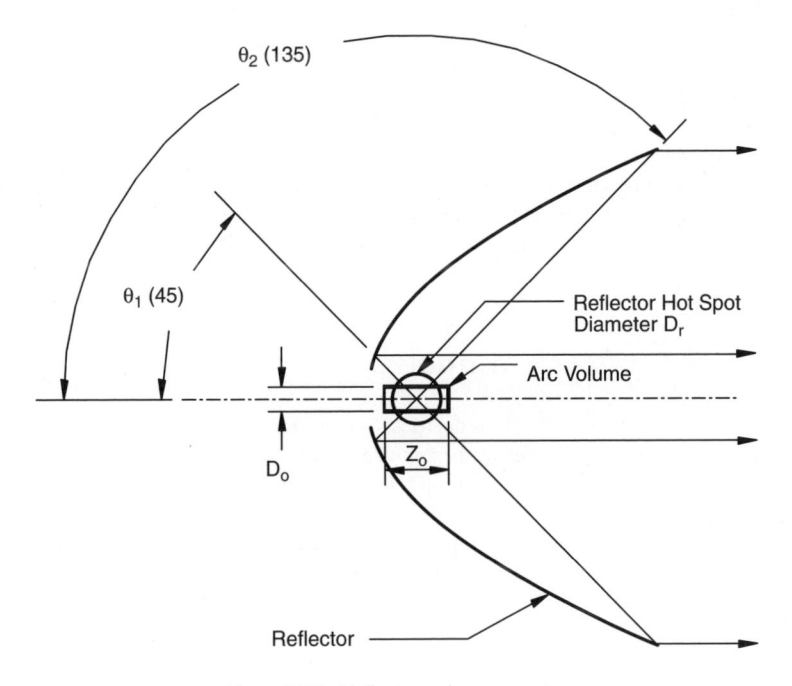

Figure A3.2 Reflector and arc geometry.

$$E = \pi Z_0 D_0 \left[\theta_2 - \theta_1 - \frac{(\sin(2\theta_2) - \sin(2\theta_1))}{2} \right] = Z_0 D_0 E_0 \tag{A3.3}$$

where:

$$E_0 = \pi \left[\theta_2 - \theta_1 - \frac{(\sin(2\theta_2) - \sin(2\theta_1))}{2} \right]$$

θ_1 is the inner collecting angle in radians,
θ_2 is the outer collecting angle in radians,
Z_0 is the length of the cylinder and
D_0 is the diameter of the cylinder.

Since the most common unit for étendue is mm²-steradian, the normal unit for Z_0 and D_0 would be millimeters. The only simplifying assumption made in this integration was that the reflector is much larger than the arc, so the inner and outer angles θ_1 and θ_2 are independent of position along the arc. The factor E_0 in this equation contains the geometric information on the reflector (θ_1 and θ_2, the inner and outer collection angles) and none of the information on the arc (length Z_0, diameter D_0). Therefore, if the same reflector is used with a different arc, the factor E_0 will not change.

For example, consider a lamp/reflector combination having arc length, diameter and collection angles of:

$\theta_1 = 45° = 0.79$ radians,
$\theta_2 = 135° = 2.36$ radians,
$Z_0 = 1.40$ mm and
$D_0 = 1.00$ mm.

This gives an E_0 value of 8.08 steradian and an arc étendue value of 11.3 mm²-steradian, when Equation (A3.3) is applied.

It must be understood that the étendue of the light beam produced by a lamp/reflector combination can be very much larger than the étendue of the arc alone. While étendue is conserved in a well corrected system, in a system with aberrations, étendue can increase dramatically. High collection, low f/# (e.g. f/0.1) conic section reflectors can increase the étendue of the beam by a factor of 10 or more compared to the étendue of the arc. The projection system designer is normally interested in the properties of the light beam, not the properties of the bare arc.

A3.5 Lamp/Reflector Model

Parameters in the model will include arc properties such as arc length, arc diameter, light distribution within the arc and total lumens produced by the arc. The model will also include reflector properties such as the inner and outer collection angles, coating efficiency and a distortion constant related to aberrations in the reflector. All these properties are well understood and readily measurable properties of an arc and a reflector, except the distortion constant. While the distortion constant could be derived from aberration theory or ray trace data on the lamp/reflector combination, here it will be derived from an empirical fit to measured data. Stewart et al. (1998) gives some details on the source of the aberrations that are lumped together by the distortion constant.

Consider a hollow cylinder of length and diameter both equal to d, with azimuthally symmetric inner and outer angles of θ_1 and θ_2. In this case, Equation (A3.3) becomes:

$$E = d^2 E_0. \tag{A3.4}$$

This is the étendue of a small volume of emitting space and it does not include the aberrations introduced by the reflector. The relationship between the arc size and the cylinder represented by the parameter d is shown in Figure A3.3. This cylindrical volume will not contain all the light from the arc unless d > Z_0

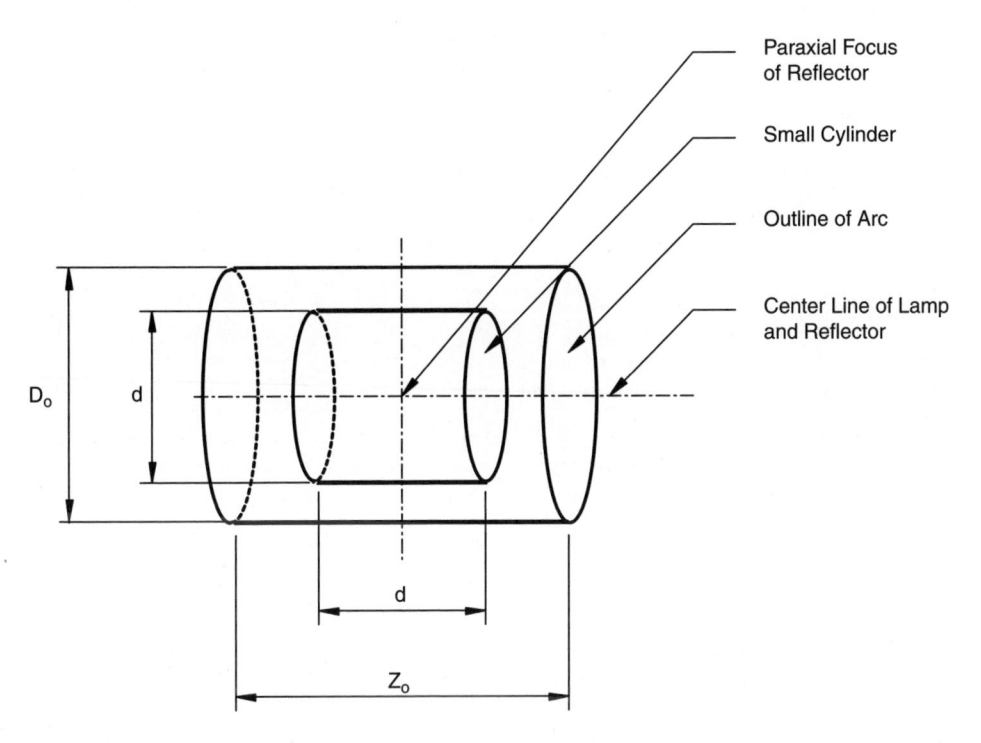

Figure A3.3 Relationship between the arc volume and the cylinder described by Equation (A3.4).

and d > D₀. Without a complete model of the arc and reflector, it is impossible to know exactly how the aberrations of the reflector will increase the étendue of the collimated beam relative to the étendue of the arc alone. In this model, it is assumed that aberrations increase étendue as a function of r^2, where r is measured from the paraxial focus of the reflector. Equation (A3.4) would then would become:

$$E = d^2 E_0 \left(1 + \frac{d^2}{D_R^2}\right) \quad (A3.5)$$

where
- E_0 is defined in Equation (A3.3),
- d is the size of the small cylinder and
- D_R is a length parameter corresponding to the reflector aberrations.

Conceptually, D_R can be thought of as the diameter of the 'hot spot' of the reflector where light emitted within the hot spot is collected efficiently. Typically, D_R is the same order of magnitude as the nominal arc size in compact reflectors suitable for use in projection systems.

Setting this equation aside for a second, let us look at the light emitted by the arc. The light from an arc lamp is not emitted from a surface, but rather is emitted from the entire volume of the arc. Let us define a luminous density function, Φ_d, in lumens/mm³. In a cylindrical coordinate system, this function would be:

$$\Phi_d = \Phi_d(r, z, \phi) = \Phi_d(r, z),$$

assuming azimuthal symmetry. Normally arcs are brightest along the centerline of the arc and decline monotonically in the radial direction to the edge of the arc. The simplest function to model this monotonic decline is a linear one. While more complex models are available that would more accurately represent the actual light distribution in typical arcs, the mathematically simple linear model seems to give satisfactory results. Therefore with the linear model, the Φ_d function can be modeled as:

$$\Phi_d(r, z, \phi) = \Phi_c\left(1 - \frac{r}{R_0}\right) \quad \text{for } r < R_0 \text{ and } |z| < \frac{Z_0}{2} \quad \text{(A3.6)}$$

$$\Phi_d(r, z, \phi) = 0 \quad \text{for } r > R_0 \text{ or } |z| > \frac{Z_0}{2}$$

where
- Z_0 is the length of the arc,
- R_0 ($D_0/2$) is the radius of the arc and
- Φ_c ($\Phi_d(0, z)$) is the luminous density on the centerline.

Unfortunately, this simple function does not adequately describe all common arcs because they often vary in light output along the length of the arc. Let us assume the arc is symmetric in z about the center, $z = 0$, and increases or decreases linearly as you go away from $z = 0$ toward the end of the arc. In this case, Equation (A3.6) becomes:

$$\Phi_d(r, z, \phi) = \Phi'_c\left(1 - \frac{r}{R_0}\right)\left[f + \frac{2(1-f)|z|}{Z_0}\right] \quad \text{for } r < R_0 \text{ and } |z| < \frac{Z_0}{2} \quad \text{(A3.7)}$$

$$\Phi_d(r, z, \phi) = 0 \quad \text{for } r > R_0 \text{ or } |z| > \frac{Z_0}{2}$$

where
- Φ'_c is the average luminous density along the centerline of the arc, and
- f is a parameter that can vary from 0 to 2.

When $f = 1$, reduces to Equation (A3.6) and the arc has uniform emission along its length. When $f = 2$, the arc has maximum emission density of $2\Phi'_c$ at the center ($z = 0$) and declines linearly to 0 emission at the ends of the arc. For $f = 0$, the arc has zero emission at $z = 0$ and increases linearly to $2\Phi'_c$ at the end of the arc, which would normally be near an electrode. For the region $0 < f < 1$, the arc emission density is low, but not zero, halfway between the electrodes and increases linearly towards the electrodes. This is a reasonable model for an AC driven lamp where there are two hot spots, one near each electrode.

This luminous density function can be integrated over r, z, and ϕ to determine the number of lumens emitted inside an arbitrary cylinder of diameter and length d centered on the axis ($r = 0$) and centered between the electrodes ($z = 0$). As long as this small cylinder is smaller than the arc, the amount of light emitted within the cylinder is:

$$\Phi(d) = \Phi_0 R G_0 \left[3\left(\frac{d}{D_0}\right)^2 - 2\left(\frac{d}{D_0}\right)^3\right]\left[f\frac{d}{Z_0} + (1-f)\left(\frac{d}{Z_0}\right)^2\right] \quad \text{for } d < Z_0 \text{ and } d < D_0 \quad \text{(A3.8)}$$

where
- $\Phi(d)$ is the lumens emitted from inside the small cylinder,
- Φ_0 is the total number of lumens emitted by the arc,
- R is the reflectivity of the coating on the reflector,
- G_0 is the geometric collection efficiency of the reflector,
- Z_0 is the total length of the arc and
- D_0 is the diameter of the arc.

The geometrical collection efficiency G_0 represents the proportion of the light emitted by the arc that actually strikes the reflector and is collected into the output beam.

COMPARISON OF MEASURED DATA TO THE MODEL

The general case including large values of d is:

$$\Phi(d) = \Phi_0 RG_0 \left\{ 3 \left[Min\left(1, \frac{d}{D_0}\right) \right]^2 - 2 \left[Min\left(1, \frac{d}{D_0}\right) \right]^3 \right\}$$

$$\times \left\{ f \left[Min\left(1, \frac{d}{Z_0}\right) \right] + (1-f) Min\left[1, \left(\frac{d}{Z_0}\right)^2 \right] \right\} \quad (A3.9)$$

where Min(a,b) is the function that takes the smaller of a or b. These Min functions replace the two inequalities of Equation (A3.7). As would be expected, for $d > D_0$ and $d > Z_0$, the total lumens value is just

$$\Phi = \Phi_0 RG_0,$$

or the total lumens of the lamp corrected with the reflector coating efficiency and the geometric collection efficiency.

Equation (A3.5) gives the étendue of the optical beam containing the light collected from within the small cylinder of diameter d and Equation A3.9 gives the lumens emitted from within that cylinder and contained in the beam. Taken together, these equations now characterize the lumen output of the lamp/reflector combination as a function of étendue.

In the present form, the equations are not very convenient to use because of the presence of the parameter d, which has little meaning to a system designer. However, for a given system étendue E, equation (A3.5) can be solved for d, giving:

$$d = \sqrt{\frac{-D_R^2 + \sqrt{D_R^4 + \frac{4ED_R^2}{E_0}}}{2}} \quad (A3.10)$$

This expression for d (Equation (A3.10)) can then be substituted into Equation (A3.9) to give Equation (A3.11), an equation for lumens in the collected beam as an explicit function of étendue of that beam. Although this expression appears rather complex algebraically, it is easy enough to program into a spreadsheet or other computer program.

$$\Phi(E) = \Phi_0 RG_0 \left\{ 3 \left(Min\left(1, \frac{\sqrt{\frac{-D_R^2 + \sqrt{D_R^4 + \frac{4ED_R^2}{E_0}}}{2}}}{D_0}\right) \right)^2 - 2 \left(Min\left(1, \frac{\sqrt{\frac{-D_R^2 + \sqrt{D_R^4 + \frac{4ED_R^2}{E_0}}}{2}}}{D_0}\right) \right)^3 \right\} *$$

$$\left\{ f \left(Min\left(1, \frac{\sqrt{\frac{-D_R^2 + \sqrt{D_R^4 + \frac{4ED_R^2}{E_0}}}{2}}}{Z_0}\right) \right) + (1-f) Min\left(1, \left(\frac{\sqrt{\frac{-D_R^2 + \sqrt{D_R^4 + \frac{4ED_R^2}{E_0}}}{2}}}{Z_0}\right)^2 \right) \right\} \quad (A3.11)$$

A3.6 Comparison of Measured Data to the Model

The lumens vs étendue curve for a lamp/reflector combination can also be measured, using the apparatus shown in Figure A3.4. The lamp is imaged onto the plane of an adjustable aperture at the entrance of an integrating sphere calibrated in lumens. The étendue of the beam can be controlled by adjusting the aperture at the focal point of the lamp, the aperture at the condenser, or both. The aperture at the sphere controls the area of the collected beam and the aperture at the lens controls the f/#. Étendue of the collected light can then be calculated using Equation (A3.4), assuming the distance between the two apertures is much larger than the diameter of either aperture.

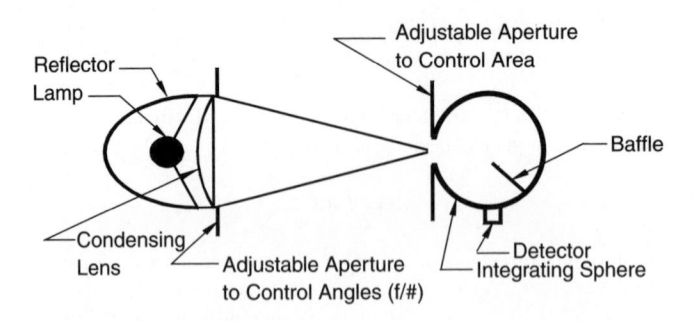

Figure A3.4 Apparatus for measuring lumens vs étendue function.

Five samples of a short arc metal halide lamp were measured in this apparatus. At any given étendue, the average and standard deviation lumen output of the five samples was computed. Étendue was changed by adjusting the aperture at the entrance port of the sphere. Model parameters for this lamp were fit to the measured collection data and geometric properties of the lamp and reflector. These parameters are shown in Table A3.1. Measured data and model data are shown for the lamp in Figure A3.5.

Table A3.1 Model parameters for sample lamps.

Reflector properties	
Type	Elliptical
Inner angle degrees	18.6
Outer angle degrees	98.1
Étendue factor (E_o) steradian	5.75
Distortion parameter (D_r) mm	1.42
Coating reflectivity (R) %	98%
Geometrical efficiency (G_0) %	58%
Arc properties	
Arc diameter (D_0) mm	1.70
Arc length (Z_o) mm	2.79
Lamp power watts	125
Lamp efficacy lumens/Watt	62
Lamp output (Φ_0) lumens	7750

As can be seen, there is an excellent match between the measured data and the simple model given by Equation (A3.11) for this lamp type. The discrepancy between the model and the average was less than 1 standard deviation at all étendues. The most visible discrepancy is at large étendue, where the model predicts a constant amount of light, once the system is no longer étendue limited, while the measured data shows a slow but steady increase in lumens with étendue. Presumably replacing the linear model of Equation (A3.7) with a Gaussian, or some other integratable function with a long tail, would improve the match at high étendue. Since systems with this high an étendue are rarely of interest, this is not felt to be a major problem.

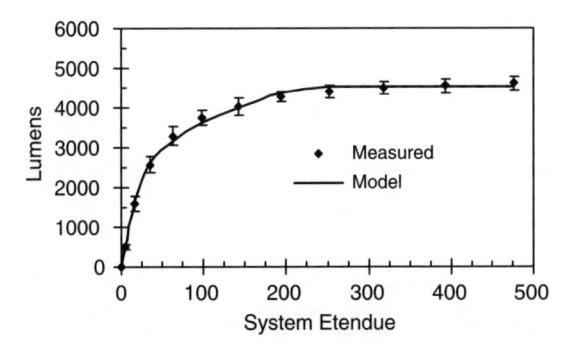

Figure A3.5 Lumen throughput vs system étendue.

The E_0 value in the table was calculated from the given inner and outer collection angles for the reflector using Equation (A3.3). Note that the angle values must be converted to radians before the equation can be used. One thing to notice about this model is that the arc length in the model that best fits the measured data is slightly larger than the gap between the electrodes: 2.79 vs. a measured 2.40 mm. It is believed that contributing factors to this include reflections off the electrodes, distortions caused by the lamp envelope and scattering at the envelope. This same effect was seen in five other lamp types when the model was constructed from the measured data.

References

Boyd, R.W. (1983) *Radiometry and the Detection of Optical Radiation*. New York: John Wiley and Sons, Inc., 89–93.
Brennesholtz, M.S. (1996) Light collection efficiency for light-valve projection systems. In M. Wu (ed.), *SPIE Proceedings* **2650**.
Stewart, C.N., Rutan, D.M., Jacobson, B.A. and Gengelbach, R.D. (1998) Metal halide lighting systems and optics for high efficiency compact LCD. In M. Wu (ed.), *SPIE Proceedings: Projection Displays IV* **3296**.

Appendix 4 Glossary

This appendix will define technical terms, jargon and acronyms related to projection displays and video in general. Words in *italics* in the definitions are also defined in the glossary.

A good general dictionary of optical terms is the *Photonics Dictionary*, reissued annually by Laurin Publishing Co.

Terms related to digital video are given in Annex A of 'Task Force for Harmonized Standards for the Exchange of Program Material as Bit Streams—First Report: User Requirements'. This paper is a joint SMPTE and EBU report, published in the *SMPTE Journal*, **106(6)**, pp. 345–405, June 1997.

The *Lighting Handbook* 8th edition, published by the Illuminating Engineering Society of North America (Mark S. Rea, Editor-in-Chief, 1993) contains an extensive glossary of lamp and lighting related terms.

Term	Definition
μm	Symbol for micron, equal to 10^{-6} meters. Sometimes μ is used alone.
1.3 K	Digital cinema with 1280 × 1024 resolution.
2 K	Digital cinema with 2048 × 1080 resolution. See *DCI* specification.
3:2 pull down	Conversion of 24 Hz movies to 30 Hz *interlaced NTSC* through the technique of displaying one movie *frame* for 3 NTSC *fields* and then displaying the next frame for 2 fields.
3LCD	A consortium setup to promote the use of 3-panel LCD projectors www.3LCD.com
4:1:1	Video that has been digitized where the horizontal and vertical resolution of the color data is half the horizontal and vertical resolution of the luminance data. The chrominance has been digitized at the same location as a luminance point.
4:2:0	Video that has been digitized where the horizontal and vertical resolution of the color data is half the horizontal and vertical resolution of the luminance data. The chrominance has also been digitized at a different location from the luminance.

Projection Displays, Second Edition M. Brennesholtz, E. Stupp
© 2008 John Wiley & Sons, Ltd

GLOSSARY

(Continued).

Term	Definition
4:2:2	Video that has been digitized where the horizontal resolution of the color data is half the horizontal resolution of the luminance and vertical resolution of the color data is the same as the vertical resolution of the luminance data. Chrominance was digitized at the same spatial location as luminance.
4:4:4	Video that has been digitized where the horizontal and vertical resolution of the color data is the same as the horizontal and vertical resolution of the luminance data and the chrominance and luminance data are digitized at the same location. Often used for RGB data, as opposed to luminance and color difference data.
4 K	Digital cinema with 4096 × 2160 resolution See *DCI*.
63° TN	An *LC* effect that allows the *twisted nematic* effect to be used in *single polarizer* reflective LCD systems.
8 K	Also called ultra-high definition. 8192 × 4320 resolution. NHK has demonstrated systems. Designed for viewing from less than 1 picture height away.
A-plate	An 'A-plate' is an optical retarder utilizing a layer of uniaxially birefringent material with its extraordinary axis oriented parallel to the plane of the layer, and its ordinary axis (also called 'a-axis') oriented perpendicular to the plane of the layer, i.e. parallel to the direction of normally incident light.
Aberration	Optical aberrations are distortions in an optical beam induced by an optical element. Some common abbreviations affecting projection systems are lateral and longitudinal chromatic aberration, spherical aberration, coma, and variation in lateral magnification.
Absorbing material	An optical material that removes some or all of the energy from a light ray passing through it.
ACATS	Advisory Committee on Advanced Television Service (USA).
Active matrix	A *matrix addressing* system where there is an electronic switch such as a *TFT* at the intersection of the *column* and *row* lines at each pixel. When this switch is activated, the signal voltage is applied to the pixel. When the switch is de-activated and the next column or row is addressed, the pixel continues to hold its voltage until readdressed. See *passive matrix* and *AM-LCD*
ADC	Analog to digital converter
Additive colors	A color system in which the *primary colors* (*red*, *green* and *blue*) are added to form *white* and other colors. Most displays and projectors are additive color. See *Subtractive color.*
Air spaced	A description of an optical element in which air (with its low index of refraction) is an inherent part of the design of the element.
Aliasing	The condition where sampling generates spatial or temporal frequencies that are in the output within the input frequency band and were not present in the input signal. This occurs when the sampling frequency is too low compared to the upper limit of the input frequency band. In a display, this may result in Moiré patterns or other artifacts.
Alignment	The orientation of the LC molecules relative the substrate.
Alignment layer	An organic or inorganic layer applied to the substrate to control the alignment of the liquid crystal molecules.
AM-LCD	*Active Matrix* Liquid Crystal Device (or Display). An *LCD* in which there is an active switch, usually a *TFT*, associated with each *pixel* to control the state of the liquid crystal material. See *Active matrix.*
Amorphous silicon	Amorphous silicon is silicon without long range order. For LCD applications, a thin layer of it is typically deposited on a glass substrate at a temperature below the melting point of the glass by *PECVD*. See *Polysilicon*.

GLOSSARY

Analog domain	The portion of a signal chain in which all signals are processed using analog techniques.
Analog format	A video format in which the information is carried by the varying amplitude, phase or frequency of a carrier wave. Examples are *Hi-Vision, PAL, SECAM, PAL+* and *NTSC*. See *Digital format*.
Analog video	Video in which the information is conveyed by a continuous variation of some parameter, such as voltage, current, carrier frequency (FM) or carrier amplitude (AM).
Analyzer	The optical element that separates the modulated from the waste light in a *light valve* projection system. See *polarizer/analyser*.
Angle of incidence	The angle between the normal to an optical surface and a light *ray* incident upon that surface.
Angle of reflection	The angle between the normal to an optical surface and a light *ray* reflected from that surface. The angle of reflection is equal to the *angle of incidence*.
Angle of refraction	The angle between the normal to an optical surface and a light *ray* that has been refracted while passing through that surface. The angle of *refraction* is related to the *angle of incidence* by *Snell's law*.
Angular color separation	Abbreviated ACS. The separation of *red, green* and *blue* light in a *LV* projector by angle instead of position.
ANSI	American National Standards Institute (http://www.ansi.org/).
ANSI lumens	The measurement of luminous flux of a projection system by averaging the *illuminance* values measured at 9 points on the screen, as prescribed by ANSI IT7.215-1992. This value is close enough to integrated *sphere lumens* to be usable for most purposes.
Answer Print	A print made from the original cinema negative. Normally viewed by directors, producers, etc. Occasionally shown at film festivals. An answer print is a 2nd generation image, compared to the fourth-generation *release print* shown in cinemas.
Anti-aliasing	Reduction of *aliasing* by modifying the *video signal*, normally with a *low pass filter*.
Aperture ratio	The fraction of a *light valve* that represents actual useful area in terms of modulating the incident light. Also called 'clear aperture', 'geometrical aperture' or 'fill factor'.
AR coating	Antireflection Coating. A coating designed to reduce the reflection at an optical surface. Without an AR coating, typically 4 % or more of the light is reflected at each glass/air interface in an optical system. With an AR coating, reflection at an interface can be reduced to 0.5 % or less. See *Dielectric coatings*.
Arc lamps	A lamp in which the light is produced by electricity passing between two electrodes through a gas or gas mixture. Also called a high intensity discharge (HID) lamp.
ARPA	See *DARPA*.
Artifact	An element visible in a displayed image that was not visible in the original image. Examples are dot crawl in *NTSC* systems, flicker in 50 Hz *PAL* systems or blockiness in *MPEG* encoded video.
ASC	American Society of Cinematographers (http://www.theasc.com/).
a-Si	*Amorphous Silicon,* sometimes also abbreviated α-Si.
Aspect ratio	The ratio of the width of a screen to the height of the screen. Can be expressed either as a ratio of whole numbers such as 4:3 or 16:9 or as decimal fractions, 1.333 or 1.778.
ATSC	*Advanced Television Systems Committee.*
Advanced Television Systems Committee	This committee supervised and evaluated the development of a digital television broadcast standard by private corporations in the US. The broadcast format developed by this committee led to the digital broadcast standard adopted by the FCC in the US.
ATTC	*Advanced Television Test Center.*

396 GLOSSARY

(Continued).

Term	Definition
ATV	Advanced TV. Also called *ATSC*. A digital broadcast standard approved by the *FCC* in the US that is scheduled to replace *NTSC* in 2009. The ATV standard allows broadcast of a single *HDTV* picture or multiple *SDTV* images, as well as non-image data. The abbreviation *DTV* is replacing ATV in common usage to describe this standard. The abbreviation ATSC is also sometimes used.
Autoconvergence	A system in which the *convergence* of *red*, *green* and *blue* is performed by special electronics built into the display.
Axicon	A cone-shaped optical element.
Back end electronics	Electronics in a projector or other display device that reformat the *video signal* to provide timing, drive levels and the number of parallel subsampled outputs required to directly drive the flat panel or other display device. The input to the back end electronics is typically the RGB signals from the *Display Processor*. See *Front End Electronics*.
Back focal length	The distance between the rearmost element of a projection lens and the image to be projected, such as the image formed by a microdisplay.
Ballast	A specialized power supply intended to run a discharge lamp, such as an *arc lamp*. The *starter* may or may not be included in the ballast.
Band pass filter	An optical filter that transmits a band of wavelengths and reflects or absorbs longer and shorter wavelengths.
Barrel distortion	A *geometric distortion* of a video image in which the edges of the image are too far from the center relative to the corners of the image. See *pincushion distortion*,
Baseband video	Video information that has no RF carrier. See *composite video*.
BD	*Blu-ray* Disc.
BDRF	Bidirectional Reflectance (distribution) Function.
BEF	Brightness Enhancement Film. A film (actually a composite made up of multiple film layers with different properties) to optimize the utilization of the light from the CCFL or other lamps in a LCD backlight unit (BLU).
Biaxial material	A *birefringent material* with three different *indices of refraction*. See *uniaxial retarder*.
Birefringence	See *birefringent material*.
Birefringent material	A material, often crystalline or polymeric, whose *index of refraction* depends on the direction of polarization of the incident light. Uniaxial material has two different indices of refraction, commonly called n_o, the ordinary index and n_e, the extraordinary index. Bi-axial material has 3 different indices of refraction.
Bit	The basic unit of digital data. A bit can only have two values, conventionally called 0 and 1.
Bit-sequential gray scale	The achievement of gray scale with a binary optical system such as the DMD or SS-FLC by displaying the highest order bit for half the available time, the next bit for quarter the time etc also called time-multiplexed gray scale.
Black	In display terms, the lowest luminance producible by the display.
Black Chrome	A Cr/CrO2 layer applied to glass as a matrix around the pixel apertures in direct view LCD with spatial color. This layer shields the TFTs from light, plus increases contrast ratio in high ambient (CrO2 applied directly to glass, Cr on top of that). Black polymer is replacing black chrome because of environmental concerns with chromium and the simpler processing of the polymer.
Black matrix	Opaque, black areas added to absorb stray light in the unused areas between active areas in a screen, microdisplay or CRT.

GLOSSARY

Blackbody radiation	The *spectral distribution* of light emitted by a hot, glowing object at a given temperature. Temperatures of interest to video people are in the range from 2700 K to 13 000 K. While this temperature range produces a wide variety of colors in x-y space, all these colors can be called 'white'. Typically the white color produced by a projector should have a *correlated color temperature* in the range of 6550 K to 9000 K.
Blue light	Light in the spectral region from approximately 400 nm to 500 nm.
Blu-ray	A high capacity optical disk storage system suitable for HDTV, supported by Sony and others.
BOM	Bill of Material. This is a complete list of parts, including the cost of those parts, needed to build a system or subsystem. Cost of Goods (COG) is also used for this same monetary figure.
Brick Sampling	A sampling process in which the samples of one row are offset by half the sampling interval compared to the samples of the rows above and below.
Brightness	The physiological and psychological response of the viewer to the physical quantity *luminance*. Brightness is not measured by an instrument: it is perceived by a human being. *Luminance* is the quantity measurable by an instrument.
Brightness uniformity	See *white field uniformity*.
Burn back	The erosion of an electrode in an *arc lamp* due to sputtering of the electrode material. This normally leads to an increase in arc size and sometimes to a motion of the arc centroid.
Burner	The discharge lamp itself, without associated reflectors, lenses or electronics.
C-plate	C-plate is a uniaxial birefringent plate with its extraordinary axis (i.e., its optic or c-axis) perpendicular to the surface of the plate and parallel to the direction of normally incident light. A C-plate is used with a LCD to correct for off-axis contrast loss.
Calligraphic scan	A calligraphic scan, also known as vector scan or stroke writing, is used to draw the characters or images to be displayed. This can be done in systems such as a CRT or a laser display where the image is created by scanning a single image-writing beam. See *Raster Scan*.
Candela	SI unit of luminous intensity. It is defined as 1/60 of the intensity of 1 cm^2 of a *blackbody* at the solidification temperature of platinum. A point source of 1 candela radiates 1 lumen into a solid angle of 1 steradian.
Cathode ray tube	Abbreviated CRT. A vacuum tube in which an electron beam strikes a phosphor which then emits light. See *raster*.
CCD	Charge coupled device. A type of integrated circuit that can be used as an image sensor. Most modern electronic cameras use CCD image sensors.
CCIR	International Radio Consultative Committee. A standards setting organization that is no longer functioning. All CCIR standards have been redesignated *ITU-R* standards.
CCITT	Consultative Committee on International Telephony and Telegraphy. An agency of the United Nations International Telecommunication Union (*ITU*), which coordinates global activities in telecommunication.
CCT	*Correlated color temperature.*
CEA	Consumer Electronics Association (http://www.ce.org/).
CEDIA	Custom Electronics Dealers and Installers Association (http://www.cedia.net/).
Center weighted lumens	Calculating the *lumens* of a projector by measuring the *illuminance* in the center of the screen and multiplying by the area of the projected image. This method is easy and nearly always gives too high a number. See *ANSI lumens* and *sphere lumens*.

GLOSSARY

(Continued).

Term	Definition
Centroid wavelength	The wavelength value where half of the light energy in a *LED* emission *spectrum* is at shorter and half the energy is at longer wavelengths. Visible LEDs are more normally specified by *dominant wavelength*.
CGA	Color graphics adapter. An obsolete IBM-PC video format with 640 by 200 pixel resolution. CGA systems could only display 16 colors.
Chiral	A general term related to handedness. In a chiral system there is a distinct difference between clockwise and counterclockwise. Circularly-polarized light behaves differently, depending on its rotation direction compared to the chirality of the material.
Chiral agent	A *chiral* material that can be added to a *liquid crystal* material to give it a preferred rotation direction.
Chiral laser	A laser utilizing the band gap associated with dyed chiral materials.
Chiral nematic	A LC phase formed by adding a chiral or optically active center to a nematic LC molecule, which in turn induces a twist in the long range order of the nematic system. Sometimes called *cholesteric* since the original implementations involved the cholesterol molecule. In display applications, this phase is bi-stable between *planar alignment* and *focal conic alignment*. See *twisted nematic liquid crystal*.
Cholesteric	See *Chiral nematic*.
Chroma or Chrominance	The encoded *color* information in the video signal. When combined with the *luma* signal, it gives a complete color picture. See Y/C, U-V.
CIE	Commission International de l'Eclairage. In English, the International Commission on Illumination. This is the international standards setting body on illumination. The CIE 1931 X-bar, Y-bar and Z-bar functions are the most commonly used definitions of 'typical' human eye response (http://www.cie.co.at/cie/home.html).
CIF	A video format used in videoconferencing systems that easily supports both NTSC and PAL signals. CIF is part of the ITU H.261 videoconferencing standard. It specifies a data rate of 30 frames per second (fps), with each frame containing 288 lines and 352 pixels per line. CIF is sometimes called Full CIF (FCIF) to distinguish it from *QCIF*.
Circularly polarized light	Light with the electric or magnetic field rotating continuously with respect to the direction of travel of the light. This rotation can be right handed or left handed. See *polarization*.
Clear aperture	See *aperture ratio*.
CMYK color system	*Cyan, magenta, yellow* and *black*, the *subtractive primary colors* used in color printing.
Codec	Compression/decompression. Commonly used in the phrase 'Video Codec' to describe the particular video compression scheme used in digital video. For example, the MPEG-4 codec is used in HDTV.
Cogging	A temporal artifact due to aliasing of moving, repetitive objects. The classic example of cogging is the spokes of a stagecoach appearing to turn backwards in a cowboy movie.
Cold mirror	A mirror that transmits infrared (*IR*) and reflects *visible* light.
Collection efficiency	(1) The fraction of the light generated by the lamp that is concentrated into the output beam. (2) The fraction of the light generated by the lamp that is focused on the light valve, ignoring any losses in the optical system except losses at the reflector and étendue related losses.
Collimated light	A light beam in which the nominal ray directions in the beam are all parallel.

GLOSSARY

Color	The differing sensations resulting from stimulation of the retina of the eye by light waves of different wavelengths.
Color coordinates	A two dimensional space established by the *CIE* used to describe colors in *colorimetry*. Two common scales are in use, *x-y* and *u'-v'*.
Color gamut	The range of colors that can be reproduced by a display. For example, the color gamut of a *RGB* monitor is typically larger than the gamut of *CMYK* inks used in printing. See *Wide color gamut display*.
Color rendering index	Abbreviated CRI. A value assigned to a lamp that depends on how closely the *spectrum* of the lamp matches the spectrum of the test source, commonly *D65*. 100 indicates a perfect match. A high color rendering index (CRI > 90) indicates that colors will appear essentially the same under the lamp illumination as under the test source.
Color uniformity	See *white field uniformity*.
Colorimetry	The science of measuring colors.
Column	In a pixelated display, pixels are normally arranged in vertical columns or staggered columns. Normally, a single conductive electrode connects every pixel in a column to the column driver.
Column driver	The electronics that drive a complete column in a display.
Coma	An optical aberration in which the center of the lens focuses off axis light at a slightly different position than the outer portion of the lens. This results in a point source imaging to a comet-like image with a small bright area and a larger, off-center tail.
Comb filter	As used in video, a 2-D filter designed to separate signals having interleaved harmonics from each other. An example is a *Luminance* and *Chrominance* separation filter in *NTSC*.
Compensation film (or foil)	A *waveplate* used with an *LCD* to correct residual *polarization* errors induced by deviations from the ideal *alignment*, e.g. correct for *pre-tilt*. See *C-plate*, *A-plate* and *O-plate*.
Composite video	Composite video is a signal that contains video information for color, brightness and horizontal and vertical synchronization. Composite video is capable of being sent over a single wire or other channel. See *baseband video*.
Compression	Image compression is used to reduce digital storage and transmission costs by eliminating redundant information. There are several standards including *JPEG* (still-frame), *MPEG* (moving pictures) and Px64 (low resolution teleconferencing). Compression techniques can cause *artifacts* in the image. See *decompression*.
Compression ratio	The ratio of the data rate required to transmit uncompressed video to the data rate required for the compressed video. Compression ratios of 70:1 or higher can still provide acceptable video with modern compression techniques.
Cone prism	See *axicon*.
Conjugate distance	The distance from a lens to the plane where the light comes to a focus. See *back focal length*.
Contouring	Visible luminance steps in a video image that occur when there are insufficient number of bits per pixel in the source *digital video*. See *posterizing*.
Contrast	The ratio between the luminances of the brightest and darkest parts of a picture. Also, the ratio between the brightest white and the darkest black it is possible to make with a projector.
Contrast enhancement	Any system designed to improve the contrast of an image. Most contrast enhancement is designed to make the dark areas darker, thereby increasing the ratio between light and dark areas.
Convergence	Alignment of the red, green and blue images accurately enough so the eye fuses them into a single full-color image. The accuracy of convergence can vary in different locations on the screen.

GLOSSARY

(Continued).

Term	Definition
Correlated color temperature	Commonly abbreviated CCT. The temperature of a *black body* object whose color is closest in *u-v color coordinates* to the color of interest. This only has meaning when a color is close to the black body curve.
CPC	Compound Parabolic Collector (or Concentrator). A collection system designed to collect light from extended sources. One application is where a *LED* is placed at the focus and a nearly collimated beam of light emerges from the CPC.
CPE	Compound Parabolic Extractor. See *CPC*.
CPL	Compact Power Light, a Philips designation for ultra-high-pressure mercury lamps capable of sustaining higher powers than the *UHP* lamp. Intended for electronic cinema and other large-venue applications.
CRI	*Color rendering index*.
Critical illumination	An illumination system in which the light source (arc or filament) is imaged on the light valve or other object to be projected. Movie projectors typically use critical illumination. See *Köhler illumination*.
Cross color and cross luminance	In a composite video signal, artifacts caused by the interaction of the luminance and chrominance signals. These artifacts can be reduced by *comb filters*.
CRT	*Cathode ray tube*.
c-Si	*Single crystal silicon*. Sometimes abbreviated *x-Si*.
Cut-off wavelength	The wavelength of an optical filter where the transmission falls rapidly through 50 % (or some other value), with low transmission at longer wavelengths and high transmission at shorter wavelengths.
Cut-on wavelength	The wavelength of an optical filter where the transmission rises rapidly through 50 % (or some other value), with high transmission at longer wavelengths and low transmission at shorter wavelengths.
CVBS	Composite Video Baseband Signal (alternatively, Chroma, Video, Blanking and Sync)
CVD	Chemical Vapor Deposition, a process technology in which chemical reactions in the vapor state result in material deposition on substrates. The technology is used in the manufacture of semiconductor and display devices.
Cyan light	Light formed by adding together the *blue* primary and the *green* primary colors of a display. Also can refer to light in a wavelength band centered around 480–495 nm. This band can be narrow or it can have a full width of as much as 200 nm.
D65	The *CIE* defined light source that closely matches the spectrum of natural daylight. The CIE 1931 color coordinates of D65 are $x = 0.313$, $y = 0.329$.
DAC	Digital to analog converter.
DARPA	Defense Advanced Research Projects Agency. A US Department of Defense agency that funds research projects for military applications.
DBS	See *Direct Broadcast Satellite*.
DCI	Digital Cinema Initiative (http://www.dcimovies.com/). This was a consortium of seven major movie studios including Disney, Fox, MGM, Paramount, Sony, Universal and Warner Bros. This consortium was formed in March 2002 to establish technical standards and business models for high performance digital cinema. The DCI standard v 1.0 was issued 27 July 2005. SMPTE later defined a series of standards and recommended practices to ensure all *DCI Compliant* projectors would work together.
DCI Compliant	A server or projector that is in compliance with DCI standards and can show DCI encrypted feature films and other material.

GLOSSARY

DCP	Digital Cinema Package. A complete, compressed group of files ready to be distributed to theaters for showing on a *DCI Compliant* projection system. The DCP is the electronic equivalent of a *release print*.
DCT	See *Discrete Cosine Transform*.
Decoder	A system that takes a specified *analog* or *digital video format* as input and produces an output stream of uncompressed *RGB* or *YUV* data.
Decompression	The reversal of the *compression* process to restore the original *video* image.
Delta arrangement	An arrangement in which the *red, green* and *blue sub-pixels* are arranged into a roughly equilateral triangle.
Depixelization	Removal of the visible *pixel* structure in an image generated by a *pixelated* display. This can be done with an optical element that increases the apparent *aperture ratio* of a pixelated microdisplay to 100 % or higher (overlap). Effectively, this is a low pass spatial filter with the roll-off starting at the pixel *Nyquist frequency*.
Desaturated color	A color made with a significant admixture of all three primary colors. Typically, these are pastel colors. See *Saturated color*.
DH-FLC	Deformable helix ferro-electric liquid crystal. A *chiral SmC LC* effect that exhibits electrically controllable *gray scale*. See *SS-FLC*.
Dichroic filter	An optical filter designed to split light into two color bands. One band is transmitted and the other reflected. Normally there is no absorption in a dichroic filter. Dichroic filters are normally made with multiple thin layers of two or more different *dielectric materials*. The word 'dichroic' is sometimes used alone to generically describe any optical filter made from multi-layer dielectric materials.
Dichroic polarizer	An absorptive *polarizer* that transmits one *linear polarization* and absorbs the other linear polarization.
Dielectric coatings	In optics, multiple layer *dielectric* coatings are used to control the behavior or light at an interface between materials with two different indices of refraction. Depending on design, these coatings can act as anti-reflective (*AR*) layers reducing reflection, mirrors to increase reflection, *dichroic filters*, or *polarizing beam splitters*.
Dielectric material	A insulating dielectric material is a material that can support an electric field with little or no conduction. Since there is no current flowing, there is no resistive heating and there is no absorption of energy from the electro-magnetic field.
Diffuser	A surface or thin object that scatters light from a smaller cone angle into a larger cone angle. A perfect reflective diffuser would scatter light uniformly over 2π steradians. A perfect transmissive diffuser would scatter light uniformly over 4π steradians, including both backscatter and forward scatter.
Digital domain	The portion of a signal processing chain in which all signals are processed using digital techniques.
Digital format	A video format in which the information is carried by a series of bits that can only take on the values 0 and 1. An example is the *Grand Alliance HDTV* format. See *Analog format*.
Digital Satellite System	A commercial system in the USA in which video signals are *MPEG* encoded and broadcast directly to homes. See *Direct Broadcast Satellite*.
Digital signal processing	The processing of signals in the *digital domain*. The abbreviation DSP is often used to indicate a Digital Signal Processor, the chip that does the processing.
Digital video	Video in which the information is conveyed by a series of *bits*.
D-ILA	Digital Integrated Light Amplifier. A LCoS system developed by JVC from the ILA technology acquired from Hughes. It uses an active matrix silicon backplane to replace the CRT-based optical drive. Newer versions are sometimes referred to as HD-ILA.

(Continued).

Term	Definition
DIN	Deutsches Institut für Normung e. V. A German standards setting organization (http://www.din.de/frames/Welcome.html).
Direct broadcast satellite	Direct broadcast from satellites to a local antenna at the display site.
Direct view	A display in which the *flat panel* or *CRT* is viewed directly, rather than projected onto a screen.
Director	Vector which describes the long molecular axis of an LC molecule.
Display processor	Electronics in digital projection and other display systems in which signal processing of the video signal is performed. The processing can include, among other actions, reformatting to adapt the signal format to the display format. These electronics typically take the decoded output from the *front-end electronics* and, after processing, provides conditioned RGB signals to the *back-end electronics*.
Dither	The simulation of a gray scale or continuous area of color by using a pattern of small dots of more than one color or brightness. Typically, no individual dot will match the color or brightness target but the human eye averages over multiple dots and will perceive the desired color and brightness. Dithering can be done in the spatial, temporal or color domains.
DLP	A Texas Instruments trade name for a projection system utilizing the *DMD*.
DLPC	Digital Light Processing–Cinema. High resolution (2K) DLP systems for electronic cinema.
DMD	Digital Micromirror Device, a reflective micro-electro-mechanical technology developed by Texas Instruments. See *DLP*.
Dominant Wavelength	The wavelength of an *LED* emission *spectrum* that corresponds to the same hue (color) as the entire emission spectrum. If a straight line is drawn from the color point corresponding to a reference spectrum is drawn through the color point of the LED to the boundary of the color diagram, the wavelength corresponding to this intersection is the dominant wavelength of the LED. The equal energy spectrum with $x = 0.333$ and $y = 0.333$ is most commonly taken as the reference point.
Double lenticular screen	A *rear projection screen* with vertical cylindrical *lenslets* on both the front and rear of the screen. The lenslets operate in pairs to provide color correction for *CRT* systems.
DPI	Dots per inch. The normal way of specifying the resolution of printed output.
DPSS or DPSSL	Diode Pumped Solid State (Laser). Also called OPSL: Optically pumped semiconductor laser.
DRAM	Dynamic *random access memory*. A semiconductor memory design that requires only one transistor per memory cell. DRAM must be periodically refreshed. During the refresh period, the data in the DRAM is not available for use. Most active matrix systems are fundamentally write-only DRAM designs. See also *SRAM*.
DSP	*Digital Signal Processing* or Digital Signal Processor.
DSS	*Digital Satellite System*.
DTV	Digital television. When used generically, this means the use of *digital formats* instead of *analog formats* for video. Often this acronym is used to specifically describe the *ATV* format that has been approved for use in the United States.
DVD	Digital Video Disk A high density industry standard that will contain enough data to encode 135 minutes of *MPEG-2 standard definition* video. The standard specifies both the encoding scheme and the disk physical properties.
DVE	Dark Video Enhanced. DLP systems with a dark green color wheel segment designed to improve the dynamic range of the system to lower gray levels.

DVI	Digital Visual Interface see www.ddwg.org
DVI-HDCP	DVI with High Definition Copy Protection encoding.
EBU	European Broadcasting Union, the governing body for European television broadcast standards (http://www.ebu.ch/).
ECB	*Electrically controlled birefringence.*
EDTV	Enhanced or extended definition TV. A backward compatible television system in which extra information is broadcast to enhance the resolution or aspect ratio of a video signal. Older sets ignore the additional information and display a standard definition TV signal. EDTV video can be either 4:3 or 16:9. *PAL+* can be considered an EDTV broadcast format. There is no accepted standard *NTSC* EDTV format. See *IDTV*.
Efficacy	Conventional symbol η. The number of *lumens* of output per watt of input electrical power for a lamp. The maximum theoretical value is 683 lumens/watt for a lamp that achieves 100 % electrical power to light conversion with all its output at 555 nm. Sometimes called luminous efficacy or, incorrectly, luminous efficiency.
EGA	Extended Graphics Adapter. An obsolete IBM-PC video format with 640 by 350 pixel resolution. EGA systems could only display 16 colors.
EHPL	Iwasaki Electric Co., Ltd trade name for its ultra-high pressure mercury lamp. See *UHP*.
EIA	Electronic Industries Association, now renamed the Electronic Industries Alliance. A trade organization that includes the full spectrum of U.S. manufacturers (http://www.eia.org/).
EIAJ	Electronic Industries Association of Japan. See *JEITA*.
Eidophor	A very high lumen output *oil film light valve* projection system, the production of which ceased after about 50 years.
Electrically controlled birefringence	A *nematic LC* phase where there is *homogenous alignment* at both substrates when there is no voltage applied to the cell, and the *directors* are parallel. In this condition, *linearly polarized* light aligned at 45° to the directors is converted to *circular polarized* light upon passing through the cell. With voltage applied, there is *homeotropic alignment* at both substrates and the polarization of light passing through the cell is not changed.
Electronic cinema	High resolution, high brightness projected video intended to provide image quality comparable to the film in a cinema projector. Electronic cinema must be an end-to-end system, including delivery and storage systems as well as projectors. See *DCI*.
EMI	Electromagnetic Interference.
Encoder	A system that takes a stream of input pictures and produces an output signal in a specified analog or digital *video format*.
Entrance pupil	In a lens or other optical system, the image of the aperture stop as seen from object space.
EPD	Electrophoretic display. Also abbreviated EPID, for electrophoretic image display or device.
Epi	Short for epitaxial. See epitaxial water.
Episcopic projection	Otherwise known as opaque projection. Light illuminates the front surface of an opaque object and is scattered into a projection lens. While this projection technique is inefficient optically, it allows projection of object or document images without any special object preparation.
Epitaxial wafer	Single crystal wafers of silicon or other semiconductor materials with a particular crystal plane, such as < 111 >, on the surface.

(Continued).

Term	Definition
ESD	Electro-static discharge. The discharge of static electricity that can potentially damage sensitive electronic circuits, including many types of *light valves* or *microdisplays*.
Étendue	The geometrical properties of the light at any surface in the optical system, without reference to light distribution within that geometry. Also known as extent or optical invariant.
Étendue limited	A projector whose total *lumen* output is reduced from the theoretical maximum by the *étendue* of one or more components.
ETSI	European Telecommunication Standards Institute.
Extent	See *étendue*.
Extinction ratio	The amount of light transmitted through a pair of polarizers when their optical axes are parallel to each other divided by the amount of light transmitted when the two optical axes are perpendicular.
f/#	Pronounced f-number, a measure of the acceptance cone angle of a lens. Often approximated by the ratio f/D, where f is the focal length and D is the diameter of the lens.
FCC	Federal Communications Commission, the US Government agency responsible for radio spectrum use.
FELC	*Ferro-electric Liquid Crystal.* Sometimes abbreviated FLC.
FELCD	An FELC-based device.
Ferro-electric liquid crystal	A bi-stable *chiral SmC LC* effect.
Field (of a lens)	The area that a projection lens projects on the screen. A field of a lens is normally circular. A microdisplay or other device typically only uses a rectangular portion of the field.
Field (video)	One *frame* of 2:1 *interlaced* video consists of two fields, one containing the even lines, the second containing the odd lines. Each field corresponds to a different time and contains 1/2 the data necessary to produce a single still image. For *non-interlaced* video, field and frame can be used interchangeably.
Field flattener	An optical element near the image or object plane whose purpose is to convert a curved image plane into a flat image plane.
Field lens	A lens used to alter the convergence or divergence of an optical beam.
Field rate	The frequency at which video *fields* are displayed.
Field sequential color	A system where the *red*, *green* and *blue* portions of a video *field* are presented sequentially, each for about 1/3 the field time. The three sub-fields are merged by the eye into a single, full color field. Sometimes abbreviated FSC.
Fill factor	See *aperture ratio*.
FIR	Finite impulse response.
Flat panel display	A display such as an LCD, plasma display or electro-luminescent device in which the display is flat and relatively thin compared to its height or width. A FPD is normally used in a *direct view* application.
Flicker	The rapid and periodic fluctuation in intensity of a light source, as perceived by the human eye. Individual fluctuations are not necessarily resolvable, but the viewer is aware the light is fluctuating.
FMV	*Full motion video.*
Focal conic alignment	A *chiral nematic* liquid crystal mode that transmits light of both circular polarizations. See *planar alignment*.

GLOSSARY 405

Focal length	The distance from the principal plane of a lens to the focal point. For a thin single lens, this is the distance from the center of the lens to the point where that lens will bring a *collimated* beam of light to a focus at a single point.
Fold mirror	A flat mirror in an optical path whose only function is to turn the optical axis of the system.
Foot-Lambert	*Lumens* per square foot when the light source is *Lambertian*. Used as a measure of the *luminance* from a *CRT* or other large area light source. Foot-lambert is obsolete and the metric unit candela/meter^2 (cd/m^2 formerly known as 'nits') should be used instead. 1 Foot-lambert = 3.426 cd/m^2.
Format	See *video format*.
FPD	*Flat panel display*.
Frame	One complete video image. A frame contains all the data for a single still image. It may consist of one or more *fields*.
Frame rate	The frequency with which video *frames* are displayed.
Frame reformatter	The electronics necessary to match the incoming signal *video format* to the display format of the *light valve*. See *display processor*.
Frame transfer addressing	An electronic addressing mode for a *light valve* in which the data for all *pixels* are reset simultaneously and all the pixels in the light valve change their optical state simultaneously.
Freeze frame	The freezing of a single image on a display out of a sequence of moving images.
Fresnel lens	A flat, faceted lens which emulates the function of a thick lens. In a rear *projection screen* the Fresnel lens is designed to act as a *field lens* and to bend the off-axis ray bundles so they are parallel or nearly parallel to the centerline of the projector.
Fresnel reflection	The reflection that occurs at any optical interface where the *index of refraction* of the materials change.
Front end electronics	Electronics that accept the incoming *video signal* from a standard video source such as a VCR, cable or terrestrial broadcast. The front end electronics then decode the data into its separate components such as luma, chroma, sync and audio. See *Decoder, Frame reformatter* and *Display processor*.
Front projection	A projection system in which the light is incident on the *projection screen* from the same side as the audience. Cinema systems, for example, are typically front projection.
Front projection screen	See *Projection screen*.
FSC	*Field Sequential Color*.
Full HD	A HDTV system with 1920 × 1080 resolution, the maximum resolution supported by the HDTV standard.
Full-motion video	A video image that moves naturally and without jerkiness. Smooth motion is possible by using from 24 to 30 frames per second. A *frame rate* fast enough to provide full motion video will not normally be fast enough to prevent *flicker*.
Gain	The gain of a projection screen is the amount of light visible in a given direction (e.g. on the centerline) compared to the amount of light that would be visible if a perfect *diffuser* were substituted for the screen.
Gamma (CRT)	The exponent, γ, in the power law relationship between the voltage in a video signal, and the cathode current in a *CRT*. Most CRT displays have a γ in the range of 2.2–2.8. See *Gamma correction*.
Gamma (video)	The exponent, γ, in the power law used to encode luminance in a video signal. The expected amount of light at the CRT screen is then proportional to the current. Most *video* signals are encoded with a γ of 1/2.2. In CRTs this is automatically decoded by the gamma of the CRT. For non-CRT displays, see *Gamma correction*.

GLOSSARY

(Continued).

Term	Definition
Gamma correction	1) Adjustment of the signal-brightness relationship of non-CRT displays to accommodate the 1/2.2 gamma encoded into normal video signals. 2) The correction of a video signal to accommodate the difference between the way the camera records the light and the way the *gamma* value specified for a given video format. This correction is often done in the camera itself.
GaN	Gallium Nitride, a material used in blue and green LEDs and laser diodes.
GCD	Gaze-Contingent Display also known as a foveal/peripheral display. A display that provides the foveal area of the retina with higher resolution images compared to the images provided to the peripheral area of the retina. Requires eye tracking and sophisticated computer software to generate the images to be displayed.
GCR	*Ghost Cancellation* Reference (signal). Reference signal for correcting errors introduced in *analog* video transmission channels.
Geometric distortion	A deviation of the reproduced picture from its intended geometry.
Ghost cancellation	In multi-path reception of *analog video* signals (e.g., reflections from big buildings) the direct and reflected signals will combine to produce ghosts on the video screen. *VSP*, with the use of the *GCR* can reduce or eliminate these ghosts.
Grand Alliance	The alliance between several US companies and US divisions of European corporations that produced the digital *HDTV* video format for the US broadcast market.
Gray scale	The ability of a *projector* or *pixel* to make a range of intensities from black to white.
Green light	Light in the spectral region from approximately 500–570 nm.
Half angle	The angle from the centerline where the apparent brightness of a projected image drops by 50%. This can be dramatically different for vertical and horizontal angles. For example, a typical consumer CRT rear projection screen will have half angles of ±45° horizontal and ±8° vertical.
Half wave plate	A thin layer of *birefringent* material aligned to introduce 1/2 wave retardation of the fast axis relative to the slow axis. On linearly polarized light starting at a 45° angle to the birefringent axis, this has the effect of rotating the *polarization* of light passing through it by 90°.
HDMI	High Definition Multimedia Interface (see www.hdmi.org). *DVI* with a smaller connector, content protection, sound and control signals added. Backward compatible with DVI, so a connector adapter is possible if all you want is video. HDMI 1.3 was released in June 2006. One main feature is 2x the bandwidth of earlier version. Also supports *xvYCC* color encoding.
HDR	High Dynamic Range
HDTV	*High Definition TV*. Sometimes abbreviated simply as HD.
HID lamp	High Intensity Discharge Lamp. See *Arc Lamp*.
High definition TV	Consumer television video signal formats that produce significantly better image qualities than a *standard definition* format. All current and proposed HDTV formats use a 16:9 *aspect ratio*. Most current and proposed HDTV systems are not backward compatible with existing standard definition receivers without conversion electronics. See *IDTV, EDTV*.
High temperature polysilicon	*Polysilicon* produced by depositing silicon at high temperatures (typically > 1000 °C) frequently using CVD, producing multigrain Si with long range order. Since processing temperatures can exceed 1000 °C, a glass substrate cannot be used. The abbreviation HTPS is sometimes used to refer to LCD microdisplays made with this material.

Highlights	The lightest part of an image, normally white or nearly white.
Histogram	The statistical distribution of levels in a video signal from black to white.
Hi-Vision	The *analog HDTV* format used in Japan.
HMD	Head-mounted display. Alternatively, helmet-mounted display. A display small enough to be mounted on a person's head or helmet. It may occupy only a small portion of the field of view, allowing normal vision or it may occupy the full field of view, allowing the display to replace the real world as the image source for vision.
HOE	*Holographic Optical Element.*
Holographic optical element	A hologram designed to replace a conventional optical element. HOE's can be designed as lenses with either positive or negative focal lengths, prisms or diffraction gratings for separating colors, diffusers, projection screens or microlenses. Typically the properties of a HOE are strongly dependent on wavelength.
Homeotropic alignment	The *directors* of the *liquid crystal* molecules are oriented perpendicular or nearly perpendicular to the surface of the substrate confining the LC. This alignment is typically seen in the *nematic phase*.
Homogeneous alignment	The *directors* of the *liquid crystal* molecules are oriented parallel or nearly parallel to the surface of the substrate confining the LC. This alignment is useful in *nematic phase* display applications.
Hot mirror	A mirror that transmits *visible* light and reflects infrared (*IR*).
Hot restrike	The reignition of an *arc lamp* after the arc is extinguished and before it has a chance to cool completely.
HUD	Head-up display. A display visible to a pilot or driver that is visible while he is looking out the window.
HVS	Human Visual System.
IARC	Internal angularly reflective coating. A multilayer dielectric coating to be applied to the inside of projection CRTs that would increase the light output of the tubes by 30–100 %.
ICC	International Color Consortium see http://www.color.org/.
IDTV	Improved (or Intermediate) definition TV. A television system intermediate between *Standard Definition* and *High Definition*. IDTV is not a well defined standard, and normally refers to a signal processing method at the receiver and/or the source that improves the resolution beyond that of standard definition in a method that is backward compatible with standard definition receivers. Normally no additional information is broadcast in an IDTV signal. See *EDTV*.
IEC	International Electrotechnical Commission (http://www.iec.ch). An international standards organization based in Switzerland that establishes standards for technical equipment.
IEEE	Institute of Electrical and Electronics Engineers (http://www.ieee.org/).
Ignitor	See *Starter*.
ILA	Image Light Amplifier. A trade name used by Hughes/JVC to describe their optically addressed *light amplifier* technology. See D-ILA.
Illuminance	The density of light incident upon a surface. 1 lumen distributed evenly over 1 square meter will produce an illuminance of 1 lux. Conventional symbol E_v.
Illumination path	The set of optical elements that produces the desired *light body* at the plane of the *light valve* in a projection system.
Image compression	Reducing the bit rate and bandwidth needed to transmit a video signal by removing data redundancy and data that codes for parts of the image that would not be visible to the human eye. See *Compression* and *decompression*.

GLOSSARY

(Continued).

Term	Definition
Index of refraction	The ratio of the speed of light in a vacuum to the speed of light in a denser medium. This property controls (among other things) the behavior of light at surfaces between two different mediums, for instance the surface between glass and air. Typically, the index of refraction of an optical glass or plastic is in the range of 1.45–1.8.
Indium tin oxide	A transparent, conducting material often used as a transparent electrode in a *LCD* or other *flat panel device*. Commonly abbreviated ITO.
Industrial Technology Research Institute	A governmental agency that supports research and development into display and other industrial technologies in Taiwan. Abbreviated ITRI.
Infitec	Infitec is a trade name of INFITEC GmbH., and is short for interference filter technology. This wavelength-multiplexed 3D technology grew out of research at DaimlerChrysler AG Germany.
InGaAlP	Indium Gallium Aluminum Phosphide is one material used to make Red & Orange *LED*s.
Integrator	An optical element, or pair of elements, designed to produce a *light body* with spatially uniform or near uniform illumination of a *light valve* or *microdisplay*.
Interlace	A video format in which the video *frame* is divided into two *fields* for 2:1 interlace. Even numbered lines are displayed in the first field, odd number lines are displayed in the second field. Typically, this is used to reduce the signal bandwidth by showing full frames at 25 Hz (*PAL*) or 30 HZ (*NTSC*). Higher interlace ratios, e.g. 3:1 or 4:1 are possible, but rarely used. In a *progressive* system (1:1 interlace), every line is displayed every frame and the field rate is equal to the frame rate.
Interpolate	To estimate values between two known values. For example, to display a low resolution image on a high resolution display, the additional pixels must be interpolated. Interpolation may also used to generate the missing lines in interlace to progressive conversion or the missing frames in a 50 Hz to 60 Hz conversion.
Invar	A metal alloy that has a coefficient of thermal expansion very close to zero.
Interpositive	Abbreviated IP. A positive print made from the original cinema negative. The IP is a 2nd generation image. See *answer print* and *release print*.
IR	Infra-red. This light has a longer wavelength than visible light and is invisible to the human eye. IR tends to heat up *light valves* and other optical components without contributing to the total lumen output of a projector. IR can be minimized with a *hot* or *cold mirror*.
IS&T	The Society for Imaging Science & Technology (http://www.imaging.org/).
ISDN	Integrated Services Digital Network.
ISO	International Standards Organization (http://www.iso.ch/).
Isotropic phase	A phase exhibited by most *LC* materials at elevated temperature in which there is no long range order to the *directors* of the individual LC molecules. The isotropic phase does not exhibit *birefringence* or optical activity.
ITO	*Indium tin oxide*.
ITRI	*Industrial Technology Research Institute* (Taiwan)
ITU	International Telecommunications Union. United Nations regulatory body working for television standardization (http://www.itu.ch/).
ITU-R	International Telecommunications Union – Radiocommunication Bureau.
ITU-R-601	A standard definition digital video standard based on 8-bit and 13.5 MHz sampling for the luminance channel of the video.
ITU-R-709	A CCIR recommendation on the gamma correction and colorimetry of video signals for digital video.

GLOSSARY

JEITA	Japan Electronic and Information Technology. The Japan Electronics and Information Technology Industries Association (JEITA) was formed on 1 November 2000, through the merger of the Electronic Industries Association of Japan (EIAJ) and Japan Electronic Industries Development Association (JEIDA). JEITA is a new industry organization in Japan with activities covering both the electronics and information technology (IT) fields. See http://www.jeita.or.jp/english/index.htm
JPEG	Joint Photographic Experts Group, a defining body of the CCITT. Acronym is most commonly used for an *image compression* scheme that divides a still image into blocks and compresses with *discrete cosine transforms*. See *MPEG*.
JPEG-2000	The *compression* scheme used by *DCI-compliant* video material.
Judder	A temporal artifact associated with moving images that produces an appearance of jerky motion in the image. This occurs most commonly when the image is sampled at a low frame rate such as 24 or 30 Hz and converted to a higher frame rate for display by frame repeat without motion interpolation.
Kell factor	Originally a fudge factor designed to relate the horizontal analog resolution to the vertical sampled resolution of a CRT. Now sometimes used more generally to relate digital *resolution* and *pixel* count to analog *MTF*. The Kell factor indicates how close the highest *spatial frequency* contained in the input image can come to the *Nyquist* frequency of the display or other sampling system without the introduction of *artifacts*.
Keystone	A *geometrical distortion* in which the top (or bottom) of the image is wider than the bottom (or top) of the image and the sides are not parallel to each other.
Köhler illumination	An illumination scheme for projectors in which the light source (arc or filament) is imaged onto the entrance pupil of the projection lens. Normally, the pupil of the illumination system is also imaged onto the light valve or other object to be projected. Slide projectors typically have Köhler illumination. The term Köhler illumination is sometimes used incorrectly to describe any illumination scheme that is not *critical illumination*.
L/PH	Lines per Picture Height see *TVL/PH*.
Lambertian distribution	An angular distribution of light which appears equally bright regardless of the angle from which it is viewed.
Laser	Light Amplification by Stimulated Emission of Radiation. A light source that produces a highly coherent and collimated beam of a single wavelength. Laser light is normally *polarized*.
Laser Diode	A laser diode is a laser where the active medium is a semiconductor p-n diode junction. Laser diodes are sometimes referred to as injection laser diodes or by the acronyms LD or ILD.
Law of refraction	See *Snell's law*.
LBO	Lithium Triborate (LiB$_3$O$_5$) A non-linear material used for frequency doubling IR laser beams to produce blue and green visible light outputs.
LC	*Liquid Crystal*.
LCD	Liquid crystal device (or display). A device utilizing *LC* material to modulate light.
LCOS	Liquid Crystal on Silicon.
LED	Light emitting diode. A semiconductor p-n junction emitter of incoherent light. See *Laser diode*.
Legacy	A term commonly used by the computer and consumer electronics industries to indicate prior generations of equipment, software or standards.
Lenslet	A small lens that is one element in an array of similar lenses. Generally larger than a *micro-lens* and smaller than conventional lenses.

(Continued).

Term	Definition
Lenticular screen	A *projection screen* designed to spread the light from the *projector* over a wide horizontal angle without spreading it over a large vertical angle. While most lenticular screens are for *rear projection*, they may also be designed for *front projection* applications.
LEP	Light emitting polymer. An organic material that emits light when a current is passed through it. See *Organic LED*.
Letterbox format	The practice of broadcasting a 16:9 image in a 4:3 format by adding black bars to the top and bottom of the picture. This practice normally wastes a quarter of the video bandwidth and reduces resolution when done to an *NTSC* image. See *PAL+*.
Light amplifier	An optically addressed *light valve*.
Light body	The patch of light created by the illumination source in the plane of the *light valve*. The light body is normally larger than the LV in order to reduce alignment and uniformity problems.
Light engine	The subsystem in a projector which converts the incoming electrical signal to the optical output. This normally includes the complete optical path, electronics that drive the display device and the mechanical supports. The term might also include the necessary power supplies and lamp *ballast*. The term would not normally include the other electronics such as tuners, *decoders*, the outside case or the *projection screen*.
Light valve	Abbreviation LV. A device to spatially modulate light to produce an image. The light is generated externally by a lamp. In current usage, the term *microdisplay* is used more commonly than light valve.
Lighthouse	The structure that houses the lamp and at least some of the collection optics.
Limiting étendue	The *étendue* of the component, or combination of components, that has the lowest value of any étendue in the projection system. This is the component that limits the light throughput in *étendue limited* systems
Line, video	A single line across a display device patterned with the bright/dark information. On a *CRT* or other *raster scanned* device, a line is written with a single horizontal sweep of the electron beam.
Linearly polarized light	The restriction of the electric or magnetic field in electromagnetic radiation (light) to one of the two possible perpendicular planes. These two planes are commonly designated s and p. Strictly speaking, s- and p-polarizations are only defined in terms of tilted surfaces. Unpolarized light is randomly oriented with respect to the two planes.
Liquid crystal	Commonly abbreviated LC. A liquid phase molecular system which has long-range order. Such systems exhibit the flow properties of liquids and the optical properties of crystals. The optical properties of liquid crystals can be changed by such outside influences as the electric field and temperature.
Lissajou scan	A scan pattern where the both the horizontal and vertical axes trace out sine-wave patterns.
Litz wire	A multi-strand wire used in higher frequency magnetic coil winding to minimize resistance increases due to the skin effect. Used in the yokes of high-resolution projection CRTs.
Lm	Abbreviation for *lumen*.
Long pass filter	An optical filter that passes all wavelengths of light that are longer than its *cut-on wavelength* and reflects or absorbs all light with a shorter wavelength.
Lossless compression	An *image compression* method in which no image data is lost during the compression process. After decompression, the image will be identical in all respects to the original image.

Lossy compression	*Image compression* schemes that discard information as the size of the file is reduced. *MPEG* and *JPEG* are lossy compression schemes. Because of the lost data, the decompressed image is not identical to the original image, but can appear identical to the original image to the human eye.
Low pass filter	A spatial or temporal filter that only passes the low spatial or temporal frequencies while attenuating the high frequencies.
Low temperature polysilicon	*Polysilicon* produced by depositing silicon as *amorphous silicon* on a substrate followed by crystallization with localized heating by a laser or some other source. 'Low temperature' refers to below the melting point of glass. Abbreviated LTPS.
Lower twisted nematic	A twisted nematic system in which the twist is less than 90°. See *TN-ECB*, which is currently the favored term.
LPF	*Low Pass Filter*.
LSB	Least Significant Bit.
LTN	*Lower (or lesser) twisted nematic*.
Luma	The encoded luminance information in a *video signal*. When combined with the *chroma* signal, a complete color image is produced. When displayed without the *chroma* information, a black-and-white image is displayed. See Y/C, U-V.
Lumen	The SI unit of luminous flux, equal to the luminous flux emitted per unit solid angle by a standard point source having a luminous intensity of one *candela*. This unit can be thought of as the photometric equivalent of watts, with the *spectrum* weighted with the *CIE 1931* $\overline{Y}$ (Y-Bar) function. A lumen is 1/683 watts of light at 555.016 nm.
Luminance	A measure of *lumens* per unit projected area per steradian. Luminance is defined to correlate as well as possible with *Brightness*, as perceived by the human eye. Conventional symbol L_V.
Luminance signal	The portion of a *video signal* that carries the *luminance* information. This portion of the signal is often abbreviated Y, as in Y/C.
Luminous efficacy	See *Efficacy*.
Luminous flux	The total amount of light (weighted for $\overline{Y}$) emitted by a source, integrated over both area and angle. Unit of measurement is the *lumen*. Conventional symbol Φ_V.
Luminous intensity	Light emitted by a source in a particular direction. Unit of measurement is the candela, or lumens per steradian. Conventional symbol I_V.
Lux	Unit of illuminance equal to 1 lumen/m^2.
LV	*Light valve*.
Magenta light	Light formed by adding together the *red* primary and the *blue* primary colors of a display.
Matrix addressing	A method of addressing a *pixelated* display where there is one (or occasionally more) conductive lines for each *column* and each *row* of the display. A pixel is addressed by simultaneously activating the column and row lines that lead to that pixel.
Media Lab	Part of *MIT*. Founded in 1985 to devise new ways for people to interact with information.
MEM	Micro-electromechanical device. A device with very small moving parts, on the order of microns. The *DMD* is the most common example of a MEM in the projection business.
Mercury	An element used in *arc lamps*, sometimes in conjunction with *xenon* or *metal halide* salts.
Metal halide	A class of *arc lamps* that typically contain *xenon*, *mercury* and the halide salts of other metals.
Metamerism	Metamerism or metameric colors refers to colors that appear identical under some circumstances but differ under other circumstances or to other viewers.

(Continued).

Term	Definition
Micro-electro-mechanical	See *MEM*.
Micro-lens	A lens in which the dimensions are small compared to the dimensions of conventional lenses. Typically, in projection systems, they would be used in large arrays with pitches that match the pixel or sub-pixel pitch of a microdisplay. Micro-lenses can be used in *angular color separation* architectures, or they can also be used to concentrate light on the *clear aperture* of a pixel.
Micro-mechanical	See *MEM*.
Micron	Symbol μ or more properly μm. 10^{-6} meters.
MII	China's Ministry of Information and Industry.
Mixed domain	The portion of a signal processing chain in which signals are processed using a mixture of analog and digital techniques.
mm	Abbreviation for millimeter, 10^{-3} meters.
MOCVD	Metal organic chemical vapor deposition, A semiconductor fabrication technique for materials like GaN used for LEDs and Lasers.
Modem	Modulator/demodulator. A device for converting digital data so it may be transmitted over an analog communication channel. Modems for PCs using telephone lines for communication convert the data to an audio tone. More generally, the word is also used for the equipment to encode digital data on any analog communication channel for transmission.
Modulation	For a sine wave signal, the ratio of the maximum amplitude minus the minimum amplitude to the maximum amplitude plus the minimum amplitude. Modulation has a maximum value of 100 % when the signal is fully modulated and a minimum value of 0 % for an unmodulated signal.
Modulation transfer function	For an imaging system, the ratio of the output *modulation* to the input modulation of a spatial sine wave input image as a function of the *spatial frequency* of the sine wave.
MOEMS	Micro-Optoelectro-Mechanical Systems see MEMS
Moiré	A macroscopic pattern that is sometimes visible in the presence two repetitive patterns in which each pattern individually may not be resolved. The spatial frequency of the Moiré pattern is the difference frequency between the two individual patterns and/or their harmonics. In projectors, Moiré can occur between any two sampling items, including the input video, the pixelated light valve, the scan lines in a raster addressed light valve or CRT, the Fresnel lens and the screen.
Motion detection	In *video signal processing*, determining what portions of an image represent moving objects, and treating those portions of the image differently than static areas.
MPEG	Motion Picture Experts Group, a defining body of the *CCITT*. The acronym is commonly used for an *image compression* scheme that divides a moving image into spatial and temporal blocks and uses motion estimation and discrete cosine transforms to remove image redundancy. Most digital video applications, including HDTV, DCI, *ATV*, Video CD and *DSS* are MPEG-based systems. See *JPEG*
MPEG-1	Older version of MPEG used mostly to encode standard definition video. Originally intended only for progressive video, this standard is commonly used for interlaced video as well.
MPEG-2	Newer version of MPEG used to encode HDTV.
MPEG-4	*MPEG*-4 is a standard used primarily to compress audio and visual digital data for applications such as HD-DVD and Blu Ray disks.

GLOSSARY 413

MPR-II	A standard originally proposed by the Swedish Department of Labor, which set maximum levels of electromagnetic radiation emitted by monitors.
MPRT	Moving picture response time, a method of evaluating motion artifacts in LCD-TV.
MQW	Multiple Quantum Well LED or diode laser construction.
MSB	Most Significant Bit.
MTF	*Modulation Transfer Function*
Mura	Mura is a Japanese word meaning blemish that has been adopted in English to provide a name for imperfections of a display pixel matrix surface that are visible when the display screen is driven to a constant gray level. Mura defects appear as low contrast, non-uniform brightness regions, typically larger than single pixels.
NA	*Numerical Aperture.*
NECSEL	Novalux Extended Cavity Surface Emitting Laser **http://www.novalux.com**
Nematic phase	The simplest LC phase, exhibiting the least amount of order. In this phase the directors of all the LC molecules are arranged substantially parallel to each other but there is no other long-range order.
Neutral density	An optical component such as a filter, mirror or lens coating, whose transmission or reflection is the same for all wavelengths of visible light.
NHK	Japanese Broadcasting Corporation. Japan's sole public broadcaster (http://www.nhk.or.jp/english/).
NIH	Not invented here.
Nit	A measure of *Luminance*. 1 Nit $= 1$ Candela$/$m^2. Candela$/$m^2 (cd$/$m^2) is the preferred nomenclature for this unit and is used throughout this book. See also *Foot-Lambert*.
nm	Abbreviation for nanometer, 10^{-9} meters. Used throughout this book as a measurement of the wavelength of visible light.
Non-interlaced video	See *progressive scan.*
NSH	Ushio designation for a DC-driven ultra-high pressure mercury lamp. See *UHP*.
NTSC	National Television Standardizing Committee. This committee established the first color TV broadcast standard in 1953 that was backward-compatible with the existing monochrome broadcast standard. NTSC used as initials today normally refers to the signal format, not the committee that established it. This NTSC broadcast standard, with minor modifications, has remained in service in the US and many other countries but will be phased out in the US in 2009 by the ATSC digital standard. The current definition of NTSC is in *ANSI/SMPTE* 170M-2004.
Numerical aperture	Abbreviated NA. The sine of the half-vertex angle of the largest cone of *rays* that can enter or leave an optical element, times the index of refraction. Generally measured at the object or image point in a system. Related to the f/# by the equation: f$/\#= 1/$2NA.
Nyquist frequency	The maximum frequency that can be represented in sampled systems, a frequency of 1/2 the sampling frequency.
O-plate	An 'O-plate' is an optical retarder utilizing a layer of a positively birefringent (e.g., liquid crystal) material with its principal optical axis oriented at an oblique angle with respect to the plane of the layer.
Object colors	The color perceived by the human eye when an optical *spectrum* from a test source is reflected by a real object. It is also possible to talk about transmitted object colors, for transparent objects. Object colors will vary not only with the spectral properties of the object, but with the spectrum of the incident light.
OCB	Optically Compensated Birefringence. A LCD effect that is faster switching than the TN effect and can be used in reduced motion artifact or color sequential LCD systems. Requires very good optical compensation films in order to get acceptable contrast.

414 GLOSSARY

(Continued).

Term	Definition
Oil film light valve	A *light valve* that uses a thin liquid film in a vacuum. Writing on this film with an electron beam distorts the free surface and the surface distortion modulates the light by diffracting or refracting it. There are currently no commercially available oil film light valve projectors.
OLED	*Organic light emitting diode.*
Opaque projection	See *Episcopic projection*
OPO	Optical Parametric Oscillator, a non-linear optical technology used to produce wavelengths other than second-harmonic wavelengths from an IR source laser.
OPS	Optically Pumped Semiconductor (Laser), more commonly known as *DPSS*, diode pumped solid-state laser.
Optical invariant	See *Étendue*.
Optical thickness	The equivalent optical path length in a vacuum to a path in a medium. The optical thickness in a medium with an index of refraction n and a geometrical length of d is nd.
Organic light emitting diode	Abbreviated OLED. A device made of organic material that emits light when a current is passed through it. It contains a diode junction between an electron-conducting region and a hole-conducting region. The electron-hole recombination at the junction emits the light.
OSD	On screen display. The displaying of auxiliary information, such as VCR setup instructions, on the main video display.
Overfill	Producing an image too large to fit on the screen so some of the image is lost. Called overscan on a *raster* driven *CRT*.
Overscan	See *Overfill*.
PAL	Phase Alternating Lines. TV color system used in many areas including Western Europe (except for France). The field rate for PAL in Europe is 50 Hz with 625 lines per screen. PAL uses a similar transmission method as NTSC but, with the color information switched 180 degrees on alternate scan lines. The subcarrier frequency is 4.43 MHz. PAL broadcasts typically have fewer color artifacts than *NTSC* broadcasts. A 60Hz version of PAL is used in Brazil.
PAL+	Phase Alternating Line Plus. An *IDTV* format, backward compatible with *PAL*, with 16:9 aspect ratio. Use is made of a *letterbox* transmission with the active picture height reduced to 430 lines. In the black bars between successive fields of video, 'Vertical Helper Information' is transmitted. This 'helper signal' is used in the receiver to reconstruct the original 576 active lines.
PALC	*Plasma Addressed Liquid Crystal.* A FPD technology using electrical discharge to change the local voltage applied to the liquid crystal rather than an active-matrix. This technology never reached production.
Passive matrix	A *matrix addressing* system where the voltage difference between the *column* and *row* lines is applied directly to the pixel. When the next column or row is addressed, the drive voltage is no longer applied to the pixel. See *active matrix*.
Patch test	A *CRT* light output test that only illuminates a portion of the screen.
PBS	*Polarizing Beam Splitter.*
PCB	Printed Circuit Board.
PCM	Pulse code modulation. The most common way of converting an analog source into a digital form. This works by taking samples of the continuously varying analog signal at regular intervals. At each sampling point a number is generated to represent the size of the signal.
PCS	*Polarization Conversion System.*

GLOSSARY

PDLC	See *Polymer Dispersed Liquid Crystal*.
PDP	Plasma Display Panel.
PECVD	Plasma Enhanced Chemical Vapor Deposition. See *CVD*.
Pel	Another name for *Pixel*.
Phosphor	Generic name for the class of substances that exhibit luminescence when bombarded with an electron beam or *blue* or *UV* light. Phosphors are deposited on the inner surface of a *CRT* screen and are excited into luminescence by the electron beam to produce a picture. Phosphors are also used in plasma, electro-luminescent and field-emission displays, white LEDs and fluorescent lamps.
Photometric	A light measurement scaled with the *CIE* 1931 weighting functions. Photometric units are intended to correlate with human perception. See *Radiometric*.
Pincushion distortion	A *geometric distortion* of a video image in which the corners of the display are too far from the center, compared to the edges. See *Barrel Distortion*.
PIP	Picture-in-picture. Placing a smaller video image on a television screen such that it covers a portion of a larger one. See *POP*.
Pixel	An independent picture element. As used in this book, a pixel has full color and gray scale and may contain more than one *sub-pixel*. A pixel cannot show any spatial detail.
Pixelated device	A microdisplay in which the active area is divided into a finite number of *pixels*, each of which can show *color* and *gray scale* but no spatial detail.
Planar alignment	A *chiral nematic* liquid crystal mode that transmits light of one circular polarization and reflects light with the other circular polarization, e.g. transmits right circularly polarized light and reflects left circularly polarized light. Sometimes called *cholesteric* since the original implementations involved the cholesterol molecule. See *Focal conic alignment*.
Plasma Addressed Liquid Crystal	See PALC.
Plasma display	A matrix addressed gas discharge device in which the UV light generated in the gas is used to stimulate phosphors which then emit the visible light. Used in direct view applications only.
PMMA	Polymethylmethacrylate (Acrylic) (poly-methyl meth-acrylate) Moldable optical plastic $n = 1.49$
Polarization	The direction of the electric or magnetic field in light with respect to some external axis. Most light sources produce unpolarized light (random polarization). After polarization, the light can be *linear polarized*, *circularly polarized* or elliptically polarized.
Polarization conversion system	An optical element, or combination of optical elements, that performs *polarization recovery* by splitting the incident light into two orthogonally polarized beams and rotating the direction of polarization of the normally unusable polarization so that it has the same orientation as the desired polarization.
Polarization recovery	Light from *arc lamps* normally consists of s- and p-*polarized light* in equal amounts. Normal *TN-LCD* and most other LCD based *light valves* normally can only use one of these polarizations. *Polarization* recovery converts s into p light (or vice versa) so most of the output of the lamp can be used by the light valves.
Polarization utilization	The fraction of the light from the lamp that is used when the light is polarized (if needed) for a light valve.
Polarized light	Light in which the (linear) *polarization* has been restricted to a single spatial orientation, often by discarding all light that does not correspond to the desired polarization. Both orthogonal polarizations can be present if they have an unvarying phase relationship to each other. For phase shifts $<\pi/2$, the light is called elliptically polarized; for $\pi/2$ shift, the light is called *circularly polarized*.

(Continued).

Term	Definition
Polarizer	An optical element that changes unpolarized light into polarized light
Polarizer/analyzer	In *LCD light valves* that require *polarized light*, there must be a *polarizer* before the LV to provide the polarized light and a second polarizer after the light valve to eliminate the unwanted light so it will not reach the screen. This second polarizer is often called the analyzer. In *single polarizer reflective* LVs, the same polarizer performs both functions.
Polarizing beam splitter	An optical element designed to split light into the s- and p- *polarizations*. One polarization is reflected, the other is transmitted. There should be no absorption in a polarizing beam splitter.
Polymer dispersed liquid crystal	Abbreviated PDLC. A *liquid crystal* device in which the liquid crystal material is dispersed into small spherical droplets in a matrix of a transparent polymer. This system is either transparent or scattering, depending on the alignment of the liquid crystal material. PDLV devices can operate in un-polarized light.
Polysilicon	Poly-crystalline silicon. Silicon formed on a substrate with multiple crystal grains. These grains are small and have no ordering between grains.
POP	Picture-out-of-picture. The display of a second image on a television screen in such a way that it does not cover any of the main image. See *PIP*.
Posterizing	Similar to *contouring*, except it is visible as steps in the color of a video image. This occurs when an insufficient number of bits is used to transmit the color information.
PPLN	Periodically Poled Lithium Niobate. A non-linear optical material that will generate visible light as the sum or difference frequency from two input laser beams.
Pre-tilt	A condition in which the directors of the LC molecules immediately adjacent to the substrate are tilted slightly instead of being parallel or perpendicular to the substrate.
Primary colors	The three (or more) fundamental colors in a display, from which all other colors are made. Typically *red*, *green* and *blue* are used as primary colors in three-primary projection systems.
Prism	A optical element with more than two flat sides used to manipulate light. The thickness of a prism is typically comparable to the lateral dimensions and plays a role in the optical functioning of the element.
Progressive scan	A *video format* in which every *field* contains all the information to produce a single still image, i. e., a field contains both odd and even *lines*. Also called non-interlaced video. See *interlaced video*.
Projection screen	A viewing screen designed to show projected images. In a front projection screen, the light is incident from the viewer's side. In a rear projection screen, the light is incident from the opposite side of the screen from the viewer.
Projector	A piece of equipment designed to project an image onto a screen.
PSCT	Polymer stabilized *cholesteric texture*. A chiral nematic LC texture stabilized by a polymer additive to the LC mixture. A PCST display can be switched between *planar* and *focal conic alignments*.
p-Si	*Polysilicon*. This abbreviation is also used for silicon with dopants to make it a p-type conductor.
PTV	Projection television.
P-VIP	Osram trade name for its ultra high pressure mercury lamp. It is a near-clone of the Philips *UHP* and is built under license.
PWM	Pulse width modulation.
QCIF	Quarter Common Intermediate Format (176 pixels by 144 lines, 30 fps; 1.22 : 1). See *CIF*. While the format was originally designed for videoconferencing over phone lines, it has also been used by cell phones and other small displays, where it is being displaced by *QVGA*.

GLOSSARY

QQVGA	Quarter-Quarter *VGA*, 128 × 160 pixels; 1.33:1 *aspect ratio*. Intended for low cost applications similar to *QVGA*.
Quarter wave coating	A *dielectric AR coating* designed to reduce reflection. The *optical thickness* of the coating is equal to 1/4 wavelength of the incident light.
Quarter wave plate	A thin layer of *birefringent* material aligned to introduce a quarter wave retardation of the fast axis relative to the slow axis. On *linearly polarized* light starting at a 45° angle to the birefringent axis, this has the effect of converting the linearly polarized light to *circularly polarized* light.
Quincunx sampling	*Brick sampling*.
QVGA	Quarter Video Graphics Array (320 by 240 pixels; 1.33:1 *aspect ratio*), a display format used in small displays such as camera viewfinders or cell phones.
QXGA	Quad *XGA* (eXtended Graphics Adaptor) 2048 × 1536 pixels, 4:3 *aspect ratio*.
Radiometric	Measurement of an optical beam based on energy, without taking the *CIE* 1931 weighting functions into account. See *Photometric*.
RAM	Random access memory. Memory in which any data location can be accessed at any time. See *SRAM, DRAM*.
Raster	Creating an image by writing a series of closely spaced horizontal or vertical lines on a CRT or other scanned device.
Ray	The path followed by an idealized light beam of infinitesimal diameter and divergence. A ray may be of any intensity, or its intensity may vary along its length.
Real time	The processing of data at a high enough data rate such that the processing does not limit the data rate of the input or output channels. See *full motion video* and *RTP*.
Rear projection	A projection system in which the light is incident on the screen from the opposite side as the audience. Consumer *CRT* and *microdisplay* projection systems are typically rear projection.
Rear projection screen	See *Projection screen*.
Red light	Light in the spectral region from approximately 600 nm to 700 nm.
Reflective LV	A *light valve* (microdisplay) in which the incident light from the lamp and the output modulated light are on the same side of the device.
Refraction	The bending of a transmitted ray of light at an interface in which there is a different *index of refraction* on each side. See *Snell's law*.
Release Print	A print made from the internegative (IN) intended to be distributed to movie theaters. The release print is a fourth-generation image. See *Answer Print*.
Resolution	The amount of detail that a *projector* is capable of providing in an image. Resolution of a *pixelated device* is often quoted as a pixel count while resolution of non-pixelated devices is described by the *MTF*. See *Kell Factor*.
RGB	A video system where separate channels carry the red, green, and blue video signals. In French, RVB (Rouge, Vert, Bleu). When combined at the display, they make up a complete color image. The color quality of the final image depends upon the total number of bits used in the signal. (e.g., 4-bit, 8-bit, 16-bit or 24-bit). At 24-bit resolution (8 bits per color), the image is considered 'true-color' and when displayed on a high resolution display, is photo-realistic. Some demanding applications require still higher bit counts such as 10, 12 or even 16 bits per color.
Ripple ratio	The ratio of the amplitude of the fundamental term to the DC term in the temporal Fourier series describing the light output of a system.
RoHS	Restriction of Hazardous Substances A 2003 European Union directive to eliminate 6 hazardous materials from electronic equipment, including lead, mercury, cadmium, hexavalent chromium, polybrominated biphenyls (PBB) and polybrominated diphenyl ethers (PBDE).

418 **GLOSSARY**

(Continued).

Term	Definition
Row	In a pixelated display, pixels are normally arranged in horizontal rows. Normally, a single conductive electrode connects every pixel in a row to addressing circuitry.
RP	Rear Projection.
RTP	*Real time* protocol. End-to-end transport for real time data such as audio and video. All broadcast video formats, whether analog or digital, must be real time.
Saturated color	A color made from one or two *primary colors*, with very little of the third primary added. See *Desaturated color*.
Scalability	A feature of certain encoding formats that allows the same content to be displayed at higher quality using more powerful processors while still supporting minimal-level quality on low-end devices.
Scattering material	A material that randomly changes the direction of light *rays* passing through it. Scattering material may have little or no *absorption*.
Screen door effect	The visible perception of an image being broken down into individual *pixels*, even when those pixels are not fully resolvable. The screen door effect is the spatial equivalent of *flicker*, since the mathematics and perceptual issues are very similar. See *depixelization, ripple ratio*.
SDTV	*Standard definition television*. Sometimes abbreviated as SD.
SECAM	Sequence Couleur à Mémoire (which translates as color sequence in memory) The standard definition video format used in France, eastern Europe and the former Soviet Union. SECAM uses alternating lines of an FM modulated color carrier to carry *U* and *V chroma*. SECAM *decoders* are significantly more complex than *NTSC* and *PAL* decoders.
SED	Spectral Energy Distribution. See *Spectrum*.
Self converged systems	A system in which *convergence* of *red*, *green* and *blue* occurs automatically due to the design of the projector. Most single panel architectures fall into this category.
SFG	Sum Frequency Generation. A technique to use a non-linear material to add two laser beams of different frequencies to get the sum-frequency. Used to generate UV and visible light from IR lasers.
SHG	Second Harmonic Generation A technique to use a non-linear material to frequency-double a laser beam to get light of 1/2 the wavelength. Used to generate UV and visible light from IR lasers.
Short arc lamp	A general term for an *arc lamp* in which the light is all emitted from a volume that is no larger than a few mm in its largest dimension.
Short pass filter	An optical filter that passes all wavelengths of light that are shorter than its *cut-off wavelength* and reflects or absorbs all light with a longer wavelength.
SHP	Phoenix Electric Co., Ltd. trade name for its ultra-high pressure mercury lamp. See *UHP*.
SID	Society for Information Display (http://www.sid.org/).
Simulator	A training system for airplanes, tanks, ships, etc. that shows the operator video images of a (simulated) outside world. Typical simulators use several high performance projection systems.
Simulcast	The simultaneous transmission of the same video signal, possibly coded in differing standards, over separate transmission channels.
Single crystal silicon	Silicon made and annealed in such a way that the entire volume of the silicon is a single crystal. Often abbreviated as x-Si or c-Si. See *Polysilicon* and *Amorphous Silicon*.

Single polarizer reflective	A reflective *LCD* system in which there is one *polarizer* between the LCD cell and the illumination source. Most reflective LCD light valves are single polarizer systems. See *Twin Polarizer reflective*.
Six-axis positioner	A mechanical system to hold a device such as a *light valve* and provide precision control over its location in space. Control is provided over three spatial axes, normally x, y and z and three angular orientations, sometimes designated θ, φ and ψ.
Sm	See *Smectic Phase*.
SmC	Smectic C phase. See *Smectic Phase*.
SmC*	Smectic C* phase. See *Smectic Phase*. This phase supports ferroelectric display modes. See *SS-FLC*.
Smectic phase	Abbreviated Sm. A *LC phase* in which all molecules are arranged in distinct layers. The *directors* of the molecules in each layer are parallel to the directors of all other molecules in the same layer. In the Smectic phase (Sm), the directors are also perpendicular to the interfaces between adjacent layers. In the Smectic C phase (SmC) the directors are tilted with respect to the interfaces between layers. In the Smectic C* phase, (SmC*), there is an azimuthal twist in the director as you go from one layer to the next.
SMPTE	Society of Motion Picture and Television Engineers (http://www.smpte.org).
Snell's law	Also called the law of *refraction*. The law relating the *angle of incidence* before an interface (θ_0), the *angle of refraction* after an interface (θ_1) and *index of refraction* before and after the interface (n_0 and n_1) interface: $n_0 \sin(\theta_0) = n_1 \sin(\theta_1)$. The incident *ray*, the refracted ray and the normal to the surface all lie in the same plane.
Spathic polarizer	A *polarizer* that does not change the *étendue* of a beam passing through it. A *polarization conversion system* is non-spathic.
Spatial frequency	For a sine wave object, spatial frequency is $2\pi/\lambda$, where λ is the wavelength of the sine wave. Informally, spatial frequency is often expressed in cycles/mm, (or some other unit of length) without the 2π. See *TV Lines*.
Spatial light modulator	Abbreviated SLM. Another term for a *light valve or microdisplay*. This term is often used to describe binary light valves that either transmit or block light, without *gray scale*. Spatial light modulator is the favored term in the optical computing industry.
Spectral colors	The color perceived by the human eye when only one wavelength in the *spectrum* is present.
Spectral distribution	See *spectrum*.
Spectral utilization	The fraction of the light from the lamp that is used when the light is broken down into *red*, *green* and *blue* components. The different colors can be distributed spatially (e.g. with three-panels), temporally (e.g. *field sequential*) or angularly (e.g. *angular color separation*).
Spectrum	pl. Spectra. The energy distribution function within an optical wavelength range. *Visible light* is in the 380–780 nm spectral range. Longer wavelengths are *IR* light, shorter wavelengths are *UV* light.
Sphere lumens	Measurement of the output of a projector by capturing all the light emerging from the projection lens into a calibrated integrating sphere. This is the scientifically accurate way to measure projector output in most cases.
SPIE	SPIE-The International Society For Optical Engineering, The former name of the society and source of the initials was the 'Society of Photo-Optical Instrumentation Engineers' (http://www.spie.org/).
SRAM	Static random access memory. A semiconductor memory architecture requiring 6 or more transistors per memory cell. SRAM memory cells do not require refreshing to maintain their information state. See *DRAM*.
SSCT	Surface-stabilized *cholesteric texture*. A *chiral nematic LC texture* stabilized by surface alignment. A SSCT display can be switched between *planar* and *focal conic alignments*.

GLOSSARY

(Continued).

Term	Definition
SS-FLC	Surface-stabilized ferro-electric liquid crystal. A bistable *Smectic C** alignment that is induced by surface forces. This results in a fast switching effect with memory. The basic effect does not support gray scale, although gray-scale has been achieved in research demonstrations. Normally, if this effect is used in an application that requires gray scale this must be achieved by *sub-pixelization* or *bit-sequential* methods. See *Smectic phase*.
SSL	Solid State Light (or Lamp or Lighting). Category includes LEDs and many lasers.
Standard definition television	The image resolution achievable with conventional transmissions standards, such as *NTSC, PAL,* or *SECAM*. Abbreviated SDTV or sometimes just SD.
Starter	A circuit to supply the high voltage required to ignite an *arc lamp*. Often integrated into the lamp *ballast*. Also called an ignitor.
Steradian	A measure of solid angle. 4π steradians cover all directions originating from a point in 3-D space.
STN	Super twisted nematic. A twisted nematic alignment with the twist >90°. This alignment is not generally used in projection applications.
Sub-NTSC	A projector in which the *resolution* is insufficient to display all the information content of a *NTSC* video signal.
Sub-pixel	An independently controllable optical modulator that does not produce all the information necessary for a *pixel*. In the most common situation, one sub-pixel will modulate one of the *primary colors*, taking three sub-pixels to make a full color pixel.
Subtractive color	A color display system, such as film, in which the primary colors are *Cyan* (C), *Magenta* (M) and *Yellow* (Y). Two colors used together get the primary colors. Cyan + magenta = *blue*, magenta + yellow = *red*, yellow + cyan = *green*. In printing, black is added as a fourth primary (abbreviated K in the CMYK system) to ensure black printing is truly black, not gray.
SVGA	Super *VGA*. A higher-resolution PC video format than VGA. 800×600 *pixels*. SVGA may use higher frame rates than 60 Hz, depending on the monitor and the video card in the PC.
SWP	Short Wavelength Pass, More commonly called *Short Pass filter*
SXGA	Super *XGA*. A higher resolution PC video format than XGA. 1280×1024 *pixels*. SXGA may use higher frame rates than 60 Hz, depending on the monitor and the video card in the PC. Note that the image must have a 5:4 *aspect ratio* to display SXGA with square pixels, while *VGA, SVGA* and *XGA* images have a 4:3 aspect ratio display for square pixels.
Synchronization	Often written or spoken as 'sync'. In an *analog TV format*, information in the signal that tells the *CRT*-based receiver when to start a new *line* (H Sync) or a new *field* (V Sync). For non-*raster scanned* devices (e.g., *active-matrix*) sync signals must be processed into the correct form for the device. Sync information must be regenerated from *digital TV formats*.
System palette	The collection of colors a video display card in a computer can process for simultaneous display. The number varies from 16 colors in a CGA or EGA display to 16 million colors (2^{24}) in full color displays.
TAC	Triacetyl Cellulose a material used in polarizing film/filters manufacturing.
Talaria	An *oil film light valve* developed by General Electric, no longer in production.
TDM	Time division multiplex. See *Bit sequential gray scale*.
TEC	Thermoelectric Cooler.

GLOSSARY

Telecentric illumination	An optical arrangement in which the chief *rays* incident at each location on the *light valve* or other optical element are parallel to each other. This puts the pupil at infinity, as seen by that element.
Teletext	Low bandwidth digital data transmitted during the *VBI*. A teletext-compatible receiver is required to separate this data from the video information for display as text.
Television lines	See *TV lines*.
Texture	The visible appearance of *liquid crystal* under a *polarized light* due to the *alignment* of the LC. Texture and alignment are sometimes used interchangeably.
TFT	Thin-film transistor. A transistor made in a thin film of a semiconductor material, normally on an insulating substrate such as glass or quartz.
Throw ratio	The ratio of the distance from the projection lens to the screen to the width of the image at the screen.
TI	Texas Instruments Corporation, Dallas, Texas.
TIFF	Tagged image file format. A file format used to represent black-and-white, grayscale and color in bit-mapped images, particularly those produced by scanners.
Timebase correction	The use of a local, stable clock to restore the correct time relationships in a video signal that has been generated by a system with a varying time base. The most common device requiring time base correction is a video tape recorder in which the tape does not move at a perfectly constant speed.
TIR	*Total internal reflection*.
TMDS	Transition-minimized differential signalling.
TN-ECB	*Twisted-nematic* electrically-controlled *birefringence*. An *ECB* cell where the *directors* at the top and bottom substrates are not parallel. In normal usage, the term TN-ECB is only used to describe cells where the twist is less than 90°. Many TN-ECB alignments are usable in single polarizer reflective mode panels. See *Lower twisted nematic*.
TN-LC	See *Twisted nematic liquid crystal*.
TN-LCD	See *Twisted nematic liquid crystal device*.
Total internal reflection	Abbreviated TIR. The total reflection of a light *ray* at an optical surface going from the high *index of refraction* to the low index side. Total internal reflection occurs when the ray strikes the surface at an angle greater than the critical angle, which is the angle which produces a *refracted* angle of 90°. Rays striking at a smaller angle are partial transmitted, partially reflected.
Transcoder	A system to perform *transcoding*, i.e. convert one *video format* to another, e.g. digital TV to *PAL*. May include both spatial and temporal *interpolation*, e.g. PAL to NTSC.
Transcoding	Conversion of one transmission *format* (e.g. one of the ATV formats) to another format (e.g. NTSC), preferably without completely decoding the source format prior to the generation of the target format.
Transmissive device	A *light valve* or *microdisplay* in which the incident light from the lamp and the modulated light are on opposite sides of the device.
Transparent material	A material that has little or no *absorption* and no *scattering* for visible wavelengths.
Trapezoid distortion	A form of *geometric distortion* which results in the image resembling a trapezoid rather than a rectangle. In projectors, this occurs when the projection screen and the light valve are not parallel.
Trichroic Prism	A trichroic prism combines 3 beams of light (RGB normally) into a white beam, or splits a white beam into 3 different colors.
True lumens	See *Sphere lumens*.

(Continued).

Term	Definition
TVL	Television lines. A single dark or bright part of a test pattern. Two TV lines are the equivalent of 1 cycle. A system with a resolution of 1 line pair/mm or 1 cycle/mm would have a resolution of 2 TVL/mm.
TVL/PH	Television lines (*TVL*) per picture height. The number of horizontal or vertical TV lines visible in a horizontal or vertical length equal to the height of the image. If 180 pairs of alternating horizontal dark and light bars are just barely discernible in an display, the display is said to have a resolution of 360 TVL/PH. Vertical resolution is always referenced to picture height. While horizontal resolution is normally also referenced to picture height it may also be referenced to picture width.
TVL/PW	Television lines per picture width. See *TVL/PH*.
Twin polarizer reflective	A reflective LCD system in which there are two polarizers, one between the LCD cell and the illumination source and the other between the LCD cell and the mirror. LCD watches are normally twin polarizer reflective systems.
Twist angle	The angle through which the orientation of the *director* is rotated in traversing a *twisted-nematic liquid crystal*.
Twisted-nematic liquid crystal	Abbreviated TN-LC. A *nematic LC alignment* in which a twist of the *directors* is observed as one passes through the LC material from one substrate to the other. The twist is about 90° in typical transmissive LCD *light valves*. The pitch of this twist is substantially greater than for a *chiral nematic* LC and can be induced by substrate orientations alone.
Twisted-nematic liquid crystal device	TN-LCD. A *LCD* that uses *TN-LC* in an arrangement in which the *polarization* of the light is rotated, typically about 90°, when no voltage is applied to the LCD cell (*homogeneous alignment*), and the polarization is not rotated when a voltage is applied (*homeotropic alignment*). The abbreviations TN, TN-LC and TN-LCD are frequently used interchangeably.
Two polarizer reflective	See *Twin polarizer reflective*
Two-element screen	A rear *projection screen* that is a combination of a *Fresnel lens* and a *double lenticular screen*. The double lenticular element corrects for color shifts in a *CRT* projection system and gives a wide horizontal viewing angle. The region between the *lenslets* on the front of the double lenticular element is frequently covered with *black matrix*, for *contrast enhancement*.
u'–v'	See *Color coordinates*.
UDTV	*Ultra definition TV.*
UHM	Panasonic (Matsushita) trade name for its version of the ultra-high pressure mercury lamp. See *UHP*.
UHP	Philips trade name for AC-driven ultra-high pressure mercury lamps in the 100 W to 400 W range.
Ultra definition TV	*Digital video format* under development in Japan by many companies, including Nippon Television Network Corp. and Fuji Television, with about 2000 TVL resolution.
Uniaxial retarder	A *birefringent material* that has two different *indices of refraction*, n_o and n_e. See biaxial material.
USDC	United States Display Consortium (www.usdc.org/).
Useful screen area	The area of a display which can be viewed when video or data is displayed. In CRT data displays, it is the area of the visible raster. In CRT TV displays, it is the area of the opening in the bezel which is placed over the tube. TV receivers are typically scanned with larger rasters than the bezel area. See O*verfill*.

USPL	US Precision Lens, Cincinnati, Ohio. A major supplier of projection lenses and other optics for video projection. Now a part of the 3M Corporation.
USXGA	Ultra *SXGA*. Normally refers to a display with 1600 × 1280 pixels.
UV	Ultra-violet. Light in which the wavelength is shorter than the wavelengths of visible light. The photons of UV carry enough energy to break chemical bonds and damage microdisplays and other optical components. Organic compounds are particularly vulnerable to damage by UV. UV can be eliminated with special UV filters or mirrors.
u-v	An obsolete *color coordinate* system related to the u'-v' system. Nowadays only used in the calculation of correlated color temperature.
U-V	Encoded *chroma* information in a video signal. Different from u-v or u'-v' *color coordinates*.
UXGA	Ultra-*XGA*. A display with 1600 × 1200 pixels.
Vari-focus lens	A variable *focal length* lens. Different from a *zoom lens* in that it does not remain in focus while the focal length is changed.
VBI	Vertical blanking interval. In an *analog TV signal*, the time allowed by the signal for a *CRT* electron beam to move from the bottom of the picture to the top and begin the next *field*. This interval is used to provide *synchronization* information from the signal source to the receiver. Other signals, such as the *GCR* and *teletext* data, can also be transmitted during this interval.
VCSEL	Vertical Cavity Surface Emitting Lasers
VDE	Verband Deutscher Elektrotechniker A German standards setting organization. http://www.vde.com/
VECSEL	Vertical External Cavity Surface Emitting Laser.
Veiling glare	Light scattering in optical components resulting in light intended for brighter parts of the image appearing in the darker parts. A result is the checkerboard *contrast* being less than the full-on/full-off device contrast.
VESA	*Video Electronics Standards Association*. A consortium of manufacturers formed to establish and maintain industry-wide standards for video cards and monitors. http://www.vesa.org/
VGA	Video graphics adapter. A common resolution in IBM-PC compatible computers. The VGA standard has 640 × 480 pixels and is non-*interlaced*, with $\geq$ 60 Hz frame rate. As an analog format, the number of displayable colors was controlled by the video card in the PC, not the display. Now generally supplanted in new computers by *SVGA* or higher resolution formats.
Video format	The specified sequence, amplitudes and encoding scheme for a video signal. Three common video formats are *PAL*, *NTSC* and *VGA*. Video formats can be *analog* or *digital*.
Video rate addressing	A display addressing scheme in which each pixel needs to be addressed at the *field rate* of the *video format*: 1/60 second (16.7 mS) for *NTSC* or 1/50 second (20 mS) for *PAL*. CRT data displays may have video rates $\geq$ 72 Hz. Addressing schemes that do not address each pixel at the field rate are not normally acceptable for *full motion video*.
Video-CD	An industry standard format for putting *MPEG-1* video and multi-channel sound on a CD-ROM. Video-CD was never a popular format in the US, Europe or Japan.
Vignetting	In an optical system, the gradual reduction of image luminance as the off-axis angle increases, resulting from the partial obscuration of the light bundle by limiting apertures of optical elements within the system.
Visible light	Light that can be perceived with the human eye. See *spectrum*.
VOD	Video on demand. A system to provide video to subscribers at any requested time, rather than according to a pre-planned schedule.
VR	Virtual reality.
VSP	Video signal processing. The application of *DSP* techniques to video signals.

(Continued).

Term	Definition
Waveplate	A thin layer of *birefringent* material intended to modify the polarization of light. Two commonly used wave plates in projection systems are *quarter-wave plates* and *half-wave plates*.
Well corrected optical element	An optical element that induces little or no *aberration* into an optical beam passing through it or reflecting off it.
White	See *black body radiation*.
White field uniformity	When a uniform white video signal is displayed, the image should be of uniform color and luminance. White field uniformity is a quantitative or qualitative evaluation of the spatial deviations from the target *white* field.
Wide band HDTV	An *HDTV* system without *compression* or other bandwidth reduction techniques. This system would offer the viewer the same image quality as is available in the television studio but it carries unacceptable bandwidth and storage capacity penalties.
Wide color gamut display	A display with an unusually large color gamut. The phrase is often used to refer to a display with a color gamut larger than the HDTV color gamut defined in ITU Recommendation 709.
WVGA	Wide *VGA* A microdisplay with 854×480 pixels. This is a 16:9 version of the VGA format.
Xenon	A noble gas often used for filling *arc lamps*, either alone or in combination with other elements.
XGA	Extended Graphics Adapter. A higher resolution PC video format than *SVGA*, with 1024×768 *pixels*. XGA displays may use higher frame rates than 60 Hz, depending on the monitor and the video card in the PC.
x-Si	*Single crystal silicon.*
xvYCC	Extended YCC Colorimetry for Video Applications. An extended color encoding system that is backward compatible with digital HDTV MPEG encoding and would enable *wide color gamut* displays.
x–y	See *Color coordinates*
Y/C	Luminance (Y) / Chrominance (C) signal. A *video format* in which the black and white (*luminance*) and color (*chrominance*) video signals are encoded separately.
YAG:Ce	YAG is a crystalline material (garnet) made from yttrium, aluminum and oxygen and doped with cerium. It is used as a phosphor with blue LEDs to generate yellow light and produce a white-emitting LED. This yellow light include weak green and red emission plus strong yellow. With the lack of sufficient red and green, the white light produced by the blue LED plus YAG:Ce has a relatively low CRI.
Yellow light	Light formed by adding together the *red* primary and the *green primary colors* of a display. Also, can refer to light in a wavelength band centered around 580–590 nm. This band can be narrow or it can extend from as low as 520 nm to as high as 770 nm or higher.
YNF	Yellow Notch Filter. Filter to remove the yellow from a Xenon or UHP lamp to improve the color gamut.
YUV	A video format encoding *color* using two *chrominance* values (U and V) and one *luminance* value (Y). An alternative to *RGB* coding, it is more efficient in transmitting video signals. Luminance is the brightness value of the signal. Chrominance is coded as U = (blue signal − Y), and V = (red signal − Y). Green signals can be extracted using only these two signals. This method of coding, while more complex than RGB, uses 50% or less of the bandwidth required by RGB coding.
Zoom lens	A variable *focal length* lens. The lens remains in focus as the focal length is changed. Used to produce a variable image size on a *projection screen* without the need to move the projector.

Index

Note: The entry or use of a word in the glossary, page 393 is not included in this index. Please consult the glossary separately for any term that is not understood

A-plate, *see* Microdisplay, compensation
Aberration, 82, 99–100, 132, 208, 258, 273–4, 278, 320, 384, 386–7
    coma, 82, 87–90, 114–15, 236
Acousto-optic modulator, 8, 227–8
Active matrix, 7–8, 29–43, 45, 53–4, 249–51, 284–7
    amorphous Si, α-Si, a-Si TFT, 32, 35–7
    addressing, 32, 249–51
    charging currents, 34–5
    crystalline Si, c-Si, x-Si, 40–2
    diode-like devices (two terminal), 42–3
    effects of leakage, 33–4, 356
    poly-Si TFT, 38–9
    MOS transistors in x-Si, 8, 30, 40–2, 57–8
    parasitic capacitance, 36–8
Adaptive processing, 343–4
Additive color, 182, 233–4, 287–8, 373–4
Aliasing, *see* Artifacts, spatial
Alignment, LC, 43–8, 51–3
    homeotropic, 46, 49, 85
    homogeneous, 46
    layers, 45–7, 51–2, 91, 244
AM-LCD, *see* Active-matrix
Amorphous silicon, *see* Active-matrix
Analyzer, *see* Polarizers
Angle of incidence, *see* Snell's law

Angle of reflection, *see* Snell's law
Angle of refraction, *see* Snell's law
Angular color separation, *see* One light valve projector
ANSI lumens, *see* Lumen
Anti-reflection coating, 71–2, 82–4, 127, 306–8
Aperture
    clear, 45, 65, 285–6, 353, 355
    lens, *see* Lens, aperture
    ratio, 39, 41, 247, 270, 285–7, 351–5
Applications, 8–16
    auditorium, 12–14, 137
    consumer, 7–12, 17
    electronic cinema, 15–16, 57, 279, 322–3
    pico-projector, 8–9, 136
    presentation, 8–10, 12–15
AR coating, *see* Anti-reflection coating
Arc lamp, *see* Lamp
Artifacts, Spatial, 347–55
    aliasing, 348–3
    depixelization, 355
    moiré, 155, 348–53
    sampling, 316–18, 348–9
    screen door effect, 62, 353–5
Artifacts, Temporal, 265, 356–62
    color sequential, 61, 249, 259, 356, 360–2
    flicker, 12, 54, 177, 228, 265, 356–8, 359

426 INDEX

Artifacts, Temporal (*Continued*)
   judder, 358–60
   motion blur, 358–60
   MPRT, 359
   smear, 357–60
a-Si, α-Si, *see* Active matrix
Aspect ratio, 9–10, 14, 22, 76, 80, 257, 279

Back focal length, *see* Lens, back focal length
Ballast, *see* Lamp
Band pass filter, *see* Filter
Barrel distortion, *see* Cathode ray tube projector, distortion
Beam splitter, *see* Polarizing prism
Birefringent material, 114, 116–17, 120, 126–7, 186
Black matrix, 150, 159
Blackbody radiation, 234–5, 375
Brewster's angle, 74–5, 109–14, 117, 122
Brightness, 325
   see also Luminance
Brightness uniformity, *see* Luminance, uniformity
Burner, *see* Lamp

C-plate, *see* Microdisplay, compensation
Candela, *see* Photometry
Cathode Ray Tube (CRT), 1, 2, 4, 8, 11, 17–26, 141, 243, 343
   cooling, 19, 21, 217–19, 224
   curved faceplate, 18, 25–6, 219–21
   gamma, 342
   phosphors and emission spectra, 19–21, 332, 334, 336, 379
   resolution, 22–5, 315–16
Cathode ray tube projector, 2–4, 163, 217–25
   convergence, 220–3, 320–1
      automatic, 222–3
   distortion, 221–2
   luminance, 18–19, 141, 223–5, 329–30, 340–1
   one lens, 220–1
   resolution, 315–18, 320–1
   screens, 147, 152–7
   three lens, 132, 152–7, 218–20
CCT, *see* Correlated color temperature
Center weighted lumens, *see* Lumen
Cholesteric, 43–4, 85
Chromaticity coordinates, *see* Color coordinates
CIE 1931, 2° Matching Function, 325–26, 372–373, 375, 381
Coatings, 75–94
   antireflective, *see* Antireflection coatings
   conductive, *see* Indium tin oxide
Cogging, *see* Artifacts, Temporal
Coherence, 162–6, 355
Cold mirror, *see* Mirrors

Collection system, 25–6, 121–3, 206–12, 271–84, 296, 383–91
   compound, 206–8
   CPC, 209–12
   constant magnification, 208–9
   LED, 192, 209–12
   non-imaging, 206, 209–12, 228
   parabolic, 206–10
   refractive, 208–9
Collimated light, 97, 99, 202, 208
Color, Colorimetry, 181–3, 332–41, 371–81
   breakup, *see* Artifacts, temporal
   CIE 2° matching function, 325–6, 372–3, 375, 381
   chromaticity diagram, 332–3, 375
   coordinates, 372–80
   correcting screen, 172–5
   correction to white, 269–70, 289–93
   difference formulas, 375–8
   filter, *see* Filter, color
   gamut, 181–3, 194, 254, 256, 264, 322, 332–40
   gamut size, 181–3, 322, 337–40
   Just noticeable difference (JND), 375–7
   MacAdam ellipses, 336–7
   measurement, 340–1, 377–80
   multiple primary, 181, 252, 389, 322, 337, 339–40, 361
   real colors, 333–4
   saturation, 84, 193, 333
   temperature, 86, 291, 309, 335, 375
      see also Correlated color temperature
   separation, 4, 234, 238, 244–6, 263–4, 270, 294–6, 305, 308
   sequential, *see* One light valve projector; Artifacts, Temporal
   uniformity, 151–3
   wheel, 178, 251–3, 255, 259, 264, 287, 289–90, 293–5, 306–10, 361
   white point, 334–5
      see also Color temperature
Colorimeter, 340, 378–80
Colorimetry, *see* Color
Column, 7–8, 29–33, 42, 62, 249
Coma, *see* Aberration, Coma
Compensation films, 91, 126–8, 285
Conjugate distance, 148–9
Contrast, 9, 47–53, 61–2, 71, 105–6, 111–12, 126–8, 142–3, 154–62, 241–2, 318–20, 322, 330–2
   ambient, 143
   ANSI checkerboard, 331–2
   full-on/full-off (Sequential), 9, 330–1
   sensitivity function, 312–13, 324, 354, 357
Convergence, 14, 22, 132–5, 218–23, 234–8, 320–1
   automatic, 223
Correlated color temperature, 309, 335, 375
Cos^4 law, 141–2
Counter-electrode, 30, 33, 53, 251

INDEX

Critical illumination, 95–7, 281
CRT, *see* Cathode ray tube
c-Si, *see* Single crystal silicon

D65, 291
   see also Correlated Color Temperature
Depixelization, 355
DH-FLC, *see* Liquid crystal
Dichroic
   filter *see* Filters
   polarizer *see* Polarizers
Diffusion, *see* Scattering
Digital Micromirror Device (DMD) 57–62, 89–90, 249–53, 259, 286, 306–10
   contrast, 61–2
   gray scale, 60–1
   operation, 58–60
   pixel design, 57–8, 61–2, 286
Director, 43–5, 47, 51, 126
Distortion, 134–5, 137–40, 206–7, 220–23, 229, 236, 321, 386–9
D-ILA, *see* Microdisplay, LCoS
DLP, *see* Digital Micromirror Device
DMD, *see* Digital Micromirror Device
Double lenticular screen, *see* Screens
Duty cycle, 54, 227, 251–6, 264, 287, 359
   see also LED, duty cycle

EBU, *see* Color gamut
ECB, *see* Liquid crystal
Edge matching, 146–7
Efficacy, 19, 170–3, 175, 192–4, 198–200, 303, 390
Efficiency
   collection, 100, 177, 192, 271–81, 283–4, 383–91
   color correction, 289–93, 309–10
   color separation, 270, 294–6, 303, 308
   component, 72–3
   light valve, 270, 284–6, 296–7
   modulation, 286–7
   photometric, 72
   polarization, 106–9, 118–25
   wavelength conversion, 188–9, 192–3
   see also Efficacy
Eidophor, 217, 260–1
Electrically controlled birefringence, *see* Liquid crystals, ECB
Electronic cinema, 15–16, 41, 243, 262–5, 279, 322–3
Étendue, 271–81, 283–4, 296, 383–91
   component, 62, 96, 99, 175, 191–2, 205–6, 208
   conserving transformation, 278–9
   cylindrical surface, 385–6
   definition, 271–3
   flat surface, 273–4
   integrator, 99–104, 281, 283–4
   limiting, 274–8
   lumens vs. étendue, 274–81, 383–91
   ratio, 277
   usable, 279–81
Extent, optical, *see* Étendue
Extinction ratio, 105–7, 120, 125

f/#, *see* Lens, aperture
FED, *see* Field emission device
FELC, *see* Liquid crystal, ferroelectric
Ferroelectric liquid crystal, *see* Liquid crystal, ferroelectric
Field emission device, 26
Field flattener, 132
Field rate, 60, 259, 263–5, 356–8, 361–2
Field sequential system, 60, 178, 249–59, 356, 360–1
Fill factor, *see* Aperture ratio
Film, 7, 12, 15–16, 136, 182, 262–3, 322–4
Filter, 73–89, 91–3, 240–2, 244–5, 263–4, 287–8, 296–8
   absorptive, 84–5, 244–5
   band pass, 79, 84
   dichroic, 75–84
   electrically tunable, 85–6
   infra-red, 91–2
   long pass, 84, 91, 155, 287–8
   short pass, 84, 287–8
   ultra-violet, 84, 91
Flicker, *see* Artifacts, Temporal
f-number (f/#), *see* Aperture, lens
Focal length, 97, 102, 134–7, 141–2, 147–9
Fold mirror, 87–8, 93, 144, 150
Foot-lambert, 12–13, 327, 368–9
Format, video, 14–15, 269, 342, 358
Frequency doubling, *see* Wavelength conversion
Fresnel
   lens, *see* Lens, Fresnel
   reflection, *see* Reflection, Fresnel
Front projection, 9–12, 142–3, 145–6, 159–62, 340–1

Gain, *see* Screens
Gamma, 342, 371
Geometric distortion, *see* Distortion
Gooch-Tarry equation, *see* Liquid crystal
Grating Light Valve, *see* Light valves, Micro-electromechanical
Gray scale, 52, 60–1, 66, 178, 233, 293, 313, 342, 362, 371

Half angle, 137, 273–5
HDTV, *see* High definition TV
HID lamp, *see* Lamps
High definition TV, 5, 9, 14–15, 22, 183, 335–8, 342
High temperature polysilicon, *see* Active matrix, Poly-Si TFT
Highlights, 225
Hot restrike, *see* Lamps
Human visual system (HVS), 311–13, 323–6, 333–4

428   INDEX

Ignitor, *see* Lamps
Illuminance, 51, 223–4, 365–9
Illuminant E, 72, 191
Illumination path, 93, 139–42, 166, 235, 237
   see also Integrator
Index of refraction, 73–6, 78–9, 83, 89, 94, 188, 259
Indium tin oxide (ITO), 54, 93–4, 284–5
Integrating sphere, 199–200, 203, 215, 326–9, 389–90
Integrator, 94–105, 121–4, 279, 281–4
   aberration correcting, 100
   étendue, 96, 281–4
   fiber optic, 103
   laser, 102–3
   lenslet, 96–100, 283–4
   lumen throughput, 281–4
   rod, 100–2, 122–4, 284
IR, *see* Filters, infra-red
Isotropic phase, 43–4
ITO, *see* Indium tin oxide

Judder, *see* Artifacts, Temporal

Kell factor, 318
Kelly Chart, *see* Color, Chromaticity diagram
Keystone, *see* Cathode ray tube projector, Distortion
Koehler illumination, 95–6

Lambertian distribution, 144, 368
Lamp, 72, 169–80, 198–206
   angular distribution, 201–2
   arc, 170, 173, 177, 195, 203, 206
      see also Lamp, high intensity discharge
   arc jump, 177–8
   ballast, 177–8, 212–13
   distribution, 201–2
   efficacy, 170–1, 172–3, 175, 192–4, 198–200
   electrodeless, 179–80
   Emission spectra, 170–80, 200–1
   Emission lines, 170, 175, 200–1, 308
   Envelope, 170–4, 176–7, 194, 203, 207, 391
   High intensity discharge (HID), 170–80, 212–13
   Hot restrike, 213
   Ignitor, *see* Lamp, ballast
   Lifetime, 202–6
   Mercury, 174–8, 338–9
   Metal-halide, 172–4, 179–80
   Short arc, 170, 174–5, 202, 285, 305, 390
   Tungsten-halogen, 178–9
   UHP *see* Lamp, mercury
   Xenon, 159, 171, 199
Laser, 8, 180–90
   area emitting, 186
   diode pumped solid state (DPSS), 185–6, 188–9
   edge emitting, 183–6, 188
   fiber, 186–7

IR, 188–9
MOPA, 187
Optically Pumped Semiconductor (OPS), *see* Laser, Diode pumped solid sate
projector, 8, 62–8, 102–3, 225–30, 253–4, 259
safety, 189–90
scanning, 66–8, 225–30
speckle, 162–6
   see also Speckle
vertical cavity, 183–5
wavelength, 181–3, 187–9
   see also Wavelength conversion
Law of refraction, *see* Snell's law
LC, *see* Liquid crystal
LCD, *see* Liquid crystal device
LCoS, *see* Microdisplay, liquid crystal on silicon
LED, 190–8
   collection system, 102, 209–12
   drive, 195–8
   dominant wavelength, 191, 194, 196–8
   duty cycle, 196
   efficacy, 192–4, 198–200
   étendue, 191–2, 195, 227
   haitz's law, 191–2
   linearity, 196–8
   thermal, 194–5
Lens, 131–42, 147–51
   aperture, 25
      see also Lens, Pupil
   back focal length, 132–4
   cathode ray tube, 132–3
   $Cos^4 q$ law, 141–2
   condenser, 208–9, 390
   field, 97, 102, 104, 126, 132–4, 147–8
   fresnel, 147–51, 220
   lenslet integrator, 96–100
   microlens, 39, 245–6, 355
   offset, 137–40
   plastic, 134, 224
   projection, 131–42
   pupil, 61, 95–6, 140–2, 162–3, 206, 252
   relay, 255
   reversed telephoto, 101–2, 134, 241–2, 296
   throw ratio, 136–7, 141–2
   vari-focus, 136
   wide-angle, 136–7
   zoom, 136–7
Lenticular screen, *see* Screens
Light amplifier, 243
   see also Light valve
Light amplifier projector, 243
Light body, 96–102, 282–4

Light collector, 25, 206–12, 271–81, 383–9
 compound, 206–7
 constant magnification, 208
 elliptical, 206–7
 LED, 209–12
 measurement, 199–200
 non-imaging, 209–12
 parabolic, 206–7
 refractive, 208–9
Light-emitting diode, *see* LED
Light engine, 2, 217–20, 238, 245, 269
Light guide, 104–5, 124, 211–12
Light pipe, *see* Light guide
Light-valve
 active-matrix, 7–8, 29–43, 45, 53–4, 249–51, 284–7
 e-beam addressed, 261
 light-amplifier, 243
 liquid crystal light-valve, 29–54
 micro-electromechanical
  diffractive, 62–6, 249, 262
  GEMS, grating electro-mechanical system, 66
  GLV, grating light-valve, 62–6, 249
  GxL, 65–6
  see also Digital Micromirror Device
 reflective, 41, 52–4
 scanning mirror, 66–8
 transmissive, 41, 45–52
 see also Microdisplay
Light-valve projector, *see* One light-valve, Two light-valve and Three light-valve projectors
Liquid crystal, 29–54, 249–51
 alignment layer, 45, 91, 244
 cells, 45–9
 cell gap, 40–1, 45, 52
 clearing temperature, 44
 cholesteric, 44
 DH-FLC, 51–2
 direct view, 1–2, 35–6, 359
 ECB, 51–2
 ferroelectric, 51–2
 Gooch-Tarry equation, 47
 hybrid aligned, 51
 inversion, 53–4
 nematic, 44, 46–9, 51–3
 PDLC, 50–1, 262
 response times, 49, 358–60
 smectic phase, 43–4, 51
 twisted nematic, 46–9, 53
 VAN, vertically-aligned nematic, 52–3
Liquid crystal device, *see* Liquid Crystal
Long pass filter, *see* Filters, long pass
Low pass filter, *see* Filters, short pass

Low temperature polysilicon, *see* Active matrix, Poly-Si TFT
Lumen, 365–9
 ANSI, 11, 327–8
 center weighted, 329
 patch test, 225, 329–30
 sphere or true, 199–200, 204–6, 275–9, 226–7
 throughput, 301–10
 variation, 296–8
 see also Photometry
Luminance, 365–9
 brightness, 11, 16, 25, 144, 227, 312, 325–6, 330
 requirements, 10–13
 uniformity, 96, 101–5, 142, 147, 210, 227, 239, 281–2, 327, 379
 see also Photometry
Luminous
 Efficacy, *see* Efficacy
 flux *see* Photometry
 intensity, *see* Photometry
LV, *see* Light valve

MacAdam Ellipses, 336–7, 338
MacNeille prism, *see* Polarizing prism
Markets, 7–16
Matrix-addressing, *see* Active-matrix
MEM, *see* Micro-electromechanical
Metal halide, *see* Lamp, metal-halide
Metric, image quality, 323–5, 359
 spatial standard observer (SSO), 324
 speckle contrast, 162–3, 324
 moving picture response time (MPRT), 359
Microdisplay
 Addressing, 8, 31–2, 58–61, 249–58, 262
  Address and flash, 254–5
  Reset addressing, 250–1
  Row at a time, 249–50
 aging, 175, 244
 compensation, 53, 126–8, 285
 DMD *See* Digital Micromirror Device
 liquid crystal, transmissive, 29–39, 41, 45–52, 234
 liquid crystal on silicon, 40–2, 53–4, 239
 see also Light valve
Micro-electromechanical, *see* Light valve
Microfilter, 244–5, 246–8, 295, 301–4
Mirrors, 8, 40–1, 62–8, 86–9
 cold, 92–3
 front surface, 87
 IR (Hot), 92–3
 micromirror, *see* Digital Micromirror Device
 scanning, 8, 66–8, 226–30
 second surface, 87, 88
 UV, 91

Modulation transfer function (MTF), 142, 314–23
  convergence, 320–1
  CRT, 22–5, 134, 315–16
  film systems, 322–3
  optical systems, 318
  raster-scanned systems, 315
  sampled systems, 246–8, 316–18
  screens, 322
Moiré, 347–53, 355
Monochrome, 17, 371
Motion compensation, 360
MTF, see Modulation transfer function
Multiple primary colors (MPC), 181, 252, 289, 337–40, 361

NA, see Lens, aperture
Nematic phase, see Liquid crystal
Neutral density, 72–3, 125, 270, 288, 297
NTSC, 5, 14–15, 295
  see also Color gamut
Numerical aperture, see Lens, aperture
Nyquist frequency, 246–47, 316–18, 322, 348–9

O-plate, see Microdisplay, compensation
Oil film light valve, see Light valves, e-beam address
One light-valve projectors, 243–58, 301–4, 306–10
  angular color separation, 245–6
  color field sequential, 249–58, 259
    addressing, 249–51
  color wheel, 251–3, 306–10
  color drum, 255–7
  microfilter, 244–5, 246–8
  resolution, 246–8
  scrolling color, 256–8
  three lamp sequential, 253–4
Optical invariant, see Étendue
Overfill, 282–3, 304–7

Patch test, see Lumens
PBS, see Polarizing beam splitter
PCS, see Polarization, Conversion system
PDLC, see Liquid crystal
Performance requirements, 8–11, 337
Phosphor, see Cathode ray tube
Photometry, 365–9
Pico-projector, 2, 8–10, 14, 254
Pincushion distortion, see Cathode ray tube projector, Distortion
Pixel, 7–10, 14–15, 29–42, 57–67, 251, 285–7, 313
  count see Resolution
  sub-pixel, 244–8, 301–4, 313
Plasma
  direct view, 2, 7
  lamp, see Lamp, electrodeless

Polarization, 105–28
  compensation, 53, 126–8, 285
  conversion system (PCS), 121–4, 280–1
  efficiency, 105–6, 108
  half-wave plate, 121–3
  recycling, 123–4
  retarder, 53, 86–7, 126–8
Polarizer, 105–28
  absorptive, 106–8
  beam splitter, see Polarizing prism
  birefringent multilayer reflective polarizer, 114–16
  brewster plate, 109–10
  dichroic, 105
  dye, 107
  extinction ratio, 105–6
  iodine, 106–7
  MacNeille prism, see Polarizing prism
  reflective, 109, 117–20
    see also Polarizing prism
  type K, 107–8
  type H, 106–7
  wire grid (Moxtek), 117–20
Polarizing prism, 111–17
  Bertrand-Feussner prism, 116–17
  Glan prism, 120
  imaging path, 124–6, 239–40
  MacNeille, 111–14, 122–3
  Nicol prism, 120
  3M, see Polarizer, Birefringent multilayer reflective polarizer
Polymer dispersed liquid crystal, see Liquid crystal, PDLC
Polysilicon, 38–9
Primary colors, 2–4, 161, 181, 249, 252, 256, 263, 289, 294, 322, 333, 335–9
Prism, 76, 82, 89–91, 103–5, 111–17, 124–6, 256–8
  combining dichroic, 82, 237–9
  cone, 103–5
  embedded dichroic, 82
  polarizing, 111–17, 120, 124–6, 239–42
  TIR, 89–91, 306
Projection ratio, see Lenses, Throw ratio
Projection screen, see Screen, projection
Pull-down, 3:2, 358–60
p-Si, see Polysilicon

Quarter wave coating, see Anti-reflection coating
Quartz, 38, 94, 127, 171–3, 176, 178, 194, 203

Radiometry, 141, 365–9
Raster, 226–7, 315–16
  CRT, 18, 22–3, 134–35, 221–5, 315–16
  laser projectors, 8, 66–8, 225–9
  scan, 226–7

INDEX

Rear projection, 10–12, 132–5, 142–4, 146–58, 162–4, 219–20, 341
   thin, 150–1
Rear projection screen, *see* Screen, projection
Reflection, Fresnel, 72–5
Reflective light valve, *see* Light valves
Reflector, *see* Light collector
Refraction, 73–5, *see* Snell's law
   index of, 73, 78–9, 83, 94, 114, 117, 188
Resolution, 9–10, 246–8, 312–23, 358
   cathode ray tube, 22–5, 320–2
   electronic cinema, 322–3
   pixel count, 9–10, 14, 313
   pixelated systems, 316–18
   projection screen, 147–59, 163–64, 322
   Spot size *See* resolution, Cathode ray tube
   see also Modulation transfer function
Retarder, 53, 86–7, 126–8
   form birefringent, 127–8
   half wave, 121–3
   quarter wave, 85, 123–4, 126, 210
Ripple ratio, 356–7
Row, 7–8, 31–3, 39–43, 54, 60, 249–58, 313
RPTV, *see* Rear Projection

Saccadic motion, *see* artifacts, temporal
Sample-and-hold, 32, 358–9
Scattering, 23, 50–1, 77, 163, 203, 272–4, 322, 331–32
   lambertian, 51, 144, 368
Schlieren systems, 259–62
   laser, 62–6
   PDLC, 50, 262
   see also Eidophor, Talaria
Screen door effect, *see* Artifacts, spatial
Screen, projection, 11–14, 18–19, 142–67, 218–20, 263, 322
   beaded, 159
   diffusing, 165, 322
   double lenticular, 152–5
   fresnel lens, 147–51
   front projection, 142–44, 159–62
      contrast enhancing, 160–2
   gain, 11–12, 144–7, 224, 340–1
   MTF, 322, 348
   polarization preserving, 263
   rear projection, 142–4, 147–59
   single lenticular, 157–9
   TIR, 155–7
Shape conversion, 278–9
Short arc lamp, *see* Lamp
Short pass filter, *see* Filter
Single crystal silicon, 8, 29, 40–2
Single light valve projector, *see* One light valve projector
Six-axis positioner, 134, 234–5, 320
Smectic phase, *see* Liquid crystal

Snell's law, 73–5, 89
Speckle, 131, 157, 162–6, 322, 324
   see also Laser, speckle
Spectral distribution, 19–20, 72–3, 170–80, 190–4, 200–1, 288–9, 334
   see also Lamps, emission spectra
   neutral density, 72–3, 125, 270, 288–9, 297
   measurement, 78, 340–1, 378–9
Spectroradiometer, 78, 200, 340–1, 378–9
Spectrum (Spectra), *see* spectral distribution
Sphere lumens, *see* Lumens
Starter, *see* Lamps
Stereoscopic 3D, 16, 262–5
Sub-pixel, 244–8, 301–4, 313
Systems
   angular color separation *see* One light valve projector
   field sequential *see* One light valve projector
   one-lens CRT, 220–1
   One panel *see* One light valve projector
   Schlieren *see* Schlieren systems
   three lens CRT, 4, 218–20
   three panel *see* Three light valve projector
   two panel *see* Two light valve projectors

Talaria, 99, 135, 206–7, 260–1
Telecentric illumination, 79–80, 96, 99–100, 126, 210, 238
Texas Instruments (TI), *see* Digital Micromirror Device
TFT, *see* Active matrix
3D, *see* Stereoscopic 3D
Three light-valve projectors, 234–43
   aberrations, 236
   equal path, 235–6, 238, 304–5
   three lens, 242–3
   unequal path, 237–8, 239
Throughput, *see* Lumen
Throw ratio, 136–7, 141–2
TIR, *see* Total internal reflection
TN-ECB, *see* Liquid crystal, ECB
TN-LC, *see* Liquid crystal, twisted-nematic
Total internal reflection, 89
   angular separation, 89–91
   see also Prism, TIR
Trapezoid distortion, 137–9, 221–3
True lumens, *see* Lumen
Twisted nematic liquid crystal, *see* Liquid crystal, twisted nematic
Two light-valve projector, 234, 259

VAN, *see* Liquid crystal, vertically aligned nematic
Vari-focus lens, 136–7
Veiling glare, 318–20
VESA, 311
Vignetting, 25–6, 140, 224, 246, 296

Wave plate, *see* Compensator or Retarder
Wavelength, *see* Colorimetry, Laser, LED
Wavelength conversion, 187–9
   periodically polled lithium niobate (PPLN) 185–6, 188
Weber-Fechner law, 313

White point, 188, 254, 270, 290–4, 332–6, 378
   see also Correlated color temperature

x-Si, *see* Single crystal silicon

Zoom lens, 132, 136–8